Exploring proteins

Exploring proteins: a student's guide to experimental skills and methods

Nicholas C. Price and **Jacqueline Nairn**

OXFORD
UNIVERSITY PRESS

Great Clarendon Street, Oxford OX2 6DP

Oxford University Press is a department of the University of Oxford.
It furthers the University's objective of excellence in research, scholarship,
and education by publishing worldwide in

Oxford New York

Auckland Cape Town Dar es Salaam Hong Kong Karachi
Kuala Lumpur Madrid Melbourne Mexico City Nairobi
New Delhi Shanghai Taipei Toronto

With offices in

Argentina Austria Brazil Chile Czech Republic France Greece
Guatemala Hungary Italy Japan Poland Portugal Singapore
South Korea Switzerland Thailand Turkey Ukraine Vietnam

Published in the United States
by Oxford University Press Inc., New York

First published 2009

British Library Cataloguing in Publication Data
Data available

Library of Congress Cataloging in Publication Data
Data available

Typeset by Graphicraft Limited, Hong Kong
Printed in Great Britain on acid free paper by
Ashford Colour Press Ltd, Gosport, Hampshire

ISBN 978-0-19-920570-7

For Margaret, Jonathan, Naomi and the memory of Rebekah, and
Stuart, Robert, and David

Foreword

The aim of this book is to introduce and illustrate the application of molecular principles and numerical concepts to the biosciences, with particular reference to the study of proteins. It is our belief (shared by many other researchers and teachers in the field) that these concepts have not been emphasized sufficiently over recent years, with the result that many students and graduates experience major problems in understanding at the molecular level how proteins play their biological roles. As we move from an era dominated by genomics to one where proteomics becomes the object of attention, it becomes increasingly important to address these issues.

This book is not intended to be a teaching or research manual, but aims to develop self-confidence in being able to understand the way proteins behave and the basis of the methods used to separate, identify, and characterize them. A number of the practical activities could be used in undergraduate teaching laboratories and some protocols for these are included in an Online Resource Centre for teachers. Other experiments may involve the use of specialized apparatus, but the analysis of the data obtained is still a very valuable activity; specimen sets of data are provided to give students practice in such analysis. In cases where laboratory methods are to be used, the specimen data sets provide the opportunity to practise the analysis part of the laboratory exercise, and will also enable a student to see whether the results actually obtained in the laboratory are 'going to plan'.

The book is in three sections. The first section (Chapters 1 to 4) provides a summary of essential concepts and tools. Although much of this material may have been encountered previously, it is absolutely essential that a good understanding of these concepts is achieved in order to allow the rest of the book to be used properly. In the second section (Chapter 5), we describe in outline the information sought about proteins and the appropriate approaches involved. The third section (Chapters 6 to 11) illustrates data acquisition and handling in practice. In this section there are frequent references to the appropriate essential concepts, helping them to be reinforced. The section concludes by explaining how experimental work should be reported.

In each chapter there is a list of key concepts covered at the start of each subsection; at the end of the subsection there is a self-test feature, which includes questions to be attempted. Solutions to these questions are given at the end of each self-test box and ancillary information is provided in marginal notes. Many of the chapters have an extensive set of problems; solutions are provided in the Online Resource Centre associated with the book.

This book aims to be challenging but extremely rewarding!

Acknowledgements

We would like to thank our many colleagues in Glasgow, Stirling, and elsewhere (particularly Sharon Kelly, Colin Kleanthous, Douglas Lamont, Adrian Lapthorn, Tim Palmer, and James Milner-White) for helpful discussions and advice about various topics covered in this book.

Danielle McGinlay checked and refined many of the experimental protocols which are in the Online Resource Centre for teachers associated with this book; her efforts helped to make several of these experiments into worthwhile practical exercises.

Jonathan Crowe at Oxford University Press was a constant source of encouragement and advice; his input has improved the book enormously.

Chemical structures were prepared using ChemDraw Standard 8.0. Protein structures were generated using the open-source software PyMOL (DeLano, W. L. *The PyMOL Molecular Graphics System* (2002) on the World Wide Web at http://pymol.sourceforge.net/), using data from the Protein Data Bank at http://www.rcsb.org

Finally we would like to thank our families for their forbearance during the protracted gestation of this book.

Table of contents

A note on units

In this book we have adopted the 'superscript and power' system for indicating the units of quantities rather than the 'slash line' approach. Thus, a concentration would be quoted in terms of mol L^{-1} or mg mL^{-1} rather than as mol/L or mg/mL, and an energy change in terms of kJ mol^{-1}, rather than kJ/mol. Both types of system are widely used in scientific books and reports, so it is important for the reader to be aware of them. One of the major advantages of the 'superscript and power' system is that it is much easier to keep track of units, especially when quantities are multiplied or divided, and thus to check, for example, that the units are the same on the two sides of an equation. When reporting the values of quantities orally, mol L^{-1}, for example, would be said as 'moles per litre' and J K^{-1} mol^{-1} as 'joules per degree Kelvin per mole'. The subject of units is discussed in more detail in Chapter 3, section 3.3.

A

The basic concepts and tools

Success depends on a good understanding of the contents of this section. Remember the crucial test for understanding an idea is 'Can I explain it to a colleague so that he/she understands it?' If the answer is 'no' then don't attempt to move on until you have mastered that idea. Once you have properly understood a concept, it will be much easier to learn and should therefore be available for recall on any later occasion.

Conceptual toolkit: the molecular principles for understanding proteins

1

Introductory note

This chapter is intended to provide a concise summary of the principles with which you should be familiar in order to understand the structures and functions of proteins. If you have already studied chemistry in some depth, you may well not need to study this chapter in detail; a good test would be to see how well you can tackle the problems at the end of the chapter. If you are in any doubt, however, it is probably worthwhile taking the time to read it carefully and master its contents!

If you wish to read more about the molecular principles outlined in this chapter you will find that there is good coverage of these in most of the standard textbooks on biochemistry or introductory chemistry, such as Berg *et al.* (2007), Crowe *et al.* (2006), Jones (2005), Mathews *et al.* (2000), Price *et al.* (2001), and Voet *et al.* (2006)

Aims of this chapter

Proteins are complex three-dimensional structures with a very wide variety of functions in nature. This chapter is designed to help you acquire or consolidate your understanding of the structure and function of molecules, with a particular focus on those relevant to the study of proteins. It is not meant to be a 'small-scale' chemistry course, but to build on a basic knowledge of the Periodic Table of the elements and the ways in which ionic and covalent compounds can be formed. It starts with the amino acids as the basic building blocks of proteins and then deals with the various levels of protein structure. The forces involved in stabilizing the three-dimensional structures of proteins and in the interactions between proteins and other binding partners are described. At the end of this chapter, there is a set of chemical structures that you should aim to learn, as they will help to deepen your understanding of the ways in which proteins behave. There is also a set of problems that you can use to check your understanding of the principles in this chapter.

1.1 The different aspects of protein structure

KEY CONCEPT

- Understanding the four levels at which protein structure can be defined

To understand how proteins function it is necessary to know their structure. Whereas for small molecules such as ethanol, benzene, or glucose we can gain very useful insights into the properties from simple two-dimensional representations of their covalent structure, the same cannot be said for the much more complex molecules of proteins, which contain many thousands of atoms. In this case we have to extend the description of structure into three dimensions.

It is convenient to discuss the structure of proteins at four levels:

A. *Primary structure*, which refers to the *sequence of amino acids* in the polypeptide chain, i.e. the covalent structure of the protein. This includes any post-translational modifications such as glycosylation, phosphorylation, etc.

B. *Secondary structure*, which refers to the *local folding* of the polypeptide chain, such that segments of the chain may form helices, strands of sheet or turns.

C. *Tertiary structure*, which refers to the *long-range folding* of the polypeptide chain so that portions of the chain that are remote in terms of sequence are brought close together in space.

D. *Quaternary structure*, which refers to the association of the individual polypeptide chains (or subunits) in a multi-subunit protein.

Before we discuss these aspects of structure we should review the properties of the amino acids, which represent the building blocks of proteins.

1.2 The constituents of proteins, the amino acids

KEY CONCEPTS

- Knowing the general structure of an amino acid and understanding its stereochemistry
- Knowing the structures of the 20 common amino acids found in proteins
- Understanding the basis of amino acid classifications
- Explaining the behaviour of amino acids in chemical terms
- Understanding the properties of water as a solvent

There are 20 different amino acids commonly found in proteins. Of these one (proline) is actually a secondary amino acid (sometimes termed an imino acid). The general structure of an amino acid is shown in Fig. 1.1, where NH_2– is the amino group, $-CO_2H$ is the acid group, and R is the side chain. The central C atom to which R is attached is referred to as the α-carbon atom (denoted C_α). If the side chain consists of a chain of C atoms, these are usually referred to by Greek letters

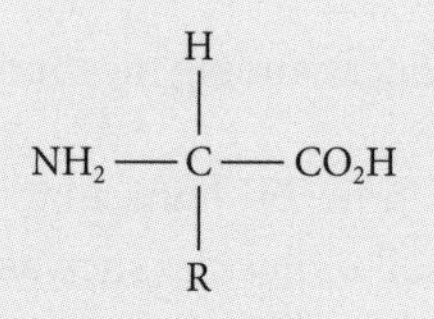

Fig. 1.1 General structure of an amino acid.

Fig. 1.2 The zwitterionic form of an amino acid in which the carboxyl group has lost a proton and the amino group has gained a proton.

$$^{+}NH_3 - \underset{\displaystyle R}{\overset{\displaystyle H}{C}} - CO_2^-$$

going away from the α-carbon atom, i.e. β, γ, δ, ε. Thus the side chain of glutamic acid (Glu) (see section 1.2.1) contains a γ-carboxyl group and of lysine (Lys) contains an ε-amino group.

In reality, in aqueous solution at pH values around neutrality, the amino group gains a proton and the acid group loses a proton, giving rise to the so-called zwitterionic form, shown in Fig. 1.2.

It will greatly help your understanding of proteins if you can learn the names, abbreviations, and chemical structures of the amino acids, i.e. of their side chains, because then you will be able to account for the structure and properties of these proteins in molecular terms. A list of the full structures of the amino acids is included in the Compendium of Structures in section 1.8.

The zwitterionic (from the German *zwitter*, meaning hybrid) form of an amino acid is one which is neutral overall but carries both a positive and a negative charge.

Many of the amino acids were named for what now seem rather obscure reasons, e.g. glycine comes from the Greek *glykeros* (sweet) because pure glycine was found to have a sweet taste. Leucine is named from the Greek *leukos* (white) because it was first isolated as a white crystalline solid; at the time whiteness was considered to be the defining property of a pure compound. Serine is derived from the Greek *serikon* and Latin *sericum* (silk) because it is found in significant quantities in the hydrolysis products of silk.

It should be noted that the classification of a few of the amino acids (particularly Cys and Gly) is difficult; they may be placed in different categories in other books.

1.2.1 The variety of amino acids

The amino acids are listed in alphabetical order in Table 1.1, with details of some of their important properties.

1.2.2 Classification of the amino acids in terms of polarity

The different amino acids can be classified in a number of ways, but probably the most useful in terms of understanding protein structure is that based on the polarity of the side chain. This can be thought of as a measure of whether the side chain promotes (polar) or discourages (non-polar) solubility in water. The polar amino acids are often further sub-divided into charged and uncharged side chains.

Non-polar side chain

Ala, Gly, Ile, Leu, Met, Phe, Pro, Trp, Val

Polar, uncharged side chain

Asn, Cys, Gln, Ser, Thr, Tyr

Polar, charged side chain

Arg, Asp, Glu, His, Lys

Table 1.1 Properties of the amino acids commonly found in proteins

Name	Abbreviation		Mass[a]	pK_a of side chain[b]	Frequency of occurrence (%)[c]
	Three-letter	One-letter			
Alanine	Ala	A	71.08	–	7.83
Arginine	Arg	R	156.19	12.5	5.35
Asparagine	Asn	N	114.10	–	4.18
Aspartic acid	Asp	D	115.09	3.9	5.32
Cysteine	Cys	C	103.14	8.4	1.52
Glutamine	Gln	Q	128.13	–	3.95
Glutamic acid	Glu	E	129.12	4.1	6.64
Glycine	Gly	G	57.05	–	6.93
Histidine	His	H	137.14	6.0	2.29
Isoleucine	Ile	I	113.16	–	5.91
Leucine	Leu	L	113.16	–	9.64
Lysine	Lys	K	128.17	10.5	5.93
Methionine	Met	M	131.20	–	2.38
Phenylalanine	Phe	F	147.18	–	4.00
Proline	Pro	P	97.12	–	4.83
Serine	Ser	S	87.08	–	6.86
Threonine	Thr	T	101.11	–	5.42
Tryptophan	Trp	W	186.21	–	1.15
Tyrosine	Tyr	Y	163.18	10.5	3.06
Valine	Val	V	99.13	–	6.71

[a]The mass in Da (Daltons) is given minus that of water. The molecular mass of a protein can be obtained by adding the values shown for the masses of the constituent amino acids plus that of one molecule of H_2O (18.02). This mass would be adjusted if necessary for the effect of any post-translational modifications, e.g. the addition of one phosphate group would add 79.98 Da.

[b]The pK_a value is that of the side chain in the free amino acids; in a protein the value of the pK_a can be markedly influenced by the precise environment of the side chain. The pK_a values of the carboxyl and α-amino groups for most amino acids are typically about 2.2 and 9.5, respectively.

[c]The frequency is taken from the occurrence of the amino acids in all sequences deposited in the Swiss-Prot database (release 49.0 7 Feb 2006; representing 75438310 amino acids in 207132 sequences).

The sign and magnitude of the free energy change indicates the tendency of a process or reaction to occur (see Chapter 4, section 4.1). A process with a large positive free energy change will have little tendency to proceed; conversely a process with a large negative free energy change will have a high tendency to proceed.

A quantitative measure of polarity can be provided by a so-called hydrophobicity scale. There are several ways in which this can be set up; that due to Engelman *et al.* (1986) corresponds to the free energy change for transfer of a given amino acid in an α-helix from the interior of a membrane to an aqueous medium (Table 1.2). In the table, the amino acids are ranked in order from most non-polar to most polar.

1.2.3 General properties of the amino acids

In this section we shall explore two general properties of amino acids before looking at specific chemical characteristics in the next section.

Table 1.2 A hydrophobicity scale for amino acids

Amino acid	Transfer free energy (kJ mol^{-1})
Phe (F)	15.5
Met (M)	14.2
Ile (I)	13.0
Leu (L)	11.7
Val (V)	10.9
Cys (C)	8.4
Trp (W)	7.9
Ala (A)	6.7
Thr (T)	5.0
Gly (G)	4.2
Ser (S)	2.5
Pro (P)	−0.8
Tyr (Y)	−2.9
His (H)	−12.5
Gln (Q)	−17.1
Asn (N)	−20.1
Glu (E)	−34.3
Lys (K)	−36.8
Asp (D)	−38.5
Arg (R)	−51.4

1.2.3.1 Stereochemistry

Since the α-carbon atom has four different substituents (except in the case of Gly for which R = H), all amino acids, except Gly, will display chirality. Only one of the enantiomers (mirror image forms) occurs in proteins; this is the L form of the amino acids (see Fig. 1.3). The D/L system for designating the stereochemistry of the amino acids is based on the D and L forms of glyceraldehyde as a reference compound, i.e. related to other compounds by a series of reactions of known stereochemistry. The more recent and absolute *R*/*S* system for describing the stereochemistry of chiral compounds is based on ranking the atoms or groups attached to the asymmetric carbon atom primarily in terms of atomic number and then establishing the direction of rotation of the ranked groups when looking along the bond from the carbon atom to the lowest ranked group (see Fig. 1.4). Almost all the L-amino acids have the *S* configuration at the α-carbon atom.

The side chains of Ile and Thr each contain a second chiral centre; these can be readily designated in the *R*/*S* system as *S* and *R* for Ile and Thr, respectively.

D-amino acids are found occasionally in nature, for example D-Ala and D-Gln in the peptidoglycan component of some bacterial cell walls, and D-Phe in gramicidin S, a cyclic decapeptide synthesized by the bacterium *Bacillus brevis*. Peptides containing these amino acids are made by special enzyme systems distinct from the normal ribosome-based protein synthesizing machinery.

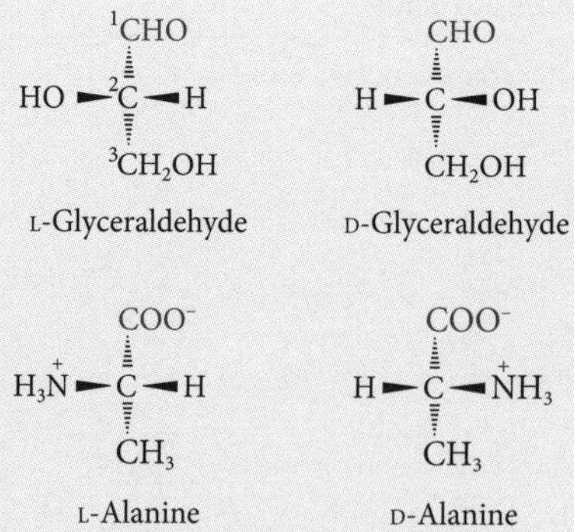

^{1}CHO
HO — ^{2}C — H
$^{3}CH_2OH$
L-Glyceraldehyde

CHO
H — C — OH
CH_2OH
D-Glyceraldehyde

COO^-
$H_3\overset{+}{N}$ — C — H
CH_3
L-Alanine

COO^-
H — C — $\overset{+}{N}H_3$
CH_3
D-Alanine

Fig. 1.3 The D/L system for denoting the stereochemistry of amino acids. The upper panels show the structures of L- and D-glyceraldehyde, which is the reference compound; the lower panels show the L and D forms of alanine. The solid arrows show atoms pointing towards the reader; the dotted arrows show atoms pointing away from the reader. The L and D forms (enantiomers) of each compound are mirror images of each other and are non-superimposable.

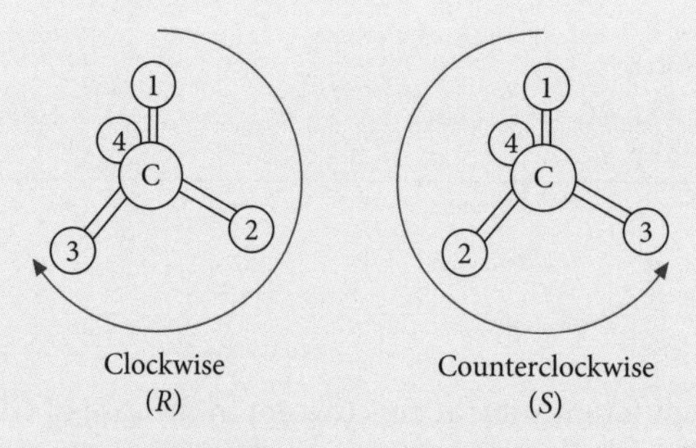

Fig. 1.4 The *R*/*S* system for denoting the stereochemistry at a chiral carbon atom. In this system the atoms linked to the carbon atom are ranked in terms of priority (1 being the highest ranked and 4 the lowest). When viewed looking from the carbon atom to atom 4, the direction in which the other atoms are ranked 1→2→3 is either clockwise (denoted *R*) or counterclockwise (denoted *S*). The priorities of some common substituent groups are $-OCH_2X > -OH > -NH_2 > -COOH > -CHO > -CH_2OH > -CH_3 > -H$.

1.2.3.2 Ionization

In Table 1.1, the average pK_a values of the α-amino and carboxyl groups of amino acids are given (9.5 and 2.2, respectively). These allow us to see why at neutral pH amino acids will exist in the zwitterionic form (NH_3^+–CHR–CO_2^-), since at pH 7 the amino group is below its pK_a and will thus be protonated, and the carboxyl group will be above its pK_a and thus be deprotonated. The actual charge carried by an amino acid at any particular pH will also depend on the nature and state of ionization of the side chain (see Table 1.1).

The isoelectric point (denoted by pI) of an amino acid is the pH at which it carries no net charge, i.e. where there is an exact balance of positive and negative charges. At the pI, the amino acid will not move in an electric field. For an amino acid with no ionizable side chain, the pI is equal to the average of the pK_a values of the carboxyl and α-amino groups (e.g. for Gly, pI = (2.4 + 9.8)/2 = 6.1). For an amino acid with an ionizable side chain, the pI is equal to the average of the two pK_a values on either side of the zwitterionic form. Thus for Asp, where the pK_a values for the α-carboxyl, the side chain carboxyl, and the α-amino group are 2.0, 3.9, and 9.9, respectively, the pI = (2.0 + 3.9)/2 = 2.95. For a protein, the actual pI will depend on the balance between the numbers of the acidic and basic amino acids in that protein.

The calculation of the pI value for amino acids with ionizing side chains is discussed in Problems 1.6 and 1.7 at the end of this chapter.

KEY INFO 

It should be remembered that the ionization states of the side chains of amino acids in proteins can be significantly different from those of the free amino acids. For example, the pK_a of Glu 35 in lysozyme is raised from 4.1 in the free amino acid to about 6.0 because the side chain is in a rather non-polar environment. This has the effect of favouring the neutral (protonated) form rather than the negative (deprotonated) form. Conversely, the pK_a of Tyr 28 in type II dehydroquinase is lowered from 10.5 in the free amino acid to about 8.0 by the presence of a neighbouring Arg side chain. The positive charge on the Arg will favour the deprotonated form of the Tyr side chain relative to the protonated form. In the case of Ser 195 of chymotrypsin, the pK_a is lowered dramatically (by over 7 units) due to the charge relay system. It is therefore always important to examine the actual environment of an amino side chain in a protein before assessing its likely function.

1.2.4 Chemical characteristics of the amino acids

It is convenient to discuss the various amino acids as groups which display similar chemical features. Again, knowing the structures of the side chains involved will help to understand the molecular basis of the property.

1.2.4.1 Aliphatic side chains (Ala, Gly, Ile, Leu, Val)

These side chains are chemically unreactive. They are important in hydrophobic interactions (section 1.7.4) with other non-polar side chains or parts of other molecules.

Because of its small side chain, Gly has a large degree of conformational flexibility, which is important when the three-dimensional structures of proteins are considered (section 1.4).

Aliphatic compounds are those in which carbon atoms are linked in straight or branched chains, rather than rings.

1.2.4.2 Aromatic side chains (Phe, Tyr, Trp)

These side chains absorb radiation in the near-ultraviolet (the absorbance maxima for Phe, Tyr, and Trp are 258 nm, 275 nm, and 280 nm, respectively).

The side chain of Phe is chemically unreactive but can participate in hydrophobic interactions (section 1.7.4).

Aromatic compounds are those which contain benzene-type rings. Their enhanced stability arises from the delocalized π-electron systems.

An electrophile is a species or part of a molecule with a positive or partial positive charge which will seek an electron-rich species or centre. Examples include the Zn^{2+} ion or the C atom in the >C=O bond. In the case of tyrosine, reagents that bring about electrophilic substitution include I_2 and the NO_2^+ (nitronium) ion.

Thyroxine, an iodinated derivative of tyrosine, is a hormone which is produced in the thyroid gland. It stimulates the synthesis of specific proteins and is essential for normal growth and development. A number of isotopes of iodine (particularly ^{129}I and ^{131}I) are radioactive and can be dangerous because they will be concentrated in the thyroid gland, leading to thyroid cancer.

A nucleophile is a species or a centre in a molecule with a negative or partial negative charge which will seek an electron-deficient species or centre. Examples include OH^- ions and the side chains of Cys ($-CH_2-S^-$), Lys ($-NH_2$), and Ser ($-CH_2-O^-$) with one or more lone pairs of electrons.

The side chain of Tyr is more reactive (due to the hydroxyl group) and will undergo electrophilic substitution reactions. One such example is iodination, which occurs naturally to give the hormone thyroxine. The hydroxyl group has a high pK_a (typically about 10.5) so will not be significantly ionized under normal physiological circumstances; it can, however, participate in hydrogen bonding (section 1.7.2) and be phosphorylated by the action of kinases (section 1.2.4.5).

The side chain of Trp is of limited chemical reactivity, but the N–H group can participate in hydrogen bonding (section 1.7.2).

1.2.4.3 *Basic side chains (Arg, Lys)*

These side chains are grouped together because at neutral pH they carry a positive charge and can therefore be involved in ionic interactions with negatively charged side chains (Asp and Glu) or with negatively charged parts of other molecules, such as phosphate groups.

The pK_a of Lys is around 10.5 for the free amino acid. The $-NH_3^+$ group is unreactive, but the deprotonated form ($-NH_2$) can act as a powerful nucleophile.

The pK_a of Arg is around 12.5, and the guanidine part of the side chain is therefore essentially positively charged under all physiological conditions.

1.2.4.4 *Acidic side chains (Asp, Glu)*

For both free amino acids, the pK_a values of the side chains are around 4.0, and hence would generally be negatively charged at neutral pH. These side chains can be involved in ionic interactions with positively charged side chains, or in binding to metal ions such as Ca^{2+} or Zn^{2+}.

1.2.4.5 *Hydroxyl side chains (Ser, Thr)*

The –OH group of the side chain of both amino acids can take part in hydrogen bonding (section 1.7.2). The pK_a of the side chains is very high (about 15), so they are in the protonated state under physiological conditions and act as very weak nucleophiles. However, in certain cases, such as the Ser proteases (see Chapter 9, section 9.9.3), the Ser can lose its proton by the charge relay system and become a very powerful nucleophile. Ser and Thr side chains (along with Tyr side chains, section 1.2.4.2) in a number of proteins can become phosphorylated by the action of specific enzymes (kinases); such processes are often involved in regulation of the activity of these proteins.

1.2.4.6 *Amide side chains (Asn, Gln)*

The amide group has only very weak acidic and basic properties so the side chains of these amino acids are always uncharged. They can participate in hydrogen bonding (section 1.7.2) as either donors or acceptors.

1.2.4.7 *Sulphur-containing side chains (Cys, Met)*

The sulphur atom of the Cys side chain is a highly reactive nucleophile, especially in the deprotonated state ($-CH_2S^-$). The pK_a of the side chain (8.4 for the free amino acid) is sufficiently low that at neutral pH there will be a small, but significant, fraction in the deprotonated form.

Under oxidizing conditions, two appropriately positioned Cys side chains can form a disulphide bond ($-CH_2-S-S-CH_2-$), which constitutes the amino acid cystine. Disulphide bonds are found in extracellular, secreted proteins such as antibodies, digestive enzymes, etc.

In the presence of oxidizing agents, the Cys side chain can be oxidized to a sulphonic acid ($-CH_2-SO_3H$) or the intermediate sulphenic ($-CH_2-SOH$) and sulphinic ($-CH_2-SO_2H$) acid forms. The sulphur atom in Cys can act as a ligand to a number of metal ions, e.g. Zn^{2+} or Fe^{2+}, and is a major site of inhibition of proteins by heavy metals such as Hg, Cd, or Pb.

The sulphur atom of Met is much less reactive, although it can function as a nucleophile towards adenosine-5′-triphosphate (ATP) to form *S*-adenosylMet (adoMet) in a reaction involving the release of the three phosphate groups, which is catalysed by methionine adenosyl transferase. The positive charge on the sulphur atom of adoMet makes it susceptible to nucleophilic attack, and hence adoMet is a powerful methylating agent, e.g. of cytosine bases in DNA. The side chain of Met can also be modified by strong oxidizing agents to give Met sulphoxide.

Disulphide bonds are only found in secreted proteins. In intracellular proteins (i.e. those which remain inside cells) the cysteine side chains remain in the reduced ($-CH_2-SH$) form. The oxidation of secreted proteins occurs in the lumen (interior compartment) of the endoplasmic reticulum in eukaryotic cells and in the periplasmic space of prokaryotic cells.

Oxidative damage to proteins, particularly to the sulphur-containing amino acids, is thought to be an important factor in ageing-related diseases. Reaction of Cys side chains in proteins with reactive oxygen species such as H_2O_2 or the superoxide anion can yield the sulphenic acid form. This reaction is usually reversible, but further oxidation cannot generally be reversed.

1.2.4.8 *Proline*

Proline is a special amino acid since its side chain is a cyclic ring. This structural feature means that there are rigid geometrical constraints on the peptide bonds to Pro (section 1.3.1). The side chain of Pro is chemically rather unreactive, although it can be hydroxylated in the 4 position by the action of prolyl-4-hydroxylase. 4-Hydroxyproline is found in the protein collagen, which occurs extensively in bone, skin, and connective tissue.

Prolyl-4-hydroxylase requires ascorbic acid (vitamin C) as a cofactor. A deficiency of this vitamin leads to the condition of scurvy, a disease of the skin, joints, etc. caused by disruption to the correct synthesis and assembly of collagen fibres.

1.2.4.9 *Histidine*

The pK_a of the imidazole ring is about 6.0 in the free amino acid. Thus, at near-neutral pH, there is balance of the protonated (positively charged) and deprotonated (neutral) forms. His side chains are often found at the active sites of enzymes as they can play a key role in acid–base catalysis. The neutral form of His is a powerful nucleophile and can participate in hydrogen bonding (section 1.7.2); in addition it can act as a ligand to metal ions such as Zn^{2+} or Fe^{2+}.

1.2.5 The structure of water

Since the majority of biological processes take place in an aqueous medium, it is appropriate to review the structure of liquid water. The geometry of the water molecule is well known, with the H–O–H bond angle = 104.5° and the O–H bond

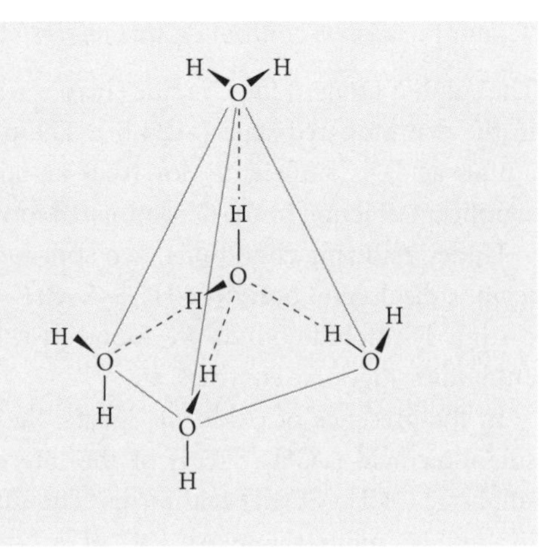

Fig. 1.5 A water molecule hydrogen bonded to its four nearest neighbours. The covalent O–H bonds are shown as solid lines; hydrogen bonds are shown as dotted lines. The four nearest-neighbour water molecules are arranged in a tetrahedral fashion. The arrangement is found in one of the most common structures of ice, where there are effectively layers of water molecules.

length = 0.096 nm. There is a separation of charge due to the higher electronegativity (electron attracting power) of the O atoms, leading to partial positive charges on the hydrogens and a partial negative charge on the oxygen (about 33% ionic character). The two lone pairs of electrons on the O atom can each act as acceptors in hydrogen bonds and the two hydrogen atoms can act as donors (section 1.7.2). Thus, each water molecule can have up to four nearest-neighbour hydrogen-bonded water molecules (Fig. 1.5). This network is found in ice, which has a significant amount of empty space and thus a relatively low density (ice floats on water).

Water is unusual in that its solid form (ice) is less dense than its liquid form. This is of considerable importance for aquatic organisms in the winter!

In liquid water, this regular hydrogen-bonded network is broken down and instead there are fluctuating clusters of water molecules with, on average, each water having about 3.5 hydrogen-bonded nearest-neighbour molecules. Because of the polar nature of the water molecule there will be strong interactions with charged or other polar species, making water an excellent solvent for these types of molecules. However, water is a very poor solvent for non-polar molecules; this is discussed further in the context of hydrophobic interactions in section 1.7.4.

1.3 The primary structure of proteins

KEY CONCEPTS

- Drawing the structure of a peptide chain
- Understanding the *cis*/*trans* isomerism of the peptide bond unit
- Understanding the information available from the primary structure of a protein, including molecular mass, extinction coefficient, isoelectric point, and any post-translational modifications

The term 'primary structure' refers to the sequence of amino acids in a protein, and is dictated by the nucleotide sequence of the gene encoding the protein, taking into account (a) the splicing out of any intervening sequences (introns) to produce the mature messenger RNA and (b) any post-translational modifications which might remove amino acids by proteolysis or modify amino acids, e.g. by addition of carbohydrate or phosphoryl groups.

1.3.1 The peptide bond

The peptide bond is formed between the carboxyl group of one amino acid and the amino group of a second amino acid, with the loss of a molecule of H_2O. By convention, a peptide chain is written such that the amino acid with the free amino group (the amino (or N)-terminal acid) is at the left-hand end of the chain and the amino acid with the free carboxyl group (the carboxyl (or C)-terminal acid) is at the right-hand end (Fig. 1.6). Peptide chains in proteins are unbranched.

The formation of a dipeptide from two amino acids is thermodynamically unfavourable. In nature, ATP is used as an energy source to activate one of the amino acids and drive the reaction to the side of synthesis.

The peptide bond itself is not adequately represented by the structure shown in Fig. 1.7, as shown by the fact that the C–N bond distance (0.132 nm) is between

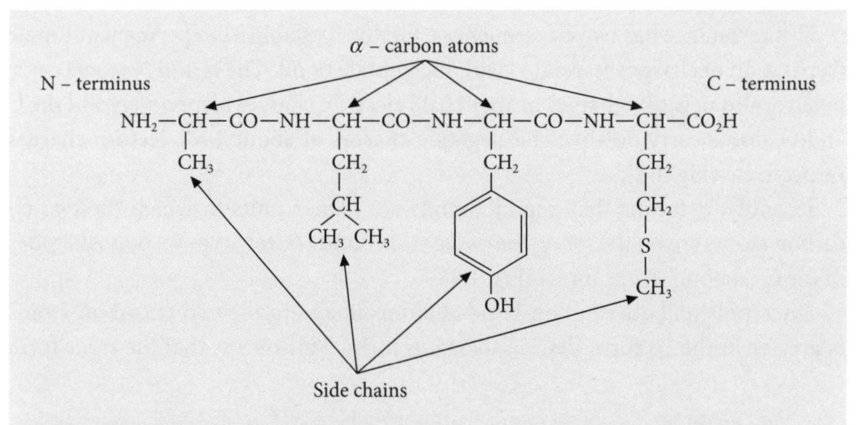

Fig. 1.6 Structure of a tetrapeptide showing the α-carbon atoms, side chains, and N- and C-termini. The tetrapeptide is written as Ala–Leu–Phe–Met (ALFM in the one-letter code). In this diagram the terminal amino and carboxyl groups are shown in the uncharged forms; in practice they would carry a positive charge and negative charge from the gain and loss of a proton, respectively.

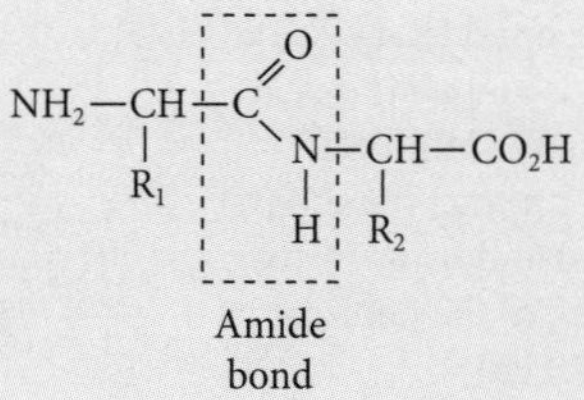

NH_2—CH—C(=O)—N(H)—CH—CO_2H (R_1, R_2)

Amide bond

Fig. 1.7 Structure of a peptide bond with the C–N bond shown as a single bond.

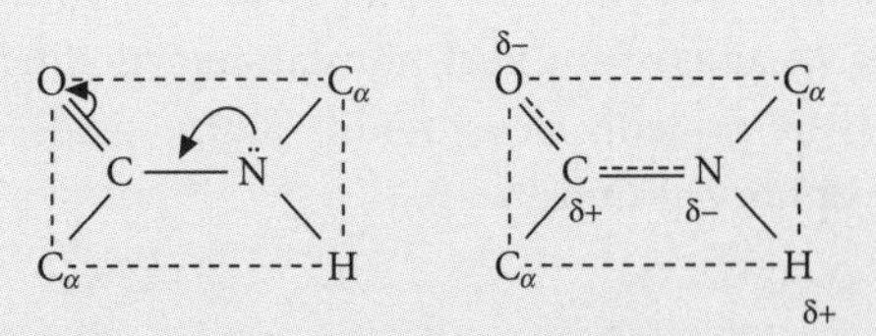

Fig. 1.8 Resonance stabilization of the peptide bond. The left-hand panel shows delocalization of the lone pair of electrons on the N atom. The right-hand panel shows the partial double bonds in the peptide bond unit and the partial charges on the C, O, N, and H atoms.

that of a C–N single bond (0.145 nm) and a C=N double bond (0.125 nm). This is usually explained by resonance stabilization, with electron density in the lone pairs on the O atom being transferred to the N atom, giving rise to partial double bond character (Fig. 1.8).

Since the N atom carries a net negative charge, there must be extensive back-donation of σ electrons from C to N to compensate for the π electron movement.

This resonance has two consequences. Firstly, it establishes a permanent dipole (separation of charge) associated with the peptide bond. The H and N atoms carry positive and negative charges of about 0.28 electron charges respectively and the C and O atoms carry positive and negative charges of about 0.39 electron charges respectively (Fig. 1.8).

Secondly, it means that peptide bonds are planar units in which the two α-carbon atoms are on the same side of the C–N bond (*cis* form) or on opposite sides of the C–N bond (*trans* form) (Fig. 1.9).

The repulsion between non-bonded atoms linked to the two α-carbon atoms is greater in the *cis* form than in the *trans* form. This means that the *trans* form

Fig. 1.9 The *cis* and *trans* forms of the peptide bond. (a) and (b) show the *cis* and *trans* forms, respectively, of the peptide bond (Xaa–Zaa), where Xaa is any amino acid and Zaa is any amino acid other than proline. (c) and (d) show the *cis* and *trans* forms, respectively, of the peptide bond (Xaa–Pro).

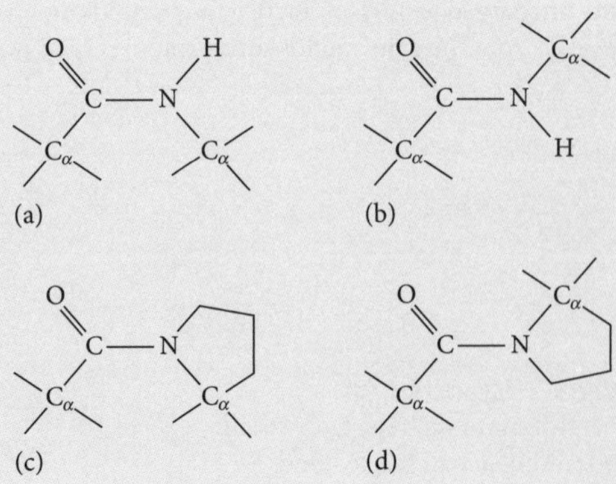

is more stable than the *cis*; in proteins it can be estimated that this is by a margin of about 20 kJ mol^{-1}.

An exception to this statement is in the case when the peptide bond is to proline (Fig. 1.9), which as we have seen is an unusual amino acid with the side chain forming a cyclic ring. This has the effect that the repulsion between non-bonded atoms is considerably reduced and the balance between *cis* and *trans* is much closer. The *trans*/*cis* ratio is now much lower (about 20), i.e. about 5% of Xaa–Pro bonds in proteins are in the *cis* form (Jabs *et al.*, 1999).

It should be noted that there is a considerable activation energy barrier (of the order of 80 kJ mol^{-1}) to rotation about the peptide bond because of its partial double bond character. This isomerization of the peptide bonds to Pro can represent one of the slow steps in the folding of proteins and there is an enzyme, peptidyl-prolyl isomerase, which can catalyse this process.

1.3.2 Information available from the amino acid sequence of a protein

Knowledge of a protein sequence gives us a good deal of information, as set out below. Further details are given in the section on bioinformatics (see Chapter 5, section 5.8).

1.3.2.1 Exact molecular mass

From the numbers of each amino acid in a protein, we can calculate the exact mass using the values given in Table 1.1 for the masses of each amino acid and adding the mass of H_2O. For example, the calculated molecular mass of the polypeptide chain of type II dehydroquinase from *Streptomyces coelicolor* is 16550.6 Da. This can be compared with the very precise measurements of mass available from mass spectrometry to confirm the identity and integrity of the protein. In this case the measured mass of the purified enzyme is 16549.7 ± 1.2 Da, in excellent agreement with the predicted value.

1.3.2.2 Isoelectric point (pI)

As mentioned in section 2.3, the pI of a protein is the pH where it carries no net charge. The way that the charge on the protein will change with pH is illustrated in Fig. 1.10.

The pI of a protein will reflect a balance between the numbers of positively charged groups (N-terminal amino group and the side chains of Arg, His, and Lys) and negatively charged groups (C-terminal carboxyl group and the side chains of Asp and Glu). The predicted pI is based on the numbers of these charged groups and their average pK_a values in proteins. If the positive groups are predominant, the pI will be high (e.g. for pig lysozyme, a value of 9.06 is predicted); if the negative groups predominate, the pI will be low (e.g. for pig pepsin, a value of 3.24 is predicted). It should be noted that these are only rough estimates of the pI value since the pK_a values of side chains in proteins can depend markedly on their environment. Nevertheless, the predicted value is a useful guide to the behaviour of the

Electrophoresis refers to the movement of a charged molecule in an electric field. Electrophoresis of proteins is often performed in a cross-linked polyacrylamide gel (see Chapter 6, section 6.2, and Chapter 8, sections 8.2.1 and 8.2.2). Ion-exchange chromatography is a widely used technique for separating proteins on the basis of the charge that they carry (see Chapter 7, section 7.2.2).

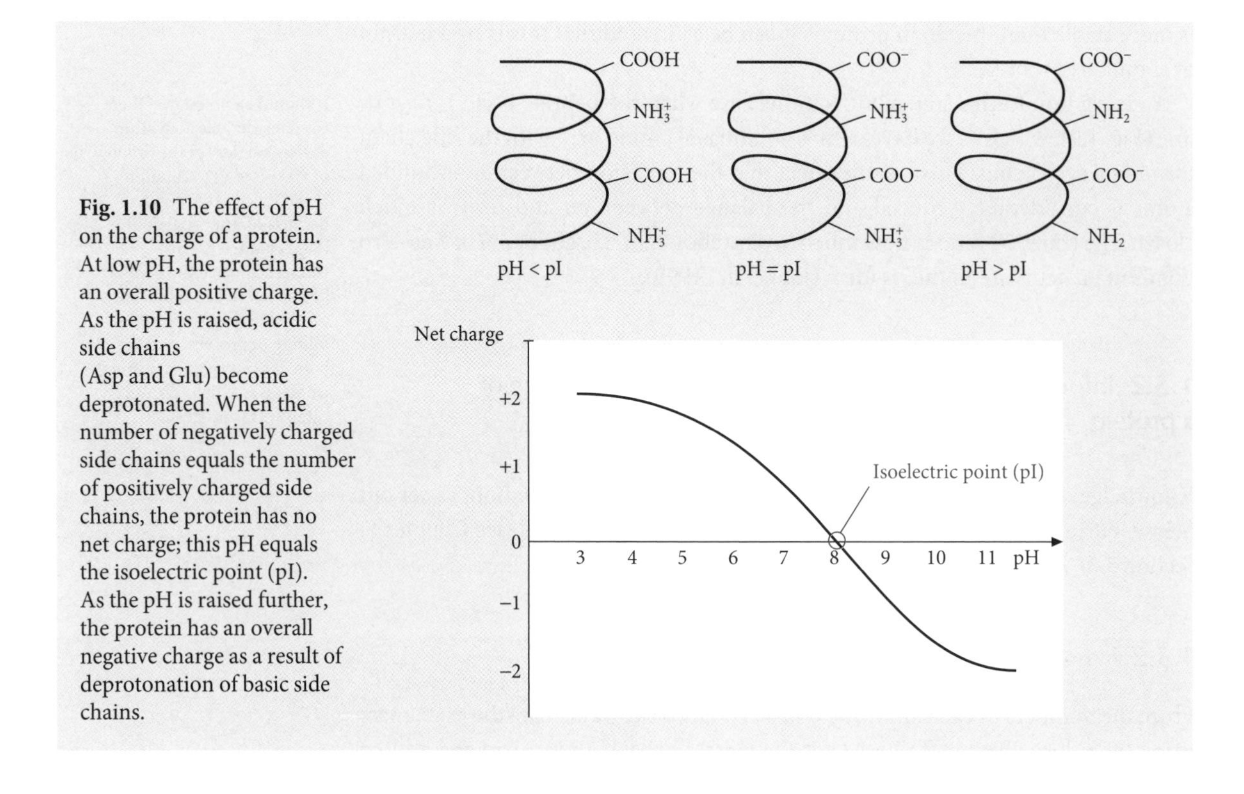

Fig. 1.10 The effect of pH on the charge of a protein. At low pH, the protein has an overall positive charge. As the pH is raised, acidic side chains (Asp and Glu) become deprotonated. When the number of negatively charged side chains equals the number of positively charged side chains, the protein has no net charge; this pH equals the isoelectric point (pI). As the pH is raised further, the protein has an overall negative charge as a result of deprotonation of basic side chains.

protein on electrophoresis (see Chapter 6, section 6.2.3) or ion-exchange chromatography (see Chapter 7, section 7.2.2), for example.

1.3.2.3 *Absorption coefficient*

The calculation of the absorption coefficient depends on the assumption that there is no other chromophore (absorbing group) present, such as the haem group in haemoglobin. If such a group is present, its contribution to the absorption at 280 nm must be added.

The absorption of radiation at 280 nm depends on the aromatic amino acids Tyr and Trp; there is also a small contribution from any disulphide bonds which may be present. The molar absorption coefficient (also known as the extinction coefficient) of the protein at 280 nm can be derived by adding together the contributions from the numbers of these different amino acids; these are based on their numbers in the protein and their known absorption coefficients at 280 nm. Division by the molecular mass of the protein gives the absorption coefficient for a 1 mg mL^{-1} solution of the protein (see Chapter 6, section 6.1.5).

1.3.2.4 *Hydrophobicity*

The balance between polar and non-polar amino acids in the protein will give a guide to the overall polarity of the protein and can point to whether it may be membrane-associated, for example. The aliphatic index is a measure of the frequency of occurrence of the bulky aliphatic side chains (Ala, Ile, Leu, and Val).

Table 1.3 Some important post-translational modifications of proteins

Type of modification	Possible effects on function
Proteolysis	Removal of targetting sequences Generation of several new products, e.g. hormones Activation of proteins, e.g. enzymes
Disulphide bond formation	Stabilization of structure of secreted proteins
Hydroxylation	Formation of hydroxy-Lys or -Pro increases the stability of the triple helix of collagen
Glycosylation	Many cell surface proteins are involved in cell–cell recognition Attachment of glycosyl-phosphatidyinositol groups anchors proteins to membranes The polar nature of proteins can be enhanced
Phosphorylation	Phosphorylation of Ser, Thr or Tyr side chains can regulate the activity of proteins, especially in signalling pathways
N-terminal acylation	Attachment of C_{14} (myristoylation) or C_{16} (palmitoylation) chains will enhance the association of the protein with membranes

1.3.2.5 Post-translational modifications

Many proteins undergo changes in covalent structure after translation of the mRNA on the ribosome. These changes are collectively termed post-translational modifications and some of the more important of these are listed in Table 1.3.

The occurrence of post-translational modifications can be deduced either by direct analysis, e.g. testing for the presence of saccharide units in the case of glycosylation, or by careful measurements of the molecular mass of the protein using mass spectrometry. If the measured molecular mass of a protein does not agree with the value calculated on the basis of the amino acid sequence, then it can be concluded that either the sequence is incorrect or (more likely) post-translational modification has occurred. From the observed mass differences it should be possible to draw firm conclusions about the types of modifications which have occurred. In the case of horse lysozyme the measured mass is 14644 Da, indicating a loss of 8 Da from the predicted value (14652 Da); this can be accounted for by the formation of four disulphide bonds from the eight Cys residues in the sequence, which corresponds to the loss of eight hydrogen atoms.

In a number of cases, extensive post-translational modification (e.g. glycosylation) may hinder the analysis of a protein by mass spectrometry, as the protein cannot readily form gaseous phase ions.

1.3.2.6 Structural and functional motifs

Analysis of the sequence can reveal the way certain parts of the protein may play distinct roles. For example, a stretch of 20 amino acids which is very likely to form a membrane-spanning (or transmembrane) α-helix (see section 1.4) can be identified by constructing a hydropathy plot in which the free energy for transferring this stretch of amino acids from the membrane to water (obtained by adding the appropriate values of the amino acids as listed in Table 1.2) is plotted as a function of the first residue in the stretch. If the free energy is above a certain threshold value (+84 kJ mol^{-1}) this indicates that a transmembrane helix is likely. Protein sequences can be scanned automatically to locate such elements, which often occur many times in a single chain, for example the G-protein coupled receptors have seven transmembrane helices and the family of glucose transporters in mammalian tissues have 12 transmembrane helices.

It should be noted that the choice of the free energy threshold for identification of a membrane-spanning element of a protein is somewhat arbitrary; the quoted value has been found to predict such elements with reasonable accuracy.

Other types of motifs that can be revealed by sequence analysis are listed below (Xaa equals any amino acid):

Targetting sequences, e.g. –Ser–Lys–Leu at the carboxyl terminus and –Lys–Asp–Glu–Leu at the carboxyl terminus act as sequences which direct proteins to the peroxisome or to be retained in the endoplasmic reticulum respectively.

Metal binding, e.g. –Cys–Xaa_4–Cys–Xaa_2–Cys– is a consensus site for Fe binding in the 2Fe–2S cluster iron sulphur proteins.

Phosphorylation sites, e.g. –Arg–Xaa_{1-2}–Ser/Thr– or –Arg–Arg–Xaa–Ser/Thr– are consensus sites for phosphorylation at the Ser or Thr by protein kinase A.

Glycosylation sites, e.g. –Asn–Xaa–Ser/Thr– (where Xaa is any amino acid but is rarely Pro or Asp) is a consensus sequence for N-glycosylation at Ser/Thr.

A commonly used tool for identifying structural motifs is Prosite (http://www.expasy.org/prosite). For further details see Chapter 5, section 5.8.5.

1.3.2.7 Sequence relationships between proteins

Analysis of the huge number of sequences in the databases has provided powerful insights into many aspects of proteins. For example, comparisons between the analogous proteins from different organisms has given clues about evolutionary relationships and helped in classification of species. When many such proteins are compared, it is seen that some amino acids are totally conserved between species, some are partially conserved (i.e. are of similar chemical types) and others are freely variable. The totally conserved residues are likely to play essential roles in the structure and/or the function of the protein. The comparisons between related, but distinct, proteins within the same organism, for example haemoglobin and myoglobin, trypsin and chymotrypsin, point to gene duplication followed by independent (divergent) evolution as the likely mechanism. Comparisons between sequences have also been used to look for clues to possible functions of proteins, or of segments of proteins. For example, analysis of the sequences of a number of steroid hormone receptor proteins shows that they consist of three domains (independent folded and functional units), namely a transactivation domain which regulates transcription, a DNA binding domain, and a hormone binding domain. For further details about the use of bioinformatics-based approaches, see Chapter 5, section 5.8.

1.4 The secondary structure of proteins

KEY CONCEPTS

- Defining the torsional (dihedral) angles in a peptide bond
- Drawing the hydrogen bonding patterns in the α-helix, β-sheet, and β-turn structures
- Explaining why proline cannot be accommodated in regular α-helix and β-sheet structures

Because of the partial double bond character of the peptide bond, the six atoms of C_α–CO–NH–C_α can be considered to lie in a plane (Fig. 1.11). Rotations around the bonds to the linking C_α atoms between peptide units will define the path traced and hence the structure adopted by the polypeptide chain. The two angles (known as torsion or dihedral angles) are defined as ϕ (phi) and ψ (psi) for the bonds between N and C_α and between C_α and C, respectively. Values of 180° refer to the fully extended polypeptide chain. Positive values of ϕ and ψ refer to clockwise rotation, negative values to counter-clockwise rotation.

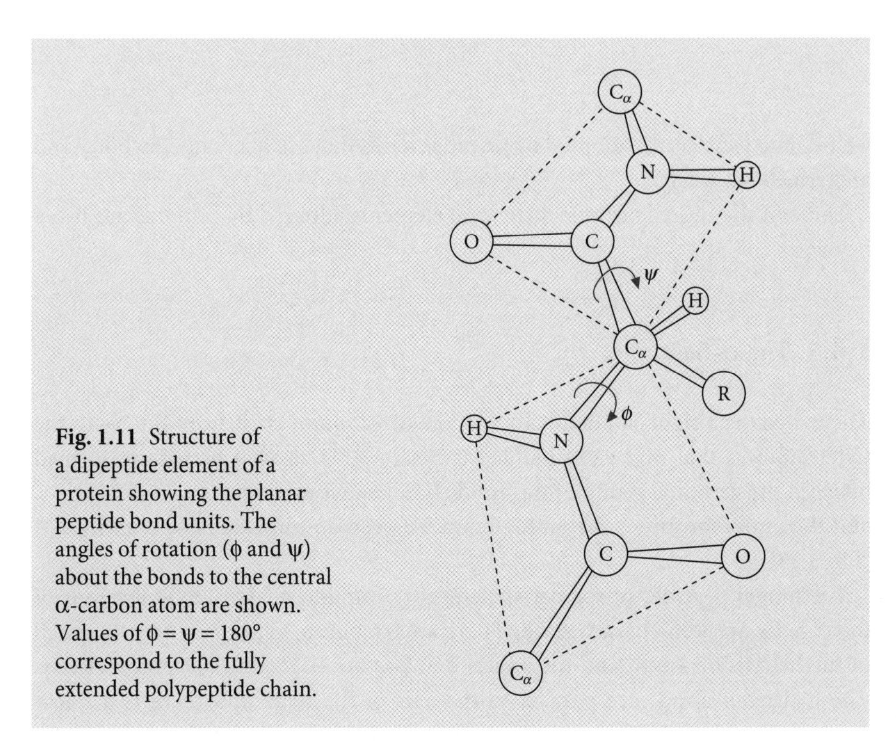

Fig. 1.11 Structure of a dipeptide element of a protein showing the planar peptide bond units. The angles of rotation (ϕ and ψ) about the bonds to the central α-carbon atom are shown. Values of $\phi = \psi = 180°$ correspond to the fully extended polypeptide chain.

Rotations about the bonds to the C_α atoms will alter the distances between non-bonded atoms of the side chains and hence the energy of the system. Calculations of the energy as a function of the angles ϕ and ψ have been made and the resulting plot (known as a Ramachandran diagram, Fig. 1.12) shows that there are certain

The precise allowed areas of the Ramachandran plot will depend on the nature of the side chains involved. Thus glycine, with its very small side chain, can occupy a considerably greater fraction of the total Ramachandran plot than can an amino acid with a bulky side chain such as valine or tryptophan.

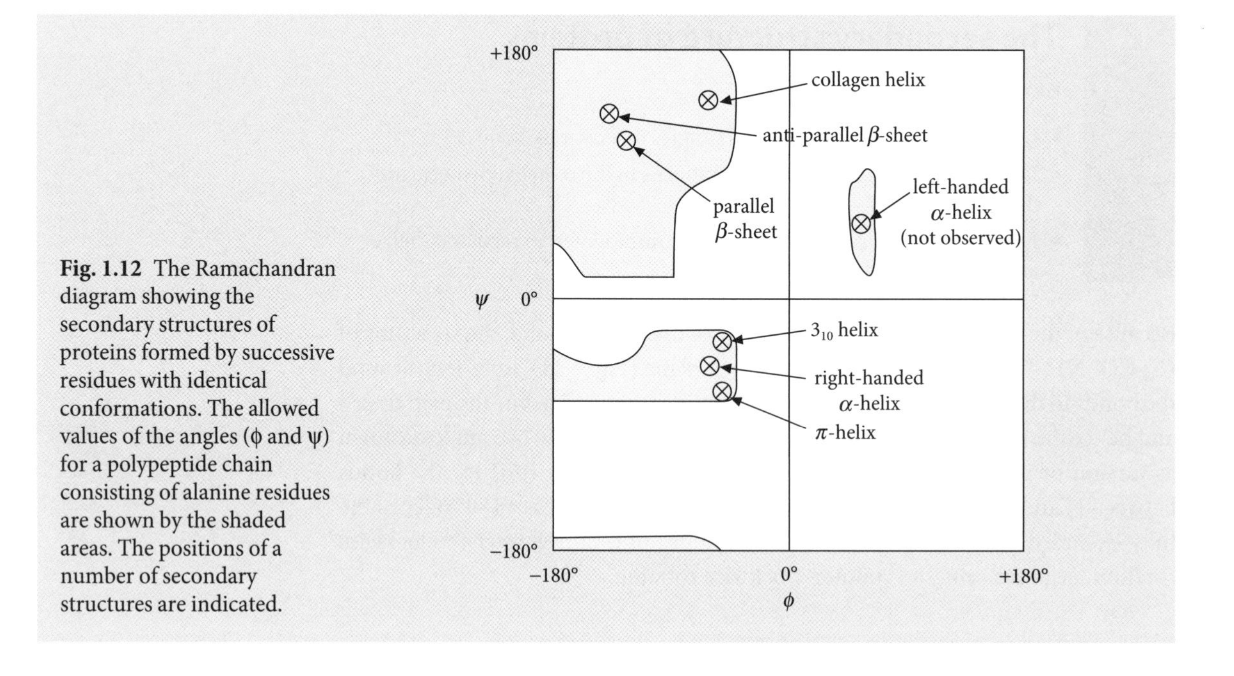

Fig. 1.12 The Ramachandran diagram showing the secondary structures of proteins formed by successive residues with identical conformations. The allowed values of the angles (ϕ and ψ) for a polypeptide chain consisting of alanine residues are shown by the shaded areas. The positions of a number of secondary structures are indicated.

well-defined structures adopted by peptide chains that are of greatest stability and are termed 'allowed'.

Some of the more common structural elements adopted by proteins are listed below.

1.4.1 The α-helix

Electrostatic interactions between the carbonyl group of amino acid n and the carbonyl groups of amino acids $n + 3$ and $n + 4$ also contribute to the stability of the α-helix. For L-amino acids, the right-handed α-helix is more stable than a left-handed helix.

The α-helix is a right-handed helix (i.e. the direction of twist from the N- to the C-terminus is that of a right-handed corkscrew). Hydrogen bonds are formed between the carbonyl group of the amide bond between amino acids n and $n + 1$, and the amino group of the amide bond between amino acids $n + 3$ and $n + 4$ (Fig. 1.13).

For model peptides of regular structure (polyamino acids) the dimensions of the α-helix are well characterized. There are 3.6 amino acids per turn, the pitch of the helix is 0.54 nm, and the angles ϕ and ψ are −57° and −47°, respectively. The hydrogen bonds are parallel to the axis of the helix and there is a linear arrangement of nuclei with the O⇌N distance 0.286 nm. There is a helix dipole with the N-terminal and C-terminal ends of the helix having positive and negative charges, respectively; this dipole can be important in stabilizing interactions between adjacent helices. Since there are 13 atoms in the ring containing the hydrogen bond (Fig. 1.13) and 3.6 amino acids per turn, an α-helix is sometimes referred to as a 3.6_{13} helix.

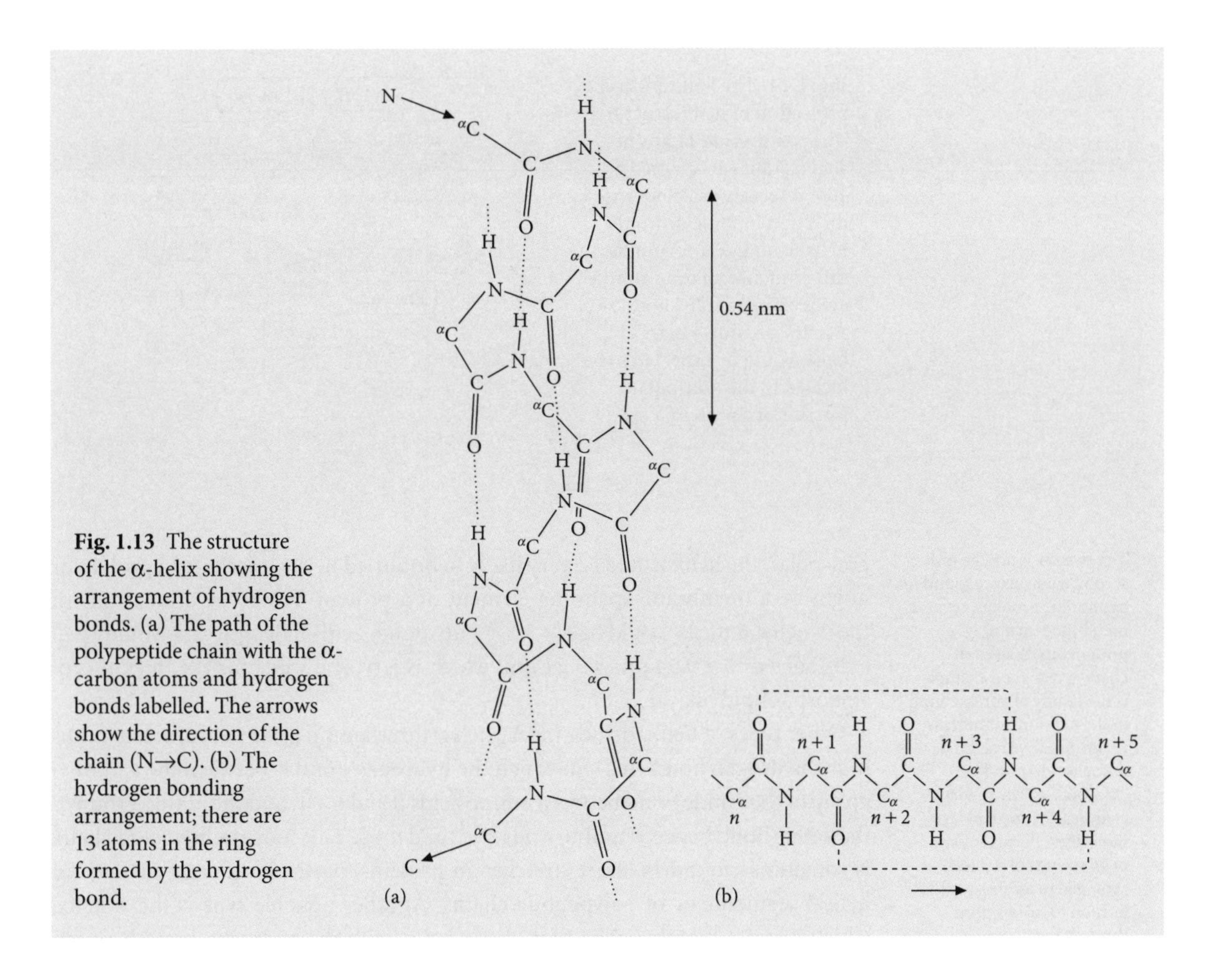

Fig. 1.13 The structure of the α-helix showing the arrangement of hydrogen bonds. (a) The path of the polypeptide chain with the α-carbon atoms and hydrogen bonds labelled. The arrows show the direction of the chain (N→C). (b) The hydrogen bonding arrangement; there are 13 atoms in the ring formed by the hydrogen bond.

Because of the steric constraints imposed by its cyclic side chain, Pro cannot be accommodated in the regular α-helix structure and will therefore either act as a helix breaker or will cause a major kink in the helix.

In proteins the α-helices are often considerably distorted from the idealized structures seen in model polypeptides with, for example, the hydrogen bond geometry less than ideal. The average length of α-helices in proteins is quite short, about 12 amino acids or 3.5 turns.

The representation of the α-helix in Fig. 1.13 does not show the side chains; these project out from the helix into the solvent. For some purposes it is useful to depict the arrangement of side chains in a projection known as the helical wheel (Fig. 1.14).

Note that with 3.6 amino acids per turn of the α-helix, the angle between successive α-carbon atoms in the helical wheel plot is 360/3.6 degrees, i.e. 100 degrees.

This might show, for example, that one face of an α-helix is polar and the other non-polar, in which case the helix is said to be amphipathic; such a helix might be involved in anchoring the protein to a membrane. If both sides of the helix were

Fig. 1.14 The helical wheel projection of helix structure. The positions of 11 amino acid residues are plotted at 100° intervals in a clockwise direction looking down the helix from the N-terminus. In this sequence all the amino acids except Ser261 and Asn267 are non-polar. This helix would be expected to be located in the non-polar interior of a protein.

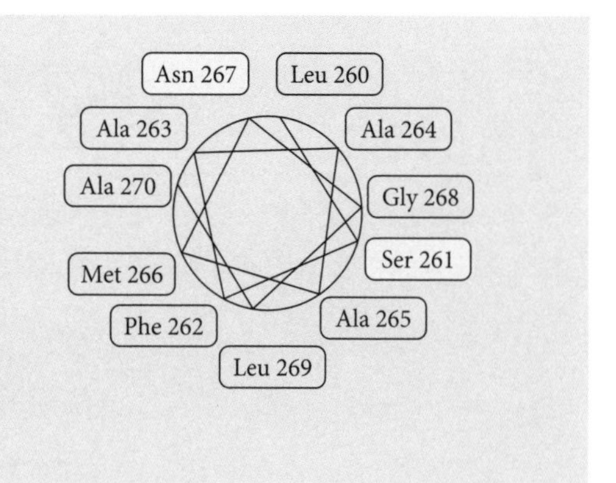

The α-helix is not the only secondary structure found in membrane proteins; several membrane-spanning proteins are β-sheet in character. A good example is the family of porins found in the outer membranes of Gram-negative bacteria, mitochondria, and chloroplasts, which form channels for small polar molecules. Porins consist of 16 strands of β-sheet arranged in an anti-parallel fashion to form a pore through the membrane.

The π-helix has been found very occasionally in proteins (Weaver, 2000).

non-polar, the helix would be very likely to be buried in the interior of a protein or might be a membrane-spanning element of a protein. From the dimensions of the α-helix, 6 turns would be $6 \times 3.6 = 21.6$ amino acids in length and would span a distance of 6×0.54 nm $= 3.24$ nm, which is a typical value for the thickness of a phospholipid bilayer.

Other types of helix include the 3_{10} helix (three amino acids per turn, ten atoms in the hydrogen bond ring), in which the hydrogen bond is between the carbonyl group of the amide bond between amino acids n and $n + 1$, and the amino group of the amide bond between amino acids $n + 2$ and $n + 3$. This more tightly folded helix is sometimes found in short stretches in proteins, particularly near the ends of helical segments or of polypeptide chains. Another possible type is the π-helix, which is more loosely coiled than the α-helix, with the hydrogen bond between the carbonyl group of the amide bond between amino acids n and $n + 1$, and the amino group of the amide bond between amino acids $n + 4$ and $n + 5$.

1.4.2 The β-sheet

Electrostatic interactions between the carbonyl groups of adjacent chains contribute significantly to the stability of β-sheet structures. In actual protein structures (rather than model compounds) there appears to be little difference between the dihedral angles for the parallel and anti-parallel sheet patterns.

A β-sheet is made up from strands of polypeptide chain in an extended structure, with the side chains of the amino acids projecting alternately above and below the plane of the sheet. The strands can run parallel or anti-parallel to each other with hydrogen bonds forming between them; the bonds are more distorted in the case of parallel strands (Fig. 1.15). The values of the angles ϕ and ψ for idealized versions of (a) parallel β-sheet are −119° and +113°, respectively, and (b) anti-parallel β-sheet −139° and +135°, respectively. In both cases, the span of the sheet is about 0.33 nm per amino acid, i.e. more than twice that of the α-helix. When the structures of proteins are analysed, it is found that sheets composed of anti-parallel strands are more common than those formed from parallel strands, with generally at least four of the latter needed to form a sheet, compared with only two of the former. There

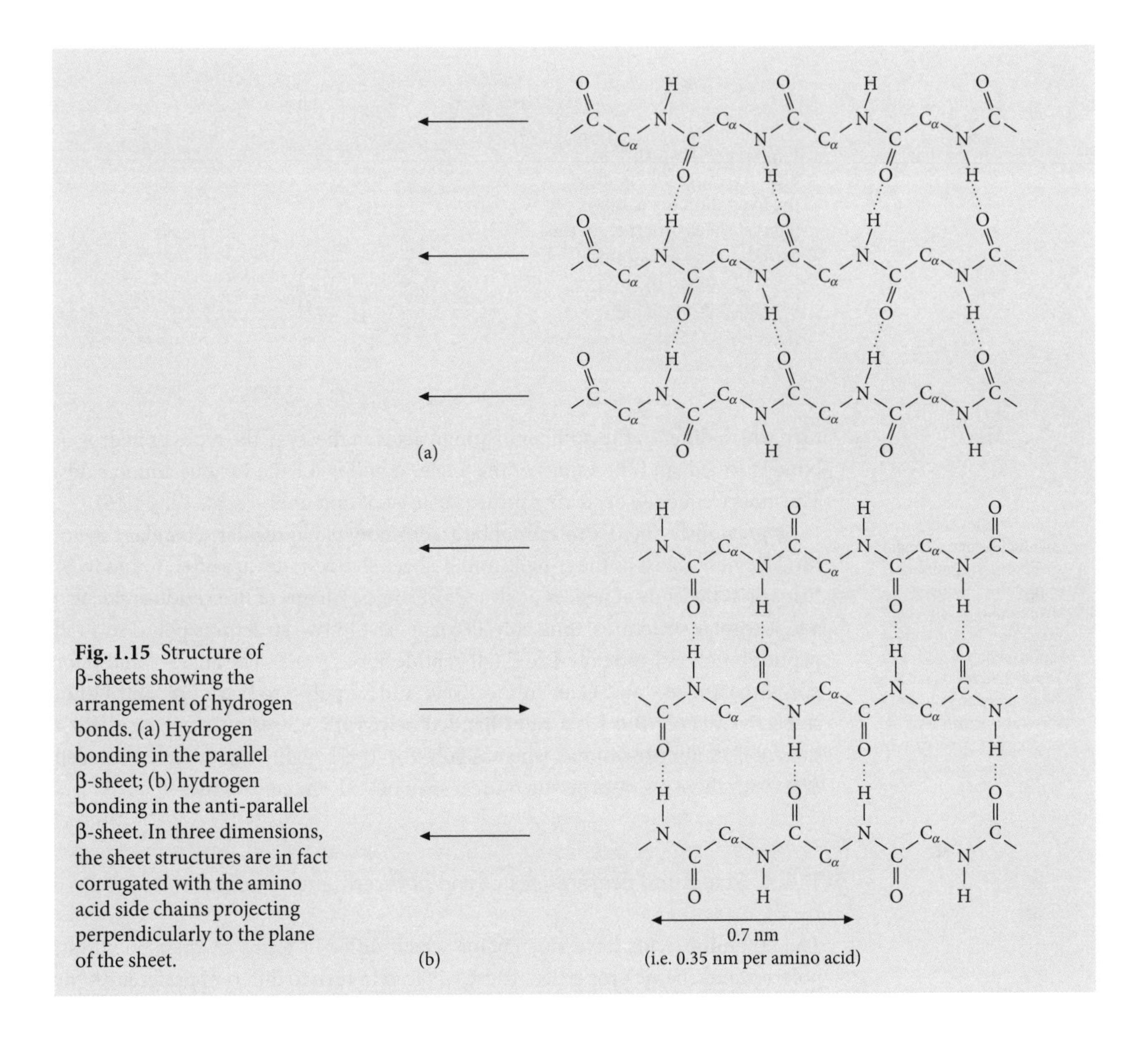

Fig. 1.15 Structure of β-sheets showing the arrangement of hydrogen bonds. (a) Hydrogen bonding in the parallel β-sheet; (b) hydrogen bonding in the anti-parallel β-sheet. In three dimensions, the sheet structures are in fact corrugated with the amino acid side chains projecting perpendicularly to the plane of the sheet.

is a right-handed twist between the strands, which appears to be a result of the electrostatic interactions between the carbonyl groups of adjacent peptide bonds in each strand; these are much more favourable with a right-handed than a left-handed twist. The average length of β-strands in proteins is about six amino acids.

1.4.3 Other structural features in proteins

In addition to α-helices and β-sheets, the most abundant structural element found in proteins is the β-turn. A β-turn is a sequence of amino acids in which the polypeptide chain folds back on itself by nearly 180°, allowing proteins to adopt globular, rather than extended, shapes. There are at least eight different types of

About 15% of residues in proteins not in α-helices or β-sheets can be in β-turns (Panasik *et al.*, 2005).

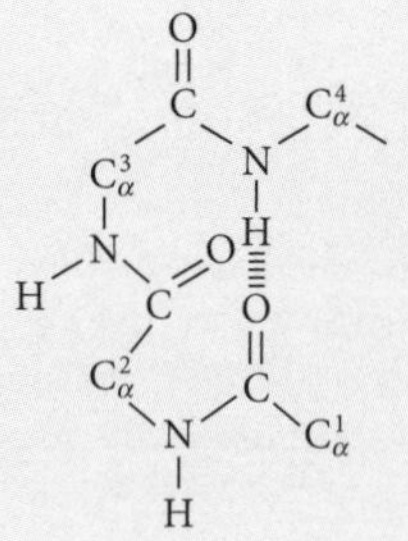

Fig. 1.16 Structure of a typical β-turn. Turns allow the formation of an anti-parallel sheet structure; in this turn there are ten atoms in the ring formed by the hydrogen bond.

turn which differ in the number of amino acids in the loop, the types of hydrogen bonds formed, and the values of the angles ϕ and ψ for the various amino acids. The most frequently occurring β-turns consist of four amino acids (Fig. 1.16).

An analysis of the crystal structures of a number of proteins of different structural types showed that a significant number (approximately 5%) of the amino acid residues could be assigned to poly-Pro II structures which were at least three amino acids in length (Sreerama and Woody, 1994).

As previously noted, Pro cannot be accommodated in regular secondary structural elements such as the α-helix and β-sheet. Pro residues are often found in β-turns or at the ends of helices or strands of sheets. Chains of Pro residues can also adopt regular structures, thus poly-Pro can exist in two structures poly-Pro I (all peptide bonds *cis*) and poly-Pro II (all peptide bonds *trans*). The angles ϕ and ψ for poly-Pro I are −83° and +158°, respectively, and for poly-Pro II are −78° and +149°, respectively. Poly-Pro I is a right-handed helix with 3.3 residues per turn and a span of 0.19 nm per residue, whereas poly-Pro II is a highly extended left-handed helix with three residues per turn and a span of 0.31 nm per residue.

1.4.4 Structural preferences of the different amino acids

The 20 amino acids have side chains which differ in terms of their bulkiness, polarity, and charge properties. These will lead in turn to different preferences for the various types of secondary structures in proteins. These preferences have been quantified in two main ways, firstly from an analysis of the structures of actual proteins and secondly from studies of the effect of incorporating different amino acids into regular polymers (e.g. introducing a Val residue into poly-Ala, which has a strong tendency to form an α-helix). From analysis of the structures of proteins, tables of preferences for the amino acids have been compiled. Met, Glu, Leu, and Ala have the highest tendency to be found in α-helices, whereas Pro, Gly, and Tyr have the lowest. Val, Ile, and Phe have the highest tendencies to be found in β-sheets; Pro and Asp the lowest. Pro, Gly, and Asp have the highest tendencies to be found in β-turns; Met, Val, and Ile the lowest. While detailed structural explanations for these preferences are not always clear, such data are useful in formulating 'rules' for predicting secondary structural features from protein sequences. It should be remembered that these rules are only guidelines; in general helices can be predicted with more confidence than sheets or turns, but even so the accuracy is relatively modest (of the order of 70%).

A commonly used tool for predicting helix, strand, and loop regions of a protein from the amino acid sequence is PHDsec (PredictProtein@EMBL-Heidelberg.de), which is stated to achieve 72% accuracy when applied to water-soluble, globular proteins.

The tertiary structure of proteins

KEY CONCEPTS

- Understanding the general principles governing the formation of the tertiary structure of proteins
- Describing how protein structures can be classified

The tertiary structure of a protein refers to its long-range folding so that parts of the polypeptide chain remote in sequence are brought into close proximity.

1.5.1 General principles

Although proteins adopt a very wide of three-dimensional structures, a number of general principles have emerged from the analysis of the many thousands of known structures of proteins. These are summarized below.

- *Close packing* Proteins generally adopt very closely packed globular structures with only a small number of internal cavities, which are generally filled by water molecules
- *Elements of secondary structure* The structural elements (helices, strands, turns etc.) are generally similar to those seen in small model compounds, but there can be deviations. For example, there is a tendency for α-helices to be distorted towards 3_{10} helices, and for strands to adopt a twist in forming sheets
- *Distributions of side chains* Non-polar side chains tend to be buried in the interior of the protein away from the solvent, and polar side chains exposed to the solvent. Polar side chains which are buried will often play key functional roles. This tendency would, of course, be reversed for membrane-spanning elements of a protein, where the non-polar side chains of a helix will project into the non-polar interior of the membrane
- *Pairing of polar groups* Polar groups which are internal are paired in hydrogen bonds; this can include carbonyl and amino groups of the main chain and pairing with internal water molecules
- *Formation of domains* Larger protein molecules tend to form structural domains, which are independent folded globular units, typically of the order of 100–150 amino acids in size. These domains are usually associated with particular functions, e.g. binding or catalysis, so that it is possible to assemble a multifunctional protein in this modular fashion.

In many cases it is possible to prepare the individual domains either by controlled proteolysis of the multi-domain protein or by expression of the appropriate part of the gene coding for that domain.

1.5.2 Classification of protein structures

A number of classification schemes have been devised to characterize the tertiary structures of proteins. These generally place the structures into one of three main categories, namely all-α, all-β, and αβ. (In some schemes, the αβ class is further divided into α/β and α + β, depending whether the helix and sheet elements

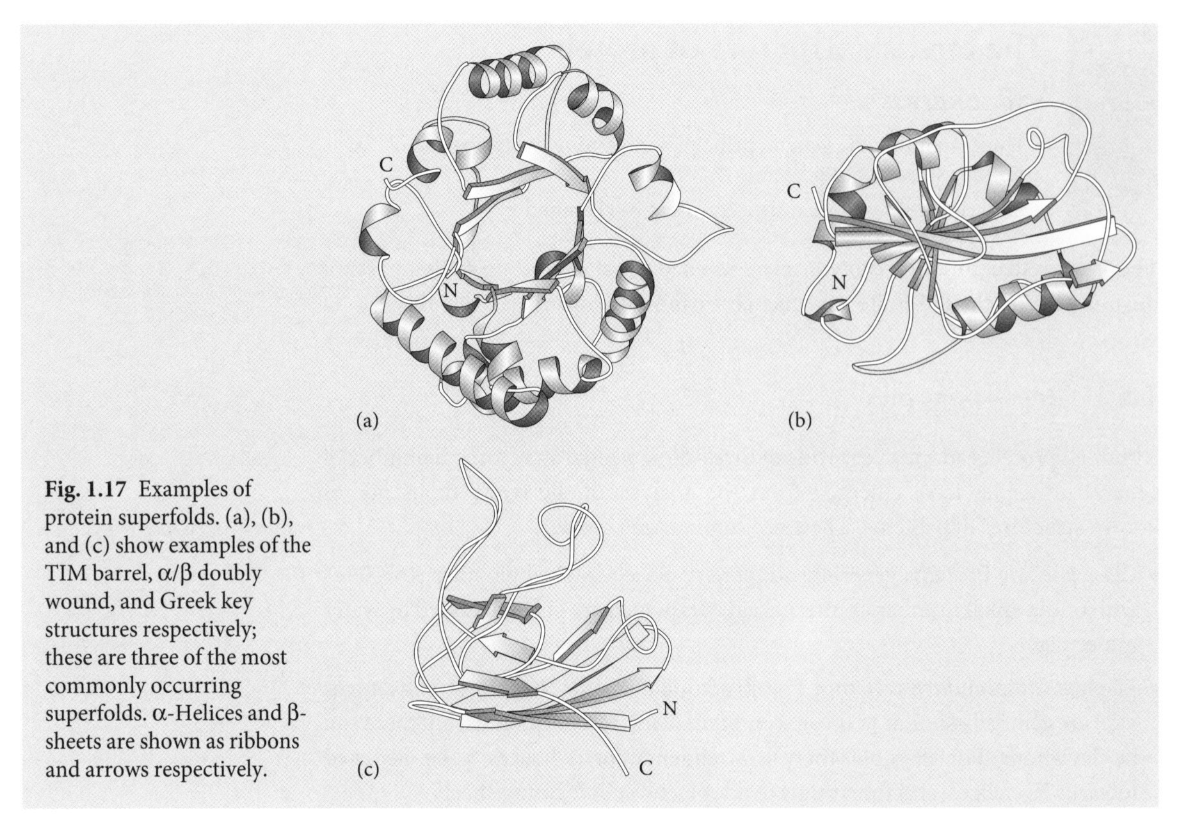

Fig. 1.17 Examples of protein superfolds. (a), (b), and (c) show examples of the TIM barrel, α/β doubly wound, and Greek key structures respectively; these are three of the most commonly occurring superfolds. α-Helices and β-sheets are shown as ribbons and arrows respectively.

alternate along the chain or are in separate regions of the sequence). One of the most useful tools for the classification of protein is the SCOP (Structural Classification Of Proteins) database, which classifies proteins at three levels: (a) protein folds, where there are similar arrangements of secondary structural elements, but no particularly close relationship in terms of function or overall structure, (b) protein superfamilies, where there are probable relationships between the proteins, as shown by similar functions, such as binding nucleotides, and (c) protein families, where there is very strong evidence for close relationships with a high (>30%) level of amino acid sequence identity between members.

The TIM barrel is named after the enzyme triosephosphate isomerase in the glycolytic pathway, the first protein structure shown to possess the repeating $(\alpha\beta)_8$ pattern.

The number of different protein folds has been estimated to be of the order of 1000. Analysis shows that only nine fold families contain proteins with no sequence or functional similarity to each other. These so-called superfolds account for over 30% of all known structures; they include the TIM barrel, α/β doubly wound, and Greek key structures (see Fig. 1.17).

1.5.3 Forces involved in stabilizing tertiary structures

The forces involved in stabilizing tertiary structures of proteins are (apart from any disulphide bonds which may be formed in extracellular proteins) of the weak non-covalent type and are discussed in section 1.7. The same types of forces are involved in the interaction of proteins with each other or with other ligands.

1.6 The quaternary structure of proteins

KEY CONCEPT

- Understanding the significance of association of polypeptide chains (subunits) in terms of the function of proteins

As noted in section 1.5, there is a tendency in polypeptide chains of more than 100–150 amino acids to form domains. For larger proteins, there is a tendency to exist as multiple subunits (polypeptide chains); this generates the quaternary structure of proteins. Proteins of molecular mass <30 kDa tend to be monomers (single subunits) and those of molecular mass >50 kDa tend to be oligomers (several subunits), although these figures should be viewed as general guidelines only. In general the arrangement of the subunits maximizes the number of inter-subunit contacts, for example in a tetrameric protein (four subunits) the preferred geometrical arrangement will be tetrahedral rather than square planar or linear. The subunits in a multi-subunit protein are generally held together by the weak non-covalent forces of the type described in section 1.7, although it should be noted that in a few cases, such as immunoglobins (Chapter 7, section 7.3.6) and bovine seminal ribonuclease, a disulphide bond is involved in holding the subunits together.

A protein consisting only of multiple copies of the same polypeptide chain is termed a 'homooligomer', e.g. lactate dehydrogenase is a homotetramer (denoted α_4). If there are different types of polypeptide chain present the protein is termed a heterooligomer, e.g. the G-proteins involved in signal transduction are heterotrimers (denoted $\alpha\beta\gamma$). Haemoglobin is a heterotetramer ($\alpha_2\beta_2$), where the α and β chains are very similar to each other. The Greek letters are used in a generic sense to distinguish different types of polypeptide chains in a protein, rather than to denote specific sequences of amino acids.

The association of several subunits in a protein can offer possibilities of novel properties to a protein. These include (a) regulation of activity, by communication between the binding sites or active sites on different subunits (e.g. in the case of O_2 binding to haemoglobin), (b) increased stability as a result of favourable packing of subunits which might be prone to unfolding as monomers, and (c) generation of novel structures such as found in chaperone proteins and proteasomes, where internal 'cavities' can be created to allow protein folding or protein degradation, respectively.

The proteasome is a large complex (1.4 MDa) that contains a barrel-shaped proteolytic unit (700 kDa) consisting of 14 copies of each of two types of subunit. The proteasome degrades proteins which have been tagged with the small protein ubiquitin for destruction.

1.7 Forces contributing to the structures and interactions of proteins

KEY CONCEPTS

- Understanding the contribution of weak forces to the structures and interactions of proteins
- Describing the basis of the principal types of weak forces (ionic interactions, hydrogen bonds, van der Waals interactions, and hydrophobic interactions)
- Knowing the range of energies involved in protein interactions

Chaperones are proteins which assist the folding of proteins in the cell by preventing unwanted interactions between hydrophobic surfaces, which would otherwise lead to the formation of protein aggregates. In the cpn60 class of chaperones, which contains 14 copies of a 57 kDa subunit, there is a large internal cavity that allows the correct folding of polypeptide chains to occur.

The amino acids in a protein are linked together by strong covalent peptide bonds (section 1.3.1). In some cases the three-dimensional structures will also be stabilized by covalent bonds, for example disulphide bonds, which can form between pairs of Cys side chains, and the pairs of Lys side chains, which (after oxidative deamination) can form strong cross-links in collagen. These covalent bonds have energies of the order of several hundred kJ mol^{-1}. However, the forces which stabilize the folded structures of the vast majority of proteins, and which underpin the interactions between proteins and other molecules, are of a non-covalent type and hence much weaker and usually short lasting. These forces can be described under the following categories.

1.7.1 Ionic (electrostatic) interactions

The effective dielectric constant (*D*) in the interior of proteins has been estimated from the effects of mutations on the p*K*a values of amino acid side chains. In the case of proteins such as cytochrome c and subtilisin, it has been estimated that *D* is in the range 30 to 60, but in staphylococcal nuclease, *D* is estimated to be 12. The value of *D* appears to depend on the extent of water penetration into the interior of the protein. In a truly non-polar region in the protein interior, *D* may be as low as about 4 (Dwyer *et al.*, 2000).

These refer to the electrostatic forces between oppositely charged groups, e.g. between negatively charged side chains (Asp, Glu) and positively charged side chains (Arg, Lys, and (under certain circumstances) His). The energy of interaction (*E*) between charges of magnitude q_1 and q_2, separated by a distance *r* is given by $E = q_1q_2/Dr$, where *D* is the dielectric constant. By definition, $D = 1$ for a vacuum; the value of D for water = 78.5. Non-polar solvents have low values of *D* (e.g. benzene, $D = 2.3$; ethanol, $D = 25$). Since the majority of charged amino acid side chains in a protein are on the surface and therefore significantly exposed to the aqueous medium, ionic interactions are generally weak (a high value of *D* means a low value for the energy), typically of the order of 5 kJ mol^{-1}. However, if ionic interactions occur in the much less polar interior of a protein (as in the case of the α-amino group of Ile 16 and the side chain of Asp 194 in chymotrypsin), the forces will be considerably stronger (possibly up to about 20 kJ mol^{-1}).

1.7.2 Hydrogen bonds

A hydrogen bond is essentially a form of electrostatic interaction involving partial charges. It occurs when an H attached to an electronegative atom (almost always O or N in biological systems) interacts with a second electronegative atom. For example, if we call these atoms A (donor) and B (acceptor), we would have the arrangement shown in Fig. 1.18, where the δ indicates a partial charge. There is an attractive force between the partial charges on the H and on the acceptor atom. This is a weak interaction (typically of the order of 5–10 kJ mol^{-1}), but when a number of hydrogen bonds are involved they can contribute significantly to local stability, as is the case in secondary structural features such as the α-helix or β-sheet (section 1.4). The particular importance of hydrogen bonds in terms of the structure and function of proteins is that they impose geometrical constraints, and thus specificity, on interactions. For maximum stability of the hydrogen bond, the three nuclei (A, H, and B) should be in a straight line and the separation between the nuclei A and B should be within fairly narrow limits (0.30 ± 0.05 nm depending on the atoms concerned). (This is how hydrogen bonds are identified in X-ray crystal protein structures, since the H atom cannot be 'seen' by X-rays as it has so little electron density compared with the heavier atoms C, N, and O). In proteins,

$$\underset{\delta-}{\text{—A}}\text{—}\underset{\delta+}{\text{H}}\cdots\cdots\underset{\delta-}{\text{B}}\text{—}$$

Fig. 1.18 Formation of a hydrogen bond. Atoms A and B represent electronegative atoms (generally O and/or N in biological systems). The partial charges on the atoms are shown.

hydrogen bonds can involve interactions between particular side chains (see section 1.2.4) or main chain atoms (the N–H or C=O groups of the peptide bond). It should also be remembered there is often a compensating mechanism whereby groups on the protein may equally well form hydrogen bonds with water, rather than with other parts of the protein, thereby reducing the energetic contribution of the interaction to the structural or binding properties of the protein.

The strengths of individual hydrogen bonds vary considerably; they are stronger if a charged group is involved, e.g. if $-NH_3^+$ acts as a donor.

1.7.3 van der Waals' interactions

This term is applied collectively to various weak forces involving interactions between dipoles (separations of charge) in molecules or parts of molecules. In addition to permanent dipoles reflecting differences in electronegativity (e.g. between C and O in the C=O bond), there are transient dipoles which arise from the fact that electrons occupy fluctuating positions in space. These dipoles can in turn induce dipoles in other molecules or groups. The van der Waals' forces include dipole$\rightleftharpoons$dipole, dipole$\rightleftharpoons$induced dipole, and induced dipole$\rightleftharpoons$induced dipole interactions. Although the forces are weak individually (of the order of 5 kJ mol^{-1}), there are usually many such interactions and thus collectively they can contribute to the stability of a protein structure or the strength of a protein–ligand interaction.

The van der Waals' forces between atoms or molecules represent a balance between the longer range attractive forces (which show a $1/r^6$ dependence, where r is a measure of internuclear distance) and the shorter range repulsive forces (which show a $1/r^{12}$ dependence).

1.7.4 Hydrophobic interactions

Water is a uniquely poor solvent for non-polar molecules or groups such as the aliphatic or aromatic side chains in proteins. As mentioned in section 1.2.5, liquid water consists of fluctuating clusters of hydrogen-bonded water molecules. When a non-polar group is introduced into this system, the water molecules tend to form more ordered hydrogen bonded 'cage' structures around the non-polar group. This process has a favourable (negative) change in enthalpy, but there is a highly unfavourable (negative) change in entropy due to the ordering of the solvent, thus the overall free energy change is unfavourable (positive) (see Chapter 4, section 4.1). The term 'hydrophobic interactions' is used to describe the tendency of non-polar molecules to be excluded from water and associate with each other, although it should be noted that such interactions are in fact very weak, being largely due to forces of the van der Waals type. For this reason, some workers prefer the term 'hydrophobic effect'.

It should be pointed out that the term 'hydrophobic' is something of a misnomer since it implies that the non-polar groups 'hate water', whereas in fact it is the water which hates them!

The hydrophobic interactions make the most important contribution in energy terms to the stability of the folded structure of a protein in water. Since there is a positive change in enthalpy associated with the exclusion of non-polar groups from water, hydrophobic interactions become weaker at lower temperatures (at least over a restricted range); this can account for the cold-induced denaturation of some proteins.

The strength of hydrophobic interactions can be altered by various agents. For example, urea, and guanidinium chloride (GdmCl) decrease the hydrophobic interactions; since they can participate in hydrogen bonds, they presumably disrupt the hydrogen-bonded network of water molecules. By contrast, the hydrophobic interactions can be strengthened by the addition of various salts; the order of effectiveness follows the Hofmeister series:

The Hofmeister series was derived in the 1880s from studies of the effectiveness of different salts in precipitating proteins from solution.

Cations $Mg^{2+} > Li^{+} > Na^{+} > NH_4^{+}$

Anions $SO_4^{2-} > HPO_4^{2-} > \text{acetate} > Cl^{-} > ClO_4^{-} > SCN^{-}$

1.7.5 Balance of energy contributions

Many of the forces involved in stabilizing the three-dimensional structures of proteins and the interactions between proteins and other molecules are relatively weak and represent a balance between competing processes. Thus, for example, hydrophobic interactions represent a balance between an unfavourable enthalpy term but a favourable entropy term, and an α-helix is stabilized by hydrogen bonding between the main chain –N–H and C=O groups, but these are formed at the expense of hydrogen bonds that might be formed to the solvent water if the protein were unfolded. The stability of the folded structure of a protein represents a very delicate balance between two opposing tendencies, namely the large (unfavourable) negative change in entropy when a compact globular structure is formed from a disordered polypeptide chain and the large (favourable) negative change in enthalpy due to the favourable interactions which occur within the folded state of the protein. For most proteins the (thermodynamic) stability of the folded state is very small, in the range 20 – 60 kJ mol^{-1} (see Chapter 8, section 8.6), compared with the entropy and enthalpy terms, each of which is of the order of several hundred kJ mol^{-1}. Of course, there may be a large kinetic barrier (activation energy) to unfolding of the folded state of a protein, but it is important to realize that, in thermodynamic terms at least, life hangs by a slender thread!

The balance of the forces can be altered by addition of certain agents. As indicated in section 1.7.4, urea and GdmCl weaken hydrophobic interactions. This is shown by the fact that different amino acid side chains are much more nearly equally soluble in 8 M urea or 6 M GdmCl than would be the case in water. Since the hydrophobic interactions make the largest contribution in energy terms to the stability of the folded states of proteins, it is easy to understand how these agents promote the unfolding of proteins.

In concentrated solutions of urea and GdmCl, proteins unfold because the entropy gain in unfolding is dominant.

Proteins can also be unfolded by exposure to extremes of pH (reflecting changes in the ionization states of amino acid side chains and hence disruption of electrostatic forces and hydrogen bonds) or temperature (reflecting alterations in the balance between polar and non-polar interactions); see also Chapter 5, section 5.6.1.

1.7.6 The range of energies involved in protein interactions

The function of the vast majority of proteins is to undertake specific interactions with other molecules (ligands). It is instructive to examine the strengths of these interactions in the context of the discussion of the forces involved. Some representative types of interactions are listed in Table 1.4, together with a note of the corresponding standard free energy change at 37°C (ΔG°_{310}), derived using the relationship $-\Delta G^\circ = RT \ln K_{eq}$ (eqn. 4.4; see Chapter 4, section 4.1).

The magnitudes of the free energy changes are such that only a relatively small number of the weak interactions (hydrogen bonds, van der Waals' forces, etc.) are required to account for them. It should be remembered, however, that many interactions involve a balance between forces, as described in section 1.7.5, thus a protein and a ligand may well each form favourable interactions with the solvent that are lost when the protein and ligand form a complex. The net free energy change can therefore be relatively small.

The values of the dissociation constants can be correlated with the functions of the complexes. For example, the extremely tight interaction between the vitamin biotin (which is produced by bacteria) and the egg white protein avidin may play a role in anti-bacterial defence for the developing chick embryo. The interactions involved in forming oligomeric proteins have to be very favourable in order to generate the stability required for a protein to survive and function under cellular conditions. Antibodies and receptors must bind their respective partners sufficiently tightly to allow subsequent events such as complement activation or signal transduction to occur. On the other hand, it is important that enzyme–substrate (and enzyme–product) interactions are not too strong since otherwise the catalytic cycle (consisting of binding of substrate(s), structural changes in the enzyme, conversion of substrate(s) to product(s), and release of product(s)) would be slowed and the enzyme would become an unacceptably inefficient catalyst.

The dissociation constant K_d for an interaction represents the ratio of the rate constants k_{off}/k_{on} (see Chapter 4, section 4.1). Since there is an upper limit for the association rate constant k_{on} (set by the rate of diffusion) of about $10^9\ M^{-1}\ s^{-1}$, an estimate can be made of the dissociation rate constant, k_{off}, and hence the life-time of the complex.

Table 1.4 Typical dissociation constants for some interactions involving proteins

Interaction	Typical dissociation constant (K_d) (M)	ΔG^0_{310} (kJ mol^{-1})*
Avidin–biotin	10^{-15}	89
Protein–protein	10^{-10}	59
Antibody–antigen	10^{-9}	53
Receptor–hormone	10^{-7}	42
Enzyme–substrate	10^{-5}	30

*The free energy values are positive as they refer to the dissociation of the complex (PL $\rightleftharpoons$ P + L), which would be unfavourable under standard state conditions (1 M concentrations of PL, P, and L). Dissociation would become progressively more favourable as the concentration of PL is lowered. In the cell the concentrations of many complexes are likely to be in the micromolar range.

1.8 Compendium of chemical structures

It is a good idea to try to learn the structures of a number of important molecules that are connected with the study of proteins; a list is given below. Once you know the structure of a compound, it will help you to understand its properties. For example, you should be able to recognize the following features of a molecule from its structure:

- Which parts are polar and which are non-polar
- Which parts may be involved in weak interactions such as hydrogen bonds, ionic bonds or hydrophobic interactions
- Which parts may be involved in ionization processes and over which pH range different charged forms may predominate
- Which parts may have important chemical roles, such as acting as a nucleophile or coordinating a metal ion.

When looking at these structures, bear in mind the following:

- Chains of carbon atoms are often represented as zig-zag lines, rather than each individual carbon atom being displayed
- In complex structures, hydrogen atoms are often omitted for the sake of clarity
- In ring structures the ring atoms are assumed to be C unless otherwise indicated (e.g. N, O, S, etc.)
- The actual charged states of compounds will depend on the prevailing pH and the pK_as of ionizing groups such as $-NH_2$ and $-CO_2H$ (see Chapter 3, section 3.7). They may not be in exactly the forms shown below.

Amino acids

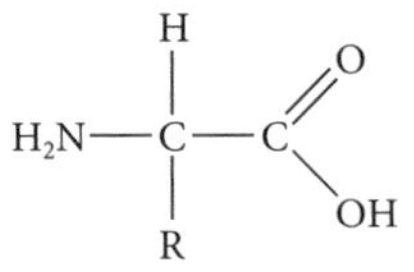

H
H_2N—C—C(=O)OH
R

The general structure of an amino acid

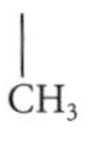

CH_3

Alanine

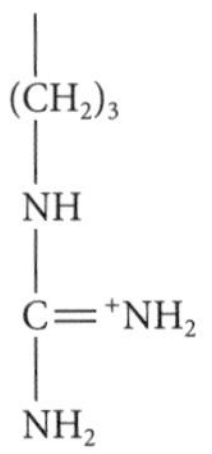

$(CH_2)_3$
NH
$C{=}^{+}NH_2$
NH_2

Arginine

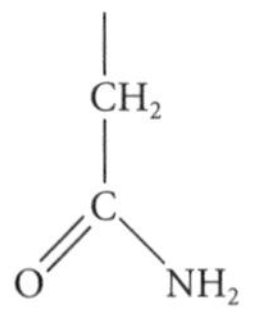

CH_2
C
O NH$_2$

Asparagine

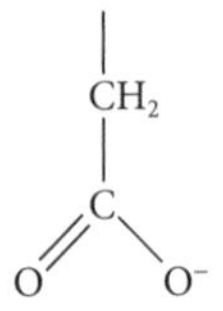

CH_2
C
O O^-

Aspartic acid

CH_2
SH

Cysteine

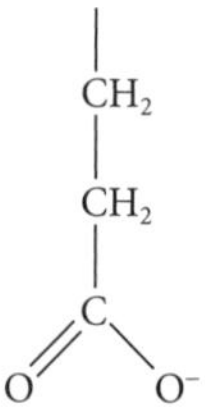

Glutamic acid

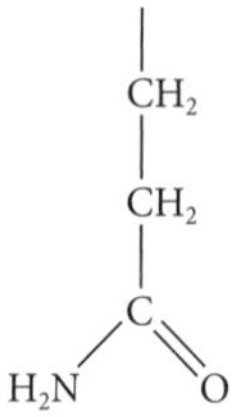

Glutamine

Glycine

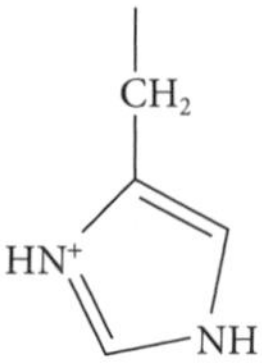

Histidine

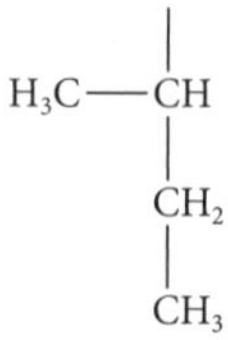

Isoleucine

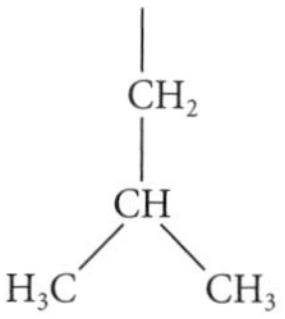

Leucine

Lysine

Methionine

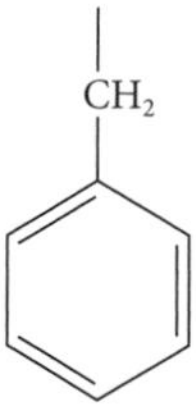

Phenylalanine

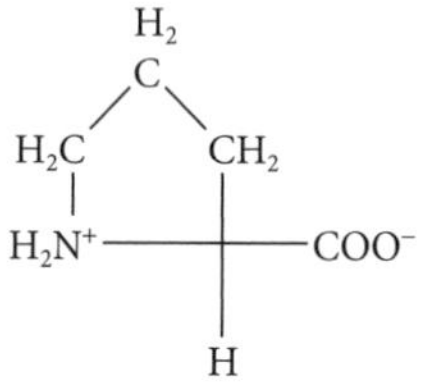

Proline

Serine

Threonine

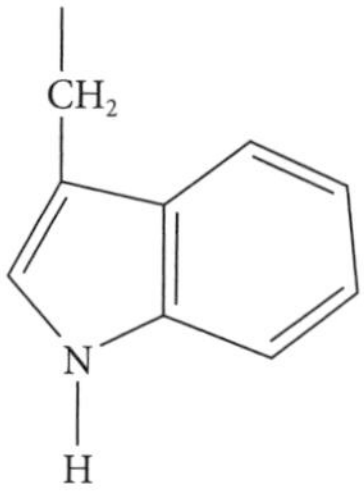

Tryptophan

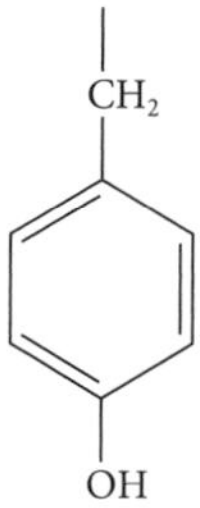

Tyrosine

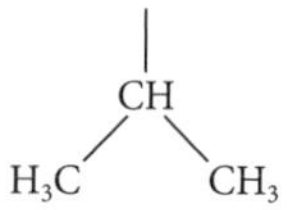

Valine

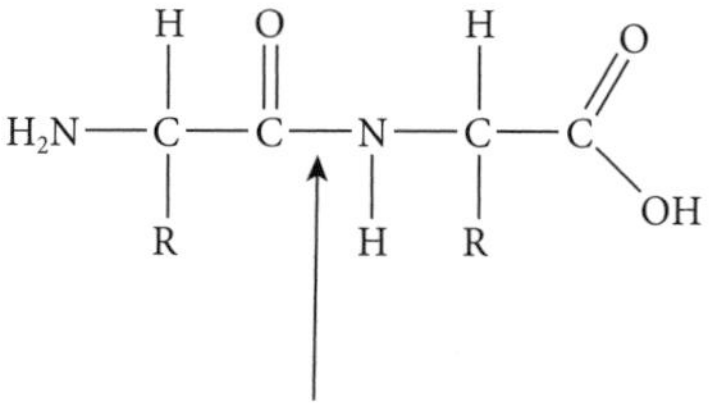

The peptide bond

Bases, nucleotides, and related compounds

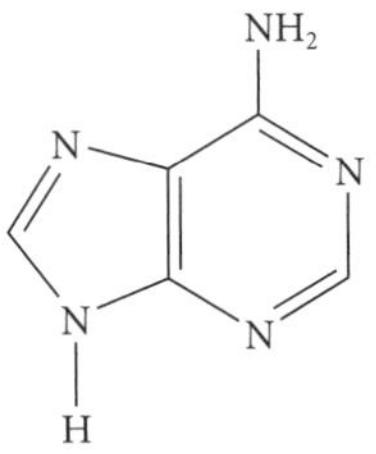

Adenine

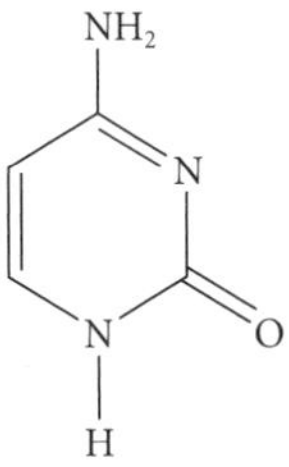

Cytosine

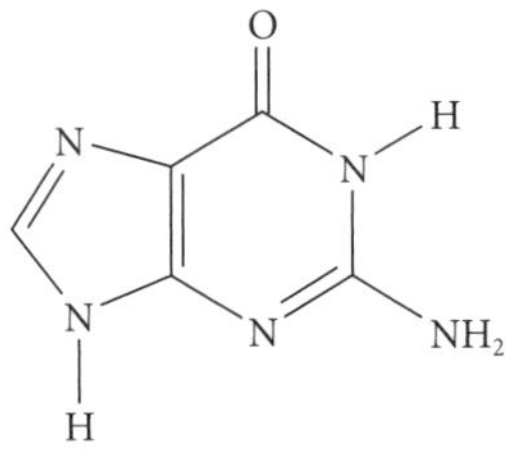

Guanine

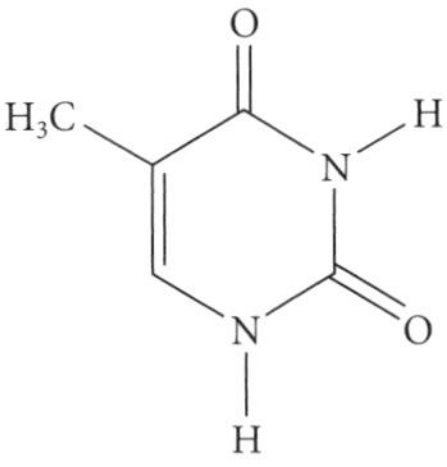

Thymine

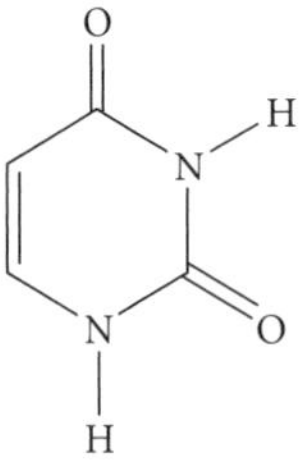

Uracil

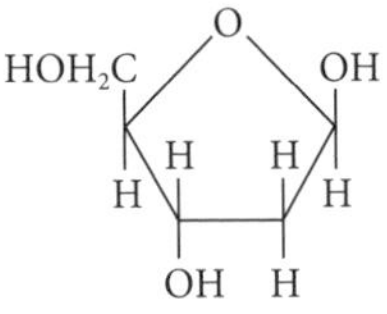

2-deoxyribose

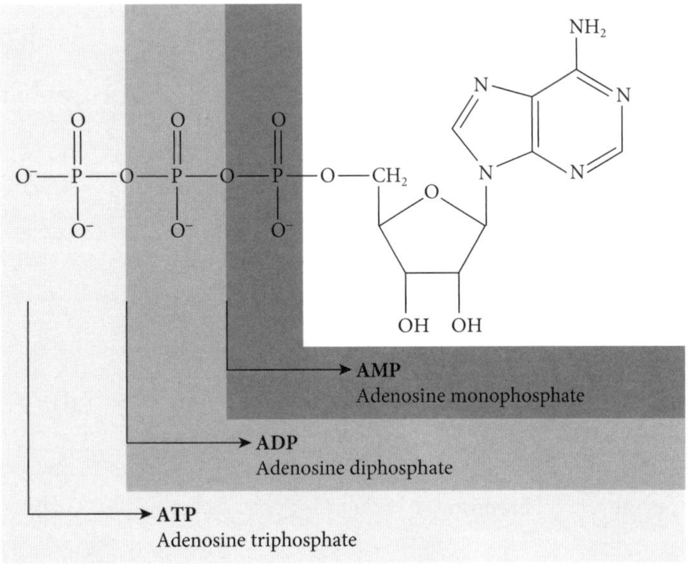

Ribose

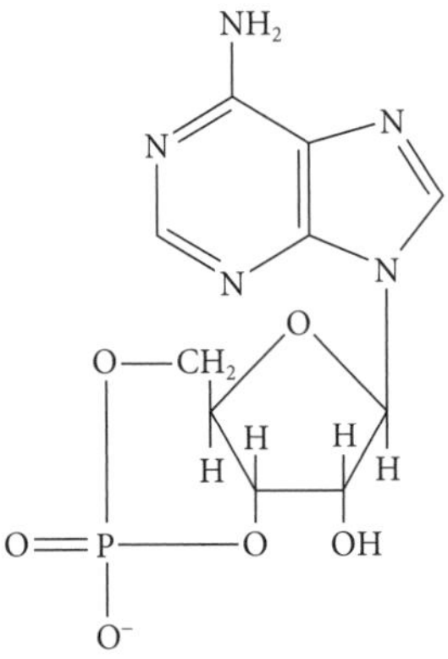

AMP
Adenosine monophosphate

ADP
Adenosine diphosphate

ATP
Adenosine triphosphate

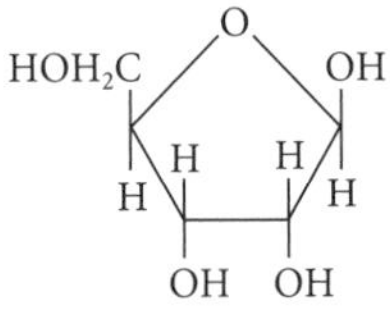

Cyclic AMP

Phosphodiester bond

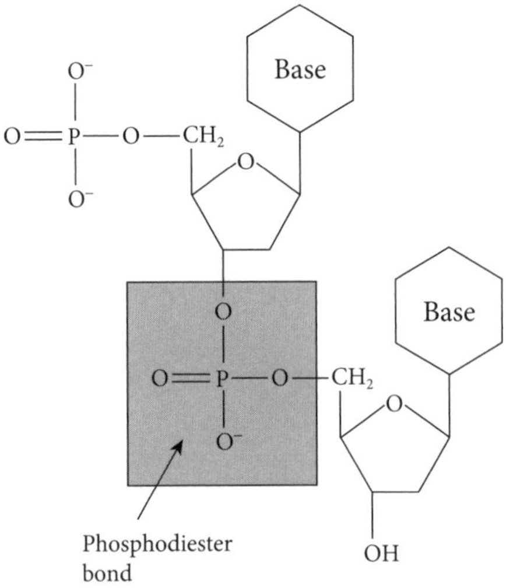

The 5′–3′ bond between nucleotides

Glycolytic intermediates

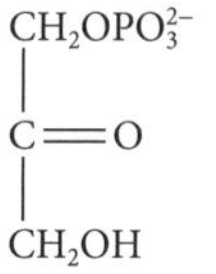

Dihydroxyacetone phosphate

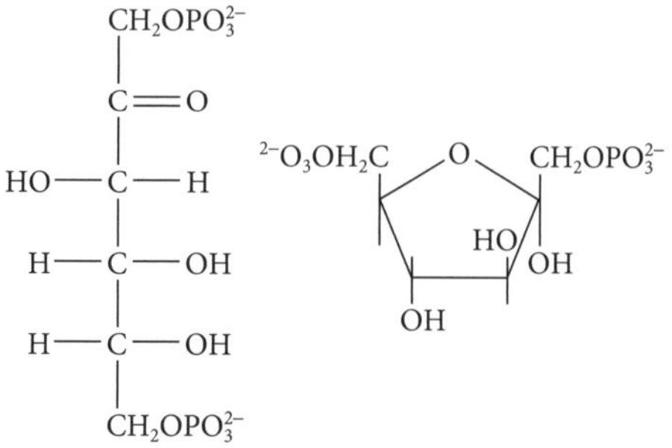

Fructose-1,6-bisphosphate

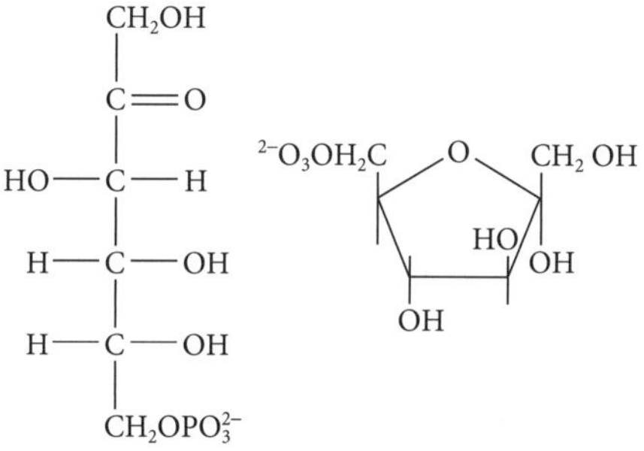

Fructose-6-phosphate

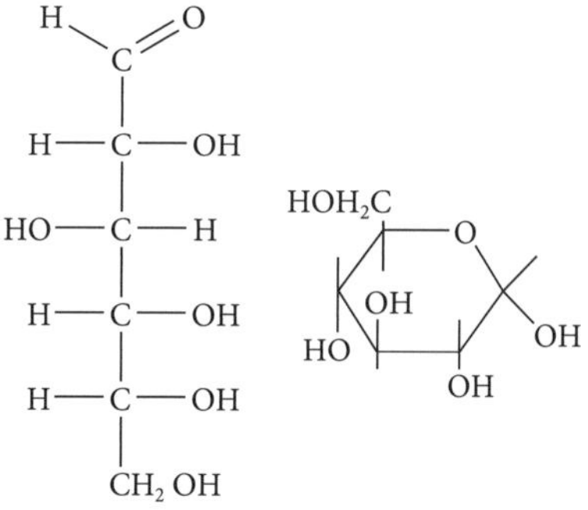

Glucose

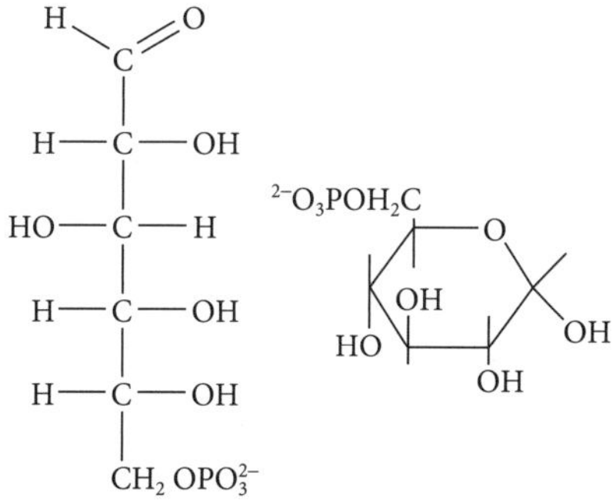

Glucose-6-phosphate

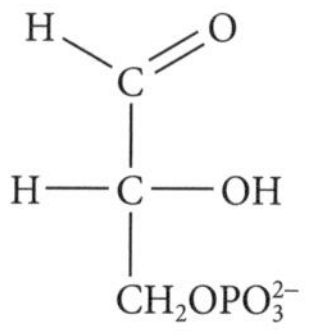

Glyceraldehyde-3-phosphate

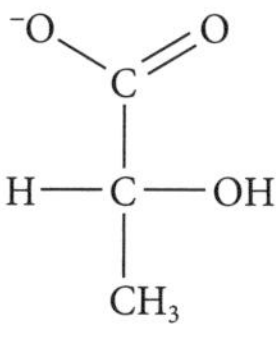

Lactate

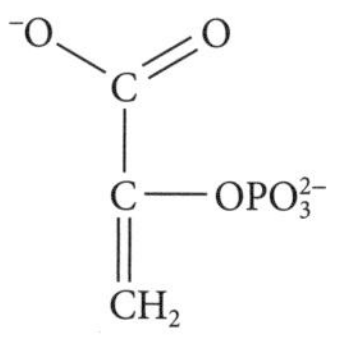

Phosphoenolpyruvate

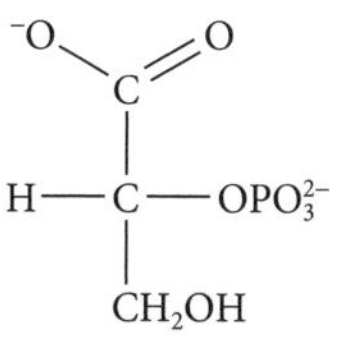

2-phosphoglycerate

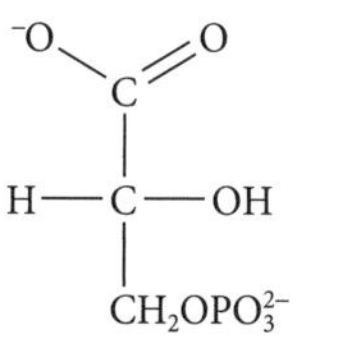

3-phosphoglycerate

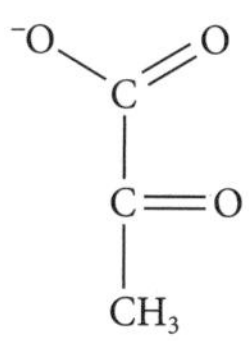

Pyruvate

Reagents

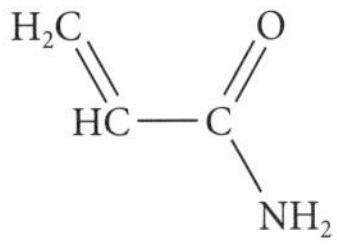

H_2C O
HC—C
NH_2

Acrylamide

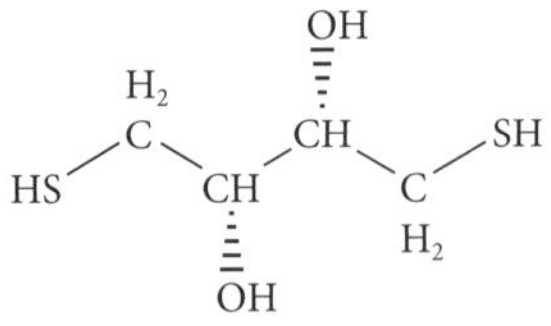

OH
H_2
C CH SH
HS CH C
H_2
OH

Dithiothreitol

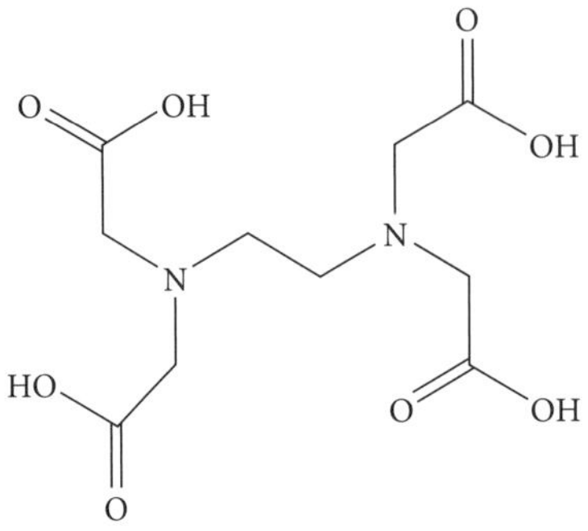

O
O OH
OH
N
N
HO
O OH
O

EDTA

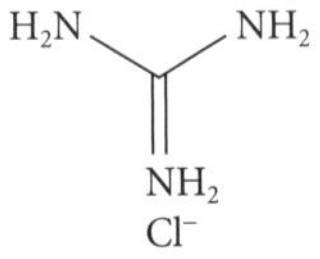

H_2N NH_2
NH_2
Cl^-

Guanidinium chloride

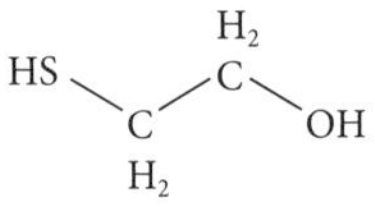

H_2
HS C
C OH
H_2

2-mercaptoethanol

O O
H_2C CH_2
N N
H H

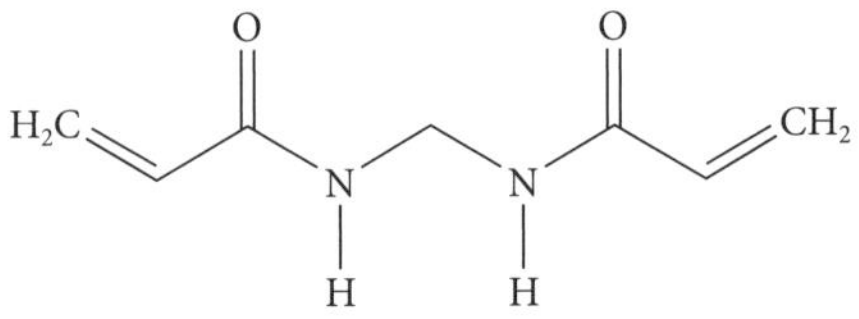

Methylene *bis*-acrylamide

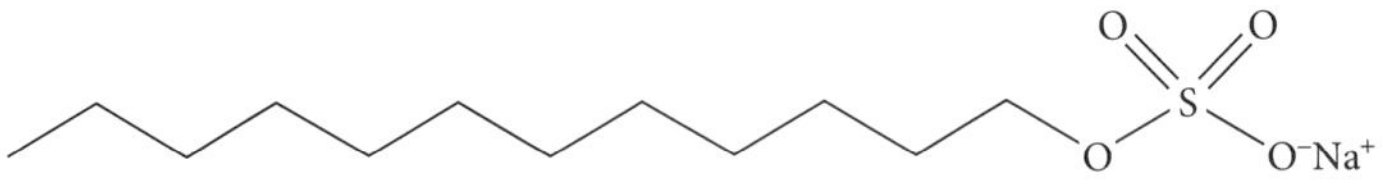

O O
S
O O^-Na^+

SDS

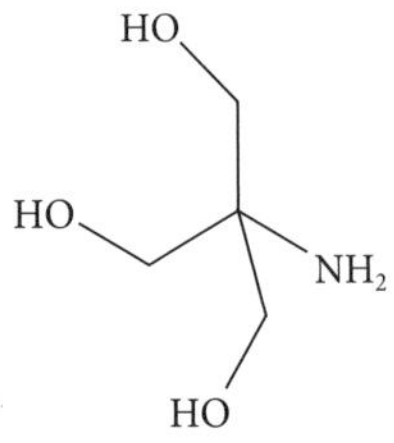

HO
HO
NH_2
HO

Tris

O
C
H_2N NH_2

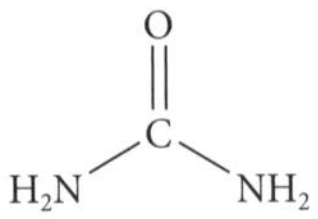

Urea

Redox cofactors

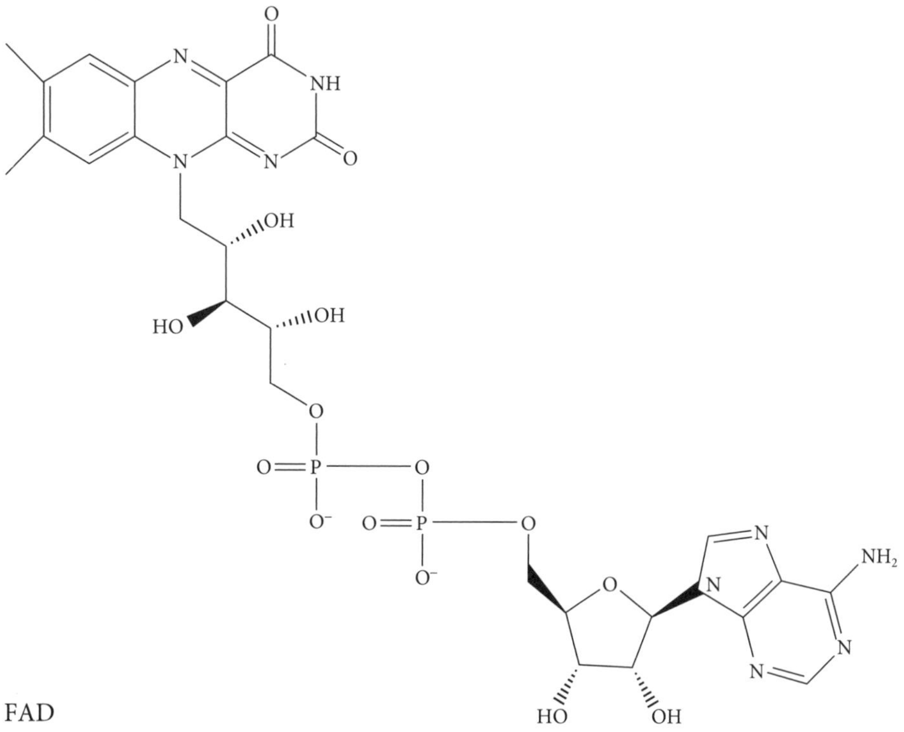

FAD

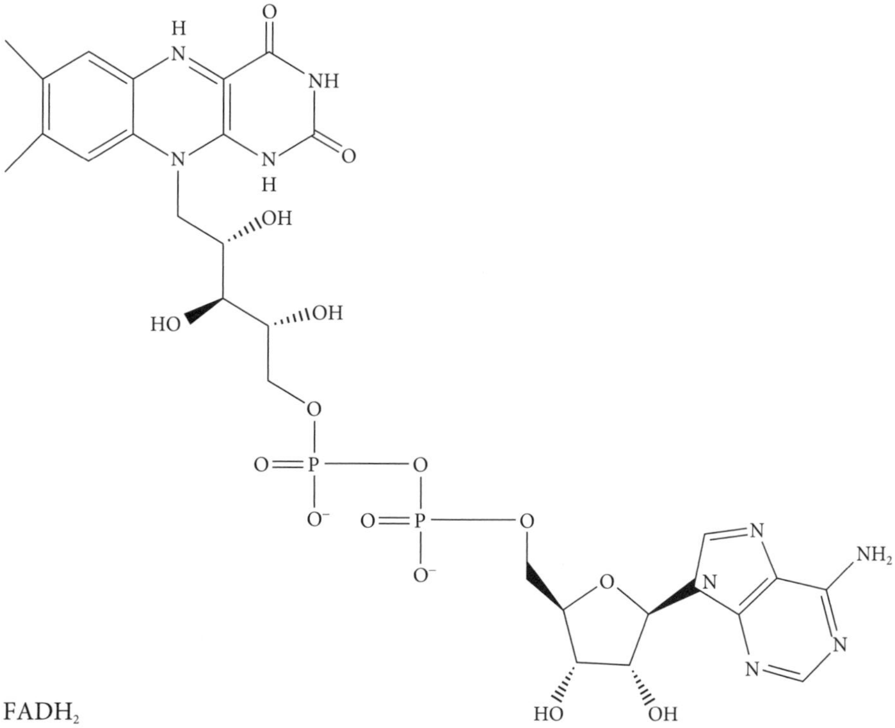

$FADH_2$

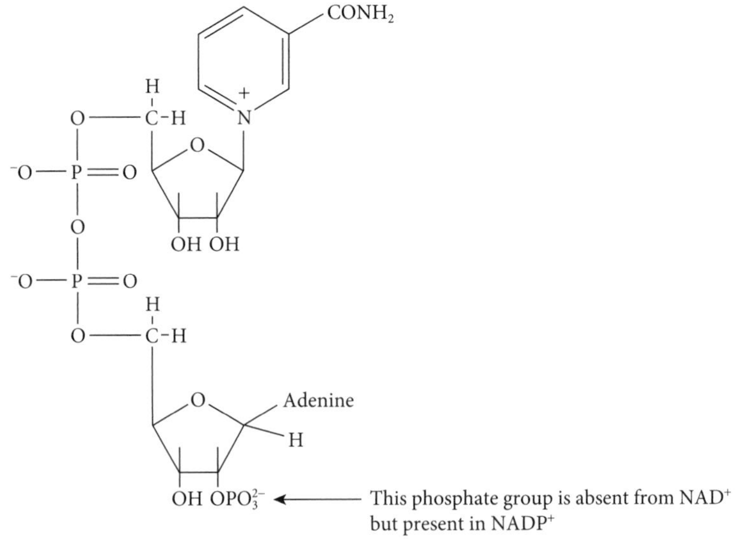

$CONH_2$

N^+

Adenine

H

$OH\ OPO_3^{2-}$ ← This phosphate group is absent from NAD^+ but present in $NADP^+$

$NAD(P)^+$

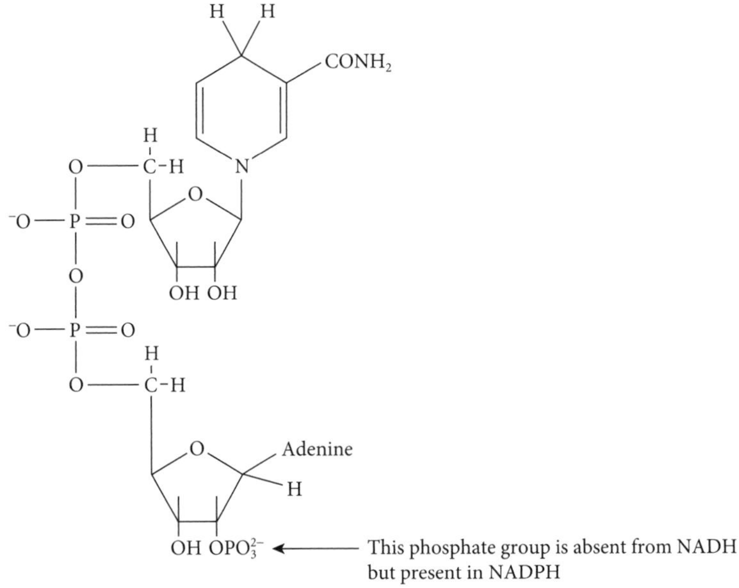

$CONH_2$

Adenine

H

$OH\ OPO_3^{2-}$ ← This phosphate group is absent from NADH but present in NADPH

NAD(P)H

TCA cycle intermediates

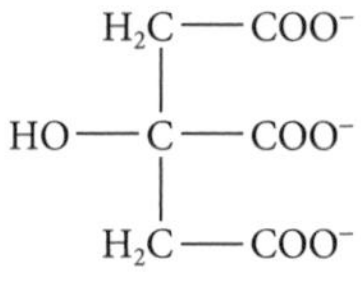

$H_2C—COO^-$
$HO—C—COO^-$
$H_2C—COO^-$
Citrate

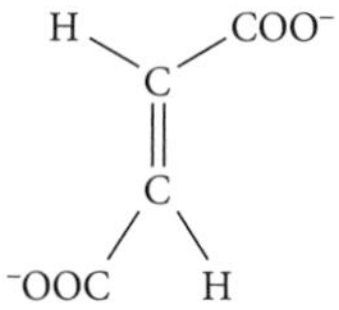

H COO^-
C
C
^-OOC H
Fumarate

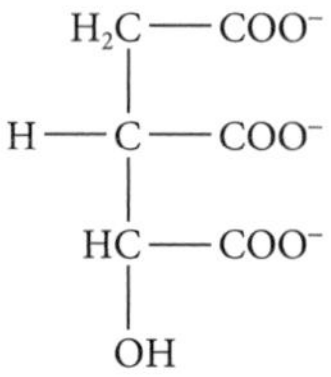

$H_2C—COO^-$
$H—C—COO^-$
$HC—COO^-$
OH
Isocitrate

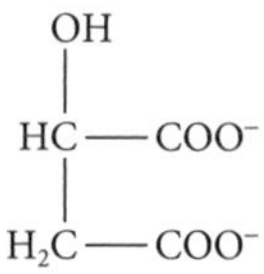

OH
$HC—COO^-$
$H_2C—COO^-$
Malate

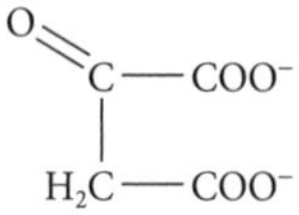

O
$C—COO^-$
$H_2C—COO^-$
Oxaloacetate

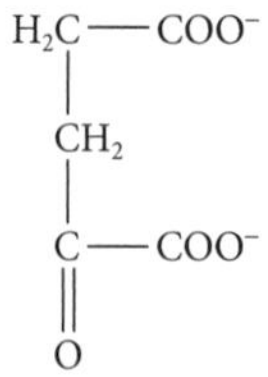

$H_2C—COO^-$
CH_2
$C—COO^-$
O
2-oxoglutarate

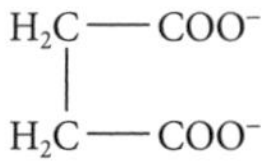

$H_2C—COO^-$
$H_2C—COO^-$
Succinate

1.9 Problems

Full solutions to odd-numbered problems are available to all in the student section of the Online Resource Centre at www.oxfordtextbooks.co.uk/orc/price/. Full solutions to even-numbered problems are available to lecturers only in the lecturer section of the Online Resource Centre.

1.1 The amino acid ornithine (which is an intermediate in the synthesis of arginine) has the side chain $-(CH_2)_3-NH_2$. What would you expect the principal chemical characteristics of this amino acid to be?

1.2 The synthetic amino acid norleucine has the side chain $-(CH_2)_3-CH_3$. How would you expect incorporation of this amino acid in place of leucine to affect the properties of a protein?

1.3 The sequence of the tripeptide glutathione is usually depicted as (γ)Glu–Cys–Gly. Draw the full chemical structure of glutathione. What might happen to glutathione under oxidizing conditions?

1.4 Calmodulin is a member of a family of small (17 kDa) proteins that can bind Ca^{2+} ions tightly. What types of amino acids might allow such proteins to achieve this?

1.5 Compared with Ca^{2+} ions, Zn^{2+} ions have a preference to coordinate with nitrogen and sulphur as compared with oxygen atoms. Which amino acids might you expect to find in Zn-binding sites in proteins?

1.6 For aspartic acid, the pK_a values for the α-carboxyl, the side chain carboxyl and the α-amino group are 2.0, 3.9, and 9.9, respectively. Draw the structures of the predominant charged forms of the amino acid at pH values 1, 3, 6, and 11. Explain why the pI is obtained by averaging the pK_a values of the α-carboxyl and the side chain carboxyl groups.

1.7 For lysine, the pK_a values of the α-carboxyl, the α-amino, and the side chain amino groups are 2.2, 9.1, and 10.5, respectively. What is the pI for lysine?

1.8 Iodoacetamide ($I-CH_2-CO-NH_2$) is a commonly used reagent to react with nucleophilic side chains in proteins. Using structural formulae, show how it reacts with the side chains of cysteine and lysine. How might you monitor the extent of modification of a protein by the reagent?

1.9 The pK_a of the side chain amino group of lysine is 10.5 in the free amino acid. Explain how the pK_a of a lysine side chain in a protein could be affected by the presence of (a) a neighbouring Arg side chain and (b) a neighbouring Asp side chain.

1.10 The following sequences of 20 amino acids occur in the protein glycophorin, which is found in the membrane of red blood cells: (a) YPPEEETGERVQLAHHFSEP and (b) EITLIIFGVMAGVIGTILLI. What would you predict about where these parts of the protein would be located?

1.11 The free energy of the *trans* form of a peptide bond is estimated to be 20 kJ mol^{-1} lower than that of the *cis* form. If a peptide bond could freely interconvert between *trans* and *cis* forms, what would be the equilibrium constant for the *trans* $\rightleftharpoons$ *cis* equilibrium at 37°C? (The gas constant $R = 8.31$ J K^{-1} mol^{-1}).

1.12 In a study of protein structures, it is found that approximately 5% of the peptide bonds to proline (Xaa–Pro) occur in the *cis* form. What is the free energy difference at 37°C between the *trans* and *cis* forms in the case of the Xaa–Pro bond?

1.13 The molar absorption coefficients at 280 nm for small model compounds of Trp and Tyr are 5690 and 1280 M^{-1} cm^{-1} respectively. The polypeptide chain of yeast alcohol dehydrogenase has a molecular mass of 36712 Da and contains 5 Trp and 14 Tyr. Calculate the molar absorption coefficient of alcohol dehydrogenase at 280 nm. (You can assume that the values for the model compounds apply to these amino acids in the protein). What would be the absorbance at 280 nm of a 0.32 mg mL^{-1} solution of the protein in a cuvette of 1-cm pathlength?

1.14 The polypeptide chain of the chaperone protein GroEL from *Escherichia coli* has a molecular mass of 57200 Da and does not contain tryptophan. The A_{280} of a 1 mg mL^{-1} solution of the purified protein in a 1-cm pathlength cuvette is 0.160. Assuming that the molar absorption coefficient at 280 nm for a small model compound of Tyr is 1280 M^{-1} cm^{-1}, calculate the number of Tyr in the polypeptide chain.

1.15 The following amino acid sequences are found in α-helical structures in three folded proteins. Use the helical wheel projection to predict where each of these helical segments may be located in the structure of the protein: (a) LSFAAAMIGLA (citrate synthase), (b) INEGFDLLRSG (alcohol dehydrogenase), and (c) KEDNKGKSEEE (troponin C).

1.16 The calculated molecular mass of the polypeptide chain of type II dehydroquinase from *Streptomyces coelicolor* is 16550.6 Da. What is the expected mass of a mutant form of the enzyme in which Tyr 28 has been replaced by Phe? Compare your answer with the observed mass (16535.1 ± 1.4 Da). In a preparation of a second mutant in which Ser 108 is replaced by Ala, there is a significant amount of material (30% of the total) with a mass some 227 Da lower than the expected value. How could this lower molecular mass material have arisen?

1.17 Draw chemical structures to show the hydrogen bonding which could arise in each case between a Tyr side chain and (a) a second Tyr side chain, (b) the carbonyl group of a peptide bond, and (c) the α-amino group of an N-terminal amino acid.

1.18 From the X-ray structure of the complex between two proteins (X and Y), the following interactions have been identified: (a) Arg 45 of X with Glu 13 of Y, (b) Ile 53 of X with both Val 84 and Phe 85 of Y. Explain the probable basis of these interactions.

1.19 Explain how the strength of ionic interactions and hydrophobic interactions would be affected by changes in the ionic strength of the solution. Use your answers to explain how changes in ionic strength can be used to elute adsorbed proteins in ion-exchange chromatography and hydrophobic interaction chromatography.

1.20 The folded form (F) of ribonuclease T1 is estimated to be 31.4 kJ mol^{-1} more stable than the unfolded form (U) at 25°C. Calculate the equilibrium constant for the $F \rightleftharpoons U$ equilibrium. In the presence of 5.6 M urea, 58.4% of the protein is unfolded; what is the free energy difference between the two forms under these conditions?

References for Chapter 1

Berg, J.M., Tymoczko, J.L., and Stryer, L. (2007) *Biochemistry*, 6th edn. Freeman, New York, 1026 pp.

Crowe, J., Bradshaw, A., and Monk, P. (2006) *Chemistry for the Biosciences: The Essential Concepts*. Oxford University Press, Oxford, 496 pp.

Dwyer, J.J., Gittis, A.G., Karp, D.A., Lattman, E.E., Spencer, D.S., Stites, W.E., and Garcia-Moreno, E.B. (2000) *Biophys. J.* **79**, 1610–20.

Engelman, D.M., Steitz, T.A., and Goldman, A. (1986) *Ann. Rev. Biophys. Biophys. Chem.* **15**, 321–53.

Jabs, A., Weiss, M.S., and Hilgenfeld, R. (1999) *J. Mol. Biol.* **286**, 291–304.

Jones, A. (2005) *Chemistry: An Introduction for Medical and Health Sciences.* John Wiley and Sons, Chichester, 260 pp.

Mathews, C.K., van Holde, K.E., and Ahern, K.G. (2000) *Biochemistry*, 3rd edn. Benjamin/Cummings, San Francisco, California, 1186 pp.

Panasik, N., Jr., Fleming, P.J., and Rose, G.D. (2005) *Protein Sci.* **14**, 2910–14.

Price, N.C., Dwek, R.A., Ratcliffe, R.G., and Wormald, M.R. (2001) *Principles and Problems in Physical Chemistry for Biochemists*, 3rd edn. Oxford University Press, Oxford, 401 pp.

Sreerama, N. and Woody, R.W. (1994) *Biochemistry* **33**, 10022–25.

Voet, D., Voet, J.G., and Pratt, C.W. (2006) *Fundamentals of Biochemistry*, 2nd edn. John Wiley and Sons, Hoboken, New Jersey, 1130 pp.

Weaver, T.M. (2000) *Protein Sci.* **9**, 201–6.

2 The key mathematical tools

Aim of this chapter

The aim of this chapter is to explain the basic concepts involved in tackling quantitative problems. Much of it is probably already familiar to you, but it is worthwhile to go through it again to consolidate your understanding of the topics. If you wish to go into more depth, the book by Cornish-Bowden (1999) should be consulted.

2.1 Estimation of the results of calculations

KEY CONCEPTS

- Breaking down calculations involving multiplication and/or division into a series of simple steps
- Making estimates as a check on the results obtained using a calculator

The Dalton (Da) is equal to 1 atomic mass unit (approximately the mass of a hydrogen atom, but more exactly 1/12 of the mass of the ^{12}C isotope of carbon (1.66×10^{-24} g)); see Chapter 3, Appendix 3.1.

Whenever you face a numerical problem, do you automatically reach for the electronic calculator? Although these wonderful devices are almost universally used for performing calculations, it is a very good idea to develop the habit of trying to estimate the result of a calculation in advance. This gives you a check that you have actually used the calculator correctly.

For example, you might wish to relate the number of amino acids in the polypeptide chain of a protein to its molecular mass. In practice, it has been found that on average each amino acid contributes about 110 Da (Daltons) to the mass of the protein. Thus, if the number of amino acids in the chain were (for example) 260, then the molecular mass would be 260×110 Da. Rather than using the calculator, we can easily estimate the result. To a first approximation we merely multiply by 100 to give 26 000 Da, or 26 kDa (kiloDaltons). We can then add 10% of this value, to give $26 + 2.6$ kDa, i.e. 28.6 kDa.

The prefixes of units are described in more detail in Chapter 3, section 3.3. k (note small k) is an abbreviation for kilo, i.e. '1000 times'; thus 1 kilogram (kg) = 1000 grams (g).

We might wish to proceed in the reverse direction, for example if a protein is of molecular mass 40 kDa, how many amino acids are in the chain? Division of the mass (in Da) by 100 gives $40\ 000/100 = 400$ amino acids; we could then take 10% of this value (40) away to give 360 amino acids (a more accurate answer is 364 amino acids). The importance of being able to relate the molecular mass and the number of amino acids in a protein is explained in Chapter 3, section 3.1.

As a further example of this type of approach it is worth noting that the average contribution of each nucleotide to the mass of a nucleic acid is 330 Da (0.33 kDa). It should thus be relatively easy to see that the mass in kDa can be obtained by dividing the number of bases by 3 (i.e. a synthetic oligonucleotide 80 bases in length has a mass of about 27 kDa).

WORKED EXAMPLE 

The genome of the bacterium *Escherichia coli* is a circular DNA molecule with 3.4 million base pairs. What is the molecular mass of this DNA?

STRATEGY

This is a relatively simple application of the rule stated above and can be solved without use of a calculator.

SOLUTION

The number of bases is 6.8×10^6; division of this by 3 gives the mass in kDa. The mass is therefore about 2.3×10^6 kDa; this could also be expressed as 2.3×10^3 MDa or 2.3 GDa.

The definition of a molar (abbreviated M) solution is given in Chapter 3, section 3.4.1. A 1 M solution contains 1 mole (abbreviated mol; equal to the gram formula weight) of the solute in 1 litre (L) of solution.

WORKED EXAMPLE 

The molar concentration of a solution can be obtained by dividing the concentration of the solute expressed in terms of mg mL^{-1} (equivalent to g L^{-1}) by the molecular mass in Da (see Chapter 3, section 3.4.1). Estimate the molarity of a 3.5 mg mL^{-1} solution of bovine serum albumin, whose molecular mass is 66 000 Da (66 kDa).

STRATEGY

This is an example of estimating the result of a division by a large number; again it is good practice to do this without a calculator.

SOLUTION

The molarity = 3.5/66 000 M (M is the abbreviation for molar). Multiply the top and bottom of this division sum by 10^{-5} to bring the denominator to a small number (in the region of 1). Hence the molarity = $3.5 \times 10^{-5}/0.66$ M. Since 0.66 goes into 3.5 about five times, the molarity can be estimated as about 5×10^{-5} M (50 μM). A more accurate answer is 5.3×10^{-5}M (53 μM).

When multiplying numbers with powers of 10, add the powers together, e.g. $(2 \times 10^5) \times (3 \times 10^3) = 6 \times 10^{5+3}$, i.e. 6×10^8. When dividing, the powers are subtracted, e.g. $(8 \times 10^{23})/(4 \times 10^{15}) = 2 \times 10^{23-15}$, i.e. 2×10^8.

A rather different sort of problem would involve an estimation of the number of heart beats in a human lifetime. You would have to make some assumptions about a typical lifespan (say 80 years) and heart rate (say 70 beats per min).

This would give the number of beats in a lifetime as:

$$\text{Number of minutes in a lifespan} = 80 \times 365 \times 24 \times 60$$

$$\text{Hence, number of beats} = 80 \times 365 \times 24 \times 60 \times 70$$

We can estimate this by taking out the powers of 10 from each term to leave small numbers that can be easily multiplied together:

In this estimation, we have taken 1 power of 10 from 80, 2 from 365, 1 from 24, 1 from 60, and 1 from 70, making 6 powers of 10 overall.

$$\text{Number of beats} = 8 \times 3.65 \times 2.4 \times 6 \times 7 \times 10^6$$

Now we estimate the multiples in pairs (i.e. 8×3.65 is about 30; 2.4×6 is about 15):

$$\begin{aligned} \text{Number of beats} &\approx 30 \times 15 \times 7 \times 10^6 \\ &\approx 30 \times 100 \times 10^6 \\ &\approx 3 \times 10^9 \text{ (i.e. three thousand million or three billion)} \end{aligned}$$

If the calculation involved a division, we would go through the same procedure separately for the numerator and the denominator before estimating the final result of the division.

Note the convenient way of representing very large or very small numbers is by use of powers of 10. For example, it is much easier to write that 1 nanometre (nm) $= 10^{-9}$ metre (m), rather than 0.000000001 m, or that the speed of light is 2.997×10^8 m s^{-1} rather than 299 700 000 m s^{-1}.

In the heartbeat example, if we were to feed the values into a calculator, we would obtain the result that there were 2.94336×10^9 heartbeats in a lifespan. However, to state this as the answer would in this case give a completely false impression of the accuracy of the estimate. The assumptions of a lifespan of 80 years and a heart rate of 70 beats per minute are likely to be at best only reasonable approximations, and hence we should be wary about stating anything other than that there are likely to be about 3 thousand million or 3 billion heart beats in a human lifespan.

? SELF TEST

Check that you have mastered the key concepts at the start of the section by attempting the following questions without using a calculator then use the calculator to check your answers.

In ST 2.1 and ST 2.2, remember each amino acid contributes about 110 Da to the mass. Assume 100 Da and then make the small (10%) adjustment.

In ST 2.3, remember each base contributes about 0.33 kDa, i.e. three bases contribute 1 kDa. The DNA consists of two strands (base pairs), so we multiply our answer for each strand by 2 to obtain the overall molecular mass.

ST 2.4 is a good example of the manipulation of powers of 10.

ST 2.1 The molecular mass of the trypsin inhibitor protein from soya bean is 21 kDa. How many amino acids does it contain?

ST 2.2 The protein hormone insulin contains 51 amino acids. Estimate its molecular mass.

ST 2.3 Human mitochondrial DNA contains about 16 000 base pairs. Estimate its molecular mass.

ST 2.4 The human body is estimated to contain 2.5×10^{13} red blood cells, each of which contains 2.8×10^8 molecules of haemoglobin. Each molecule of haemoglobin has four binding sites for oxygen. How many molecules of oxygen can be bound by the haemoglobin in the body? If 1 mole of oxygen contains 6.02×10^{23} molecules, how many moles of oxygen does this correspond to?

Answers

ST 2.1 The number of amino acids is estimated as 190 amino acids.

ST 2.2 The molecular mass is estimated as 5600 Da or 5.6 kDa.

ST 2.3 The molecular mass is estimated as 11 000 kDa or 11 MDa.

ST 2.4 The number of molecules is estimated as 3×10^{22}; the number of moles is estimated as 0.05.

2.2 Significant figures

KEY CONCEPTS

- Expressing the value of a quantity to the stated number of significant figures
- Understanding the degree of precision appropriate for the experimental approaches employed

The number of significant figures in the quoted value of a quantity is the number of figures ignoring leading or trailing zeroes, ignoring the position of the decimal point; it provides a measure of the confidence with which that value is known. Thus, if the molecular mass of a protein is quoted as 30 kDa (i.e. 30 000 Da), this represents only 1 significant figure; we would be confident that the mass were between 25 and 35 kDa. A different technique might yield the answer to 2 significant figures, e.g. 34 kDa. The technique of mass spectrometry might give an answer of 34.503 kDa; this would represent 5 significant figures. When values are rounded off, 0–4 are rounded down, 5–9 are rounded up. Thus, to 1 significant figure, 34 would be expressed as 30; 35 would be expressed as 40.

Try to develop the skill of quoting the results to the appropriate number of significant figures. This shows that you have understood the basis of the calculation or measurement.

It is extremely important to quote the results of calculations to the appropriate number of significant figures. For example, the molecular mass of a protein can be determined by SDS-PAGE (see Chapter 8, section 8.2.1), in which the mobility of the protein on electrophoresis is compared with standard proteins of known molecular mass. If the mobility is measured to 2 significant figures (e.g. the distance travelled by a band on a gel was 5.2 cm), then the molecular mass should not be quoted to more than 2 significant figures, e.g. 35 kDa, even if the calculator display gives an answer of 34.631782 kDa.

It is very tempting to think that because a calculator gives, for example, 8 places of decimals it must somehow be accurate and authoritative. This is not the case! Of course, it is good practice to carry as much precision as possible forwards during calculations, so long as proper rounding off is performed at the end.

SELF TEST 

Check that you have mastered the key concepts at the start of this section by attempting the following question.

ST 2.5 A calculator gives the result of a calculation as 4623.708. Express this result to 1, 2, 3, and 4 significant figures.

Answer

ST 2.5 The results are 5000, 4600, 4620, and 4624, respectively.

2.3 Logarithms

KEY CONCEPTS

- Understanding what is meant by the logarithm of a number
- Deriving the values of log x, ln x, 10^x, e^x for a given value of x
- Understanding the importance of logarithms in analysing biological systems

The logarithm (abbreviated log) of a number n is the power to which the reference base number (usually 10) must be raised to give n.

Thus, $10^2 = 100$, so log 100 = 2; similarly log 100 000 = 5.

The log does not have to be an integer, thus $10^{2.7634} = 580$, so log 580 = 2.7634.

From the rules regarding powers during multiplication and division:

$$10^a \times 10^b = 10^{a+b}, \text{ so } \log(a \times b) = \log a + \log b$$

$$10^a/10^b = 10^{a-b}, \text{ so } \log(a/b) = \log a - \log b$$

$$a^2 = a \times a, \text{ so } \log(a^2) = \log a + \log a = 2\log a; \text{ in general } \log(a^n) = n \log a$$

Leonhard Euler was an 18th-century Swiss mathematician who made major contributions to many areas of mathematics, even after he became totally blind. Natural logarithms (to base e) arise in the branch of mathematics known as integral calculus (the area under the curve $y = 1/x$ from $x = e$ to $x = 1$ equals 1).

An alternative reference base number for logarithms is the Euler number, e (equal to 2.71828 . . .). Logarithms to base e are known as natural logarithms and generally denoted by ln (though you will see $\log_e$ used in some books).

Now ln 10 = 2.303, so in general ln x = 2.303 log x.

If you enter various numbers into your calculator and use the log and ln keys you should be able to get a feel for the behaviour of logarithms. You will discover the following key properties of logarithms.

KEY INFO

KEY PROPERTIES OF LOGARITHMS

- **The log of 1 = 0 (this is because $10^0 = 1$)**
- **The log of a number between 0 and 1 is negative (an example of this is pH, see section 2.3.1)**
- **The log of a number greater than 1 is positive**
- **Negative numbers do not have logarithms; if you try to find the log of –4 for example, you will get an error message.**

You should also learn to use the anti-logarithm or inverse logarithm keys (10^x and e^x for logarithms to base 10 and e, respectively). If you know the logarithm of a number, you can use these keys to evaluate the number. For example, if $x = 4.702$, $10^x = 50\,350.1$, and $e^x = 110.17$.

Because logarithms are expressions of the power to which a number is raised, we can use the log x and 10^x functions on the calculator to work out squares, square roots, cubes, and cube roots, etc. of numbers. For example, if we wished to work out the cube root of 983, we take the log of 983 (2.9926), divide this by 3 (0.9975), and then take 10^x of this number (9.943). To work out 7.52 cubed (7.52^3), take the log of 7.52, multiply it by 3, and then take 10^x of the result (to give 425.26). We can also use this approach to deal with non-integral powers of numbers. For example, $7.52^{0.28}$ can be shown to be equal to 1.759. The self-test question **ST 2.8** at the end of this section provides an application of this type of calculation.

The use of logarithms makes it possible to compress what can be a huge numerical range. Some applications of logarithms in biology are described in sections 2.3.1–2.3.6.

2.3.1 Acid-base behaviour and the pH scale

Acidity is quantitatively defined by the concentration of protons (H^+ ions) present in a solution. The $[H^+]$ varies enormously in living systems. Thus, after a meal, the $[H^+]$ in the stomach is typically about 0.03 M (30 mM), whereas in the duodenum it is around 0.00000001 M (1×10^{-8} M or 10 nM). Inside the lysosome (a subcellular organelle concerned with degradation of macromolecules), the $[H^+]$ is usually 0.00003 M (3×10^{-5} M or 30 μM).

In order to handle this huge range of numbers, the pH scale is used as a measure of acidity. pH is defined by eqn. 2.1:

$$\text{pH} = -\log\,[H^+] \tag{2.1}$$

The stomach contains a very strong solution of hydrochloric acid (HCl). This would degrade the stomach wall if it was not protected by a layer of mucus. Ulcers arise if this mucus layer is damaged, e.g. by aspirin or other drugs, excessive alcohol, smoking, etc.

Thus, in the stomach the $\text{pH} = -\log\,(0.03) = -(-1.52) = 1.52$
In the duodenum, the $\text{pH} = -\log\,(1 \times 10^{-8}) = -(-8) = 8$
In the lysosome, the $\text{pH} = -\log\,(3 \times 10^{-5}) = -(-4.52) = 4.52$

As we shall see in Chapter 3, section 3.7, an analogous system is used to denote the strengths of acids, employing the term pK_a (equal to $-\log K_a$, where K_a is the dissociation constant of the acid).

Another illustration of logarithmic scales to show a very large range of concentrations is the formation plot, used to depict the binding of a drug to a receptor, for example (see Chapter 4, section 4.3.1).

2.3.2 Variation of reaction rates with temperature

The rates of reactions increase dramatically with temperature as a greater proportion of the reactants possess the energy necessary to surmount the activation energy barrier for reaction to occur. The equation derived by Arrhenius (eqn. 2.2;

see Chapter 4, section 4.2.3) to describe the variation of the rate constant of the reaction (k) with absolute temperature (T) is:

Svante Arrhenius was a Swedish physical chemist who made important contributions in the 19th and 20th centuries to the theory of reaction rates and to understanding the behaviour of ionic solutions.

$$k = Ae^{-Ea/RT} \qquad 2.2$$

where A is the pre-exponential factor, E_a is the activation energy for the reaction, and R is the gas constant (8.31 J K^{-1} mol^{-1}).

We shall see in section 2.5.2 how eqn. 2.2 can be transformed to plot data conveniently.

2.3.3 First-order processes and bacterial growth

The decay of radioactive isotopes or the decrease in the concentration of drugs in the blood plasma normally follow first-order kinetics, according to eqn. 2.3:

$$[A]_t = [A]_0 e^{-kt} \qquad 2.3$$

where $[A]_t$ and $[A]_0$ are the concentrations at time t and at zero time, respectively, and k is the rate constant for the reaction. We shall see in section 2.5.2 how eqn. 2.3 can be transformed to plot data conveniently.

Bacteria with a plentiful supply of nutrients will grow in an exponential (or logarithmic) fashion; thus if, say, the generation time (the time for cell growth and division to provide to daughter cells) were 30 min, and we start with 100 cells in a culture, then after 30 min there will be 200 cells, after 60 min, 400 cells, and after 10 h, 1.0486×10^8 (i.e. $2^{20} \times 100$) cells. At this rate, after 20 h there would be no less than 1.0995×10^{14} (i.e. $2^{40} \times 100$) cells! Of course, the culture will eventually run out of nutrients and the numbers will level off. The period of rapid growth is known as the log phase. A plot of the log of the number of cells against time in this phase can be used to determine the generation time ($t_{1/2}$) of the bacterial culture; the slope of this plot is equal to (log 2)/$t_{1/2}$, i.e. 0.301/$t_{1/2}$.

The rapid growth in the number of bacteria makes it very important to try to achieve 100% killing to combat disease. That is why it is advisable to complete a prescribed course of antibiotics, for example.

✓ WORKED EXAMPLE

The number of bacterial cells in a culture (where there is a plentiful supply of nutrients) increases from 1.2×10^5 to 5.8×10^5 over 120 min. What is the generation (doubling) time for the bacteria under these conditions?

STRATEGY

The two data points can be used to calculate the slope of the plot of log (number of cells) against time. This can be used to calculate $t_{1/2}$.

SOLUTION

The slope of the plot is 0.684/120 min^{-1}, i.e. 0.0057 min^{-1}. Thus, $0.301/t_{1/2} = 0.0057$, from which $t_{1/2} = 0.301/0.0057$ min = 52.8 min.

2.3.4 Molecular mass calibration graphs

Molecular masses of proteins are often estimated by the techniques of gel filtration and SDS-PAGE (see Chapter 8, sections 8.2.1 and 8.2.3). The former method is usually carried out under conditions where a protein retains the three-dimensional structure required for activity (i.e. native conditions) and therefore can be used to estimate the mass of the intact protein. The latter is performed under denaturing conditions and almost invariably will yield the mass of the constituent polypeptide chains of the protein. In both cases, the behaviour of protein being analysed is compared with those of standard proteins of known molecular mass, and calibration graphs are constructed. These are log molecular mass vs. elution volume (gel filtration) and log molecular mass vs. mobility (SDS-PAGE).

SDS-PAGE is an abbreviation for sodium dodecylsulphate-polyacrylamide gel electrophoresis. It is a technique that measures the mobility of a protein in an electric field in the presence of SDS which is a detergent. It can be used to give a good estimate of the molecular mass of a protein as well as the degree of purity of a protein preparation.

2.3.5 Spectrophotometry

As we shall see in Chapter 3, section 3.6, in many cases measurement of the absorption of light by a solution provides a convenient way of determining its concentration. The quantity measured is known as the absorbance (A), which is defined by the equation $A = \log (I_0/I_t)$ where I_0 and I_t are the intensities of incident and transmitted light. The logarithmic nature of this relationship has important practical consequences for the accurate determination of concentrations (see Chapter 3, section 3.6).

2.3.6 Energy changes and equilibrium constants of reactions

The standard free energy change in a reaction (ΔG^0) is related to the equilibrium constant for the reaction (K_{eq}) by eqn. 2.4:

$$\Delta G^0 = -RT \ln K_{eq} \qquad 2.4$$

where R is the gas constant (8.31 J K^{-1} mol^{-1}) and T is the absolute temperature.

The nature of this equation means that the value of K_{eq} will change logarithmically with changes in ΔG^0; at 310 K (37°C) each change of about 5.9 kJ mol^{-1} will lead to a 10-fold change in the value of K_{eq}. This point is discussed further in Chapter 4, section 4.1.

The free energy change of a reaction under standard state conditions (ΔG^0) is discussed further in Chapter 4, section 4.1. The Greek letter Δ (capital delta) is used to mean 'the change of'; G is the symbol for free energy (denoted as G in honour of Josiah Willard Gibbs, an American 19th-century physical chemist. The superscript zero indicates that the change in free energy is under standard state conditions (see chapter 4, section 4.1).

? SELF TEST

Check that you have mastered the key concepts at the start of this section by attempting the following questions.

Use a calculator to perform the following calculations.

ST 2.6 Find the values of log x and ln x for the following values of x: 0.018, 0.632, 1.589, 29.97, 8713

ST 2.7 If log x and ln x have the values −3.72, −1.59, 0.033, 1.15, 4.858, what are the values of x?

ST 2.8 The resting heart rate (H in beats min^{-1}) for mammals has been found to vary with body mass (m in kg) according to an empirical relationship $H = 202/(m^{0.25})$. Use this equation to estimate the heart rate for the following animals: elephant (6000 kg), white rhinoceros (2500 kg), lion (220 kg), human (75 kg), domestic cat (5 kg), rat (0.5 kg).

ST 2.8 provides an application of the use of logarithms to evaluate powers of numbers. Using more extreme examples, we could estimate that the heart rate of a blue whale (100 000 kg) is 11 beats min^{-1} and that of a small shrew (0.003 kg, i.e. 3 g) is 863 beats min^{-1}. These values are in line with measured values for these parameters. It is worth noting that the heart of a blue whale is the size of a modest saloon car and the aorta is large enough for an adult human to crawl along!

Answers

ST 2.6 The values of log x are −1.745, −0.199, 0.201, 1.477, 3.940, respectively; the values of ln x are −4.017, −0.459, 0.463, 3.400, 9.073, respectively.

ST 2.7 The values of x are: (log x) 1.905×10^{-4}, 0.0257, 1.079, 14.125, 72 111 respectively; the values of x are: (ln x) 0.0242, 0.204, 1.034, 3.158, 128.77, respectively.

ST 2.8 The heart rates (beats min^{-1}) are: elephant, 23; rhinoceros, 29; lion, 53; human, 69; cat, 135; rat 241.

2.4 Reciprocals

KEY CONCEPTS

- Understanding what is meant by the reciprocal of a number
- Using reciprocals to evaluate a number of important parameters such as V_{max}, K_m, K_d, and E_a from appropriate graphs

The reciprocal of a number is 1 divided by that number; thus the reciprocal of 8 is 0.125 and the reciprocal of 0.02 is 50. Use the $1/x$ button on the calculator to calculate reciprocals and to explore this function.

Calculations of reciprocals are required in a number of situations, for example:

- In the Lineweaver–Burk plot of enzyme kinetic data and the subsequent calculations of the parameters K_m and V_{max} (see Chapter 4, section 4.4)
- Calculating the K_m or K_d from a the slope of an Eadie–Hofstee or a Scatchard plot, respectively (see Chapter 4, section 4.4)
- Interconverting association and dissociation constants for binding processes (see Chapter 4, section 4.3.1)
- In the Arrhenius plot where the x-axis of the plot is $1/T$ (T is the absolute temperature) (see Chapter 4, section 4.2.3).

SELF TEST ?

Check that you have mastered the key concepts at the start of this section by attempting the following questions.

Use a calculator to perform the following calculations

ST 2.9 From a graph, $1/V_{max}$ is found to be 0.0235 min μM^{-1}. What is the value of V_{max}?

ST 2.10 From the same graph, $-1/K_m$ is found to be -0.0065 μM^{-1}. What is the value of K_m?

ST 2.11 The value of K_a for a binding process is 4.53×10^4 M^{-1}. What is the value of K_d, given that $K_a = 1/K_d$?

Answers

ST 2.9 The value of V_{max} is 42.6 μM min^{-1}.

ST 2.10 The value of K_m is 154 μM; note that the minus signs on each side of the equation cancel out, so that K_m is a positive number.

ST 2.11 The value of K_d is 2.208×10^{-5} M, or 22.08 μM.

In **ST 2.9–2.11** note that when taking reciprocals, the units are also inverted.

In **ST 2.9–2.11** note that the prefix μ (micro: small Greek letter mu) means '10^{-6} times', i.e. 1 μg = 10^{-6}g.

In **ST 2.10** note that K_m is effectively a concentration (see Chapter 4, section 4.3.3), so it must be a positive number.

2.5 Testing hypotheses

KEY CONCEPTS

- Understanding the equation $y = mx + c$ for a straight line graph, and being able to derive the slope and intercept of this graph
- Rearranging simple equations into the form $y = mx + c$

Biochemistry and related subjects, e.g. molecular cell biology, aim to provide explanations of the behaviour of biological systems based on physical laws. The aim is to produce a hypothesis or model that can be tested against experimental data. An important aspect of the process is to derive an equation and then test the experimental data against this equation, usually by means of an appropriate plot. Once a model is verified, the equation can be used to predict the outcome of an experiment under a new set of conditions. If the data do not support the model, it may well be necessary to change it to accommodate the data. This section will deal with the way in which we analyse data so as to confirm that they obey proposed models. Section 2.6 will give a brief outline of some important statistical concepts, which allow us to assign the degree of confidence with which we can make such statements.

2.5.1 Dependent and independent variables

In a graph, the convention is that the x-axis (abscissa) is used to plot the variable that the experimenter varies (e.g. time, concentration of substrate, etc.). This is the *independent variable.*

Guidelines for plotting graphs are given in Chapter 11, section 11.2.9. Most of the points made also apply to graphs generated by computers.

The y-axis (ordinate) is used to plot the quantity that is then observed (e.g. concentration of product formed, rate of reaction, etc.). This is the *dependent variable.*

A very important relationship between y and x is given by the equation of a straight line (eqn. 2.5):

$$y = mx + c \tag{2.5}$$

where m is the slope (gradient) of the line and c is the intercept of the line on the x-axis.

It is very important when plotting data to make sure that the points on the graph actually correspond to the numerical values of the data points. This is a particular problem with certain computer-based graphics programs such as Excel, which will not automatically plot data points with the correct uniform scale on the x-axis; the advice is to look carefully at the plot and see whether it corresponds to what you intend. You should also be able to calculate the value of the slope (change in the value of y divided by the change in value of x) and express it in the correct units. Finally, you should be able to look at an equation and recognize what terms could represent the y-axis and x-axis values, remembering that the slope must be a constant (or a combination of terms that are constant).

Some typical straight line plots which might be obtained are shown in Figs. 2.1–2.3.

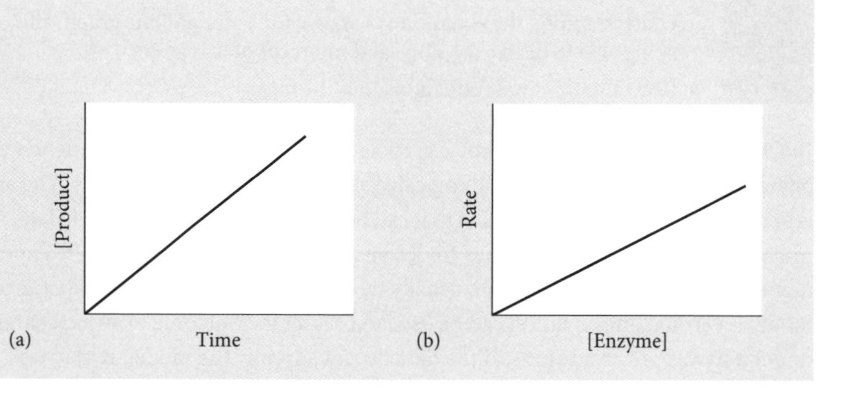

Fig. 2.1 Examples of straight-line graphs through the origin. (a) Concentration of product formed against time in a reaction (straight line through the origin); (b) the rate of reaction against the concentration of enzyme added, where there is no significant rate in the absence of enzyme. The equation for the line is $y = mx$, where m is the gradient.

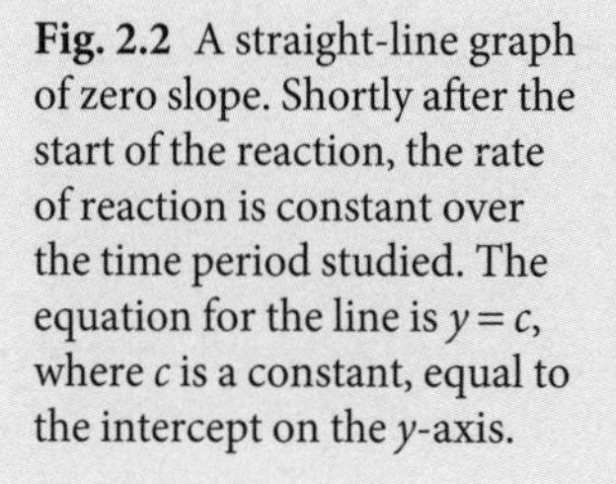

Fig. 2.2 A straight-line graph of zero slope. Shortly after the start of the reaction, the rate of reaction is constant over the time period studied. The equation for the line is $y = c$, where c is a constant, equal to the intercept on the y-axis.

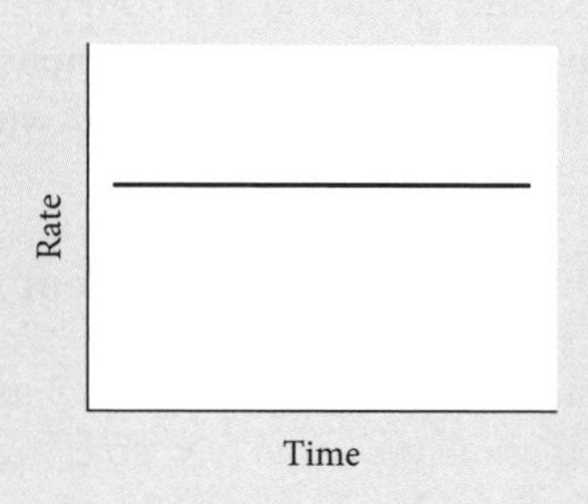

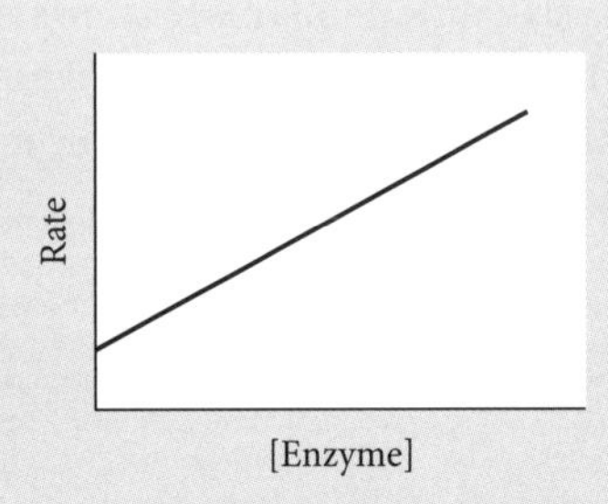

Fig. 2.3 A straight-line graph with a non-zero intercept on the y-axis. The rate of product formation is plotted against the concentration of enzyme for the case where there is a significant blank rate in the absence of enzyme. The equation for the line is $y = mx + c$, where m is the gradient and c is the intercept on the y-axis.

Fig. 2.1(a), which depicts the concentration of product formed against time in a reaction, clearly shows a simple straight line relationship, with the equation $y = mx$ (m = slope). This shows that product is being formed at a constant rate; at zero time, no product is present. We would normally expect to see a graph of this type if we plotted the rate of an enzyme-catalysed reaction against the concentration of enzyme added (Fig. 2.1(b)).

Fig. 2.2, which depicts the rate of product formation of the reaction in Fig. 2.1(a) against time, is a straight line of zero slope, i.e. the rate of the reaction is constant over the time period studied. The equation for this line is $y = c$ (i.e. $m = 0$, since there is no dependence on time).

At the instant the reaction starts, the rate is very low; it takes a finite period before the so-called steady-state rate is achieved. The plot in Fig. 2.2 assumes that this 'pre-steady-state' period is very short.

Fig. 2.3, which depicts the rate of a small number of enzyme-catalysed reactions against the concentration of enzyme added, shows a straight line relationship. However, in this case there is still a significant background rate of reaction when no enzyme is present. The equation is $y = mx + c$, where m is the slope and c is the intercept on the y-axis. The intercept would correspond to the background (or blank) rate of reaction.

Fig. 2.4, which depicts the rate of reaction of an enzyme-catalysed reaction against the concentration of substrate, is clearly not a straight line. The rate of the reaction shows saturation behaviour with respect to the concentration of

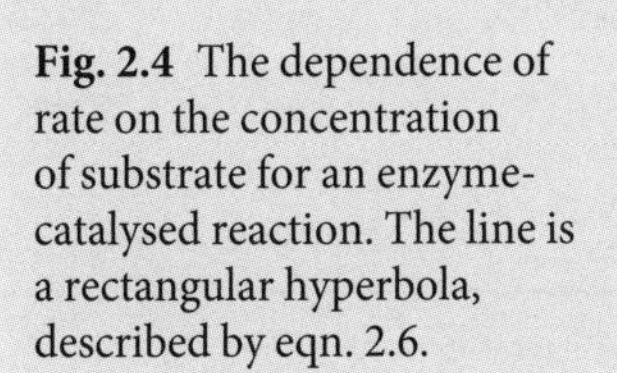
Fig. 2.4 The dependence of rate on the concentration of substrate for an enzyme-catalysed reaction. The line is a rectangular hyperbola, described by eqn. 2.6.

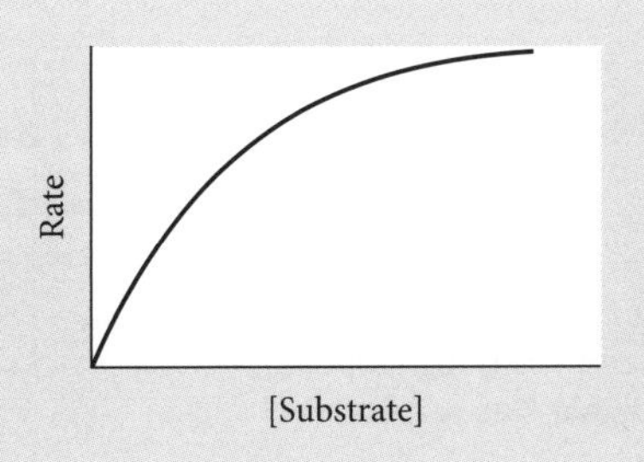

Mathematicians prefer the use of the term 'limiting value', as it indicates that a limit is approached at high concentrations of substrate. The term 'maximum value' should be used when the value can decline as the substrate concentration is increased further. However, the term V_{max} is very widely used by biochemists, so will be kept here.

substrate, i.e. it tends towards a maximum (more accurately, a limiting) value. The actual equation (eqn. 2.6) describing this type of curve is known mathematically as that for a rectangular hyperbola. It is discussed in more detail in Chapter 4, section 4.3.3.

$$v = \frac{V_{max}[S]}{K_m + [S]} \tag{2.6}$$

Equation 2.6 can be transformed in a number of ways to give a straight line relationship that would allow the validity of the equation to be tested. One of the most commonly used is the Lineweaver–Burk plot. By taking reciprocals of the terms on both sides of eqn. 2.6, we obtain eqn. 2.7:

$$\frac{1}{v} = \frac{K_m + [S]}{V_{max}[S]} \tag{2.7}$$

Dividing each term in the numerator of the right hand side of eqn. 2.7 by V_{max} $[S]$, we obtain eqn. 2.8:

$$\frac{1}{v} = \frac{K_m}{V_{max}}\left(\frac{1}{[S]}\right) + \frac{1}{V_{max}} \tag{2.8}$$

Since K_m and V_{max} (and hence also K_m/V_{max}) are constants, it follows that eqn. 2.8 is of the form $y = mx + c$, where y is $1/v$, x is $1/[S]$, m is K_m/V_{max}, and c is $1/V_{max}$.

The Lineweaver–Burk plot is very commonly used to analyse enzyme kinetic data, but it should be remembered that the reciprocal nature of the axes makes it subject to a highly non-uniform distribution of errors (see Chapter 4, section 4.4).

A plot of $1/v$ vs. $1/[S]$ is a straight line (Fig. 2.5), known as the Lineweaver–Burk plot, with the y-axis intercept $= 1/V_{max}$ (see Chapter 4, section 4.4).

Note that some plots you may obtain in the laboratory, e.g. the response of a dye-binding assay to the amount of protein added (see Chapter 6, section 6.1.1), do not necessarily conform to any simple theoretical equation and would be represented by smooth curves. Appropriate values can then be read off the calibration graphs.

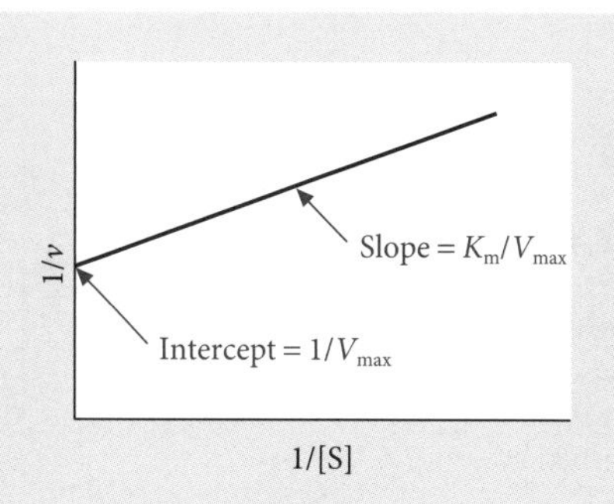

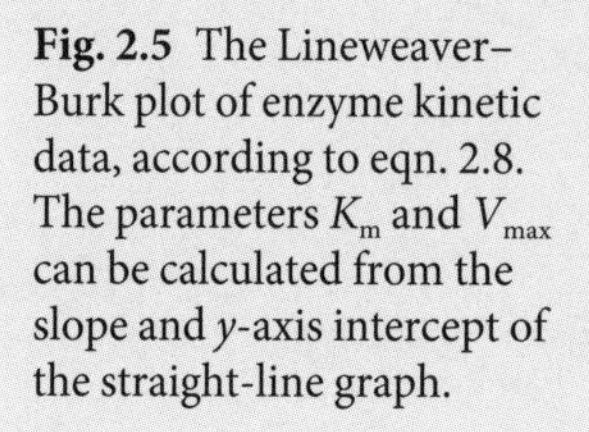

Fig. 2.5 The Lineweaver–Burk plot of enzyme kinetic data, according to eqn. 2.8. The parameters K_m and V_{max} can be calculated from the slope and y-axis intercept of the straight-line graph.

2.5.2 Rearranging equations

In many cases, it is necessary to rearrange an equation to put it into a form appropriate for plotting and subsequent analysis. In section 2.5.1, we saw how eqn. 2.6

$$v = \frac{V_{max}[S]}{K_m + [S]} \qquad 2.6$$

could be transformed (by taking reciprocals of both sides) into one (eqn. 2.8) which would give a straight line, i.e.

$$\frac{1}{v} = \frac{K_m}{V_{max}}\left(\frac{1}{[S]}\right) + \frac{1}{V_{max}} \qquad 2.8$$

There are other ways in which eqn. 2.6 can be transformed so as to give the equation of the form $y = mx + c$ for a straight line. For example, by multiplying both sides of eqn. 2.6 by $(K_m + [S])$, we obtain:

$$vK_m + v[S] = V_{max}\,[S]$$

Rearranging terms:

$$vK_m = V_{max}\,[S] - v[S]$$

Dividing each term on both sides by $[S]$:

$$\frac{vK_m}{[S]} = V_{max} - v$$

Dividing each term on both sides by K_m, we obtain eqn.2.9:

$$\frac{v}{[S]} = \frac{V_{max}}{K_m} - \frac{v}{K_m} \qquad 2.9$$

This is of the form $y = mx + c$, where y is $v/[S]$, x is v, m is $-1/K_m$, and c is V_{max}/K_m. Thus, a plot of $v/[S]$ vs. v is a straight line of slope $-1/K_m$ and a y-axis intercept of V_{max}/K_m. The intercept on the x-axis (derived by setting $v/[S] = 0$) is V_{max}. This is known as the Eadie–Hofstee plot (see Chapter 4, section 4.4).

The Eadie–Hofstee plot is a better way of analysing enzyme kinetic data than the Lineweaver–Burk plot because the distribution of errors is more uniform (see Chapter 4, section 4.4).

✓ WORKED EXAMPLE

Transform eqn. 2.6, i.e. $v = \frac{V_{max}[S]}{K_m + [S]}$ so as to give the Hanes–Woolf equation (eqn. 2.10), for which the plot is $[S]/v$ vs. $[S]$.

$$\frac{[S]}{v} = \frac{[S]}{V_{max}} + \frac{K_m}{V_{max}} \qquad 2.10$$

STRATEGY

The approach is to rearrange the equation so as to be able to separate terms in the $y = mx + c$ form.

SOLUTION

By multiplying both sides of eqn. 2.6 by $(K_m + [S])$, we obtain:

$$vK_m + v[S] = V_{max}[S]$$

Rearranging terms:

$$V_{max}[S] = v[S] + vK_m$$

Dividing each term on both sides of the equation by V_{max}:

$$[S] = \frac{v[S]}{V_{max}} + \frac{vK_m}{V_{max}}$$

Dividing each term by v:

$$\frac{[S]}{v} = \frac{[S]}{V_{max}} + \frac{K_m}{V_{max}} \qquad 2.10$$

Because K_m and V_{max} are constants, this equation is of the form $y = mx + c$, with $y = [S]/v$ and $x = [S]$. A plot of $[S]/v$ vs. $[S]$ will be a straight line with an intercept on the y-axis of K_m/V_{max} and a slope of $1/V_{max}$. The intercept on the x-axis is $-K_m$.

The Hanes–Woolf plot is a better way of analysing enzyme kinetic data than the Lineweaver–Burk plot because the distribution of errors is more uniform (see Chapter 4, section 4.4).

Other examples of rearranging equations to produce straight line graphs include:

The Arrhenius equation for variation of reaction rate constant with temperature

The equation is:

$$k = Ae^{-Ea/RT} \qquad 2.2$$

where A, E_a, and R are constants. (The Arrhenius equation has been mentioned in section 2.3.2 and is discussed in more detail in Chapter 4, section 4.2.3.) Taking the natural logarithms of both sides of eqn. 2.2, we obtain eqn. 2.11:

$$\ln k = \ln A - \frac{E_a}{RT} \qquad 2.11$$

Remember that the natural logarithm of $e^x = x$, and that the logarithm of the product of two numbers is sum of the logarithms of the numbers (see section 2.3).

Thus, a plot of $\ln k$ vs. $1/T$ is a straight line of slope $-E_a/R$, from which E_a can be calculated (see Chapter 4, section 4.2.3).

The equation for a first-order process

Equation 2.3 describes a first-order process (see section 2.4.3 and Chapter 4, section 4.2.1):

$$[A]_t = [A]_0 e^{-kt} \quad 2.3$$

where $[A]_0$ and k are constants.

Taking the natural logarithms of both sides of eqn. 2.3, we obtain:

$$\ln [A]_t = \ln [A]_0 - kt \quad 2.12$$

A plot of $\ln [A]_t$ vs. t is a straight line of slope $-k$, yielding the rate constant directly (see Chapter 4, section 4.2.1).

SELF TEST ?

Check that you have mastered the key concepts at the start of this section by attempting the following question.

ST 2.12 The equation for the osmotic pressure (P) exerted by a solution of a protein whose molecular mass equals M and of concentration c g L^{-1} is given by:

$$P/RTc = 1/M + Bc$$

where R, T, and B are constants. How would you determine M from a suitable graph?

Answer

ST 2.12 A plot of P/RTc vs. c will have a y-axis intercept of 1/M; M is obtained by taking the reciprocal of this intercept.

In ST 2.12 the equation is already in the form $y = mx + c$, where c is plotted on the x-axis and P/RTc on the y-axis. For a given protein, M (and hence $1/M$) will be constant.

2.6 Some basic statistics

KEY CONCEPTS

- Defining the mean, median, and mode of a distribution curve
- Defining the mean and standard deviation of a normal distribution curve
- Testing the difference of two means using the Student's t function
- Testing for correlation between variables; linear and non-linear regression

It is important to realize that virtually all the statements we make in an experimental science are statistical ones. We may be very confident, for example that falling out of an airplane at an altitude of 6500 m without a parachute will be fatal, or at a more mundane level, that administration of a statin-type drug (such as simvastatin)

In 1942, Lt I.M. Chisov, a Soviet pilot survived after ejecting at over 6500 m from his Ilyushin 4 plane when his parachute failed to open. Although he sustained significant injuries, he was back in the cockpit a few months later.

The statin drugs work by inhibiting a key enzyme involved in the biosynthesis of cholesterol. This can lead to a significant reduction (up to 50%) in blood cholesterol levels, and thus reduce the risk of suffering a heart attack.

will lead to the lowering of blood cholesterol levels, but this is not always the case for every individual. We need some way of estimating the degree of confidence with which we can make statements; this is the realm of statistics. The coverage of this topic for the molecular biosciences is much less than would be needed for the environmental and ecological sciences, principally because in the former we usually perform experiments in which we vary the important parameters (concentration, temperature, pH, etc.) in a systematic way to test some accepted theory or model. In contrast, in the more complex relationships in ecology we may have to consider the effects of many variables at the same time. This would require a much more detailed statistical approach to establish significant correlations between parameters, which could then be investigated in detail to derive the causal mechanisms involved (for example, how A influences B, and subsequently C). Statistics is also useful for establishing the degree of confidence with which we can quote the value of an experimentally measured or derived parameter such as the amount of protein in a solution, the rate constant of a reaction, or the Michaelis constant for the substrate of an enzyme. For many applications of statistics, it is common to use a 95% significance threshold, i.e. that we can be 95% confident about a certain outcome, but in some cases it may be important to be at least 99% confident.

2.6.1 Distributions of variables

The starting point for our discussion is the way that the values of parameters can be distributed. For example, if we were to measure the speeds at which vehicles were proceeding along an autobahn in Germany (where there is no official speed limit in rural areas), we might find that most of the vehicles were at speeds in the range 90–110 kph (roughly 55–70 miles per hour) but there would be some lorries going slower than this, and some high-performance cars going at speeds of 150 kph or higher. When plotted as a graph with the value of the speed (or rather the range of speeds, such as between 90 and 92 kph) on the x-axis and the number of vehicles measured as being within that range on the y-axis, we might obtain a distribution curve of the type as shown in Fig. 2.6.

There are three important values associated with a distribution curve. The *mean*, or more strictly the arithmetic mean, ($\bar{x}$) is defined as the average of the values (x_1,x_2,x_3,x_4, etc.) of the parameter plotted on the x-axis. This is defined mathematically by eqn. 2.13:

$$\bar{x} = \frac{x_1 + x_2 + x_3 \ldots + x_n}{n} = \frac{\Sigma(x)}{n} \quad 2.13$$

The symbol Σ is the Greek capital letter sigma (S).

where n is the number of values of the parameter in question (x in this case), and Σ means 'the sum of the values'.

The *median* is defined as the middle value of the parameter, i.e. that value with as many values above it as below it. If we have an even number of values, the

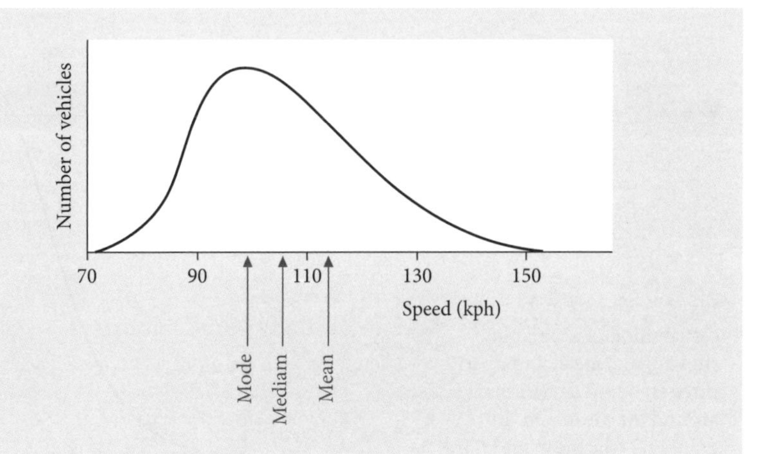

Fig. 2.6 A skewed distribution showing the hypothetical distribution of vehicle speeds along a German autobahn. The mode, median, and mean of the distribution are indicated.

median is the numerical average of the middle two values. For example, if the median of the values 10.2, 10.7, 11.0, 11.1, 11.5, 11.6, 11.9, 12.2 would be the average of 11.1 and 11.5, i.e. 11.3. The median can be given with its quartiles. The first quartile value has $^1/_4$ of the values below it, the third quartile value has $^1/_4$ of the values above it. The inter-quartile range contains the middle $^1/_2$ of the values.

The *mode* is that value which occurs most commonly. The term mode is valuable in describing a distribution of variables which cannot be ranked (e.g. eye colour) and distribution which might show two peaks (this would be termed bimodal).

The mean, median, and mode of the distribution of vehicle speeds are indicated in Fig. 2.6.

2.6.2 The normal distribution

One particularly important type of distribution is known as the normal distribution in which the values of a continuous variable (i.e. one which can take any value, rather than just discrete values) are distributed symmetrically around the mean value. This is shown in Fig. 2.7. This would apply, for example, to physical characteristics such as height or weight, or to examination scores when measured for a suitably large sample size of the population. However, of more importance in the present context is that it also describes the distribution of values of experimental measurements subject to random variations. These would include, for instance, properties of samples taken from individual organisms that have been chosen to be well matched, or replicate determinations of some property of a sample from one particular source. Because of the symmetrical nature of this distribution, the values of the mean, mode, and median all coincide.

The word 'normal' refers to the mathematical form of the distribution curve; it does not mean 'expected' or 'typical'.

A normal distribution is characterized by the values of the mean and the standard deviation.

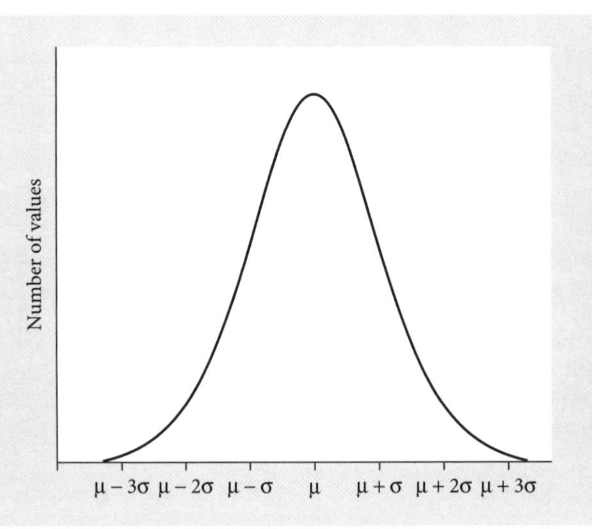

Fig. 2.7 The normal distribution of a variable showing the mean and up to three standard deviations around the mean. On the graph, μ is the mean and σ is the standard deviation.

The population mean (μ) is defined by eqn. 2.14:

$$\mu = \frac{x_1 + x_2 + x_3 \ldots + x_n}{n} = \frac{\Sigma(x)}{n} \qquad 2.14$$

where x_1, x_2, etc. are the individual values of the property and n is the number of values in the population.

In practice, we are rarely able to study the entire population so we study the properties of a sample of the population; the sample mean ($\bar{x}$) has already been defined by eqn. 2.13:

$$\bar{x} = \frac{x_1 + x_2 + x_3 \ldots + x_n}{n} = \frac{\Sigma(x)}{n} \qquad 2.13$$

where n is the size of the sample.

The standard deviation (*SD*, also designated as σ) is defined for a population by eqn. 2.15:

$$SD = \sqrt{\frac{\Sigma(x - \mu)^2}{n}} \qquad 2.15$$

The symbol σ is the Greek small letter sigma (s).

where x is an individual value of the property, μ is the population mean, and n is the number of values in the population.

The introduction of the term $n - 1$ rather than n into the expression for the standard deviation is difficult to explain in simple terms, but it gives a better shape to the distribution curve. For values of $n > 20$, there is only a small (<3%) difference between the two expressions.

For a sample taken from the entire population, the standard deviation (*SD*) is defined in an analogous fashion, except that the term $n - 1$ is introduced into the denominator, as shown in eqn. 2.16:

$$SD = \sqrt{\frac{\Sigma(x - \bar{x})^2}{n - 1}} \qquad 2.16$$

The value of the standard deviation relative to that of the mean gives an indication of how tightly the values are grouped around the mean. In terms of experimentally derived values this would indicate the degree of confidence we had in stating the value of the given parameter, e.g. molecular mass. (See the discussion on significant figures in section 2.2.) From the mathematical equation for a normal distribution, it is found that 68.2% of the total area under the curve is within 1 standard deviation of the mean, 95.4% of the total area is within two standard deviations of the mean and 99.7% within three standard deviations of the mean. (The areas of a standard normal distribution are given in the table in Appendix 2.1 at the end of this chapter.) For example, the average height of adult males in the UK is 178 cm with a standard deviation of 5 cm; thus, 95% of the male population is between 168 and 188 cm in height. There is thus only a 1 in 20 probability (which we express using the symbol p, i.e. $p = 0.05$) that the height of a male will fall outside that range. For example, there would be a less than 1 in 500 probability ($p < 0.003$) that a male is 195 cm tall.

The coefficient of variation (CV) is often used to describe the degree of variability of a population. It is defined as: $CV = 100\ (SD/\bar{x})\%$, where $\bar{x}$ and SD are the mean and standard deviation of the population, respectively.

The way in which the sample mean ($\bar{x}$) might vary from the (true) population mean (μ) is described by the term standard error of the mean (SEM) which is defined by eqn. 2.17.

$$SEM = \frac{SD}{\sqrt{n}} \tag{2.17}$$

Clearly, the larger the sample size n, the smaller the value of the SEM.

From the properties of normal distributions:

$\mu \pm 1.96\ SEM$ will include 95% of the sample means

$\mu \pm 2.58\ SEM$ will include 99% of the sample means

WORKED EXAMPLE 

The operation of a pipette was checked by repeatedly dispensing and weighing volumes of water. The volume on the pipette was set at 1 mL, and the following volumes (mL) were dispensed in succession: 0.932, 0.927, 0.948, 0.937, 0.918, 0.929, 0.940, and 0.942. What is the mean and standard deviation of these values? Comment on the reliability of the pipette.

STRATEGY

We use eqns. 2.13 and 2.16 to evaluate the mean and standard deviation, respectively.

SOLUTION

The mean value is 0.934 mL and the standard deviation is 0.0096 mL. From the properties of the normal distribution 99.7% of the values would be within the range 0.905 to 0.963 mL, which is significantly different from the nominal value of 1.000 mL. Thus, we can conclude that the pipette is *precise* (i.e. it delivers volumes which are reproducibly close to each other, with a low standard deviation), but it is *not accurate* (i.e. it is not sufficiently close to the true, or required, value). If the experiment had given a mean of 1.002 mL with a standard deviation of 0.0096 mL, the pipette would be both *precise* and *accurate.*

The words 'accurate' and 'precise' are often used interchangeably; it is important to appreciate their correct scientific usage, as in this example.

Having looked at the way in which statistics can be used to describe the distributions of variables, we shall now briefly consider two important applications of statistics in drawing conclusions from such distributions.

2.6.3 Testing the difference between two means

The large numbers of patients required for the later stages of drug trials is a major factor in the cost of developing new drugs. It is estimated that each new drug would have cost several hundred million dollars to bring to market.

A very common use of statistics is to decide whether a change in a parameter is significant. For example, does the administration of a certain drug lead to a significant reduction in blood pressure, or are any changes observed merely due to chance? A trial may be set up with matched pairs of patients half of whom are given the drug and the other half given a dummy 'placebo' which is the control. The blood pressure data are collected and presented in the form of a mean and standard deviation for each group. In order to be able to draw reliable conclusions, the sample sizes should be as large as possible; indeed the trials of new drugs usually involve at least several hundred patients. In each group (drug and placebo) there will be a range of values of blood pressure, each with its own mean and standard deviation (Fig. 2.8).

The way we usually proceed is to test the so-called 'null hypothesis', that is that there is no real difference between the mean values for the two groups (i.e. that the drug does not really cause any effect) and that any difference observed reflects random variations between individuals. Testing this hypothesis would certainly be important for the trial of a new drug, since there is an onus on the company to prove that the new drug is more effective than any existing treatments.

We first calculate the standard error of the difference (SE_d) between the two means, according to eqn. 2.18:

$$SE_d = \sqrt{\frac{SD_1^2}{n_1} + \frac{SD_2^2}{n_2}} \qquad \textbf{2.18}$$

where SD_1 and SD_2 are the standard deviations of the two groups (of sizes n_1 and n_2, respectively).

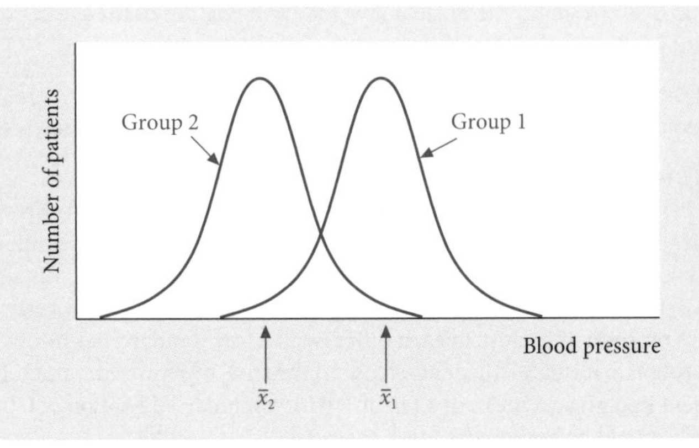

Fig. 2.8 Distribution curves for the blood pressure data for two groups of patients. The mean values for the two groups are indicated ($\bar{x}_1$ and $\bar{x}_2$).

We then use the t function (more properly known as the Student's t function) to assess the significance of the differences. The t function is defined by eqn. 2.19:

$$t = \frac{(x-\mu)\sqrt{n}}{SD} \quad 2.19$$

It shows a distribution around a mean value similar to the normal distribution, but is more appropriate for smaller sample sizes.

In terms of testing the differences between two means, the appropriate definition of t is given by eqn. 2.20:

The t function was defined by William Gosset in 1908. His employers, Guinness Breweries, requested him to publish his work under a pseudonym, so he chose the name 'Student').

$$t = \frac{|\bar{x}_1 - \bar{x}_2|}{SE_d} \quad 2.20$$

where $\bar{x}_1$ and $\bar{x}_2$ are the two sample means and $|\ |$ means 'irrespective of sign'.

The probability (p) that the two sample means are identical can be deduced from the properties of the t function for the appropriate number of degrees of freedom (equal to $n_1 + n_2 - 2$). The values of the t function are given in the table in Appendix 2.2 at the end of this chapter.

In a small-scale drug trial, the sample mean values of the diastolic blood pressures of the drug and placebo groups were 122.5 and 110.3 mm, respectively. There were 20 patients in each group. The standard deviations for the two groups were 20.5 and 18.1 mm, respectively. Do the data show (at the 95% confidence level) that the drug has an effect on the blood pressure?

STRATEGY

We calculate the standard error of the difference (eqn. 2.18), and from that the value of t (eqn. 2.20). Reference to the table of t values allows us to reach a conclusion about the significance of any change.

SOLUTION

The value of $SE_d = 6.11$ mm. The value of $\bar{x}_1 - \bar{x}_2 = 12.2$ mm. Hence $t = 1.997$. Reference to the table in Appendix 2.2 shows that t is below the entry value (2.02) for 95% confidence. Hence, we cannot reject the null hypothesis and must conclude that the drug has not been shown to have an effect. Since t is quite close to the entry value, it would probably be worthwhile extending the test to include more patients; this may well increase the value of t significantly.

2.6.4 The correlation coefficient and linear regression

The term 'correlation' refers to how strongly two variables are related. For example, if we were to plot a scatter diagram showing the shoe sizes of individuals against their height, we would expect to see a positive relationship between the two (tall

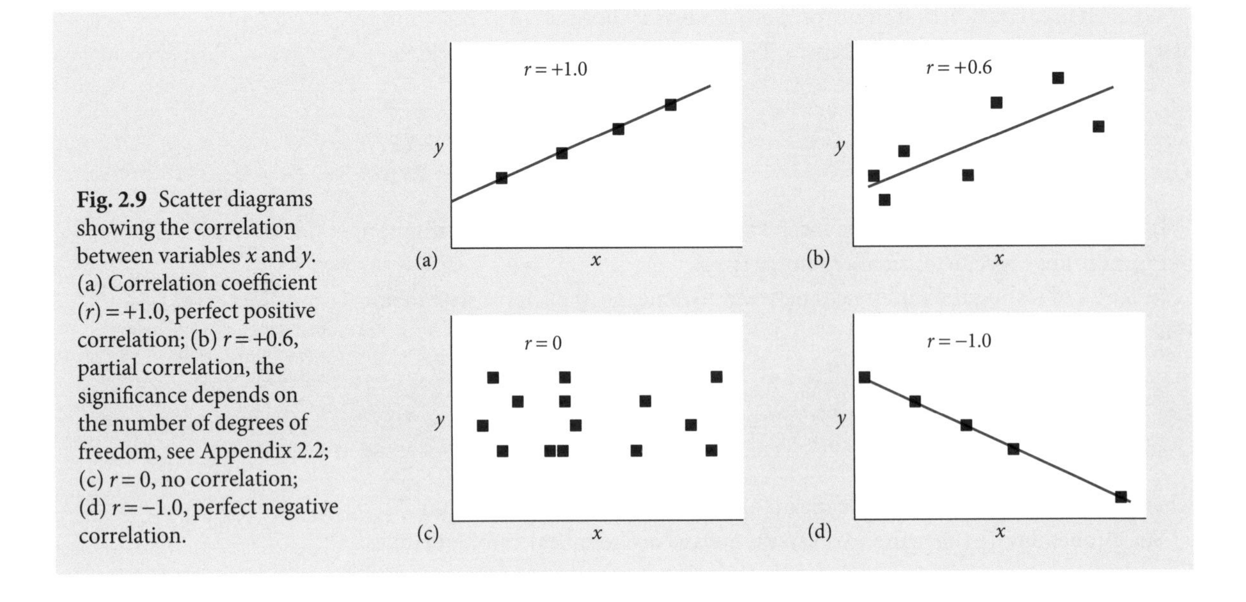

Fig. 2.9 Scatter diagrams showing the correlation between variables x and y. (a) Correlation coefficient $(r) = +1.0$, perfect positive correlation; (b) $r = +0.6$, partial correlation, the significance depends on the number of degrees of freedom, see Appendix 2.2; (c) $r = 0$, no correlation; (d) $r = -1.0$, perfect negative correlation.

people generally have large feet). We can define the correlation coefficient, r, for the variables x and y by eqn. 2.21:

$$r = \frac{\Sigma(x-\bar{x})(y-\bar{y})}{\sqrt{\Sigma(x-\bar{x})^2\Sigma(y-\bar{y})^2}} \qquad 2.21$$

where $\bar{x}$ and $\bar{y}$ are the means of the values of x and y, respectively, in the data set..

Values of r can range from −1 (perfect negative correlation) to +1 (perfect positive correlation). Some examples of scatter diagrams and the associated values of r are shown in Fig. 2.9.

The value of r which indicates a significant correlation between two variables depends on the number of (x,y) data points we have (strictly speaking on the degrees of freedom (n), which equals the number of (x,y) data points −2). Values of r which are used to establish a correlation are listed in the table in Appendix 2.3 at the end of this chapter. For example, we could say (at the 95% confidence level) that two variables are positively correlated if $r \geq 0.754$ $(n = 5)$ or $r \geq 0.576$ $(n = 10)$. Drawing this sort of conclusion is important if one is trying to establish a correlation between two variables before trying to propose a mechanism for a causal relationship.

In the molecular biosciences it is more likely that we are investigating the validity of a model and are testing experimental data against that model. If we are testing an equation where we would expect a straight line relationship (see section 2.5.2), then we can plot the appropriate parameters on a graph to check that the equation and hence the model are obeyed. In an ideal world (perfect data), all the points would fall on the straight line and determination of the slope and

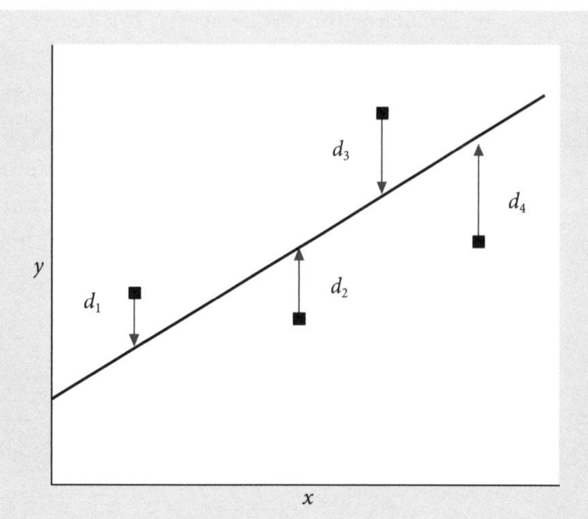

Fig. 2.10 Least-squares line for the data points shown by filled squares. The deviations from the straight line are indicated by d_1, d_2, etc.

intercept would be trivial. In practice, due to errors in measurements, it is likely that the points would be scattered around a straight line. The determination of the best straight line is known as linear regression and the line produced is known as the regression line of y on x. The most widely used method to do this is the least-squares method, in which the sum of the squares of the differences between the calculated and observed values of y at each value of x is minimized (Fig. 2.10).

Note in Fig. 2.10 that the squares of the differences are used so that deviations below the line and above the line both contribute to the total deviation.

The equation for the least-squares fit straight line is $y = mx + c$, where the slope m is given by eqn. 2.22:

$$m = \frac{\Sigma(x - \bar{x})(y - \bar{y})}{\Sigma(x - \bar{x})^2} \quad \textbf{2.22}$$

Knowing the values of m, $\bar{x}$, and $\bar{y}$, the value of the y-axis intercept c can be calculated from eqn. 2.23:

$$\bar{y} = m\bar{x} + c \quad \textbf{2.23}$$

Once the best straight line has been determined, it can be used to predict values of y for given values of x. The standard error of the estimate of y (s_e) is given by eqn. 2.24:

$$s_e = \sqrt{\frac{\Sigma(y - \bar{y})^2}{n - 2}} \quad \textbf{2.24}$$

where n is the number of (x,y) points.

We can then establish a 95% confidence band around the regression line, which will be marked by two lines (one 1.96 times s_e above the line, the other 1.96 times s_e

below the line). Clearly, it is better if one is using this approach within the measured range of the values of x (interpolation) than outside this range (extrapolation). In the latter case, it would be important to establish (or necessary to assume) that the model and equation were valid outside the range.

The values of the slope and intercept may then be used to derive parameters such as the rate constant or activation energy of a reaction, or the number of binding sites on a protein for a given ligand. It is possible to calculate the standard errors in the estimates of the slope (s_m) and the y-axis intercept (s_c) using eqns. 2.25 and 2.26, respectively.

$$s_m = \sqrt{\left(\frac{1}{n-2}\right)\left(\frac{n\Sigma(y^2) - (\Sigma(y))^2}{n\Sigma(x^2) - (\Sigma(x))^2} - m^2\right)} \qquad 2.25$$

$$s_c = s_m\sqrt{\frac{\Sigma(x^2)}{n}} \qquad 2.26$$

WORKED EXAMPLE

The rate (v, in units of µM min^{-1}) of an enzyme-catalysed reaction was studied as function of substrate concentration ([S], in units of µM). The data were analysed by the Hanes–Woolf plot (eqn. 2.10) in which [S]/v is plotted against [S]. The following values were obtained:

[S]/v	13.2	18.0	18.2	22.9	25.3	27.0	30.7
[S]	10	20	30	40	50	60	70

Calculate the correlation coefficient for the plot of [S]/v vs. [S], and use linear regression to calculate the best straight line.

STRATEGY

This is an application of eqns. 2.21, 2.22, and 2.23. It is a good idea to draw up a table to calculate the various terms required for these equations.

SOLUTION

[S]/v is designated as y and [S] as x. The values of $\bar{y}$ and $\bar{x}$ are 22.19 and 40, respectively. The values of $\Sigma(y-\bar{y})^2$ and $\Sigma(x-\bar{x})^2$ are 220.03 and 2800, respectively. The value of $\Sigma(x-\bar{x})(y-\bar{y})$ is 776. From this, using eqn. 2.21, $r = 0.9886$; this is highly significant correlation ($p < 0.001$ for 5 degrees of freedom, i.e. the number of (x,y) data points (7) −2). The slope (m) and y-axis intercept (c) of the least-squares line are 0.277 and 11.1, respectively. Further analysis using eqns. 2.25 and 2.26 shows that $s_m = 0.029$ and $s_c = 1.3$.

2.6.5 Non-linear regression

Although linear regression is a very useful method, there are many occasions when the relationships between variables cannot be expressed in terms of a simple straight line equation, or where such a relationship could cause problems. (One example of the latter is the Lineweaver–Burk rearrangement of the Michaelis–Menten equation (see section 2.5.1). The reciprocal nature of the parameters plotted ($1/v$ and $1/[S]$) means that there is a highly non-uniform distribution of errors over the range of values, so that in determining the best straight line by the least-squares method, the greatest weight is given to the points at high $1/[S]$, i.e. low $[S]$, which are associated with the greatest experimental errors).

In cases where a straight line relationship does not hold, it is possible to use non-linear regression, where the data are fitted to more complex equations, often involving higher power dependence on x (e.g. x^2, x^3, etc.). Most fitting procedures involve the use of complex numerical algorithms and are most conveniently performed by computers. One way of assessing the overall quality of the fit is by evaluating the normalized root mean square deviation (NRMSD), which is defined by eqn. 2.27:

$$NRMSD = \sqrt{\frac{\Sigma(y_{obs} - y_{cal})^2}{\Sigma(y_{obs})^2}} \qquad 2.27$$

where y_{obs} and y_{cal} are the observed and calculated (according to the fitting equation) values of y, at each specified value of x. The NRMSD can take values ranging from 0 (perfect fit) to 1 (no fit whatsoever); generally, values less than 0.1 are considered satisfactory.

There are many programs commercially available for the direct fitting of enzyme kinetic data to theoretical models such as that described by the Michaelis–Menten equation. The majority of these use the Levenberg–Marquardt algorithm, which employs an iterative approach to find the values of the parameters in the chosen model which give the best fit to the experimental data as judged by the sum of the squares of the differences being minimized. Initial trial values of these parameters are either supplied or guessed and these are then varied in an incremental fashion and the effect on the goodness of fit assessed. In the Levenberg–Marquardt approach, the sizes of the incremental changes can be automatically adjusted according to how well the values of the parameters are converging towards their final values.

In the case of enzyme kinetic data, these direct fitting procedures can be used to fit the data (v as a function of $[S]$) directly to the Michaelis–Menten equation (see Chapter 4, section 4.4). The program will produce estimates of the parameters K_m and V_{max}, together with the standard errors of the estimates in these quantities. Ideally, the errors should be ≤5% of the values of the parameters. Low values of these errors give confidence that the equation (and the model on which it is based) is obeyed, and that the measurements are not subject to excessive random errors.

A low error value does not, however, exclude the possibility of a systematic error. For example, if a stock solution of substrate had been made up at the wrong concentration, then the value of K_m would be incorrect, even if there were no errors in pipetting or measurement of rates.

As well as the NRMSD, a useful further check on the appropriateness of the analysis of the data is to look at the so-called pattern of residuals (the differences between the calculated and observed values of the y parameter at each value of x). When these differences are plotted against the values of the x parameter, there

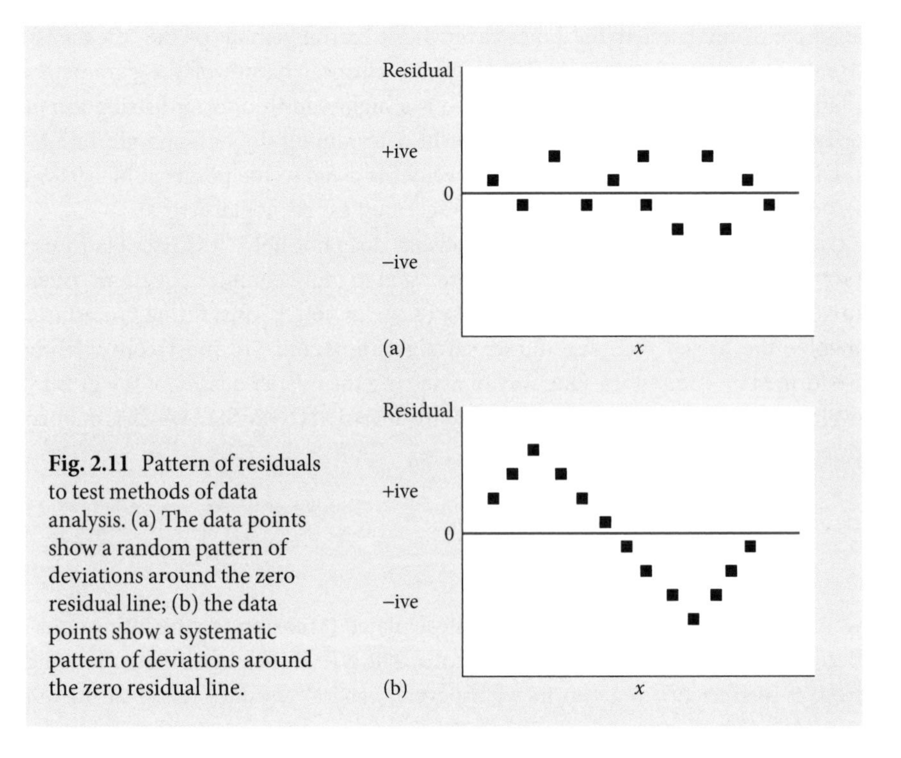

Fig. 2.11 Pattern of residuals to test methods of data analysis. (a) The data points show a random pattern of deviations around the zero residual line; (b) the data points show a systematic pattern of deviations around the zero residual line.

should be a random pattern around the line $y = 0$ (Fig. 2.11(a)). If there is a systematic pattern of residuals (as, for example in Fig. 2.11(b)), this indicates that the method of analysis is inappropriate, and should be altered to yield a random pattern of residuals.

? SELF TEST

Check that you have mastered the key concepts at the start of this section by attempting the following questions.

With respect to ST 2.13, the current guidelines for blood cholesterol levels are that they should be kept below 5.2 mM, although the distribution between various lipoprotein complexes is also important.

ST 2.13 The following values of blood cholesterol (mM) were found in a sample of eight healthy females: 4.3, 4.1, 5.8, 5.0, 3.9, 5.5, 5.2, and 4.9. What is the median, mean and standard deviation of these values? What is the 95% confidence limit for the population mean (μ)?

ST 2.14 A group of 21 healthy students undertook a glucose tolerance test in which they fasted overnight and then ingested 75 g glucose. Before taking the glucose, the blood glucose levels of the group had a mean value of 4.73 mM (SD = 0.48 mM). 30 min after taking the glucose, the blood glucose levels had a mean value of 7.95 mM (SD = 1.63 mM). After further 90 min, the levels had a mean value of 5.25 mM (SD = 1.53 mM). Are the levels at 30 and 120 min significantly different from that at the start?

ST 2.15 A Hanes–Woolf plot (eqn. 2.10) used to analyse a set of enzyme kinetic data obtained at eight values of substrate concentration showed a correlation coefficient (r) of 0.669. What would you recommend to the investigator who produced the data?

Answers

ST 2.13 The values are: median, 4.95 mM; mean, 4.84 mM; standard deviation, 0.68 mM; 95% confidence limit for μ, 4.37–5.31 mM.

ST 2.14 Comparing 30 min and start values, $t = 8.68$; this gives $p < 0.01$, i.e. the null hypothesis can be rejected with >99% confidence. Comparing 120 min and start values, $t = 1.49$; this gives a p value between 0.2 and 0.1 ($0.2 > p > 0.1$); i.e. the null hypothesis cannot be rejected with at least 95% confidence. Thus, the 30 min value is significantly higher than the start value, but the 120 min value is not.

ST 2.15 The value of r is below the value required for 95% confidence of a positive correlation. It would not therefore be appropriate to use the plot to try to obtain reliable values of the kinetic parameters (K_m and V_{max}) for the enzyme. It would be sensible to try to improve the experimental technique and to obtain more data points.

With respect to **ST 2.14**, the current guidelines are that the fasting blood glucose levels should be in the range 3.3–6.1 mM; 30 min after the ingestion of glucose the level should be below 11.1 mM; after further 90 min, the level should have dropped to below 7.8 mM. Fasting blood glucose levels greater than 7.8 mM, and greater than 11.1 mM at the 120 min point indicate diabetes.

References for Chapter 2

Cornish-Bowden, A. (1999) *Basic Mathematics for Biochemists*, 2nd edn. Oxford University Press, Oxford, 221 pp.

Appendix

Appendix 2.1 Table of areas of a standard normal distribution

The entries in the table show the proportion of the total area under the curve which lies between $x = 0$ and the actual value of x. The areas for negative values of x are obtained by symmetry.

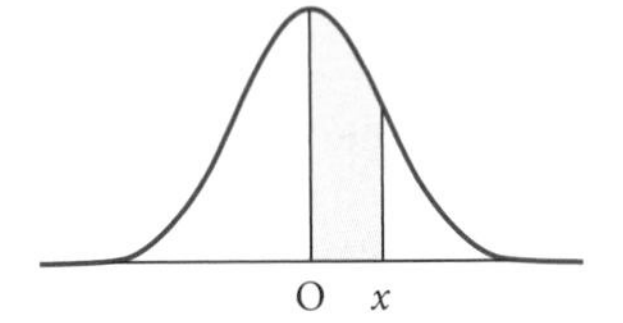

x	.00	.01	.02	.03	.04	.05	.06	.07	.08	.09
0.0	.0000	.0040	.0080	.0120	.0160	.0199	.0239	.0279	.0319	.0359
0.1	.0398	.0438	.0478	.0517	.0557	.0596	.0636	.0675	.0714	.0753
0.2	.0793	.0832	.0871	.0910	.0948	.0987	.1026	.1064	.1103	.1141
0.3	.1179	.1217	.1255	.1293	.1331	.1368	.1406	.1443	.1480	.1517
0.4	.1554	.1591	.1628	.1664	.1700	.1736	.1772	.1808	.1844	.1879
0.5	.1915	.1950	.1985	.2019	.2054	.2088	.2123	.2157	.2190	.2224
0.6	.2257	.2291	.2324	.2357	.2389	.2422	.2454	.2486	.2517	.2549
0.7	.2580	.2611	.2642	.2673	.2703	.2734	.2764	.2794	.2823	.2852
0.8	.2881	.2910	.2939	.2967	.2995	.3023	.3051	.3078	.3106	.3133
0.9	.3159	.3186	.3212	.3238	.3264	.3289	.3315	.3340	.3365	.3389
1.0	.3413	.3438	.3461	.3485	.3508	.3531	.3554	.3577	.3599	.3621
1.1	.3643	.3665	.3686	.3708	.3729	.3749	.3770	.3790	.3810	.3830
1.2	.3849	.3869	.3888	.3907	.3925	.3944	.3962	.3980	.3997	.4015
1.3	.4032	.4049	.4066	.4082	.4099	.4115	.4131	.4147	.4162	.4177
1.4	.4192	.4207	.4222	.4236	.4251	.4265	.4279	.4292	.4306	.4319
1.5	.4332	.4345	.4357	.4370	.4382	.4394	.4406	.4418	.4429	.4441
1.6	.4452	.4463	.4474	.4484	.4495	.4505	.4515	.4525	.4535	.4545
1.7	.4554	.4564	.4573	.4582	.4591	.4599	.4608	.4616	.4625	.4633
1.8	.4641	.4649	.4656	.4664	.4671	.4678	.4686	.4693	.4699	.4706
1.9	.4713	.4719	.4726	.4732	.4738	.4744	.4750	.4756	.4761	.4767
2.0	.4772	.4778	.4783	.4788	.4793	.4798	.4803	.4808	.4812	.4817
2.1	.4821	.4826	.4830	.4834	.4838	.4842	.4846	.4850	.4854	.4857
2.2	.4861	.4864	.4868	.4871	.4875	.4878	.4881	.4884	.4887	.4890
2.3	.4893	.4896	.4898	.4901	.4904	.4906	.4909	.4911	.4913	.4916
2.4	.4918	.4920	.4922	.4925	.4927	.4929	.4931	.4932	.4934	.4936
2.5	.4938	.4940	.4941	.4943	.4945	.4946	.4948	.4949	.4951	.4952
2.6	.4953	.4955	.4956	.4957	.4959	.4960	.4961	.4962	.4963	.4964
2.7	.4965	.4966	.4967	.4968	.4969	.4970	.4971	.4972	.4973	.4974
2.8	.4974	.4975	.4976	.4977	.4977	.4978	.4979	.4979	.4980	.4981
2.9	.4981	.4982	.4982	.4983	.4984	.4984	.4985	.4985	.4986	.4986
3.0	.4987	.4987	.4987	.4988	.4988	.4989	.4989	.4989	.4990	.4990

Appendix 2.2 Table of values of Student's *t* function

The first column lists the number of degrees of freedom, *n*. The other columns show the probabilities (p) for t to be greater than the values listed. The values of p for negative values of t are obtained by symmetry.

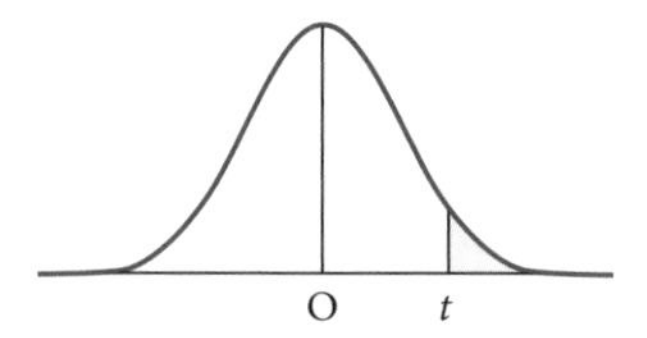

n \ p	.10	.05	.025	.01	.005
1	3.078	6.314	12.706	31.821	63.657
2	1.886	2.920	4.303	6.965	9.925
3	1.638	2.353	3.182	4.541	5.841
4	1.533	2.132	2.776	3.747	4.604
5	1.476	2.015	2.571	3.365	4.032
6	1.440	1.943	2.447	3.143	3.707
7	1.415	1.895	2.365	2.998	3.499
8	1.397	1.860	2.306	2.896	3.355
9	1.383	1.833	2.262	2.821	3.250
10	1.372	1.812	2.228	2.764	3.169
11	1.363	1.796	2.201	2.718	3.106
12	1.356	1.782	2.179	2.681	3.055
13	1.350	1.771	2.160	2.650	3.012
14	1.345	1.761	2.145	2.624	2.977
15	1.341	1.753	2.131	2.602	2.947
16	1.337	1.746	2.120	2.583	2.921
17	1.333	1.740	2.110	2.567	2.898
18	1.330	1.734	2.101	2.552	2.878
19	1.328	1.729	2.093	2.539	2.861
20	1.325	1.725	2.086	2.528	2.845
21	1.323	1.721	2.080	2.518	2.831
22	1.321	1.717	2.074	2.508	2.819
23	1.319	1.714	2.069	2.500	2.807
24	1.318	1.711	2.064	2.492	2.797
25	1.316	1.708	2.060	2.485	2.787
26	1.315	1.706	2.056	2.479	2.779
27	1.314	1.703	2.052	2.473	2.771
28	1.313	1.701	2.048	2.467	2.763
29	1.311	1.699	2.045	2.462	2.756
30	1.310	1.697	2.042	2.457	2.750
40	1.303	1.684	2.021	2.423	2.704
60	1.296	1.671	2.000	2.390	2.660
120	1.289	1.658	1.980	2.358	2.617
∞	1.282	1.645	1.960	2.326	2.576

Appendix 2.3 Table of critical values of correlation coefficient

The value of p shown is the probability that the absolute value of r exceeds the value shown in the table. Thus, for 8 degrees of freedom, the probability that r exceeds 0.632 when its true value is 0 (no correlation) is 0.05.

The number of degrees of freedom is the number of (x,y) data points −2.

THE CORRELATION COEFFICIENT

Degrees of freedom	*Value of p*				
	0.10	0.05	0.02	0.01	0.001
1	0.9877	0.99692	0.99951	0.99988	0.9999988
2	0.900	0.950	0.980	0.990	0.999
3	0.805	0.878	0.934	0.959	0.991
4	0.729	0.811	0.882	0.917	0.974
5	0.669	0.754	0.833	0.875	0.951
6	0.621	0.707	0.789	0.834	0.925
7	0.582	0.666	0.750	0.798	0.898
8	0.549	0.632	0.715	0.765	0.872
9	0.521	0.602	0.685	0.735	0.847
10	0.497	0.576	0.658	0.708	0.823
11	0.476	0.553	0.634	0.684	0.801
12	0.457	0.532	0.612	0.661	0.780
13	0.441	0.514	0.592	0.641	0.760
14	0.426	0.497	0.574	0.623	0.742
15	0.412	0.482	0.558	0.606	0.725
16	0.400	0.468	0.543	0.590	0.708
17	0.389	0.456	0.529	0.575	0.693
18	0.378	0.444	0.516	0.561	0.679
19	0.369	0.433	0.503	0.549	0.665
20	0.360	0.423	0.492	0.537	0.652
25	0.323	0.381	0.445	0.487	0.597
30	0.296	0.349	0.409	0.449	0.554
35	0.275	0.325	0.381	0.418	0.519
40	0.257	0.304	0.358	0.393	0.490
45	0.243	0.288	0.338	0.372	0.465
50	0.231	0.273	0.322	0.354	0.443
60	0.211	0.250	0.295	0.325	0.408
70	0.195	0.232	0.274	0.302	0.380
80	0.183	0.217	0.257	0.283	0.357
90	0.173	0.205	0.242	0.267	0.338
100	0.164	0.195	0.230	0.254	0.321

Calculations in the molecular biosciences

3

Part 1: Characterization of biological molecules

The next two chapters will explain the main types of calculations involved in studying proteins. In this chapter, we shall first of all deal with some general principles important in any calculations, and then concentrate on how to study the individual components of reactions. This includes the preparation and dilution of solutions and how spectrophotometry can be used to check concentrations in some cases. The behaviour of acids, bases, and buffers is described and this is followed by a discussion of the specific activities of proteins and the manner in which their purification is recorded. Chapter 4 will concentrate on the analysis of a number of processes and reactions of biological interest.

At the end of this chapter, there are several problems for you to check your understanding of the material covered.

If you wish to explore the topics covered here in more detail, the books mentioned in Chapter 1 can be consulted.

3.1 The golden rules for successful calculations

KEY CONCEPTS

- Working out the strategy for a calculation
- Setting out and explain the working
- Making sure the number of significant figures and the units are correct

Whenever you do a calculation, make sure that the answer to each of the following questions is 'yes'. This will not absolutely guarantee success but will go a very long way towards that goal.

- Have I worked out the correct strategy for the calculation?
- Is my working clearly set out and explained?
- Is the value obtained reasonable?
- Is it reported to the appropriate number of significant figures?
- Is it quoted in the correct units?

The aim of this chapter is to help you carry out calculations successfully in a number of topics covering the purification and characterization of proteins.

We can expand these points as follows:

Have I worked out the correct strategy for the calculation? Before starting to write down your solution, make sure that you have worked out how you will proceed from the information supplied in the problem you are trying to solve to the point where you can calculate the final answer. If the path is not immediately obvious, it can be a good idea to think about what you can deduce from the information and then try to think of what might be required immediately before the final step. Working from both ends in this way can often help to define the strategy.

Is the working clearly set out and explained? It is very important to break down the calculation into steps and use words to explain. Do not just write down numbers without explanation. If you do, you will not convince anyone else reading your work that you know what you are doing. If you make a small slip in the first step and then this is (correctly) carried through subsequent steps, you should obtain credit for this; if you have not explained what you are doing and the final result is incorrect it is highly unlikely that you will obtain any credit at all.

Is the value obtained reasonable? Sometimes it is not clear what sort of value might be expected, although it is usually possible to check that the outcome of the final calculation step is of the right order of magnitude (see Chapter 2, section 2.1). However, in many cases it is possible to think about whether the value is reasonable.

It is difficult to overestimate the importance of the need for this check.

For example, one would expect to find small integral values for the number of ligand-binding sites on a protein, e.g. one substrate binding site per polypeptide chain. Values of 0.0009 or 926 sites per polypeptide chain are very unlikely and suggest that a serious error has been made in the calculation, probably involving either the use of incorrect units or mistakes in handling powers of 10. A second example would be if you were asked to use some experimental chemical modification data to calculate the number of cysteine side chains in a protein of molecular mass 30 kDa. Before starting, one can make a useful estimate as follows. A 30 kDa protein will contain about 30 000/110 amino acids, i.e. 273 amino acids (see Chapter 2, section 2.1). There are 20 different kinds of amino acids in proteins, so if they occurred with equal frequency (which in practice they do not, but that is not important here), there would be about 14 of each type. Thus, we might expect to find about 14 cysteines in each molecule of the protein. In fact, cysteine is among the rarer amino acids so in most proteins of this size the actual value would be less than 14. The key point is that a value of, for example, 538 would clearly be incorrect. In addition, a value of 0.25 would be incorrect since the value should be either zero or an integral value such as 1, 2, etc.

The average frequency of occurrence of amino acids in proteins is given in Chapter 1, Table 1.1. Leucine and alanine occur most frequently (9.6% and 7.8%, respectively); cysteine (1.5%) and tryptophan (1.2%) occur least frequently. These are only average figures and do not necessarily apply to any particular protein, for example the small protein metallothionein that plays a role in storing a number of heavy metal ions consists of about 30% cysteine residues.

Is the appropriate number of significant figures used? This has been dealt with in Chapter 2, section 2.2; remember to use the display in the calculator with caution when writing down the final answer.

Is the answer quoted in the correct units? The subject of units of quantities is discussed in section 3.3. You should get into the habit of not only quoting the correct units of the final answer, but also quoting the units at the various stages of the calculation. This will act as a guide to the direction of the calculation process, and will also help to convince anyone else that you know what you are doing!

SELF TEST ?

The key concepts at the start of the section can be checked at the end of section 3.3.

3.2 Magnitudes of quantities

KEY CONCEPTS

- Appreciating the range of sizes of quantities used in describing biological systems
- Performing calculations involving powers of 10

Molecular biology and biochemistry are quantitative sciences; the values of quantities we discuss range from the very large (numbers of molecules, Avogadro's number, frequency of radiation, etc.) to the very small (dimensions of cells, organelles and molecular complexes, concentrations of hormones, etc.); the range of dimensions of biological objects is illustrated in Fig. 3.1. This range of magnitudes means that you need to develop confidence in handling powers of 10 and the prefixes used for quantities (section 3.3).

Fig. 3.1 shows the advantage of using a logarithmic scale. For example, if we wished to represent the range of dimensions from 10 nm to 100 m on linear scale graph paper with a grid where each 1 mm represents 10 nm, then to represent 100 m, we would need a piece of paper 10^4 km long (more than the distance from London to Los Angeles!).

SELF TEST ?

The key concepts at the start of the section can be checked at the end of section 3.3.

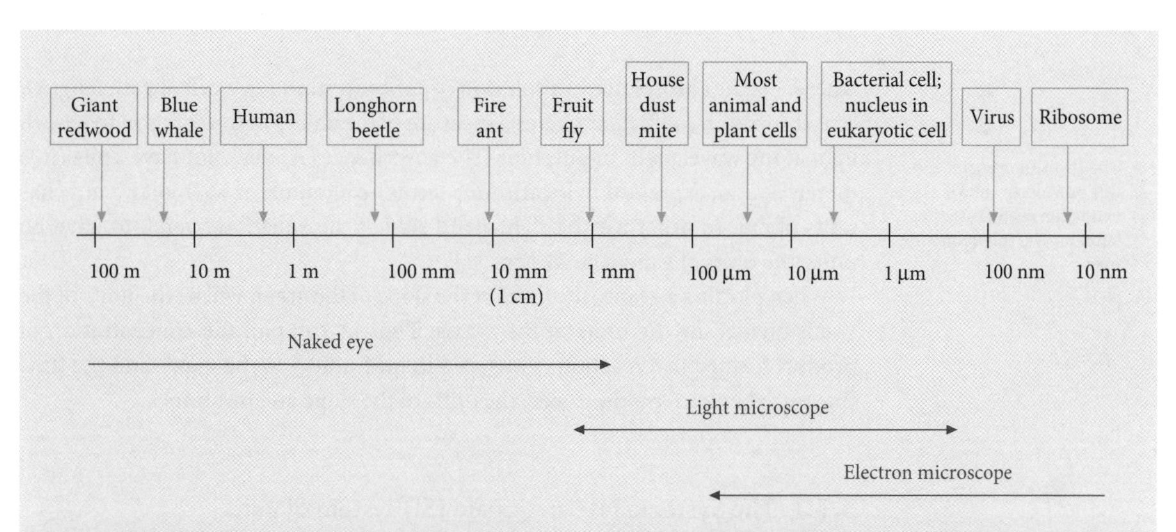

Fig. 3.1 The dimensions of biological objects. The scale is logarithmic, with each division representing a factor of 10. The dimensions associated with each object are for guidance only as there will be a range of values in each case. Individual atoms and small molecules would occupy the zone between 0.1 nm (100 pm) and 1 nm. The lower arrow bars indicate the observation ranges for the naked eye, light microscope, and electron microscope. Highly sophisticated electron microscopes can resolve objects down to about 0.2 nm (200 pm) in size.

3.3 Units of quantities

KEY CONCEPTS

- Knowing the SI prefixes for units and interconvert them as appropriate
- Using correct units in equations and graphs

Virtually all quantities we come across have units, for example mass (e.g. kg, kDa), length (e.g. mm, nm), concentration (e.g. g L^{-1}, mol L^{-1} (or M)), time (e.g. ms), etc. It is essential to state these units in any description of a system or during calculations. The only exceptions arise where the quantity represents a ratio of two values, e.g. the absorbance (A) of a solution is defined in terms of the ratios of intensities of two light beams (incident on the solution and transmitted by the solution) and therefore does not have units.

It should be pointed out that equilibrium constants (see Chapter 4, section 4.1) raise an apparent contradiction; strictly speaking they are dimensionless since all the terms in the definition refer to ratios to the standard state for that component. However, since for solutes, the values (of concentrations) are each expressed relative to a standard state of 1 M concentration (see Chapter 4, section 4.1), equilibrium constants are usually quoted with units of M to indicate this.

It is important to refer to a 1 M solution for calculation purposes, i.e. a K_d of 50 μM should be expressed as 50×10^{-6} M (or 5×10^{-5} M).

Remember that in any equation, the units must balance on the two sides. For example, in spectrophotometry the Beer-Lambert law (eqn. 3.1; see section 3.6) states that:

$$A = \varepsilon \times c \times l \tag{3.1}$$

where A is the absorbance of a solution of concentration c in a cell of path length l. The absorption coefficient ε gives a measure of the ability of the solution to absorb light at the wavelength in question. The absorbance (A) does not have units (it is merely a ratio, expressed in logarithmic terms), c has units of M (molar), and l has units of cm. In order for the right-hand side of the equation overall to have no units, the units of ε must be M^{-1} cm^{-1}.

The absorption coefficient was previously known as the extinction coefficient; the latter term is still frequently used.

When plotting a graph, the units of the slope of the graph will be the units of the y-axis divided by the units of the x-axis. Thus, if you plot the concentration of product formed in a reaction (measured in mM units) on the y-axis and the time (measured in min) on the x-axis, the units of the slope are mM min^{-1}.

3.3.1 The Système Internationale (SI) system of units

The SI system of units is based on the metre (m), kilogram (kg), second (s) quantities and units derived from these. Information on these units and units derived from them as well as the values of physical constants is included in Appendix 3.1 at the end of this chapter. It uses prefixes to bring numerical values of quantities into the 'easy to scale' range, i.e. between 0.1 and 100; thus 800 nm = 0.8 μm;

2000 μm = 2 mm, etc. The most important prefixes involve factors of 1000, so we can set up a scale as shown in Table 3.1:

Table 3.1 The prefixes used in the SI system of units

Prefix	G, giga	M, mega	k, kilo	Base unit	m, milli	μ, micro	n, nano	p, pico	f, femto
Numerical value	10^9	10^6	10^3	1	10^{-3}	10^{-6}	10^{-9}	10^{-12}	10^{-15}

We only need to go outside the range shown on very rare occasions; for interest 10^{12} is known as Tera (abbreviated 'T') and 10^{-18} is known as atto (abbreviated as 'a').

One example of the use of the prefix T would be in denoting the frequency of radiation. Thus, red light of wavelength 700 nm has a frequency of 4.3×10^{14} Hz (hertz) or 430 THz.

It is very important indeed for you to be able to convert quantities between the different prefixed units rapidly and reliably, for example to show that 5000 nm = 5 μm and that 0.012 μmol = 12 nmol.

Some intermediate prefixes are retained for convenience, e.g. c (centi, 10^{-2}), as in cm; d (deci, 10^{-1}), as in dm (note 1 litre = 1 dm^3). For measures of volume there would be an inconveniently large factor of 10^9 between 1 m^3 and 1 mm^3, hence the use of 1 dm^3 and 1 cm^3 (1 cm^3 = 1000 mm^3; 1 dm^3 = 1000 cm^3).

WORKED EXAMPLE 

The length of the C–N bond in the peptide bond unit is 0.132 nm. Express this in terms of μm, pm and also in terms of Angstrom units (Å), which are commonly used to describe protein structures (1 Å = 10^{-8} cm).

STRATEGY

This is a straightforward application of the interconversion of prefixes in the SI system of units.

SOLUTION

1 nm = 10^{-3} μm, so 0.132 nm = 1.32×10^{-4} μm = 132 pm = 1.32 Å

The Ångstrom unit (Å) is not an SI unit, but is widely used by structural biologists; most bond lengths are in the range 1–2 Å (0.1–0.2 nm).

WORKED EXAMPLE 

The molecular mass of lysozyme is 14 300 Da, and its density is 1.4 g mL^{-1}. Assuming that the lysozyme molecule can be regarded as spherical, calculate its radius. Avogadro's number is 6.02×10^{23} mol^{-1}. The volume of a sphere of radius r is given by $(4\pi/3)r^3$.

STRATEGY

This problem involves calculating the mass of an individual molecule of lysozyme and hence its volume and radius. It is very important to keep track of the units and powers of 10.

SOLUTION

The mass of 1 mol lysozyme is 14 300 g (see section 3.4).

Hence the mass of 1 molecule = 14 300/(6.02×10^{23}) g = 2.375×10^{-20} g.

Since volume = mass/density, the volume of 1 molecule = 1.697×10^{-20} mL (cm^3).

So $(4\pi/3)r^3 = 1.697 \times 10^{-20}$, hence $r^3 = 4.05 \times 10^{-21}$, therefore $r = 1.59 \times 10^{-7}$ cm, or 1.59 nm (note 1 cm = 10^{-2} m; 1 nm = 10^{-9} m).

Thus, the radius of a lysozyme molecule = 1.59 nm.

In this calculation, the cube root can be readily derived by using logarithms (see Chapter 2, section 2.3).

? SELF TEST

Check that you have mastered the key concepts at the start of sections 3.1–3.3 by attempting the following questions.

ST 3.1 A pipetting device is stated to be capable of delivering 2.5 mm^3 portions of a solution. Express this volume in terms of mL and μL.

ST 3.2 A Scatchard plot is often used to analyse binding data. In this plot, $r/[L]_{free}$ is plotted against r (r is a ratio and therefore has no units; $[L]_{free}$ is the concentration of free ligand and in the case in question has units of μM). What are the units of the slope of the graph?

ST 3.3 Calculate the radius of a bacterial ribosome which can be assumed to be spherical. The molecular mass of the ribosome is 2.7 MDa and the density is 1.4 g mL^{-1}. Avogadro's number is 6.02×10^{23} mol^{-1}. The volume of a sphere of radius r is given by $(4\pi/3)r^3$.

Answers

ST 3.1 The volume delivered = 2.5×10^{-3} mL or 2.5 μL.

ST 3.2 The units of the slope of the graph are μM^{-1}.

ST 3.3 The radius of the ribosome = 9.1 nm.

Attempt Problem 3.1 at the end of the chapter.

3.4 The concepts of moles and molarity

KEY CONCEPTS

- Distinguishing clearly between moles and molarity
- Knowing the molecular mass of the solute, interconverting concentrations expressed in terms of mass/volume and in terms of molarity

It is important to note that the term mole can refer to molecules, atoms, subatomic particles, photons, etc.

One mole (abbreviated mol), which is the gram formula mass of a substance, contains Avogadro's number (6×10^{23}, or more accurately 6.022×10^{23}) of entities (which can be molecules, atoms, ions, etc.).

Substances react with one another in terms of moles, not in terms of masses. Thus, in the hydrolysis of urea:

$$CO(NH_2)_2 + H_2O \rightarrow CO_2 + 2NH_3$$

1 mol of urea (60 g) reacts with 1 mol water (18 g) to give 1 mol carbon dioxide (44 g) and 2 mol ammonia (2×17 g, i.e. 34 g).

We can also refer to 1 mol of atoms (e.g. H atoms or Cl atoms) or ions (e.g. H^+ ions or electrons); in each case this will mean the Avogadro number of that species.

3.4.1 Molarity of solutions

Virtually all biochemical reactions take place in solution. (A solution consists of one or more *solutes* dissolved in the *solvent*). A 1 molar (M or mol dm^{-3}) solution contains 1 mol of the solute in 1 litre (L or dm^3) of solution.

KEY RELATIONSHIP !

number of moles = concentration (expressed in M) × volume (expressed in L)

WORKED EXAMPLE 

A certain solution of NADH has a concentration of 0.6 mM. How many moles of NADH are present in 0.5 mL of this solution?

STRATEGY

This is an application of the relationship: number of moles = concentration × volume

SOLUTION

1 L of the solution contains 0.6 mmol. Hence 0.5 mL contains 0.3×10^{-6} mol, i.e. 0.3 μmol.

NADH (nicotinamide adenine dinucleotide, reduced form) is a biological reducing agent, which participates in redox reactions catalysed by dehydrogenases, such as lactate dehydrogenase (pyruvate + NADH + H^+ $\rightleftharpoons$ lactate + NAD^+). The structure of NADH is given in Chapter 1, section 1.8).

If the molecular mass of the solute is X dalton (Da), then 1 mol = X g (e.g. for glucose $C_6H_{12}O_6$ the gram formula mass would be 180 g).

A 1 M solution therefore contains X g solute L^{-1} (for glucose, this would be 180 g L^{-1}).

With practice, you should be able to work out rapidly how many mol of a solute are present in a certain volume of solution of specified molarity. Most biochemical reactions involve small volumes of dilute solutions, so we are likely to be dealing with μmol, nmol or even pmol quantities of solute.

We can readily work out the concentrations of solutions as follows:

A solution containing 1 g solute of molecular mass X per litre would be $(1/X)$ M (for glucose this would be (1/180) M = 5.56×10^{-3} M).

For biochemistry, g and L are large quantities; mg and mL (or even μg and μL) are much more likely to represent the scale of experiments undertaken and to take account of the high costs of many of the reagents involved.

KEY RELATIONSHIP !

In terms of concentration, 1 μg μL^{-1} = 1 mg mL^{-1} = 1 g L^{-1}.

Thus, we can always calculate and obtain the molar concentration of a solution by dividing the mg mL^{-1} (or g L^{-1} or μg μL^{-1}) concentration by the molecular mass (in Da).

Thus, a 0.2 mg mL^{-1} solution of glucose = 0.2/180 M = 1.11×10^{-3} M (1.11 mM).

BSA is an abundant protein in the serum of cattle blood. It can bind a number of non-polar substances and serves to transport fatty acids in the blood stream.

A 0.15 mg mL^{-1} solution of bovine serum albumin (BSA) (molecular mass 66 000 Da) = 0.15/66 000 M = 2.273 μM.

 **WORKED EXAMPLE**

The concentration of protein required for an nuclear magnetic resonance experiment is 0.5 mM. If the molecular mass of the protein is 27.5 kDa, what concentration is required in terms of mg mL^{-1}? If the sample volume is 0.4 mL, what mass of protein is required?

STRATEGY

This is a relatively straightforward calculation, provided that you keep track of the units and powers of 10.

SOLUTION

A 1 M solution of protein contains 27 500 mg mL^{-1}. Thus a 0.5 mM solution contains $0.5 \times 10^{-3} \times 27\,500$ mg mL^{-1} = 13.75 mg mL^{-1}. In 0.4 mL there would be 13.75×0.4 mg, i.e. 5.5 mg.

It should be noted that for large molecules such as proteins a 1 M solution is physically unrealistic; it would be impossible for instance to dissolve 27 500 mg (27.5 g) protein in buffer to give a 1 mL solution. Nevertheless, the concept of a 1 M solution represents a useful basis for calculations.

In a number of cases, it is convenient to describe the concentration of a solute in a solution in terms of % weight/volume (abbreviated as w/v); this would represent the grams of solute per 100 mL solution. If two liquids are being mixed (e.g. ethanol and water) it can be convenient to express the concentration of ethanol in terms of percentage volume/volume (abbreviated as v/v).

 **WORKED EXAMPLE**

The structure of acrylamide is given in Chapter 1, section 1.8. When acrylamide is polymerized and cross-linked it gives a gel which provides a convenient support medium for electrophoresis in which convection is minimized. The monomer (but not the polymer) is a powerful neurotoxin and must be handled with great care.

Acrylamide, which can be polymerized to make a gel widely used in electrophoresis, is available as an aqueous solution at 30% (w/v). The molecular mass of acrylamide is 71.1 Da. What is the concentration of this solution in molar terms?

STRATEGY

This is again a relatively straightforward calculation provided the correct units are used throughout the calculation.

SOLUTION

A 30% (w/v) solution corresponds to 300 g L^{-1}. A 1 M solution of acrylamide would be 71.1 g L^{-1}. Hence, the concentration is 300/71.1 M = 4.22 M.

A final note of caution: it is *absolutely essential* that you appreciate the distinction between the *amount* of a solute in a solution (measured in *moles*) and its *concentration* (measured in *molar*). You should be able to work out that, for example, 1 mL of a 1 mM solution contains 1 μmol and that 0.02 mL of a 0.7 μM solution contains 14 pmol.

For some calculations it is appropriate to work in terms of mol, for others it may be best to work in terms of M. You must never slip from one to the other during a calculation without being aware of what you are doing. If you label the quantities with the correct units as you go along, you should avoid making mistakes.

SELF TEST ?

Check that you have mastered the key concepts at the start of the section by attempting the following questions.

ST 3.4 How many moles of solute are present in 2.5 mL of a solution whose concentration is 1.35 mM?

ST 3.5 A solution contains 2.1 pmol of a given protein in 0.5 μL. What is the concentration of the solution?

ST 3.6 A mass spectrometer requires 0.2 ng protein in a volume of 2.5 μL. If the molecular mass of the protein is 34 kDa, what is the concentration of the solution in molar terms?

ST 3.7 The molecular mass of a small peptide is 791 Da. What would be the concentration of a 5 mM solution of the peptide in terms of mg mL^{-1}?

Answers

ST 3.4 The amount of solute = 3.375×10^{-6} mol or 3.375 μmol.

ST 3.5 The concentration = 4.2 μM.

ST 3.6 The concentration = 2.35 nM.

ST 3.7 The concentration = 3.955 mg mL^{-1}.

Attempt Problems 3.2 to 3.8 at the end of the chapter.

3.5 Preparation and dilution of stock solutions

KEY CONCEPTS

- Working out how to prepare stock solutions of the required concentration
- Calculating how to dilute these solutions appropriately for particular applications

The accurate preparation and dilution of solutions is essential for accurate work in the molecular biosciences. The following sections should give you guidance as to how to carry out these operations successfully.

3.5.1 Stock solutions

In experimental work, we often make use of 'stock solutions' of compounds which have either been prepared to a high degree of accuracy in the laboratory or purchased from a reputable supplier. These stock solutions are usually much more concentrated than our 'working solutions' (in actual experiments) and must be diluted appropriately before use.

The preparation of stock solutions is usually carried out as follows. Suppose that you wished to prepare 5 mL of a 5 mM stock solution of NADH. You would need to know the molecular mass of NADH; inspection of the catalogue reveals that the disodium salt hydrate of NADH (which is the form you have purchased) has a molecular mass of 709.4 Da. A 1 M solution of NADH would be 709.4 g L^{-1}, i.e. 709.4 mg mL^{-1}. Thus, a 5 mM solution would be $5 \times 10^{-3} \times 709.4$ mg mL^{-1}, i.e. 3.55 mg mL^{-1}.

If we wished to make up exactly 5 mL of this solution, we would have to weigh out 5×3.55 mg, i.e. 17.75 mg NADH on an appropriate balance. This could take some considerable time. It is much more likely that we require about 5 mL of the stock solution, in which case we can adjust the volume of solvent (water, buffer, etc.) so that the concentration is correct. Thus, if we weighed out 17.43 mg NADH, we would add 4.91 mL solvent to obtain the correct concentration. (In practice, we would probably use measurements of light absorbance to confirm the concentration of this stock solution, as explained in section 3.6. This would be particularly important in those cases where a substance may have absorbed moisture, which often happens to powders stored at low temperatures.)

Many biological compounds are stored at low temperatures because they are unstable at room temperature. When taking bottles from the fridge or freezer it is important to allow them to warm up to room temperature before opening, otherwise water vapour will condense on the inside of the bottle.

In some cases, particularly of very concentrated stock solutions, there can be significant volume changes when the solvent is added to the solute. For example, if we wished to prepare 10 mL of an 8 M solution of the denaturant guandinium chloride (GdmCl) of molecular mass 95.5 Da, we can readily calculate that we would require 95.5×8 mg mL^{-1}, i.e. 764 mg mL^{-1} GdmCl. For 10 mL, this would be 7.64 g GdmCl. If we were to put this into a graduated tube and add 10 mL of water or buffer, the final volume would be close to 14 mL and the solution would not be of the required concentration. Instead, we would carefully add the solvent, making sure that the GdmCl was dissolved and bringing the volume of the solution up to the 10 mL mark. It should be noted that 8 M is close to the solubility limit of GdmCl, so it takes some time to ensure that it is all dissolved. (Since GdmCl does not have a convenient absorption band, the concentration of the solution can be checked by density or refractive index measurements, by reference to standard tables of values).

The dissolution of many substances including urea and GdmCl is endothermic (i.e. takes in heat), so it is often necessary to warm solutions gently to aid dissolution.

3.5.2 Dilution of stock solutions

Many procedures involve the preparation of working solutions from more concentrated stock solutions which have either been purchased or made up in the laboratory. It is very important to be able to do this accurately.

Two basic points are:

A. The amount of solute cannot be increased or decreased by any dilution procedure (matter cannot be created or destroyed!), and

B. The *amount* of solute = the *volume* of solution multiplied by its *concentration*.

Thus if:

C_S = concentration of stock solution

V_S = volume of stock solution taken

and

C_W = final concentration of working solution

V_W = final volume of working solution

Then we can derive the following important equations:

KEY RELATIONSHIP !

concentration of stock × volume of stock = final concentration × final volume

$$\text{i.e. } C_S \times V_S = C_W \times V_W \quad (3.2)$$

from which it follows that:

volume of stock = (final concentration × final volume)/concentration of stock

$$\text{i.e. } V_S = \frac{C_W \times V_W}{C_S} \quad (3.3)$$

The application of eqn. 3.3 is illustrated in the worked examples below.

Note the importance of relating the volume of the stock solution to the final (total) volume of the solution (eqn. 3.2), not to the volume of buffer or solvent added to dilute the stock solution.

WORKED EXAMPLE 

How much of a stock solution of ATP (4.0 mM) in a given buffer should be taken to give a concentration of 0.1 mM in a working solution of total volume 1.0 mL?

STRATEGY

This is an application of eqn. (3.3); it is important to ensure that the units are the same on both sides of the equation.

SOLUTION

In this case, $C_S = 4.0$ mM, $C_W = 0.1$ mM, $V_W = 1.0$ mL

Hence from eqn. (3.3) $V_S = (0.1 \times 1.0)/4.0$ mL $= 0.025$ mL (25 μL)

The remaining volume (0.975 mL or 975 μL) would be made up of the appropriate buffer.

WORKED EXAMPLE

Stock solutions of ATP and $MgCl_2$ are 4.0 and 20 mM, respectively, in a given buffer. What volumes of ATP, $MgCl_2$ and buffer would you add together to produce a working solution of final volume 2.0 mL, in which the concentrations of ATP and $MgCl_2$ were 0.5 and 1.5 mM, respectively?

STRATEGY

This calculation involves the application of eqn. (3.3) to each solute separately. Again it is important to ensure that the correct units are used on both sides of the equation.

SOLUTION

We must consider the components separately.

For ATP, $C_S = 4.0$ mM, $C_W = 0.5$ mM and $V_W = 2.0$ mL

Hence from eqn. (2.3) $V_S = (0.5 \times 2.0)/4.0$ mL $= 0.25$ mL

For $MgCl_2$, $C_S = 20$ mM, $C_W = 1.5$ mM and $V_W = 2.0$ mL

Hence from eqn. (2.3) $V_S = (1.5 \times 2.0)/20$ mL $= 0.15$ mL

Thus, the working solution would consist of 0.25 mL ATP, 0.15 mL $MgCl_2$ and $(2.0 - (0.25 + 0.15))$ mL, i.e. 1.6 mL buffer.

SELF TEST

Check that you have mastered the key concepts at the start of the section by attempting the following question.

In ST 3.8 remember to use eqn. 3.3 to calculate the volume of each stock solution required. The volume of buffer required to make up the total volume is calculated by subtracting the stock volumes added from the required volume of the working solution.

ST 3.8 You are provided with stock solutions of an enzyme (2.5 mg mL^{-1}), its substrate (5 mM), and urea (8 M), all in a given buffer. You wish to make up a solution of volume 0.5 mL with the following final concentrations: enzyme, 0.1 mg mL^{-1}; substrate 0.25 mM; urea, 2 M. What volumes of the stock solutions and buffer would you require?

Answer

ST 3.8 The volumes of the solutions required are: enzyme, 0.02 mL (20 μL); substrate, 0.025 mL (25 μL); urea 0.125 mL (125 μL); buffer, 0.33 mL (330 μL).

Attempt Problems 3.9 to 3.15 at the end of the chapter.

3.6 Absorbance measurements

KEY CONCEPTS

- Using the Beer–Lambert law to calculate concentrations from absorbance data
- Knowing the range of absorbance values where measurements are most reliable

Measurements of the absorption of light by a solution (spectrophotometry) can be an extremely convenient way of determining the concentration of that solution, provided that (a) the solute has a suitable absorption band, and (b) you have access to a properly calibrated instrument (spectrophotometer).

The fundamental equation governing spectrophotometry is the Beer–Lambert law (eqn. 3.1):

$$A = \varepsilon \times c \times l \qquad 3.1$$

where A is the absorbance at a particular wavelength, ε is the absorption coefficient (a measure of the degree of absorption of light at that wavelength), and l is the path length of the cuvette (or cell) that holds the solution of concentration c. From this equation we can see that at a given wavelength and at a constant path length, A is proportional to c.

We have already seen this equation (section 3.3), where it was noted that the units of ε will depend on the units of c and l, since overall, each side of the equation has no units. This is because A is defined by eqn. 3.4 as:

$$A = \log\left(\frac{I_0}{I_t}\right) \qquad 3.4$$

where I_0 is the intensity of the incident light, i.e. the light striking the cuvette, and I_t the intensity of the transmitted light, i.e. the light leaving the cuvette.

WORKED EXAMPLE 

What percentage of the incident light is transmitted after passage through solutions which have absorbance values of (a) 0.05, (b) 0.3, (c) 1.0 and (d) 2.0?

STRATEGY

This is an application of eqn. 3.4 for each value of A by putting the value of A into the calculator and use the 10^x function to obtain (I_0/I_t). The reciprocal of this value gives (I_t/I_0).

SOLUTION

The value of (I_t/I_0) at each value of A is multiplied by 100 to give the per cent of the incident light that is transmitted. Thus, the per cent of transmitted light is (a) 89.1%, (b) 50.1%, (c) 10% and (d) 1%.

The logarithmic nature of A is not always readily appreciated because most modern instruments merely give a digital display of the absorbance. *However, it is particularly important to note that you should always aim to measure absorbance values in the range 0.1–1.0. Despite what the manufacturers of the instrument may claim, absorbance values significantly above 1.0 are not usually reliable.* You can bring absorbance values into the appropriate range either by an accurate dilution of a concentrated solution, or sometimes by using a cuvette of different path length. Measurements will be most accurately made when the absorbance is about 0.3.

The absorbance at a given wavelength is often denoted by a subscript; thus A_{280} is the absorbance at 280 nm. The absorption coefficient would be similarly abbreviated, e.g. ε_{280}.

If we know (or can look up) the absorption coefficient for a substance at a particular wavelength, we can use the Beer-Lambert law to determine the concentration of a solution, as illustrated below in the worked examples below.

WORKED EXAMPLE

The absorption coefficient of NADH at 340 nm is 6220 M^{-1} cm^{-1}. You have made up a 2.5 mM stock solution of NADH (see section 3.5.1). How would you check its concentration spectrophotometrically?

STRATEGY

The key point is to make sure that the absorbance to be measured is in the range for reliable measurements (0.1–1.0).

SOLUTION

A 2.5 mM solution of NADH in a cuvette of 1 cm path length would, according to the Beer-Lambert law, have an absorbance at 340 nm (denoted as A_{340}) of $6220 \times 2.5 \times 10^{-3} = 15.6$. This is far too high to measure, so we need to perform an accurate dilution. If the solution were diluted 50-fold, e.g. by adding 0.02 mL to 0.98 mL buffer, the A_{340} should be $15.6/50 = 0.312$, which is within the required range for measurement. The spectrophotometer should be adjusted so that the absorbance of buffer alone is set at zero. (Note that this dilution procedure would be appropriate for a cuvette of 1 mL capacity with a path length of 1 cm; if we were using a cuvette of 3 mL capacity, we would add 0.06 to 2.94 mL buffer.)

When performing dilutions it is important to ensure that the correct pipettes are used and the solutions are completely mixed.

WORKED EXAMPLE

A solution consists of a mixture of NADH and its oxidized form NAD^+. NADH and NAD^+ have equal absorbances at 260 nm (absorption coefficient 18 000 M^{-1} cm^{-1}). NADH absorbs at 340 nm (absorption coefficient 6220 M^{-1} cm^{-1}); NAD^+ does not absorb at this wavelength. In a cuvette of 1 cm path length, the absorbance of the solution is 0.933 at 260 nm and 0.135 at 340 nm. What are the concentrations of NADH and NAD^+ in the solution?

STRATEGY

We can use the A_{260} to determine the sum of [NADH] and [NAD^+], since both absorb equally at this wavelength. The A_{340} can be used to determine the [NADH].

SOLUTION

From the A_{260}, the sum of [NADH] and [NAD^+] is 0.933/18 000 M = 51.8 μM. From the A_{340}, [NADH] = 0.135/6220 M = 21.7 μM. Hence by difference [NAD^+] = 30.1 μM.

Spectrophotometry provides a convenient way of measuring the concentration of a protein solution. Virtually all proteins absorb radiation at 280 nm due to the aromatic amino acid side chains (principally tyrosine and tryptophan) they contain. The absorption coefficients of different proteins at 280 nm will vary because proteins differ in terms of the exact proportions of these amino acids they contain. For example, the absorption coefficient of BSA (a very commonly employed protein standard) at 280 nm (ε_{280}) is 0.66 (mg mL^{-1})$^{-1}$ cm^{-1}, whereas that of hen egg white lysozyme is 2.65 (mg mL^{-1})$^{-1}$ cm^{-1}, reflecting the greater proportion of aromatic amino acids in the latter.

The relationship between the absorption coefficient of a protein at 280 nm and the content of aromatic amino acids is discussed in Chapter 6, section 6.1.5.

For proteins, the absorption coefficients are often referred to concentrations in terms of mg mL^{-1} (i.e. w/v) rather than in terms of molarity. If molarity were to be used as the basis for expressing concentrations, not only would the quoted absorption coefficients be inconveniently large (see worked example below), but in addition there also could be ambiguities about the molarity in the case of proteins with multiple polypeptide chains. Does the molarity refer to the individual polypeptide chains or to the intact protein? This should be clearly stated.

WORKED EXAMPLE 

The molecular mass of lysozyme is 14.3 kDa. What is the molar absorption coefficient of the protein at 280 nm, given that the A_{280} of a 1 mg mL^{-1} solution in a cuvette of 1 cm path length is 2.65?

STRATEGY

This calculation relies on evaluating the properties of a 1 M solution of the protein, and scaling the absorbance accordingly.

SOLUTION

A 1 M solution of lysozyme would be 14 300 mg mL^{-1}. Using the Beer-Lambert law, the A_{280} of this solution in a 1-cm pathlength cuvette is $2.65 \times 14\,300 = 37\,900$. Thus, the absorption coefficient of lysozyme is 37 900 M^{-1} cm^{-1}.

Many biochemical compounds, including proteins, nucleic acids and small molecules such as ATP and NADH, have convenient absorption bands. If at all possible, it is a good idea to use spectrophotometry to check the concentrations of solutions.

WORKED EXAMPLE 

The successful ligation of a DNA fragment (insert) into a cloning vector (DNA) requires a vector/insert ratio of 1/10 in the ligation reaction. Following the careful preparation of a 3.5 kilobase pair (kbp) vector and a 1.0 kbp insert, A_{260} measurements were made to determine the concentration of DNA in each preparation, using the relationship that a value of 1.0 for the A_{260} in a cuvette of 1 cm path length corresponds to 50 μg mL^{-1} DNA. If the A_{260} values of the vector and insert preparations are 0.80 and 0.60, respectively, calculate the volumes of the two preparations required for a ligation reaction which contains 20 fmol vector and 200 fmol insert.

STRATEGY

The DNA concentrations of vector and insert are worked out in terms of μg mL^{-1}, and hence mg mL^{-1}. The molecular masses of the two samples of DNA can be estimated from the number of bases (see Chapter 2, section 2.1), and this can be used to express the concentrations in molar terms. From these values, the volumes of solution which contain the required number of moles can be calculated.

SOLUTION

The DNA concentration of the vector is $(0.80/1.0) \times 50\ \mu g\ mL^{-1} = 40\ \mu g\ mL^{-1}$ (i.e. $0.04\ mg\ mL^{-1}$). Similarly, the DNA concentration of the insert $= 30\ \mu g\ mL^{-1}$ (i.e. $0.03\ mg\ mL^{-1}$).

The molecular mass of the vector can be estimated from the relationship (Chapter 2, section 2.1) that each nucleotide in a strand of DNA contributes 330 Da to the molecular mass. Thus, the molecular mass of the vector $= 3500 \times 2 \times 330$ Da $= 2.31 \times 10^6$ Da (note that we have multiplied by 2 because there are 3500 base pairs in this sample of double-stranded DNA). Similarly the molecular mass of the insert $= 1000 \times 2 \times 330$ Da $= 0.66 \times 10^6$ Da.

This worked example provides good practice in being able to interconvert the different prefixes for units in the SI system (see section 3.3.1). It also highlights the very small volumes of reagents used in many molecular biology techniques; great care should be taken in handling such small volumes.

Hence the concentration of the vector = concentration (in $mg\ mL^{-1}$)/molecular mass (in Da) $= 0.04/(2.31 \times 10^6)$ M $= 1.73 \times 10^{-8}$ M $= 17.3$ nM. The concentration of insert $= 0.03/(0.66 \times 10^6)$ M $= 45.5$ nM.

The solution of vector contains 17.3 nmol L^{-1}; this is equivalent to 17.3 fmol μL^{-1}. Thus, the volume of solution that contains 20 fmol $= 20/17.3\ \mu L = 1.16\ \mu L$.

The solution of insert contains 45.5 nmol L^{-1}; this is equivalent to 45.5 fmol μL^{-1}. Thus, the volume of solution that contains 200 fmol $= 200/45.5\ \mu L = 4.40\ \mu L$.

? SELF TEST

Check that you have mastered the key concepts at the start of the section by attempting the following questions.

ST 3.9 A solution of *N*-acetyltyrosine has an A_{280} of 0.265 in a cuvette of path length 0.5 cm. The absorption coefficient of the compound at this wavelength is $1280\ M^{-1}\ cm^{-1}$. What is the concentration of the solution?

ST 3.10 A solution of bovine insulin, when diluted 20-fold into buffer, had an A_{280} of 0.732 in a cuvette of path length 1 cm. If the absorption coefficient of insulin at 280 nm is $0.97\ (mg\ mL^{-1})^{-1}\ cm^{-1}$ and the molecular mass of the protein is 5734 Da, what is the concentration of the original solution expressed in terms of $mg\ mL^{-1}$ and in terms of molarity?

In ST 3.11, it should be noted that a nitroanilide is a good substrate for a proteolytic enzyme (protease) as it contains an amide bond, which links the amino acid units in a protein (see Chapter 9, section 9.9.3).

ST 3.11 4-Nitroaniline (4NA) is a product of the action of proteases on 4-nitroanilide substrates; the absorption coefficient of 4NA at 405 nm is $9500\ M^{-1}\ cm^{-1}$. In a cuvette of 1 cm path length, the A_{405} of a solution of 4NA whose volume is 2.5 mL is 0.27. What is the concentration of the solution and how many moles of 4NA are present?

Answers

ST 3.9 The concentration = 0.414 mM or 414 μM.

ST 3.10 The concentration = 15.1 $mg\ mL^{-1}$ or 2.63 mM.

ST 3.11 The concentration $= 2.84 \times 10^{-5}$ M or 28.4 μM; there are 7.11×10^{-8} mol or 71.1 nmol present.

Attempt Problems 3.16 and 3.17 at the end of the chapter.

3.7 Acids, bases, and buffers

KEY CONCEPTS

- Using the Henderson-Hasselbalch equation, calculate the balance of an ionization equilibrium of known pK_a at a given pH
- Understanding how buffers operate and how they can be prepared

As mentioned in Chapter 2, section 2.3.1, the pH scale provides a convenient measure of the acidity of a solution; pH = –log [H^+]. At 25°C (298K), pure water has a pH of 7.

The definition of pH means that as [H^+] increases, pH decreases. Distilled water usually contains some dissolved CO_2 (i.e. carbonic acid) from the atmosphere, so that the observed pH is lower than 7, typically in the range 5.5–6.0.

An acid can be defined as a substance which has a tendency to donate H^+ ions and a base as a substance which has a tendency to accept H^+ ions. Strong acids such as HCl are completely dissociated (to give H^+ and Cl^- ions), whereas weak acids (such as ethanoic acid) are only partially dissociated. In a similar manner, strong and weak bases will differ in terms of their readiness to accept H^+ ions. Thus, NaOH is a strong base since it is completely dissociated to yield Na^+ and OH^- ions; the latter will readily accept H^+ ions to form water.

The strength of an acid HA (which dissociates to give H^+ and A^- ions) can be expressed quantitatively in terms of its acid dissociation constant K_a, which is defined in eqn. 3.5 as:

$$K_a = ([H^+]\,[A^-])/[HA] \qquad 3.5$$

By analogy with the definition of pH, we define $pK_a = -\log K_a$.

Taking logarithms of both sides of eqn. 3.5, we obtain:

By analogy with the definition of pH, the lower the value of pK_a of an acid, the stronger acid it is.

$$\log K_a = \log [H^+] + \log ([A^-]/[HA])$$

Rearranging the terms:

$$-\log [H^+] = -\log K_a + \log ([A^-]/[HA])$$

which can be expressed as eqn. 3.6 (remembering that pH = –log [H^+] and that $pK_a = -\log K_a$):

$$pH = pK_a + \log ([A^-]/[HA]) \qquad 3.6$$

Eqn. 3.6 is known as the *Henderson–Hasselbalch equation*; it provides a simple way of visualizing the predominant species in the equilibrium as the pH varies with respect to the pK_a; this is depicted in Fig. 3.2.

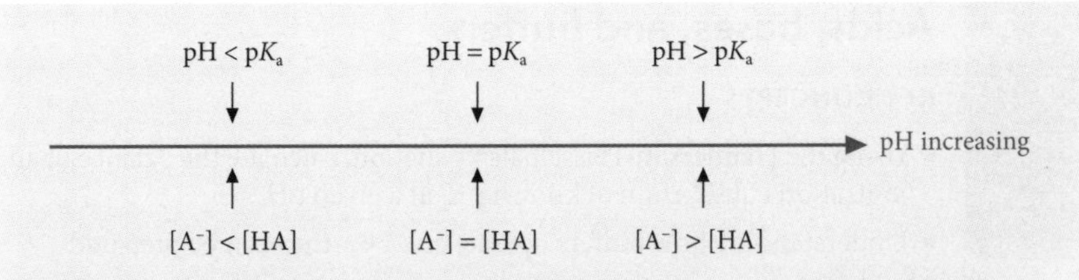

Fig. 3.2 The effect of pH on the position of the $HA \rightleftharpoons H^+ + A^-$ equilibrium. When $pH = pK_a$, exactly 50% of the HA has dissociated to give H^+ and A^-. The exact ratio of $[A^-]:[HA]$ at any pH can be calculated using the Henderson–Hasselbalch equation (eqn. 3.6).

WORKED EXAMPLE

What percentage of the acid is in the form of A^- (a) when the pH is 1 unit below the pK_a, (b) when the $pH = pK_a$, and (c) when the pH is 1 unit above the pK_a?

STRATEGY

This calculation applies eqn. 3.6; it is important to keep track of the signs and to use the 10^x key on the calculator reliably. To calculate the percentage of each form we must remember that in each case, the sum of $[A^-]$ and [HA] is 100%.

SOLUTION

From the Henderson–Hasselbalch equation (eqn. 3.6), it is found that $([A^-]/[HA])$ is 0.1, 1.0 and 10 for (a), (b), and (c), respectively. This would give the percentages of A^- as 9.1, 50, and 90.9%, respectively.

The structure of Tris is given in Chapter 1, section 1.8.

The pK_a values for some acids of biochemical interest are 4.76 for ethanoic acid (acetic acid), 7.2 for the second dissociation of phosphoric acid ($H_2PO_4^- \rightleftharpoons H^+ + HPO_4^{2-}$), and 8.1 for Tris base ($TrisH^+ \rightleftharpoons Tris + H^+$).

Since the structures and hence biological activities of many proteins are very sensitive to pH, we usually try to maintain the pH in the solution by the use of a buffer, which is a system designed to minimize the change in pH on addition of H^+ or OH^- ions. A buffer typically consists of a mixture of a weak acid (HA) and its salt (A^-); the latter is usually supplied as its sodium salt, for example. (Note that a buffer can also be prepared from a mixture of a weak base and its acid.)

We can explain the action of buffers by considering the equilibrium $HA \rightleftharpoons H^+ + A^-$. If H^+ ions are added to the system, they will combine with the A^- ions to form HA and thus effectively be removed. When OH^- ions are added, these will react with HA to generate H_2O and A^-, again resisting the change in pH. Buffers will be most effective when there are reasonable concentrations of HA and A^- present, and thus should be used within a narrow range (usually taken to be 1 pH unit on either side) of the relevant pK_a. Thus, the phosphate buffer system based on $H_2PO_4^- \rightleftharpoons H^+ + HPO_4^{2-}$ (i.e. using the salts NaH_2PO_4 and Na_2HPO_4) would be suitable for buffering in the range of pH between 6.2 and 8.2. For effective buffering the concentration of the buffer should be kept high.

WORKED EXAMPLE ✓

The Tris buffer system ($TrisH^+ \rightleftharpoons Tris + H^+$, based on the base Tris and its hydrochloride TrisHCl) has a pK_a of 8.1. What are the concentrations of the $TrisH^+$ and Tris forms at pH 8.3, if the total concentration of Tris species is 50 mM? During a reaction, 5 mM H^+ ions are formed. What is the new pH?

STRATEGY

This calculation depends on applying eqn. 3.6 to evaluate the concentrations of Tris and $TrisH^+$ and then consider the effect of adding 5 mM H^+ to the system.

SOLUTION

From the Henderson–Hasselbalch equation (eqn. 3.6), the ratio $[Tris]/[TrisH^+] = 10^{0.2} = 1.58$. Thus, [Tris] = 61.2 mM and $[TrisH^+]$ = 38.8 mM. On addition of 5 mM H^+, [Tris] will be reduced by 5 mM to 56.2 mM and $[TrisH^+]$ will be raised by 5 mM to 43.8 mM. Application of the Henderson–Hasselbalch equation shows that the new pH would be 8.21. (Note if no buffer had been present, 5 mM H^+ would give a pH of 2.3.)

WORKED EXAMPLE 

In the previous worked example, if the total concentration of Tris species is 20 mM, what would be the new pH on adding 5 mM H^+ ions?

STRATEGY

This follows the same steps as the previous example, with the addition of 5 mM H^+ having a proportionately greater effect.

SOLUTION

At pH 8.3, the concentrations of Tris and $TrisH^+$ are 12.2 and 7.8 mM, respectively. On addition of 5 mM H^+, the concentrations become 7.2 and 12.8 mM, respectively; the new pH is 7.85. Comparison with the previous example clearly demonstrates the beneficial effect of using a high concentration of buffer.

WORKED EXAMPLE 

A phosphate buffer system can be made up by mixing appropriate quantities of the two salts NaH_2PO_4 (supplied as the dihydrate $NaH_2PO_4.2H_2O$; molecular mass 156 Da) and Na_2HPO_4 (supplied as the anhydrous salt; molecular mass 142 Da). You wish to make up 1 L of a buffer of pH 7.5 with a total phosphate concentration of 0.1 M. Explain how you would make up 0.1 M stock solutions of the two salts in water and then in what proportion you would mix them to obtain the required pH. The appropriate pK_a for the $H_2PO_4^- \rightleftharpoons H^+ + HPO_4^{2-}$ ionization is 7.2.

STRATEGY

This is a straightforward application of eqn. 3.6 to calculate the ratio of the two salts required for the buffer.

SOLUTION

0.1 M stock solutions of $NaH_2PO_4.2H_2O$ and Na_2HPO_4 can be made up by weighing out 15.6 (i.e. 156×0.1) g and 14.2 (i.e. 142×0.1) g of the salts, respectively, dissolving each in water and making up to 1 L in each case. From the Henderson–Hasselbalch equation (eqn. 3.6) the ratio of $[HPO_4^{2-}]/[H_2PO_4^-]$ at pH 7.5 is 1.995 (i.e. $10^{0.3}$). Thus, to obtain 1 L of the required buffer, we need to mix 666 mL $((1.995/(1 + 1.995) \times 1000)$ mL) of the 0.1 M Na_2HPO_4 solution and 334 mL $((1/(1 + 1.995) \times 1000)$ mL) of the 0.1 M $NaH_2PO_4.2H_2O$ solution.

In practice, we could also make up the buffer in the last worked example (0.1 M sodium phosphate, pH 7.5) in a rather simpler way. The appropriate amount of the acidic form of the buffer ($NaH_2PO_4.2H_2O$) is weighed out to make 1 L of the buffer, i.e. 0.1×156 g = 15.6 g. This is then dissolved in about 900 mL water, and titrated to the required pH using concentrated (5–10 M) NaOH. Finally, the volume is made up to the required mark (1 L) with water. One advantage of this second method is that if other components are added, e.g. NaCl, they could affect the pK_a of the buffer system, which would alter the ratio of the two solutions to be added in the first method.

We would commonly use a titration method to make up Tris buffers. The required amount of the solid (Tris(hydroxymethyl)methylamine) is weighed out and dissolved in water before being titrated down to the required pH with the appropriate concentrated acid (HCl, ethanoic acid, etc.) before being made up to the required volume with water.

? SELF TEST

Check that you have mastered the key concepts at the start of the section by attempting the following questions.

ST 3.12 The pK_a of the amino group in the side chain of lysine is 10.5. What percentage of lysine is in the unprotonated (–NH_2) form at pH 8.2?

ST 3.13 The pK_a of a particular lysine side chain in a protein is 7.8. What percentage of this lysine side chain is in the unprotonated form at pH 8.2?

In ST 3.14, the Henderson–Hasselbalch equation (eqn. 3.6) is used to calculate the ratio [ethanoate]/[ethanoic acid] at the stated pH. The total concentration of the two species is 0.2 M.

ST 3.14 The pK_a of ethanoic acid is 4.76. What concentrations of ethanoic acid and sodium ethanoate would be required to make up a buffer at pH 5.2, if the total ethanoic acid and ethanoate concentration is 0.2 M?

Answers

ST 3.12 0.5% of the lysine is unprotonated.

ST 3.13 71.6% of the lysine side chain is unprotonated.

ST 3.14 The concentrations are: 0.053 M ethanoic acid and 0.147 M sodium ethanoate.

Attempt Problems 3.18 and 3.19 at the end of the chapter.

3.8 Specific activities of proteins and enzymes

KEY CONCEPTS

- Understanding how to calculate the specific activity of a protein
- Expressing the result in the appropriate units

A very important quantity when studying a protein is a measure of its biological activity, usually called an assay of its activity. In the case of a binding protein, we will want to know how much ligand it can bind and with what affinity. If the protein was able to elicit some more complex response such as cell growth, we would need to measure the ability of a given quantity of the protein to cause cell proliferation in a suitable system. In the case of an enzyme, the activity will be measured in terms of how much substrate is converted to product in a given time.

A very commonly used unit for enzyme activity is the amount that converts 1 μmol substrate to product in 1 min; in fact this is often abbreviated simply to 1 unit (U) or sometimes 1 International Unit (IU).

In the SI system, the unit of enzyme activity (the katal) is defined as the amount which converts 1 mol substrate to product in 1 s. This is, in fact, a very large amount of activity and in practice we would often be dealing with nkatal amounts.

An important property is the specific activity of the protein, which is defined as the amount of biological activity per given weight of the protein. For an enzyme, this might be measured in terms of μmol min^{-1} mg^{-1} (U mg^{-1} or IU mg^{-1}) or using the SI definition in terms of katal kg^{-1}.

WORKED EXAMPLE 

Show that for an enzyme a specific activity of 1 katal kg^{-1} corresponds to 60 μmol min^{-1} mg^{-1}.

STRATEGY

The important point here is to keep a very careful track of the units.

SOLUTION

The specific activity of 1 katal kg^{-1} corresponds to 1 mol substrate s^{-1} kg^{-1}, i.e. to 60 mol substrate min^{-1} kg^{-1}. This is 60×10^6 μmol substrate min^{-1} kg^{-1} and hence to 60 μmol substrate min^{-1} mg^{-1}. Thus, despite the fact that both the katal and the kg are large and impractical amounts of activity and protein, the SI definition of specific activity gives numbers that are of similar orders of magnitude to the more widely used definition (U mg^{-1}).

The specific activity of a protein is essentially a measure of the quality and purity of the sample and will be discussed further in section 3.9.

WORKED EXAMPLE 

The activity of the enzyme lactate dehydrogenase can be measured by monitoring the reaction in the direction pyruvate + NADH → lactate + NAD^+. NADH absorbs radiation at 340 nm (absorption coefficient = 6220 M^{-1} cm^{-1}); NAD^+ does not absorb at this wavelength. In an assay, 25 μL of a sample of enzyme (containing 5 μg protein mL^{-1}) was added to a mixture of pyruvate and NADH to give a total volume of 3 mL in a cuvette of 1-cm pathlength. The decrease in A_{340} was 0.14 min^{-1}. What is the specific activity of the enzyme expressed in terms of μmol min^{-1} mg^{-1} and in SI units (katal kg^{-1})?

STRATEGY

This calculation depends on evaluating the amount of activity in terms of μmol min^{-1} and the amount of enzyme added. It is advisable to break down the calculation into small steps and set the working out clearly.

Strictly speaking the equation for the reaction catalysed by lactate dehydrogenase should include H^+ on the left-hand side. It is assumed that the reaction is carried out in a suitable buffer, so that H^+ is often omitted.

SOLUTION

The amount of enzyme added is 0.025 × 5 μg (note that 25 μL = 0.025 mL), i.e. 0.125 μg.

The rate of the reaction (change in A_{340} min^{-1}) can be converted to a concentration change min^{-1}, i.e. 0.14/6220 M min^{-1} = 22.51 μM min^{-1}.

In a 3 mL assay mixture this corresponds to change of 22.51 × 3/1000 μmol min^{-1} = 0.0675 μmol min^{-1}.

The specific activity is thus 0.0675 μmol min^{-1} per 0.125 μg, or 0.0675 × 1000/0.125 μmol min^{-1} mg^{-1} = 540 μmol min^{-1} mg^{-1}.

In SI units this would be 540/60 katal kg^{-1}, i.e. 9.0 katal kg^{-1}.

We shall see in Chapter 4, section 4.5 how the specific activity can be used to calculate the turnover number (k_{cat}) of an enzyme, provided the molecular mass of the enzyme is known.

? SELF TEST

Check that you have mastered the key concepts at the start of the section by attempting the following questions.

ST 3.15 The rate of an enzyme-catalysed reaction as measured by the formation of product is 7.3 μM min^{-1} in a solution of volume 2 mL. What does this rate correspond to in terms of μmol min^{-1}? If 0.15 μg enzyme had been added to the reaction, what is the the specific activity of the enzyme expressed as μmol min^{-1} mg^{-1} and as katal kg^{-1}?

ST 3.16 The specific activity of a certain enzyme is 745 μmol min^{-1} mg^{-1}. How much enzyme should be added to a reaction mixture of volume 1 mL to give a rate of reaction of 12 μM min^{-1}?

In ST 3.16 the rate should be calculated in terms of mol/min from the concentration change/min and the volume of solution. Knowing the specific activity of the enzyme allows you to work out how much enzyme is needed to give the required rate.

Answers

ST 3.15 The rate corresponds to 0.0146 μmol min^{-1}; the specific activity is 97.3 μmol min^{-1} mg^{-1} (or 1.62 katal kg^{-1}).

ST 3.16 The amount of enzyme to be added is 0.0161 μg.

Attempt Problem 3.20 at the end of the chapter.

3.9 Purification tables

KEY CONCEPTS

- Defining the terms specific activity, yield and purification factor
- Calculating these quantities to draw up a purification table

A very important part of the study of proteins involves their purification. Only when purified protein is available can we determine its structure and investigate the basis of its biological activity. Traditionally, purification would have involved initially identifying a suitable source in which the protein of interest was abundant, for example yeast would be an excellent source of enzymes of the glycolytic pathway.

After breakage of the cells to provide a crude extract, various purification procedures would be applied in turn until pure, active protein was obtained. Recently, the trend has been to overexpress a recombinant version of the protein (often with some tag incorporated to facilitate purification) in a convenient host organism, such as *Escherichia coli*, yeast or insect cells, which can often be grown on a large scale using relatively simple growth media (see Chapter 5, section 5.4.2).

Before undertaking a purification process, it is essential to devise a reliable assay system for the protein in question, this will usually involve a measurement of the biological activity (catalysis, binding, etc.); see Chapter 5, section 5.3.

Whatever method of purification is used, it is important to record the progress of the procedure in a suitable way; this involves drawing up a purification table, which will record the following information:

- Step involved
- Total protein
- Total activity
- Specific activity (i.e. the activity per unit weight of protein)
- Yield (i.e. the recovery of activity expressed as a per cent of that in the initial extract)
- Purification factor (the factor by which the specific activity has increased).

It is essential for you to be able to calculate the specific activity, yield, and purification factor given data on the total protein and total activity.

To illustrate the concepts involved, consider the following data from a purification of an enzyme.

Step 1 refers to the crude extract. The total protein is 940 mg, and the total activity is 56 780 units (1 unit equals 1 μmol substrate consumed per min).

Step 2 refers to the extract which has been subjected to ion-exchange chromatography on DEAE-cellulose. The total protein in the pooled peak fractions is 53 mg and the total activity is 47 640 units.

We can then draw up our initial purification table (Table 3.2).

Of course, in a real example it is likely that there will be several more steps before the protein is completely purified.

Table 3.2 Purification table for a protein

Step	Total protein (mg)	Total activity (units)	Specific activity (units mg^{-1})	Yield (%)	Purification factor
Crude extract	940	56 780	60.4	100	1.0
Ion-exchange	53	47 640	899	83.9	14.9

The *specific activity* is obtained by dividing the total activity by the total protein at that step (e.g. 56 780/940 = 60.4 units mg^{-1}).

The *yield* refers to the recovery of activity, *not* to the recovery of protein. After all, the purpose of the purification is to get rid of contaminating proteins and retain the protein of interest. In this case, the yield at step 2 is (47 640/56 780) × 100% = 83.9%.

The *purification factor* refers to the factor by which the specific activity has increased, in this case 899/60.4 after step 2 = 14.9-fold. (Note in a multi-step purification, we can express this either in terms of the factor for each individual step or as a running total factor relative to the initial crude extract as 1.0).

Some examples of protein purification experiments are described in Chapter 7.

? SELF TEST

Check that you have mastered the key concepts at the start of the section by attempting the following question.

ST 3.17 In a purification procedure for a certain enzyme, the total protein and total activity at each step was as follows:

Step	Total protein (mg)	Total activity (units)
Crude extract	53.5	780
Ion-exchange chromatography	8.7	525
Gel filtration	1.6	490

Calculate the specific activity, yield and purification factor at each step in the procedure.

Answer

ST 3.17 The values are: specific activity, 14.6, 60.3, 306.3 units mg^{-1}; yield, 100, 67.3, 62.8%; purification factors (per step), 1.0, 4.1, 5.1).

Attempt Problem 3.21 at the end of the chapter.

3.10 Problems

Full solutions to odd-numbered problems are available to all in the student section of the Online Resource Centre at www.oxfordtextbooks.co.uk/orc/price/. Full solutions to even-numbered problems are available to lecturers only in the lecturer section of the Online Resource Centre.

3.1 Hair is composed of the protein α-keratin, which is α-helical in structure. How many turns of α-helix are produced per second in growing hair? Assume that the rate of hair growth is 18 cm $year^{-1}$ and that each turn of α-helix corresponds to a length of 0.54 nm.

3.2 How many moles of GTP are present in 2 µl of a 15 µM solution?

3.3 GTP (sodium salt) has a molecular mass of 545 Da. What mass of GTP is present in 3 mL of a 15 µM solution?

3.4 A certain analytical technique is said to be capable of determining 100 pmol of the amino acid tyrosine (molecular mass 181 Da). What mass of tyrosine does this correspond to?

3.5 The concentration of haemoglobin in red blood cells is estimated to be 250 mg mL^{-1}. Haemoglobin consists of four polypeptide chains each of molecular mass 16.5 kDa (16 500 Da); what is the molar concentration of haemoglobin chains in the red blood cell?

3.6 The molecular mass of chymotrypsin is 24 500 Da (24.5 kDa). What is the concentration of a 0.35 mg mL^{-1} solution expressed in molar terms?

3.7 A solution of aspartate aminotransferase is stated to be 13.6 µM in terms of the dimeric (two polypeptide chain) form of the enzyme (the molecular mass of each polypeptide chain is 44 kDa). What is the concentration of the enzyme in terms of mg mL^{-1}?

3.8 The enzyme phosphoglycerate mutase from *Saccharomyces cerevisiae* (polypeptide molecular mass 27 kDa) is typically assayed at a concentration of 0.05 µg mL^{-1}. What is the molar concentration of enzyme polypeptide chains in the assay?

3.9 Explain how you would make up 15 mL of a 10 M solution of urea in water. The molecular mass of urea is 60 Da.

3.10 For the preparation of a certain gel for electrophoresis, 20 mL of an 8% (w/v) acrylamide solution is required. What volume of a 30% (w/v) stock solution of acrylamide is required for this?

3.11 A stock solution of a protein in buffer is 4.3 mg mL^{-1}. How would you prepare 1 mL of a solution containing 0.2 mg mL^{-1} protein?

3.12 The concentration of glacial acetic (ethanoic) acid is stated to be 17.3 M. How would you prepare 200 mL of a 1 M solution of the acid in water?

3.13 IPTG (isopropyl-β-D-thiogalactopyranoside) is widely used as an inducer for the expression of recombinant proteins in the bacterium *E. coli*. Its molecular mass is 238.3. How much IPTG is required to make up 5 mL of a 0.5 M stock solution in water? What volume of this solution should be added to 50 mL of bacterial culture to give a final concentration of 1.2 mM?

3.14 The four deoxyribonucleotide triphosphates (NTPs), i.e. dATP, dCTP, dGTP, and dTTP, are essential components of the polymerase chain reaction (PCR) used to amplify a selected section of DNA. The NTPs are often supplied as separate stock solutions, each at a concentration of 100 mM in sterile H_2O. How would you make up a working solution (in sterile H_2O) of final volume 1 mL which contains the four NTPs each at a concentration of 2 mM?

3.15 The concentration of a stock solution of NADH is 3 mg mL^{-1}. What does this correspond to in molar terms? (The molecular mass of NADH (disodium salt) is 709 Da). What volume of this stock solution would you add to an assay mixture (of final volume 1.0 mL) to obtain a final concentration of 0.1 mM?

3.16 The absorbance of a solution of ATP was checked by performing two successive dilutions, the first 50-fold and the second 40-fold. The A_{260} of this final solution was 0.47 in a cuvette of 1 cm path length. What is the concentration of the original solution of ATP? The absorption coefficient of ATP at 260 nm is 15 000 M^{-1} cm^{-1}.

3.17 A solution of BSA is made up by weighing out 20 mg of the lyophilized (freeze dried) powder of the protein supplied and dissolving in 2 mL water. The concentration of the protein was checked by adding 0.04 mL of the solution to 0.96 mL water; the A_{280} of the resulting solution relative to a water blank was 0.237. The absorption coefficient for BSA is 0.66 (mg $mL^{-1})^{-1}$ cm^{-1}. Calculate the actual concentration of the original solution of BSA and comment on your answer.

3.18 What would be the change in pH on adding OH^- ions to a final concentration of 20 mM to a 0.1 M sodium phosphate buffer, pH 7.0? The pK_a for the $H_2PO_4^- \rightleftharpoons H^+ + HPO_4^{2-}$ ionization is 7.2.

3.19 Explain how you would make up 1 L of a 50 mM Tris/HCl buffer at pH 7.5, containing 20 mM NaCl. The molecular masses of Tris(hydroxymethyl)methylamine and NaCl are 121.1 and 58.45 Da, respectively.

3.20 A sample of malate dehydrogenase (3.2 mg mL^{-1}) was diluted 500-fold for assay. 20 μL of this solution was added to an assay system of total volume 1 mL in a cuvette of 1 cm path length. The increase in A_{340} (corresponding to the conversion malate + NAD^+ → oxaloacetate + NADH + H^+) was 0.23 min^{-1}. What is the specific activity of the enzyme expressed in terms of μmol min^{-1} mg^{-1} and in SI units? (The absorption coefficient of NADH is 6220 M^{-1} cm^{-1}; NAD^+ does not absorb at this wavelength.)

3.21 The enzyme adenylate kinase catalyses the interconversion of adenine nucleotides in the reaction AMP + ATP ⇌ 2ADP. The enzyme was purified from pig muscle by procedures including precipitation at low pH, elution from phosphocellulose by a pulse of AMP and gel filtration on Sephadex G-75. The following results were obtained starting from 1 kg muscle:

Step	Total protein (mg)	Total activity (units)*
Crude extract	72 500	413 000
Low pH	18 700	365 000
Phosphocellulose	290	223 000
Gel filtration	77	200 000

*1 unit of enzyme activity corresponds to 1 μmol substrate consumed per minute.

Complete the purification table for this procedure.

Appendix

Appendix 3.1 The SI units and physical constants

SI units are based on the metre-kilogram-second system of measurement, and are very widely accepted among scientists. SI units for various physical quantities are listed below.

Quantity	SI unit	Notes
Amount of substance	mole (mol)	This quantity contains 1 Avogadro number of the basic units (atoms, molecules etc.)
Electric charge	coulomb (C)	
Length	metre (m)	Å (Ångstrom) = 1×10^{-10} m = 0.1 nm
Mass	kilogram (kg)	Multiple units are based on g (e.g. mg) although kg is the basic unit
Molecular mass	dalton (Da)	1 Da = 1 atomic mass unit ≈ mass of 1 hydrogen atom (1.66×10^{-24} g); more exactly it is 1/12 of the mass of the ^{12}C isotope of carbon
Temperature	degree Kelvin (K)	0°C = 273.15K
Time	second (s)	
Volume	cubic metre (m^3)	1 litre (L) = 1 dm^3; 1 mL = 1 cm^3
Electric potential	volt (V) ($J\ C^{-1}$)	
Energy	joule (J) ($1\ m^2\ kg\ s^{-2}$)	1 cal = 4.18 J
Force	newton (N) ($1\ m\ kg\ s^{-2}$)	
Frequency	hertz (Hz) ($1\ s^{-1}$)	
Power	watt (W) ($J\ s^{-1}$)	
Pressure	pascal (Pa) ($1\ N\ m^{-2}$)	1atm = 101.325 kPa
Radioactive decay	becquerel (Bq) $1\ s^{-1}$	1 disintegration per second; 1 curie (Ci) = $3.7 \times 10^{10}\ s^{-1}$
Concentration	mol kg^{-1}	Moles per kg of solvent, also called molality Concentration is more usually quoted in terms of molarity (M), moles per L of solution For dilute aqueous solutions, molarity is very nearly equal to molality.

The following prefixes are used in conjunction with the SI units to bring the numerical values of quantities into a convenient range (see Chapter 3, section 3.3.1).

Prefix	G, giga	M, mega	k, kilo	Base unit	m, milli	μ, micro	n, nano	p, pico	f, femto
Numerical value	10^{9}	10^{6}	10^{3}	1	10^{-3}	10^{-6}	10^{-9}	10^{-12}	10^{-15}

Physical constants used in this book are:

Avogadro number (L)	6.022×10^{23} mol^{-1}
Faraday constant (F)	9.648×10^{4} C mol^{-1}
Gas constant (R)	8.314 J K^{-1} mol^{-1}
Planck constant (h)	6.626×10^{-34} J s
Speed of light in a vacuum (c)	2.997×10^{8} m s^{-1}

4 Calculations in the molecular biosciences

Part 2: Characterization of biological processes

This chapter continues the approach of Chapter 3, and deals with the analysis of some important types of biological processes. The first two sections deal with the energetics of processes (how far and how fast will they proceed?). These sections are followed by a detailed discussion of binding equilibria; this includes the kinetics of enzyme-catalysed reactions since many of the equations involved are similar. A final section deals with radioactivity, which is widely used to track specific compounds. As with Chapter 3, there are several problems at the end of each section and also at the end of the chapter to allow you to check your understanding of the material covered.

4.1 The energetics of processes: thermodynamics

KEY CONCEPTS

- Defining the terms ΔH, ΔS, and ΔG for a process
- Defining the standard state and biochemical standard state
- Knowing the relationship between ΔG^0 and the equilibrium constant
- Defining the mass action ratio (MAR) for a process

In biological systems we deal with two main types of processes, namely (a) chemical reactions in which covalent bonds are broken and made, so that one or more new compounds (products) are made from the starting compounds (reactants or substrates), and (b) complex formation in which two (or more) compounds can associate reversibly with one another. The latter types of processes almost always involve weak non-covalent interactions (see Chapter 1, section 1.7).

Both types of processes will eventually reach a position of equilibrium where there is no further tendency to change with time; this is, however, a dynamic state, rather than a static one. The forward and backward reactions are both proceeding, but at equal rates.

The science of thermodynamics is concerned with the overall changes in energy when processes occur; it can be used to predict how far a process will proceed, i.e. does the equilibrium lie more towards the side of the products or the reactants?

The energy changes that occur during processes can be described using different parameters.

ΔH is the *change in enthalpy* (change in heat content) of a system. This is a measure of the energy changes in breaking and making of bonds (either covalent or non-covalent) in a process. An exothermic reaction or process is one in which heat is given out by the system to its surroundings; this has a negative ΔH. An endothermic reaction or process has a positive ΔH and therefore absorbs heat from the surroundings. The units of ΔH are energy mol^{-1}, usually kJ mol^{-1}.

ΔS is the *change in entropy*. This is a measure of the change in the extent of randomness (or disorder) of a system. Thus, a process in which the number of molecules increases would be expected to have a positive ΔS. Similarly the transition from a solid to a liquid or a liquid to a gas would have a positive ΔS. The units of ΔS are energy degrees $Kelvin^{-1}$ mol^{-1}, e.g. kJ K^{-1} mol^{-1} (or because these values are usually small, J K^{-1} mol^{-1}).

Note that we use the absolute or Kelvin scale of temperature in energy calculations. 0°C is equivalent to 273 K (more precisely 273.15 K).

ΔG is the *changes in (Gibbs) free energy*. This is a measure of the work or usable energy that can be obtained from a process. A process in which free energy is liberated (exergonic) has a negative ΔG. An endergonic process has a positive ΔG. The units of ΔG are energy mol^{-1}, usually kJ mol^{-1}.

The terms enthalpy, entropy and free energy are dealt with in greater detail in many standard textbooks of chemistry and biochemistry, such as those referred to in Chapter 1.

A negative value of ΔG indicates that the process has a tendency to proceed from left to right. Although the value of ΔG will give no indication about the rate at which the process may proceed, it should be noted that in the case of the vast majority of biochemical reactions, the presence of appropriate enzymes normally ensures that reactions that are feasible energetically will proceed at a suitable rate to meet the needs of the organism concerned. The change in free energy for a process is defined by eqn. 4.1:

The rates at which processes proceed are dealt with in the section on kinetics (section 4.2).

$$\Delta G = \Delta H - T\Delta S \qquad 4.1$$

where T is the absolute temperature (degrees Kelvin, K). Thus, the change in free energy is the change in enthalpy minus the energy associated with an increase in entropy.

The values of ΔG, ΔH, and ΔS can be measured under any set of conditions. However, to provide a reference point for measurements and calculations, it is necessary to define a standard state, and the values under these conditions are then denoted by a superscript zero, i.e. ΔG^0, ΔH^0, and ΔS^0. The *standard state of a substance* refers to the substance under 1 atm pressure at the temperature in question. For almost all biochemical purposes we are dealing with solutions and the standard state of a solute is a 1 M solution of that solute. (For a gas, the standard state is the gas at 1 atm; for a solvent, it is the pure solvent). Under standard state conditions, eqn. 4.1 would then be written $\Delta G^0 = \Delta H^0 - T\Delta S^0$.

In the special case of biochemical processes where H^+ is either a reactant or a product of the reaction, it is useful to define a new biochemical standard state, where the standard states of all solutes is 1 M, but that of H^+ is 10^{-7} M (i.e. pH 7). The values of the quantities in the biochemical standard state are denoted by a superscript dash, i.e. $\Delta G^{0\prime}$, $\Delta H^{0\prime}$, and $\Delta S^{0\prime}$.

For most processes either the ΔH term or the ΔS term (strictly speaking the $T\Delta S$ term) makes the dominant contribution to the overall ΔG.

 WORKED EXAMPLE

For the combustion of glucose ($C_6H_{12}O_6 + 6O_2 \rightarrow 6CO_2 + 6H_2O$) the values of ΔG^0 and ΔH^0 at 25°C are −2872 and −2822 kJ mol^{-1}, respectively. Calculate the value of ΔS^0 and comment on its sign. Which term makes the dominant contribution to ΔG^0?

STRATEGY

This is an application of eqn. 4.1 using the standard state values of the thermodynamic parameters.

SOLUTION

We can rearrange eqn. 4.1 to give $\Delta S^0 = (\Delta H^0 - \Delta G^0)/T$. Putting $T = 298$ K, we find that $\Delta S^0 = (-2822 + 2872)/298$ kJ K^{-1} mol^{-1} = 0.168 kJ K^{-1} mol^{-1}, or 168 J K^{-1} mol^{-1}. The positive value of ΔS^0 reflects the greater number of molecules of products (12) compared with reactants (7). In this reaction, the enthalpy term makes the dominant contribution to the free energy change.

The derivation of eqn. 4.2 involves application of the laws of thermodynamics and the ideal gas law and is described in the textbook by Price *et al.* (2001).

For the process $A + B \rightleftharpoons C + D$, it is possible to derive the general eqn. 4.2 which applies under any set of conditions:

$$\Delta G - \Delta G^0 = RT \ln\left(\frac{[C] \times [D]}{[A] \times [B]}\right) \qquad 4.2$$

where ΔG is the free energy change under the given set of conditions, ΔG^0 is the free energy change when A and B in their standard states are converted to C and D in their standard states, R is the gas constant (8.31 J K^{-1} mol^{-1}) and ln is the natural logarithm.

It should be noted that the terms [C], etc. in eqn. 4.2 are usually taken to be concentrations; in fact each represents a ratio of the concentration under the given conditions to the concentration in the standard state, which for a solute is a 1 M solution under 1 atmosphere pressure. Hence, the term $\left(\frac{[C] \times [D]}{[A] \times [B]}\right)$ (or $\left(\frac{[C]}{[A] \times [B]}\right)$ for the process $A + B \rightleftharpoons C$) is dimensionless, i.e. a pure number – indeed if it were not, it would not be possible to take logarithms. However, values of equilibrium constants (K_{eq}) or MAR (Γ) (see below) are usually given in molar units to indicate that the standard state of the each component (solute) is a 1 M solution.

The term $\left(\frac{[C]\times[D]}{[A]\times[B]}\right)$ is often known as the MAR and given the symbol Γ.

It can be used to calculate the actual free energy change in a process under any given set of conditions. Thus, eqn. 4.2 can be rewritten as eqn. 4.3:

$$\Delta G - \Delta G^0 = RT \ln \Gamma \qquad 4.3$$

Γ is the Greek capital letter gamma.

A special case arises when the process is at equilibrium, i.e. when there is no overall tendency for it to proceed in either direction, although the forward and backward processes will occur. In this case, $\Delta G = 0$, and the values $[A]$, $[B]$, etc. are their equilibrium values ($[A]_{eq}$, $[B]_{eq}$, etc.). Thus, eqn. 4.2 can be rewritten as eqn. 4.4:

$$-\Delta G^0 = RT \ln\left(\frac{[C]_{eq}\times[D]_{eq}}{[A]_{eq}\times[B]_{eq}}\right) = RT \ln K_{eq} \qquad 4.4$$

where K_{eq} is the equilibrium constant for the process.

The formation and dissociation of a protein–ligand complex, PL, is discussed in detail in section 4.3.1.

We can use eqn. (4.4) to calculate K_{eq} (and hence know on which side the equilibrium lies) from the value of ΔG^0 for a process and vice versa. However, it is a very good practice to make an estimate of the value of K_{eq} you would expect before using the calculator.

4.1.1 Key relationships

In Chapter 2, section 2.3.6 we mentioned that at 310 K (37°C) each change of approximately 6 kJ mol^{-1} leads to a 10-fold change in the value of K_{eq}. So we can proceed as follows.

- If the value of ΔG^0 is positive, then the process is unfavourable as written and the equilibrium will lie on the side of the reactants, (i.e. K_{eq} will be less than 1)
- If ΔG^0 is negative, the equilibrium will lie towards the products, i.e. K_{eq} will be greater than 1
- If $\Delta G^0 = 0$, then K_{eq} would equal 1 and the process would be balanced between reactants and products).

A worked example will illustrate these principles.

WORKED EXAMPLE

The ΔG^0 for the reaction catalysed by pyruvate kinase is -31.4 kJ mol^{-1} at 310 K.

$$\text{phosphoenolpyruvate} + \text{ADP} \rightleftharpoons \text{pyruvate} + \text{ATP}$$

Estimate the equilibrium constant for the reaction.

STRATEGY

This is an application of the general principles stated above.

SOLUTION

Because ΔG^0 is negative, we know K_{eq} will be greater than 1. The value of ΔG^0 represents about five multiples of 6 kJ mol^{-1}. From this we can estimate that K_{eq} should equal about 1×10^5 (M). The accurate answer can be obtained from a calculator using eqn. 4.4., i.e. $-(-31\ 400) = 8.31 \times 310 \times \ln K_{eq}$. Hence $\ln K_{eq} = 12.19$, i.e. $K_{eq} = 1.97 \times 10^5$ (M). (Note that we have expressed ΔG^0 in terms of J, since the value of R is given in J.)

The term 'steady state' refers to the situation when the concentrations of intermediates in a series of reactions remains constant (i.e. in each case, the rate of formation equals the rate of breakdown). In living organisms, this would apply to the concentrations of metabolic intermediates in pathways such as glycolysis, the tricarboxylic acid cycle, etc.

Biological systems are not at overall equilibrium with their surroundings (otherwise they would be dead!), although a significant proportion of the several thousand metabolic reactions will be at or very close to equilibrium; this provides a justification for the study of equilibrium thermodynamics. Thus, the value of ΔG^0 for a reaction may not give a very reliable guide to the actual tendency (ΔG, driving force) of the reaction to proceed under the conditions prevailing in the living cell.

Another branch of thermodynamics has been developed to describe and analyse systems that are not at equilibrium. However, non-equilibrium thermodynamics (in which there is an emphasis on the steady state of a system) is beyond the scope of this book.

WORKED EXAMPLE

Phosphofructokinase catalyses the reaction:

$$\text{fructose-6-phosphate} + \text{ATP} \rightleftharpoons \text{fructose-1,6-bisphosphate} + \text{ADP}$$

for which the $\Delta G^{0\prime} = -17.7$ kJ mol^{-1} at 308 K. In a study of perfused rat heart, the following concentrations of metabolites were found: fructose-6-phosphate (F6P), 60 μM; ATP, 5.3 mM, fructose-1,6-bisphosphate (FBP), 9 μM; ADP, 1.1 mM. Is the reaction at equilibrium in the perfused heart and, if not, what is the value of $\Delta G'$? In the same tissue, the concentration of AMP was found to be 95 μM. Is the reaction catalysed by adenylate kinase (2ADP $\rightleftharpoons$ ATP + AMP) at equilibrium ($\Delta G^{0\prime} = 2.1$ kJ mol^{-1})?

STRATEGY

This problem depends on the application of eqn. 4.3, making sure that the concentrations are quoted in molar terms, e.g. [ATP] $= 5.3 \times 10^{-3}$ M.

SOLUTION

The MAR (Γ) for the phosphofructokinase reaction, i.e.

([FBP] × [ADP])/([F6P] × [ATP]) = $(9 \times 10^{-6} \times 1.1 \times 10^{-3})/(6 \times 10^{-5} \times 5.3 \times 10^{-3}) = 0.031$

Hence, from eqn. 4.3 ($\Delta G' - \Delta G^{0\prime} = RT \ln \Gamma$). $\Delta G'$ can be calculated as:

$$\Delta G' = -17\,700 - 8890 \text{ J mol}^{-1} = -26\,600 \text{ J mol}^{-1} = -26.6 \text{ kJ mol}^{-1}$$

Hence, the reaction is clearly not at equilibrium. For the adenylate kinase reaction, the MAR is ([ATP] [AMP])/$[ADP]^2$ = 0.416. Hence $\Delta G' = 2.1 - 2.24$ kJ mol^{-1}, i.e. −0.14 kJ mol^{-1}. Since $\Delta G'$ is very small, this latter reaction is very close indeed to equilibrium.

SELF TEST ?

Check that you have mastered the key concepts at the start of the section by attempting the following questions.

ST 4.1 For the hydrolysis of ATP to give ADP and phosphate (P_i) at 37°C, $\Delta H^{0\prime}$ and $\Delta S^{0\prime}$ are −20.5 kJ mol^{-1} and 30.6 J K^{-1} mol^{-1}, respectively. What is $\Delta G^{0\prime}$ for the reaction?

ST 4.2 The ΔG^0 for the reaction catalysed by lactate dehydrogenase (lactate + $NAD^+ \rightleftharpoons$ pyruvate + NADH) is +23.5 kJ mol^{-1} at 37°C. Estimate the value of the equilibrium constant.

In ST 4.2 use the key relationships to link the values of K_{eq} and ΔG^0.

ST 4.3 The enzyme aldolase catalyses the reaction FBP $\rightleftharpoons$ G3P + DHAP (where FBP is fructose-1,6-bisphosphate, G3P is glyceraldehyde-3-phosphate and DHAP is dihydroxyacetone phosphate). The $\Delta G^{0\prime}$ for this reaction at 37°C is 23.5 kJ mol^{-1}. The concentrations of FBP, G3P, and DHAP in resting muscle were measured as 30, 20, and 110 μM, respectively. What is the value of K_{eq} for the aldolase reaction? What is the value of $\Delta G'$ for this reaction in muscle?

Answers

ST 4.1 The $\Delta G^{0\prime}$ for the reaction is −30.0 kJ mol^{-1}.

ST 4.2 The estimate of K_{eq} is about 1×10^{-4} (M); the accurate answer is 1.09×10^{-4} (M).

ST 4.3 The values of K_{eq} and $\Delta G'$ are 1.09×10^{-4} (M) and −1.0 kJ mol^{-1}, respectively.

Attempt Problems 4.1–4.4 at the end of the chapter.

4.2 The rates of reactions: kinetics

KEY CONCEPTS

- Illustrating the energy profile for a reaction indicating the following terms: energy of reaction, activation energy, transition state, intermediate
- Distinguishing the order and molecularity of a reaction
- Stating the rate laws for zeroth, first, pseudo-first, and second-order reactions
- Deducing the units of rate constants for different orders of reactions
- Describing the dependence of reaction rate on temperature

Thermodynamics deals with the energy changes in processes and reactions and can thus indicate which reactions are feasible. However, it says nothing about the rates at which they might occur; this is the subject of kinetics. (In this section, we shall use the term 'reaction' to include chemical reactions as well as the formation of complexes between molecules, unless otherwise indicated).

The energy changes which occur during a typical chemical reaction can be depicted by an energy profile (Fig. 4.1).

In order to react, the reactants have to possess sufficient energy (the activation energy) to reach the transition state of the reaction, the state of highest energy in the profile. If the profile involves a local minimum, this would represent an

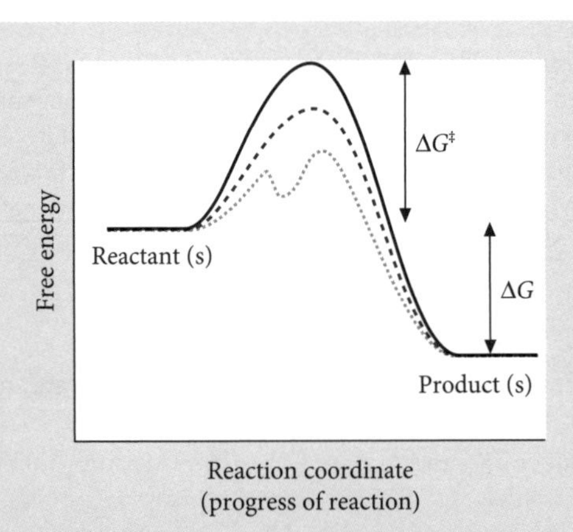

Fig. 4.1 Energy profile for a reaction. The solid line represents a typical energy profile, in which an activation energy barrier ($\Delta G^{\ddagger}$) must be surmounted. The highest point in the profile is the transition state in the reaction. The dashed and dotted lines represent possible profiles for catalysed reactions in which smaller energy barriers must be surmounted. In the case of the dotted line there is a local minimum before the transition state is reached; this represents an intermediate in the reaction. The overall free energy change in the reaction is represented by ΔG.

intermediate. The stability of the intermediate will depend on the depth of this minimum relative to the thermal energy; under favourable circumstances it may be possible to isolate and characterize the intermediate. (It will be easier to isolate the intermediate by working at lower temperatures. The lower thermal energy of the intermediate will trap it in its 'energy well').

As shown in Fig. 4.1, catalysis of the reaction will occur when a new pathway of lower activation energy is available to the reactants.

Formation or dissociation of a complex also involves surmounting an activation energy barrier, but this will usually be much smaller than for a chemical reaction where strong covalent bonds are broken and made.

There is no necessary relationship between the overall energy change for a reaction and the rate at which it will occur (which depends on the magnitude of the activation energy). Thus, a solution of ATP (which is often referred to as a high-energy compound, with a ΔG^0 of -30 kJ mol^{-1}) is stable for weeks at pH 7.0. However, in the cell there are enzymes which will catalyse reactions involving the transfer of the terminal (γ) phosphate group of ATP to a variety of acceptor molecules so that reactions occur on the time scale required by the demands of the organism.

ATP is often referred to as the 'energy currency' of the cell. There are reactions and processes which generate sufficient energy to drive the synthesis of ATP from ADP plus phosphate. Other reactions and processes can be driven towards completion by being coupled to the hydrolysis of ATP. Problems 4.2 and 4.3 at the end of the chapter deal with this concept in more detail.

Kinetics is concerned with measuring the rate of a reaction under a variety of conditions, changing parameters such as the concentrations of reactants, pH, temperature, etc. From the data acquired, we aim to formulate the rate law (i.e. determine the order of the reaction) and use this and other information (such as attempting to isolate intermediates) to propose a mechanism. The mechanism is a description at the atomic level of the elementary steps of bond breaking and making as the reaction proceeds. Knowledge of the mechanisms of chemical reactions is important to understand the basis of enzyme catalysis.

Rates can be measured in a continuous fashion (in which a property which changes during the reaction, such as absorbance or fluorescence, is monitored) or in a discontinuous fashion (in which samples are taken from the mixture at stated times, the reaction is stopped (quenched) and analysis is undertaken to determine the extent of the reaction). In addition to being much more convenient experimentally, continuous monitoring of a reaction (if it is feasible) provides many more data points for subsequent mathematical analysis of the kinetics.

4.2.1 The order of a reaction

The order of a reaction is defined as the power to which the concentration of a reactant is raised in the rate law. Determination of the order of a reaction is an important part of investigation of its mechanism; any proposed mechanism must generate a rate law consistent with the experimental observations.

Thus, for a hypothetical reaction in which x moles of A and y moles of B react to form products:

$$x\text{A} + y\text{B} \rightarrow \text{products}$$

The rate of the reaction (i.e. the rate of formation of products, or the rate of disappearance of reactants) might be given by eqn. 4.5:

$$\text{Rate} = k \times [A]^a \times [B]^b \qquad 4.5$$

where k is known as the rate constant.

The reaction is said to be of order a in A, of order b in B, and of overall order $a + b$. The order is an experimentally determined quantity and can have integral or non-integral values. For example, enzyme-catalysed reactions are of approximately first order with respect to substrate at low substrate concentrations, but the order decreases to near zero as the substrate concentration becomes saturating (see section 4.3.3).

The order of a reaction should not be confused with its molecularity, which is defined as the minimum number of species involved in the slowest step of the reaction, i.e. the minimum number of species involved in the transition state. The molecularity must therefore be a whole number.

The rate laws for reactions of different orders will be mentioned briefly and the method of analysis indicated.

Zeroth-order reactions are relatively rare. Examples in chemistry include some reactions of gases adsorbed on solid surfaces. In enzyme-catalysed reactions at very high substrate concentrations, the rate is close to zeroth order in substrate (see section 4.3.3).

Zeroth-order reactions: In this case, the rate does not depend on the concentration of the reactant A, i.e. rate of change of $[A] = -k$, where k is the rate constant. (The rate is negative because A is being consumed.) A plot of $[A]_t$ against time (Fig. 4.2) is a straight line of slope $-k$.

First-order reactions: In this case, the rate depends on $[A]$, i.e. rate of change of $[A] = k \times [A]$. Mathematical analysis shows that the concentration of A changes with time (t) as given by eqn. 4.6 (see Chapter 2, section 2.3.3):

$$[A]_t = [A]_0 \times e^{-kt} \qquad 4.6$$

By taking the natural logarithms of the terms on both sides of eqn. 4.6 and rearranging, we derive eqn. 4.7:

$$\ln [A]_t - \ln [A]_0 = -k \times t \qquad 4.7$$

where $[A]_0$ is the concentration of A at time $t = 0$, and k is the rate constant. A plot of $\ln [A]_t$ vs time (t) (Fig. 4.3) gives a straight line of slope $-k$.

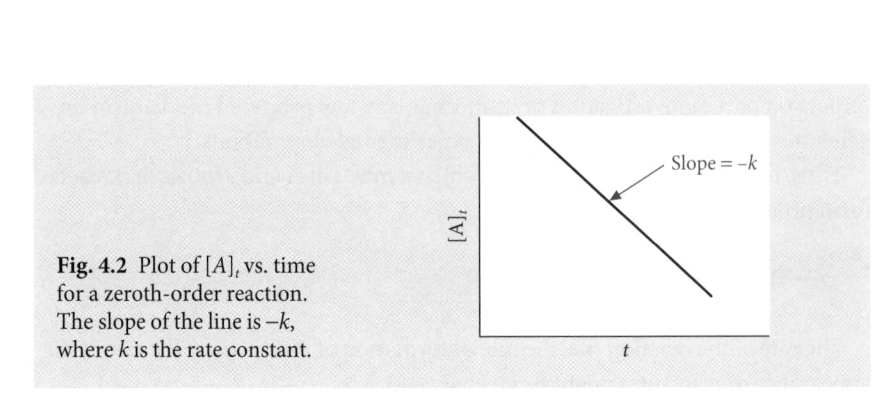

Fig. 4.2 Plot of $[A]_t$ vs. time for a zeroth-order reaction. The slope of the line is $-k$, where k is the rate constant.

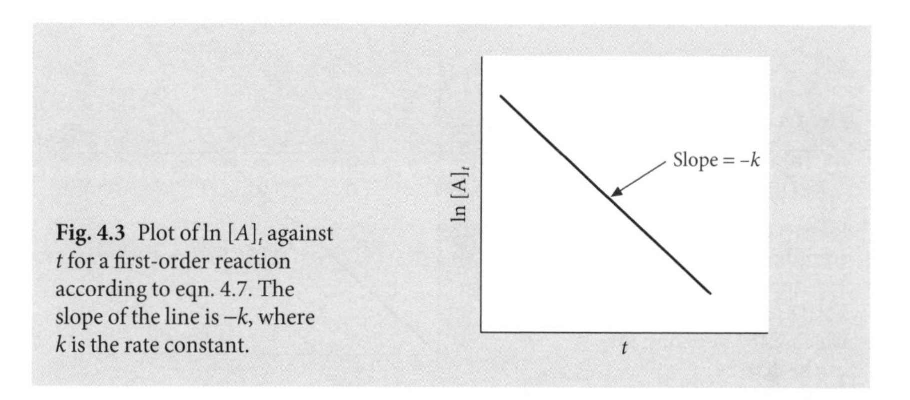

Fig. 4.3 Plot of ln $[A]_t$ against t for a first-order reaction according to eqn. 4.7. The slope of the line is $-k$, where k is the rate constant.

For a first-order reaction, the half-time $t_{1/2}$, i.e. the time taken for $[A]_t$ to decline to half its initial value, is found by setting $[A]_t = 0.5[A]_0$ in eqn. 4.6. This gives eqn. 4.8:

$$t_{1/2} = \frac{\ln 2}{k} = \frac{0.693}{k} \quad \textbf{4.8}$$

Note the half-time is independent of $[A]$, i.e. it takes as long for $[A]_t$ to fall from $[A]_0$ to $0.5[A]_0$ as it does for it to fall from $0.5[A]_0$ to $0.25[A]_0$.

The constant values of successive half-times is a very convenient diagnostic test for a first-order reaction.

Second-order reactions: If a reaction involves two reactants, A and B, whose initial concentrations are $[A]_0$ and $[B]_0$, respectively, then the concentration of product ($[P]$) formed after time t is given by eqn. 4.9:

$$\frac{1}{([A]_0 - [B]_0)} \times \ln\left(\frac{[B]_0([A]_0 - [P])}{[A]_0([B]_0 - [P])}\right) = k \times t \quad \textbf{4.9}$$

where k is the rate constant.

A plot of $\ln\left(\frac{[B]_0([A]_0 - [P])}{[A]_0([B]_0 - [P])}\right)$ vs t (Fig. 4.4) will give a straight line of slope $k \times ([A]_0 - [B]_0)$, from which k can be determined.

The analysis of second-order reactions can be quite complex, but is simplified considerably under two conditions.

Condition 1 The first condition is where the concentrations of the two reactants are equal, i.e. $[A]_0 = [B]_0$. In this case, the concentration of product P after time t is given by eqn. 4.10:

$$\frac{[P]}{[A]_0([A]_0 - [P])} = k \times t \quad \textbf{4.10}$$

so a plot of $\frac{[P]}{[A]_0([A]_0 - [P])}$ vs t (Fig. 4.5) will be a straight line of slope k.

Fig. 4.4 Plot of

$$\ln\left(\frac{[B]_0([A]_0-[P])}{[A]_0([B]_0-[P])}\right) \text{ vs. } t$$

for a second-order reaction according to eqn. 4.9. The slope of the line is $k \times ([A]_0 - [B]_0)$, from which k, the rate constant, can be derived.

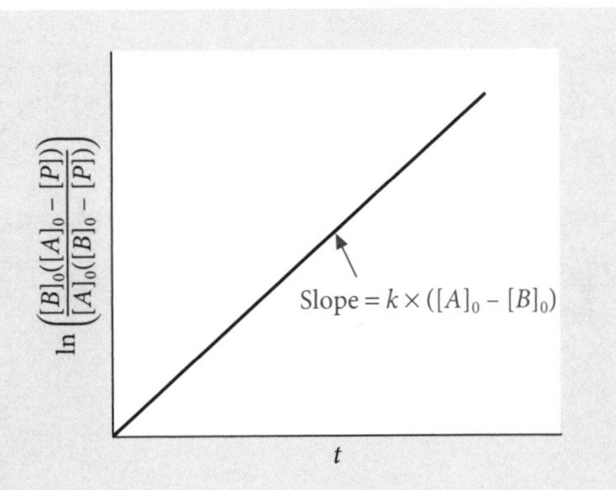

Fig. 4.5 Plot of

$$\frac{[P]}{[A]_0([A]_0-[P])} \text{ vs. } t$$

for a second-order reaction when $[A]_0 = [B]_0$ according to eqn. 4.10. The slope of the line = k, where k is the rate constant.

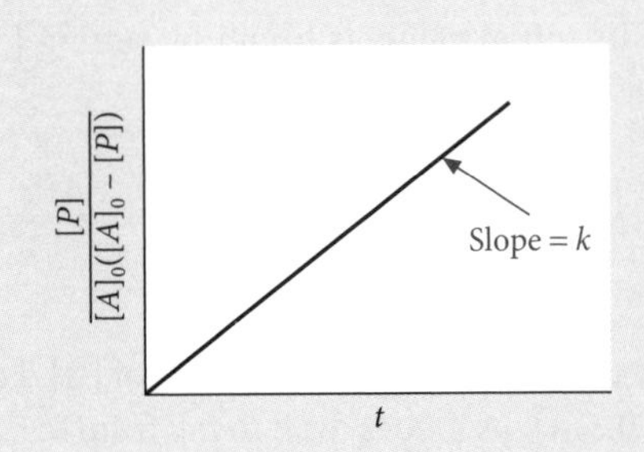

The half-time of the reaction is derived from eqn. 4.10, by putting $[P] = 0.5\ [A]_0$; this gives eqn. 4.11:

$$t_{1/2} = \frac{1}{k \times [A]_0} \qquad \textbf{4.11}$$

It is clear that each successive half-time becomes longer, i.e. it takes twice as long for $[A]$ to fall from $0.5\ [A]_0$ to $0.25\ [A]_0$ as it does from $[A]_0$ to $0.5\ [A]_0$.

✓ **WORKED EXAMPLE**

Coenzyme A (denoted CoASH) which acts as a carrier of acyl groups in the metabolism of fatty acids and carbohydrates, contains a sulphydryl group. This group reacts with 5,5′-dithiobis-(2-nitrobenzoic acid) (Nbs_2) with a 1:1 stoichiometry to give a yellow product (P), the thionitrophenolate ion which can be monitored by its absorbance at 412 nm. The data in Table 4.1 were obtained when equal concentrations (50 μM) of CoASH and Nbs_2 were reacted.

Table 4.1 Time course of the reaction between CoASH and Nbs_2

Time (min)	0	1	2	4	8	12	20	30
[P] (μM)	0	8.3	14.3	20.2	30.8	35.3	40.0	42.9

Show that the data conform to a second-order reaction with $[A]_0 = [B]_0$, and determine the second-order rate constant.

STRATEGY

This calculation involves evaluating the term $\frac{[P]}{[A]_0([A]_0 - [P])}$ at each time point to construct the appropriate plot. Use of the correct units will ensure that the slope is evaluated correctly.

SOLUTION

The plot of $\frac{[P]}{[A]_0([A]_0 - [P])}$ vs t (Fig. 4.6) is a straight line, showing that the reaction does obey second-order kinetics. The slope of the line corresponds to the second-order rate constant and is equal to 0.04 μM^{-1} min^{-1} = 4000 M^{-1} min^{-1} (note 1 $\mu M^{-1} = 10^6$ M^{-1}). This value can be seen to be reasonable by calculating the first-half life for the reaction using eqn. 4.11. Thus, $t_{1/2} = 1/k[A]_0 =$ $1/(4000 \times 50 \times 10^{-6})$ min = 5 min. Inspection of the data shows that $[P]$ has risen to half its final value (i.e. to 25 μM) between 4 and 8 min of reaction.

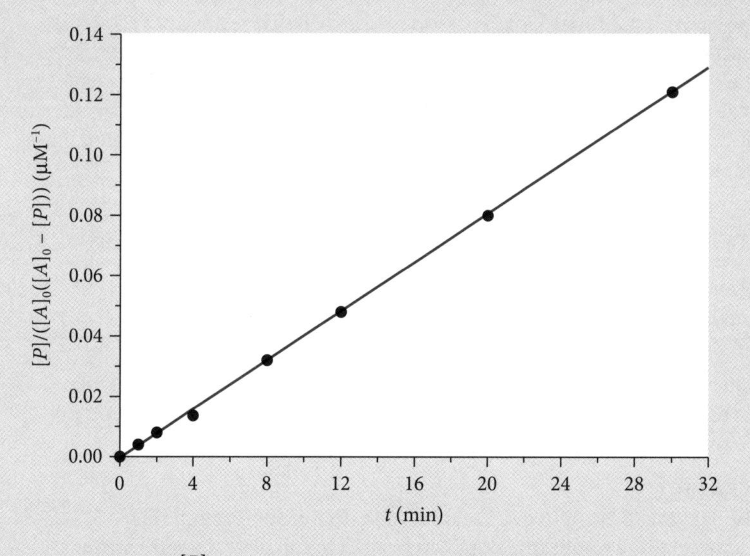

Fig. 4.6 Plot of $\frac{[P]}{[A]_0([A]_0 - [P])}$ vs. t for the reaction between CoA and Nbs_2 in the worked example according to eqn. 4.10. The slope of the line = k, where k is the rate constant.

Condition 2 The second condition under which the analysis of second-order reactions can be simplified is when the concentration of one of the reactants (say A) is much greater than that of the other (B). (In practice, this applies if the ratio $[A]_0/[B]_0$ is greater than about 20). If this applies, then the concentration of A remains essentially constant during the course of the reaction.

We can then write the rate of the reaction as eqn. 4.12:

$$\text{Rate} = k \times [A] \times [B] \quad (4.12)$$

but if $[A]$ remains effectively constant, this becomes eqn. 4.13:

$$\text{Rate} = k' \times [B] \quad (4.13)$$

where $k' = k \times [A]$.

Note it is important to express the pseudo-first-order rate constant and the second-order rate constant in the correct units (see section 4.2.2).

The reaction now behaves as a first-order reaction (it is termed a pseudo-first-order reaction) and can be analysed as such to give k', the pseudo-first-order rate constant. Division of k' by $[A]$, i.e. the concentration of the reactant in excess gives k, the second-order rate constant for the reaction.

WORKED EXAMPLE

The enzyme shikimate kinase possesses a single lysine residue per molecule. Reaction of the side chain of this lysine leads to loss of enzyme activity.
The reaction of shikimate kinase (0.1 μM) with the lysine-specific reagent 2,4,6-trinitrobenzenesulphonic acid (TNBS) (5 μM) was monitored by measuring the activity of samples withdrawn from the reaction mixture after stated times (Table 4.2).

Table 4.2 Inactivation of shikimate kinase by reaction with TNBS

Time (min)	0	2	4	6	8	10	15	20
Activity remaining (%)	100	74.1	54.9	40.7	30.1	22.3	10.5	5.0

Show that the data conform to a pseudo-first-order reaction and determine the pseudo-first-order rate constant and the second-order rate constant for the reaction of the lysine side chain of the enzyme with TNBS.

STRATEGY

The data are to be plotted in the form for a first-order reaction (Fig. 4.3); if a straight line results, the slope can be used to calculate the rate constant.

SOLUTION

A plot of $\ln([A]_0/[A]_t)$ vs t is a straight line, showing that the loss in activity is occurring in a pseudo-first-order fashion. (Note the molar ratio of TNBS to enzyme is 50). The slope of the line $= -0.15\ \text{min}^{-1}$, so the pseudo-first-order rate constant $(k') = 0.15\ \text{min}^{-1}$. This is a reasonable value since it would correspond to a half-time for the reaction of $(\ln 2)/0.15\ \text{min} = 0.693/0.15\ \text{min} = 4.62\ \text{min}$. Inspection of the data shows that 50% of the original activity has been lost between 4 and 6 min of reaction. The second-order rate constant (k) is found by dividing k' by the concentration of TNBS (which is in excess). Hence $k = 0.15/(5 \times 10^{-6})\ \text{M}^{-1}\ \text{min}^{-1} = 30\,000\ \text{M}^{-1}\ \text{min}^{-1}$. It might be noted that this rate constant is much higher than for the reaction of the free amino acid lysine with TNBS and suggests that the side chain in the enzyme must be especially reactive.

4.2.2 The units of rates and rate constants

The rate of a reaction is measured in terms of the change in concentration of a reactant or product (depending on how the reaction is being monitored) with time. The units of rate will thus be concentration × (time)$^{-1}$, e.g. mM min^{-1}, µM s^{-1} etc.

The units of a rate constant will depend on the order of the reaction concerned, and can be deduced by taking into account the fact that units must balance on both sides of an equation (see Chapter 3, section 3.3).

Thus for a first-order reaction, rate = $k \times [A]$, so in terms of units we have:

left-hand side – the units are (concentration × (time)$^{-1}$)

right-hand side – the units are ((units of k) × concentration)

hence the units of k must be (time)$^{-1}$, e.g. s^{-1}, min^{-1}, etc.

WORKED EXAMPLE ✓

What are the units of a second-order rate constant?

STRATEGY
From the rate law for a second-order reaction we can derive the units, noting that the units on the two sides of an equation must balance.

SOLUTION
For a second-order reaction, rate = $k \times [A] \times [B]$, so applying the rule about the need to balance units on both sides of the equation, the units of the second-order rate constant are (concentration)$^{-1}$ × (time)$^{-1}$, e.g. M min^{-1}.

4.2.3 The variation of reaction rate with temperature

As the temperature increases, a greater proportion of the reactant molecules have the energy necessary to surmount the activation energy barrier. This leads to a marked increase in the rate of the reaction with temperature. The equation proposed by Arrhenius (see Chapter 2, section 2.3.2) is:

$$k = A\,e^{-Ea/RT} \quad (4.14)$$

where k is the rate constant for the reaction, E_a is the activation energy, R is the gas constant (8.31 J K^{-1} mol^{-1}), and T is the temperature in degrees Kelvin. A is known as the pre-exponential factor and is related to the frequency of successful collisions between reacting molecules.

Taking natural logarithms of both sides of eqn. 4.14, we obtain eqn. 4.15:

$$\ln k = \ln A - \frac{E_a}{RT} \quad (4.15)$$

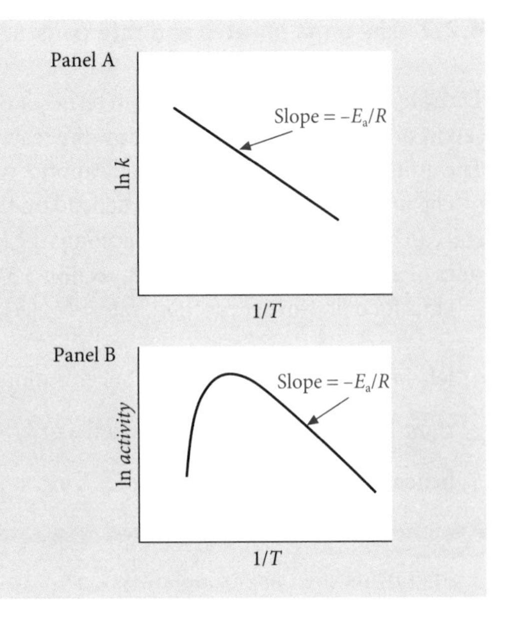

Fig. 4.7 Arrhenius plots to show the effect of temperature on reaction rate according to eqn. 4.15. (a) A typical plot for a chemical reaction. The slope of the line is $-E_a/R$, from which E_a can be derived. (b) A typical plot for an enzyme-catalysed reaction in which, above a certain temperature, there is a sharp loss of activity, reflecting the unfolding and inactivation of the enzyme.

Thus, an Arrhenius plot (ln k vs $1/T$) (Fig. 4.7) is a straight line of slope $-E_a/R$, from which E_a can be calculated.

In the case of enzyme-catalysed reactions, the rate will start to decline at higher temperatures because the three-dimensional structure of the enzyme which is required for the expression of catalytic activity will be disrupted by the loss of the weak interactions which stabilize it (see Chapter 1, section 1.7). The effect on the Arrhenius plot is shown in Fig. 4.7 (Panel B). The temperature at which the activity reaches a maximum value is often referred to as the 'optimum temperature', but this can be difficult to measure as it will depend on the length of time the enzyme is incubated at the temperature in question prior to assay. (For this reason, the term 'apparent optimum temperature' is often used.) In most cases the 'optimum temperature' of an enzyme will reflect the physiological needs of the organism in question. Some enzymes from hyperthermophilic bacteria, for example, seem to be stable for long periods at temperatures close to 100°C.

The availability of stable enzymes from thermophilic bacteria has been a key factor in their application in many processes. A good example is the use of the thermostable DNA polymerase from *Thermus aquaticus* in the polymerase chain reaction, which can amplify the amount of selected DNA by a very large factor (1 million-fold or greater).

✓ WORKED EXAMPLE

The data in Table 4.3 were obtained for the activity of a sample of mammalian lactate dehydrogenase over a range of temperatures.

Table 4.3 **The variation of lactate dehydrogenase activity with temperature**

Temperature (°C)	5	15	25	35	45	55	65
Activity (μmol min^{-1} mg^{-1})	52.5	108	212	401	488	270	40.3

Explain the dependence of activity on temperature and determine the activation energy for the enzyme-catalysed reaction.

STRATEGY

The data can be plotted according to the Arrhenius equation to give a straight line (eqn. 4.15). Deviations from a straight line would indicate that additional factors have to be considered.

SOLUTION

The Arrhenius plot (ln (activity) vs $1/T$, where T is the temperature in degrees Kelvin = °C + 273, i.e. 25°C = 298 K) is linear over the range 5–35°C. The slope of the graph = −5810 K. The slope = $-E_a/R$, so the $E_a = 5810 \times 8.31$ J mol^{-1} = 48.3 kJ mol^{-1}. At 45°C, the rate is less than predicted on the basis of this straight line and at temperatures above 45°C, the activity declines sharply. This is due to progressive unfolding of the enzyme at the higher temperatures and the loss of catalytic activity. It should be noted that where a comparison can be made between an enzyme-catalysed reaction and the corresponding non-catalysed reaction, the activation energy of the latter is considerably greater.

SELF TEST ?

Check that you have mastered the key concepts at the start of the section by attempting the following questions.

ST 4.4 At low concentrations of substrate, the rate of an enzyme-catalysed reaction is given by: rate = $(k_{cat}/K_m) \times$ [Enzyme] × [Substrate]. At high concentrations of substrate, the rate is given by: rate = $k_{cat} \times$ [Enzyme]. Given that k_{cat} and K_m are constants, what is the order of the reaction under each condition?

ST 4.5 The dissociation of an antibody–antigen complex is a first-order reaction with a rate constant of 3×10^{-3} s^{-1}. What is the half-time for this reaction?

ST 4.6 The rate of a reaction catalysed by an enzyme from an intertidal species (winkle) increases by 1.2-fold as the temperature is raised from 5°C to 15°C. What is the activation energy for this reaction?

Answers

ST 4.4 The orders of the reaction are: at low [substrate], first order in enzyme and first order in substrate, second order overall; at high [substrate], first order in enzyme, zeroth order in substrate, first order overall.

ST 4.5 The half-time = 231 s, or 3 min 51 s.

ST 4.6 The activation energy = 12.1 kJ mol^{-1}.

Attempt Problems 4.5 and 4.6 at the end of the chapter.

In ST 4.6, the relatively small dependence of the rate with temperature is key to the survival of the organism, since the temperature will fluctuate considerably during the 24-h cycle of high and low tides.

4.3 Binding of ligands to macromolecules: saturation curves

KEY CONCEPTS

- Understanding the meaning of K_a and K_d in binding processes
- Understanding the relationship between K_d and ΔG^0
- Writing the equation for hyperbolic binding to single and multiple (independent) binding sites
- Writing the Hill equation to describe cooperative binding
- Understanding the meaning of V_{max} and K_m in enzyme-catalysed reactions
- Explaining the different types of inhibition of enzyme-catalysed reactions

Many biological processes depend on the interactions between molecules, for example regulation of gene expression (protein/DNA), catalysis (enzyme/substrate), cell signalling (hormone/receptor), etc. These interactions depend very largely on non-covalent forces (hydrogen bonds, van der Waals, electrostatic, hydrophobic interactions), which are individually weak but collectively can generate high affinities and specificities between the interacting molecules (for further details see Chapter 1, section 1.7). It is important to characterize binding processes to judge their importance under physiological conditions.

4.3.1 Single binding site on a macromolecule

For many of these interactions, simple saturation behaviour is observed:

$$P + L \rightleftharpoons PL$$

where P represents a protein and L a ligand, i.e. any molecule which binds to the protein; this could include a nucleic acid, a low molecular mass molecule or another protein, for example.

For this equilibrium, we can define two equilibrium constants, K_a and K_d, which are known as the association and dissociation constants, respectively:

$$K_a = \frac{[PL]}{[P] \times [L]} \qquad 4.16$$

$$K_d = \frac{[P] \times [L]}{[PL]} \qquad 4.17$$

where $[P]$ and $[L]$ represent the concentrations of free (i.e. unbound) protein and ligand, respectively, and $[PL]$ represents the concentration of the PL complex.

K_a represents a measure of the tendency for P and L to associate to form the PL complex. The larger this tendency, i.e. the tighter the binding, the greater the value of K_a.

K_d represents a measure of the tendency for the PL complex to dissociate into P and L. The smaller this tendency, i.e. the tighter the binding, the smaller the value of K_d.

For largely historical reasons, K_a is used predominantly by chemists and K_d by molecular biologists and biochemists. The two constants are reciprocals (see Chapter 2, section 2.4) of each other, i.e. $K_a = 1/K_d$ and $K_d = 1/K_a$. The units of K_a are (concentration)$^{-1}$, e.g. M^{-1}; those of K_d are concentration, e.g. M.

Strictly speaking, K_a and K_d are dimensionless since they refer to a standard state of a 1 M solution of P, L and PL (see section 4.1). It is convenient to keep the units of M^{-1} and M for K_a and K_d, respectively, to remember that for calculations, we must always express the concentrations in molar terms.

Since the equilibrium between P, L and PL is a dynamic one, at equilibrium the rate of the association reaction (P + L $\rightarrow$ PL), i.e. $k_a \times [P] \times [L]$, must equal the rate of the dissociation reaction (PL $\rightarrow$ P + L), i.e. $k_d \times [PL]$, where k_a and k_d are the association and dissociation rate constants, respectively. Thus, we have:

$$k_a \times [P] \times [L] = k_d \times [PL] \quad (4.18)$$

Hence, by combining eqn. 4.18 with eqns. 4.16 or 4.17, it follows that:

$$K_a = \frac{k_a}{k_d} \quad (4.19)$$

$$K_d = \frac{k_d}{k_a} \quad (4.20)$$

Thus, the equilibrium constants K_a and K_d can be expressed as the appropriate ratios of rate constants for the individual association and dissociation steps.

In order to develop molecular explanations for the interactions involving proteins, it is important to be able to link the value of K_a or K_d to the energy changes involved in binding or dissociation.

For the dissociation of the *PL* complex (PL $\rightleftharpoons$ P + L), the relationship between the standard state free energy change (see section 4.1) and the value of K_d is given by eqn. 4.21 (which is analogous to eqn. 4.4).

$$-\Delta G^0 = RT \ln K_d \quad (4.21)$$

Thus, if $K_d = 1\ \mu M$ at 310 K (37°C), $\Delta G^0 = -8.31 \times 310 \times \ln (1 \times 10^{-6})\ J\ mol^{-1} =$ 35 600 J mol^{-1} = 35.6 kJ mol^{-1}. The energy change can be compared with typical values for the weak, non-covalent forces described in Chapter 1, section 1.7.

Note that K_d is expressed in terms of molar concentration, since this is referred to the standard state of the ligand, i.e. a 1 M solution.

In this example, the value of ΔG^0 is large and positive, since under standard state conditions (i.e. 1 M PL) there would be little tendency of the reaction to proceed from left to right, i.e. for PL to dissociate. Dissociation of the complex would be promoted by lowering the concentration of PL, which would be the case in living

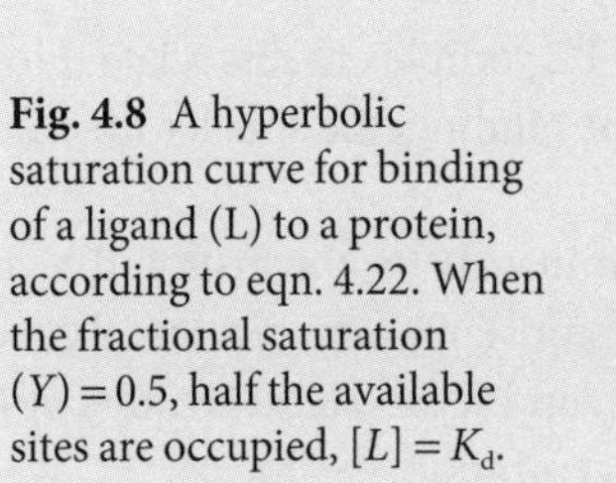

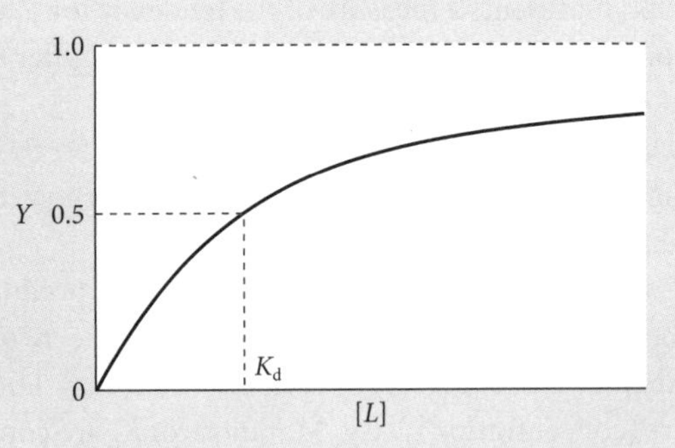

Fig. 4.8 A hyperbolic saturation curve for binding of a ligand (L) to a protein, according to eqn. 4.22. When the fractional saturation $(Y) = 0.5$, half the available sites are occupied, $[L] = K_d$.

systems, where concentrations of macromolecules are likely to be in the micromolar range, or even lower.

Remembering that at 310 K each factor of 10 in the value of K_d corresponds to approximately 6 kJ mol^{-1} (accurately 5.93 kJ mol^{-1}; see Chapter 2, section 2.3.6 and section 4.1 of this chapter) we can easily estimate the ΔG^0 involved. For example, a K_d of 10^{-9} M (1 nM) for a complex corresponds to a ΔG^0 of approximately 54 kJ mol^{-1} (accurately 53.4 kJ mol^{-1}).

It is a relatively simple task (see Appendix 4.1) to derive an expression for the fraction (Y) of the protein P which is saturated by ligand L (eqn. 4.22):

$$Y = \frac{[L]}{K_d + [L]} \qquad 4.22$$

The dependence of Y on $[L]$ according to eqn. 4.22 is described mathematically as a rectangular hyperbola (or simply a hyperbola). The binding is said to be hyperbolic in nature.

The value of Y can range from 0 to 1; a graph indicating how Y varies with $[L]$ is shown in Fig. 4.8.

Note that when half the total binding sites are occupied, i.e. $Y = 0.5$, then $K_d = [L]$. This gives a useful operational definition of K_d, namely that it corresponds to the concentration of (free) L at which half-saturation of the protein has occurred.

One of the problems about depicting the saturation curve as in Fig. 4.8 is that it is difficult to span a very wide range of ligand concentrations. For example, to go from 10% saturation to 90% saturation requires an 81-fold change in $[L]$ (from $K_d/9$ to $9 \times K_d$). The use of a logarithmic scale for $[L]$ can overcome this problem since it allows us to cover a very large numerical range of values (see Chapter 3, section 3.2). This type of plot is known as a formation plot and an example is shown in Fig. 4.9. In this figure the binding is hyperbolic (as in Fig. 4.8) but the logarithmic nature of the x-axis gives the curve a 'sigmoidal' (or elongated S-shape) appearance.

4.3.2 Multiple binding sites on a macromolecule

If instead of a single binding site on the protein, there are n sites for binding the ligand and these are assumed to be equivalent and independent of each other

$$P + nL \rightleftharpoons PL_n$$

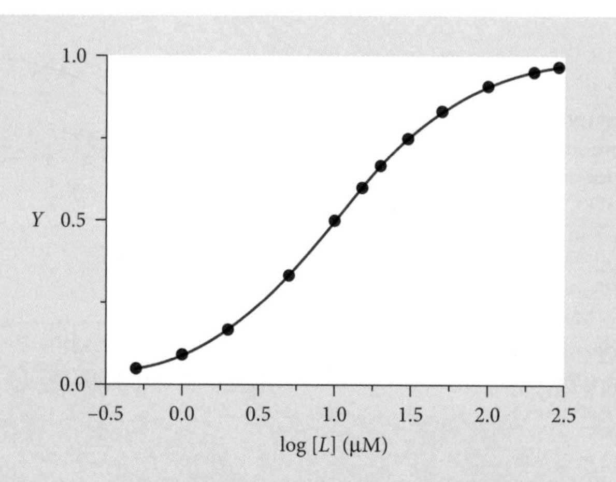

Fig. 4.9 Formation plot for binding of a ligand (L) to a protein, where $[L]$ is plotted on a logarithmic scale. The data points correspond to a hyperbolic binding model (eqn. 4.22) with $K_d = 1$ μM. Note that in this example the use of the logarithmic scale allows data over a range of nearly 1000-fold values of $[L]$ to be plotted.

then the saturation equation is analogous to eqn. 4.22, taking into account the number of binding sites; see eqn. 4.23:

$$r = \frac{n \times [L]}{K_d + [L]} \qquad 4.23$$

where r is the average number of molecules of ligand bound per molecule of protein, n is the number of binding sites, $[L]$ is the concentration of free ligand, and K_d is the dissociation constant.

A plot of r against $[L]$ is shown in Fig. 4.10. Note that r can take values ranging from 0 to n.

Substitution into eqn. 4.23 shows that when $r = 0.5n$ (i.e. when half the total sites are saturated), $[L] = K_d$. Thus, just as in the case of a single binding site, the value of

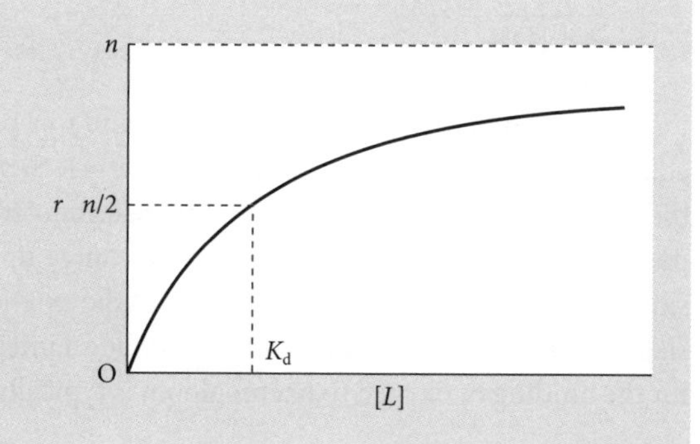

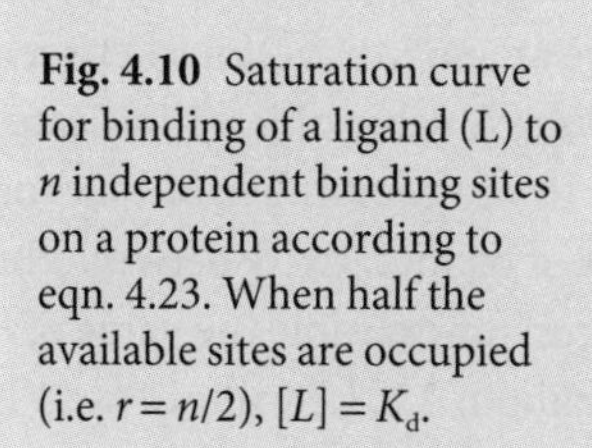

Fig. 4.10 Saturation curve for binding of a ligand (L) to n independent binding sites on a protein according to eqn. 4.23. When half the available sites are occupied (i.e. $r = n/2$), $[L] = K_d$.

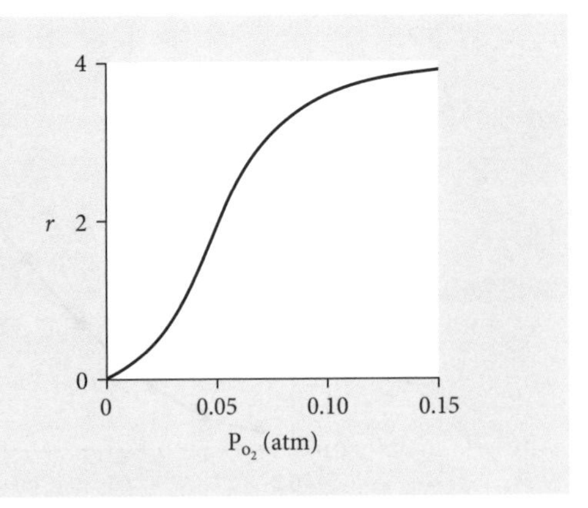

Fig. 4.11 Saturation curve for binding of O_2 to haemoglobin. At high pressures of O_2 all four binding sites in the haemoglobin molecule are occupied. The sigmoidal shape of the binding curve reflects cooperatively between the binding sites.

K_d is equal to the concentration of ligand required to bring about 50% saturation of the available sites.

Note in those cases where there are interactions between the binding sites, the shape of the saturation curve will differ. Thus, the binding of oxygen to haemoglobin (four binding sites) shows positive cooperativity, with the binding at the first site making it easier to bind to the subsequent sites. The saturation curve in this case is sigmoidal (see Fig. 4.11). This behaviour arises because of structural changes relayed between the subunits (polypeptide chains) of the protein.

Although a detailed mathematical analysis of cooperativity is beyond the scope of this book, mention should be made of the Hill equation (eqn. 4.24), which provides a relatively simple way of describing this type of behaviour. The equation is a modified form of eqn. 4.23 in which the term $[L]$ is raised to a power h, where h is known as the Hill coefficient.

Hill proposed this equation in 1910 to describe the binding of oxygen to haemoglobin.

$$r = \frac{n \times [L]^h}{K + [L]^h} \qquad 4.24$$

The Hill equation is usually expressed in terms of the fractional saturation (Y) of the binding sites by ligand, i.e. $Y = r/n$. Thus eqn. 4.24 becomes:

$$Y = \frac{[L]^h}{K + [L]^h} \qquad 4.25$$

The value of h for binding of a given ligand to a protein can be found by a suitable plot as described in section 4.4. When $h = 1$, there are no interactions between the multiple ligand binding sites on the macromolecule. A value of $h > 1$ indicates positive cooperativity. The value of h can range up to n, the number of binding sites, and its magnitude gives a measure of the extent of cooperativity between the sites. The Hill coefficient does not have to be an integer; for example the value of h for the binding of oxygen to haemoglobin is typically about 2.8. (A value of $h < 1$ is said to indicate negative cooperativity, where the binding of the first ligand molecule hinders binding of subsequent molecules.) An equation analogous to eqn. 4.25 can be derived for the analysis of enzyme-catalysed reactions, see section 4.4.

4.3.3 Enzyme kinetics

The treatment of enzyme kinetics is, in a mathematical sense, analogous to that of ligand binding considered above. However, it is considered separately because a number of the symbols used differ from those used in considering binding equilibria. Enzyme kinetics is an important topic because it is actually measuring the enzyme performing its biological role (i.e. catalysing a reaction). By studying how the rate of the reaction depends on the concentration(s) of substrate(s), and the effect of inhibitor molecules (section 4.3.4) it is possible to draw conclusions not only about the mechanism of the reaction, but also about the possibilities of designing specific inhibitors which could act, for example as therapeutic drugs.

In the simplest case, where an enzyme (E) catalyses a reaction involving one substrate (S) and one product (P), the reaction is proposed to proceed via formation of an enzyme–substrate complex which breaks down to generate product and regenerate enzyme as follows:

$$\mathrm{E} + \mathrm{S} \rightleftharpoons \mathrm{ES} \rightarrow \mathrm{E} + \mathrm{P}$$

By using either an equilibrium assumption (i.e. the breakdown of ES to give E + P is so slow as to not to perturb the $\mathrm{E} + \mathrm{S} \rightleftharpoons \mathrm{ES}$ equilibrium significantly) or the steady-state assumption (i.e. the concentration of ES remains effectively constant once the steady-state condition has been achieved), we can derive an equation (4.26) to show how the rate (or velocity, v) varies with the concentration of substrate ([S]). This is known as the Michaelis–Menten equation.

$$v = \frac{V_{max} \times [S]}{K_m + [S]} \qquad 4.26$$

where V_{max} is the maximum rate (more correctly, the limiting rate) and K_m is known as the Michaelis constant. From eqn. 4.26 it can be seen that if v is set equal to 0.5 V_{max}, $[S] = K_m$.

The derivation of the Michaelis–Menten equation is discussed in more detail in Appendix 4.1.

The Michaelis–Menten equation was derived in 1913 by Leonor Michaelis (Belgian) and Maud Menten (Canadian) using the equilibrium assumption. In 1925, Briggs and Haldane (both British) applied the steady-state approximation to derive an equation of similar form.

The plot of v vs. [S] is shown in Fig. 4.12. From this it can be seen that at low [S] the rate of reaction increases almost linearly with [S], but at high [S] the rate tends to level off somewhat and approaches a maximum (or limiting) value. This levelling off reflects the saturation of the active sites of the enzyme molecules present; no matter how much more substrate is added, the rate cannot increase further.

Strictly speaking [S] refers to the concentration of free substrate. However, the vast majority of studies of enzyme-catalysed reactions take place under steady-state conditions where the molar concentration of enzyme is very much less than

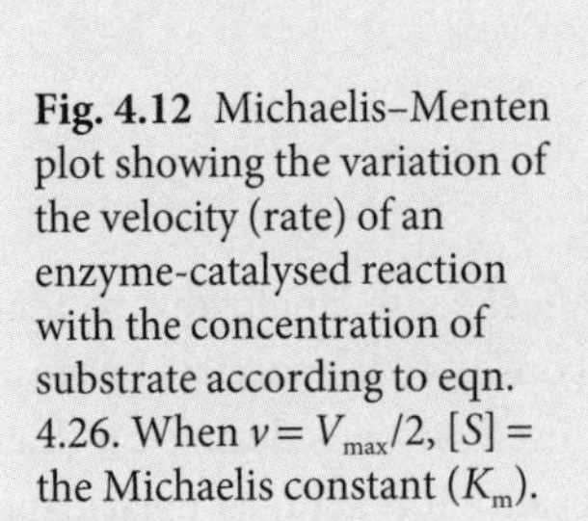

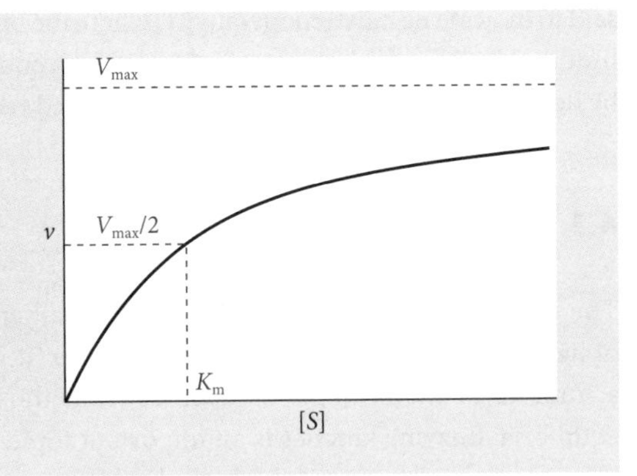

Fig. 4.12 Michaelis–Menten plot showing the variation of the velocity (rate) of an enzyme-catalysed reaction with the concentration of substrate according to eqn. 4.26. When $v = V_{max}/2$, $[S] =$ the Michaelis constant (K_m).

the molar concentration of substrate. In these circumstances, the concentration of free substrate is effectively equal to that of the total added substrate.

It should be noted that when E and S are mixed the system takes a short time to reach the 'steady state'. It is possible to observe events in the pre-steady state period using specialized apparatus; this can give additional information on the rates of individual steps in the overall reaction (see Chapter 9, section 9.9.4).

4.3.4 Inhibition of enzyme-catalysed reactions

Although a full discussion of the inhibition of enzyme-catalysed reactions is beyond the scope of this book, it is important that you are able to analyse appropriate data to determine the type of inhibition being observed in any particular case and to evaluate the strength of the interaction between enzyme and inhibitor in terms of a suitable dissociation constant. We will confine our attention to those inhibitors which bind reversibly to enzymes.

In the mechanism quoted previously:

$$\mathrm{E + S \rightleftharpoons ES \rightarrow E + P}$$

we could envisage a general scheme for the way in which an inhibitor (I) might interact with E and the ES complex.

We can consider three limiting cases for enzyme–inhibitor interaction:

A. *Competitive inhibition* The inhibitor (I) binds to E but not to ES (i.e. $K_{ESI} = \infty$ which means that the ESI complex has no tendency to form). In this case, the addition of I will pull some of E over to the EI complex. However, addition of increasing concentrations of S will eventually overcome the effect of I, since the balance will progressively swing towards formation of ES rather than EI. This is termed competitive inhibition; V_{max} remains unchanged, but K_m is raised (reflecting the higher $[S]$ to be added to overcome the effects of I). The most likely explanation for competitive inhibition is that although the inhibitor cannot undergo the reaction, it has a structural resemblance to the substrate and therefore binds at the active site, preventing the binding of S.

In quantitative terms, K_m in the presence of inhibitor is raised by a factor $\left(1 + \frac{[I]}{K_{EI}}\right)$ compared with the absence of inhibitor (K_{EI} is the dissociation constant of the EI complex). V_{max} remains unchanged.

Competitive inhibitors are widely studied to define the way that the substrate and enzyme interact, and as the basis for designing potential therapeutic drugs. Examples of the latter include angiotensin-converting enzyme inhibitors to control hypertension, and inhibitors of hydroxymethylglutaryl coenzyme A reductase (HMG CoA reductase) to lower blood cholesterol levels and thereby decrease the risk of heart attacks.

B. *Non-competitive inhibition*: The inhibitor (I) binds to E and ES with equal affinity (i.e. $K_{EI} = K_{ESI}$). In this case, both E and ES will be pulled over to the complexes EI and ESI, respectively, and the net effect will be that a certain fraction of the enzyme is complexed with I and unable to catalyse the reaction. The remaining enzyme will have the same K_m for substrate as in the absence of inhibitor. This is termed non-competitive inhibition. The inhibitor cannot bind to the active site, since it does not affect the affinity of the enzyme for substrate.

In quantitative terms, V_{max} in the presence of inhibitor is lowered by a factor $\left(1 + \frac{[I]}{K_{EI}}\right)$ compared with the absence of inhibitor. K_m remains unchanged.

Effectively a proportion of the enzyme has been removed from the system; this could also be achieved by irreversible inhibition of part of the enzyme present. The case discussed here is more properly termed pure non-competitive inhibition. There is in fact a range of possible effects of inhibitors on V_{max} and K_m, which are too complex to be dealt with in detail here.

C. *Uncompetitive inhibition*: The inhibitor binds only to the ES complex and not to E ($K_{EI} = \infty$, i.e. there is no tendency for the EI complex to form). In the presence of I, E will be pulled over to the ES complex, which will then be pulled towards the ESI complex. This will have the effect of lowering both K_m (because E is pulled over to ES) and V_{max} (because some of the enzyme in the form of ESI will be unable to take part in the catalytic reaction). This is termed uncompetitive inhibition and is extremely rare in one-substrate reactions (although several examples are found in multi-substrate reactions). The inhibition could arise from a structural change occurring on binding of substrate to enzyme which then creates a binding site for the inhibitor.

In quantitative terms, both K_m and V_{max} are lowered in the presence of inhibitor by a factor $\left(1 + \frac{[I]}{K_{ESI}}\right)$ compared with the absence of inhibitor (K_{ESI} is the dissociation constant of the ESI complex to give ES + I).

The three different types of inhibitors can be easily recognized by their effects on the kinetic plots as described in section 4.4.

? SELF TEST

Check that you have mastered the key concepts at the start of the section by attempting the following questions.

ST 4.7 A certain drug binds to a target receptor with a K_d of 5×10^{-8} M: what is the ΔG^0 for dissociation of this complex at 37°C?

ST 4.8 For an enzyme-catalysed reaction which obeys Michaelis–Menten (hyperbolic) kinetics, what fold change in [*S*] is required to go from a rate of 20% V_{max} to 80% V_{max}?

ST 4.9 For a binding process which obeys the Hill equation (eqn. 4.25), assume that $K = 1$ mM and that [*L*] changes from 0.5 to 2 mM. What change in fractional saturation occurs over this range for the cases (a) when $h = 2.0$, and (b) when $h = 3.0$?

Answers

ST 4.7 The ΔG^0 for the dissociation = 43.3 kJ mol^{-1}.

ST 4.8 A 16-fold change is required.

ST 4.9 The changes in fractional saturation are: (a) from 0.2 to 0.8; (b) from 0.11 to 0.89. These show the much greater sensitivity to changes in [*L*] compared with hyperbolic binding (eqn. 4.22), where the change is only 2-fold, from 0.33 to 0.67.

Attempt Problem 4.7 at the end of the chapter.

4.4 Analysis of binding and kinetic data

KEY CONCEPTS

- Understanding the graphical procedures used to analyse binding and kinetic data
- Expressing the parameters obtained in the correct units and checking that they are reasonable in terms of the data supplied

As we have seen for systems where there are no interactions between the binding sites, there are essentially two parameters to be determined from the experimental data, one representing the property of the fully saturated protein (V_{max} or n), and the other representing the concentration of ligand or substrate to achieve half-saturation (K_m or K_d). In order to obtain these parameters experimentally we need to make measurements of the extent of binding or the enzyme activity at different concentrations of ligand or substrate. From these data, it is usually difficult to estimate the limiting value (V_{max} or n) directly since this would involve the use of very high concentrations of ligand or substrate.

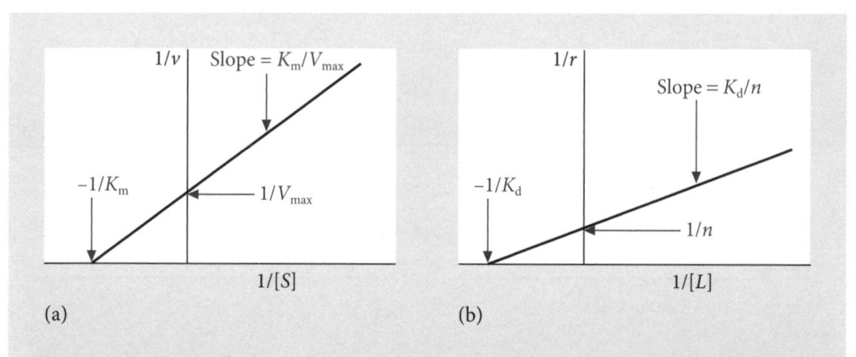

Fig. 4.13 Analysis of enzyme kinetic data (a) and ligand-binding data (b). The plots are known as Lineweaver–Burk and Hughes–Klotz plots, respectively.

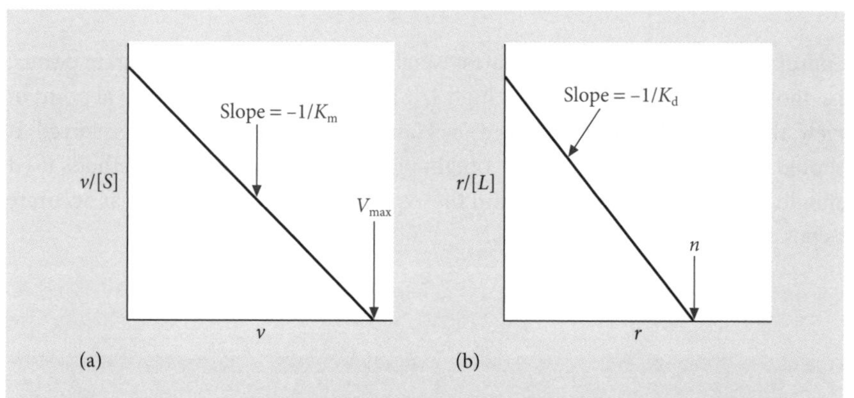

Fig. 4.14 Analysis of enzyme kinetic data (a) and ligand-binding data (b). The plots are known as Eadie–Hofstee and Scatchard plots, respectively.

The criteria for assessing the results of application of non-linear regression to binding or kinetic data are described in Chapter 2, section 2.6.5.

We can use either direct, computer-based, fitting to the saturation curve (non-linear regression; see Chapter 2, section 2.6.5) or transformations (see Chapter 2, section 2.5.2) of the saturation equations (eqns. 4.22, 4.23, and 4.26) to give linear plots, which are termed Lineweaver–Burk (Hughes–Klotz), Eadie–Hofstee (Scatchard) and Hanes–Woolf. (The names in brackets are used for the plots used to analyse binding, rather than enzyme kinetic data). From the slopes and intercepts of these plots (Figs. 4.13, 4.14, and 4.15) the parameters (V_{max} or n; K_m or K_d) can be obtained. Figure 4.16 depicts the effects of the various types of inhibitors (section 4.3.4) on the Lineweaver–Burk plots.

The Lineweaver–Burk plot is the most commonly used for analysis of enzyme kinetics; it involves plotting the two variables, or rather derivatives of them, on separate axes. However, because of the reciprocal nature of the axes, the error distribution is highly non-uniform, which makes it difficult to decide on the most

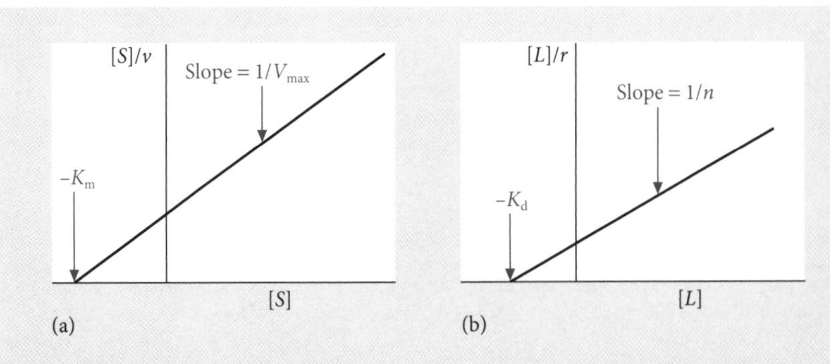

Fig. 4.15 Analysis of enzyme kinetic data (a) and ligand-binding data (b). The plots are known as Hanes–Woolf plots.

appropriate straight line. The greatest weight is given to the least accurate points, i.e. those at low $[S]$ and low v (i.e. high $1/[S]$, high $1/v$). From a statistical point of view, the Eadie–Hofstee (Scatchard) or Hanes–Woolf plots are to be preferred. It should be noted, however, that to obtain plots of good quality, the methods used should be reliable and validated and the experimental data need to be as accurate as possible.

✓ WORKED EXAMPLE

Analyse the data in Table 4.4 showing the variation of rate with substrate concentration for an enzyme-catalysed reaction, using the Lineweaver–Burk plot. State the values of K_m and V_{max} you obtain, quoting the correct units.

Table 4.4 Variation of velocity of an enzyme-catalysed reaction with substrate concentration

[Substrate] (mM)	Velocity (ΔA min^{-1})
4.0	0.035
8.0	0.054
12.0	0.066
20.0	0.080
30.0	0.090

STRATEGY

The data are manipulated to be in the form required for the Lineweaver–Burk plot ($1/v$ vs. $1/[S]$). It is important to choose appropriate scales for the plot so that the intercepts and slope can be read off easily and accurately.

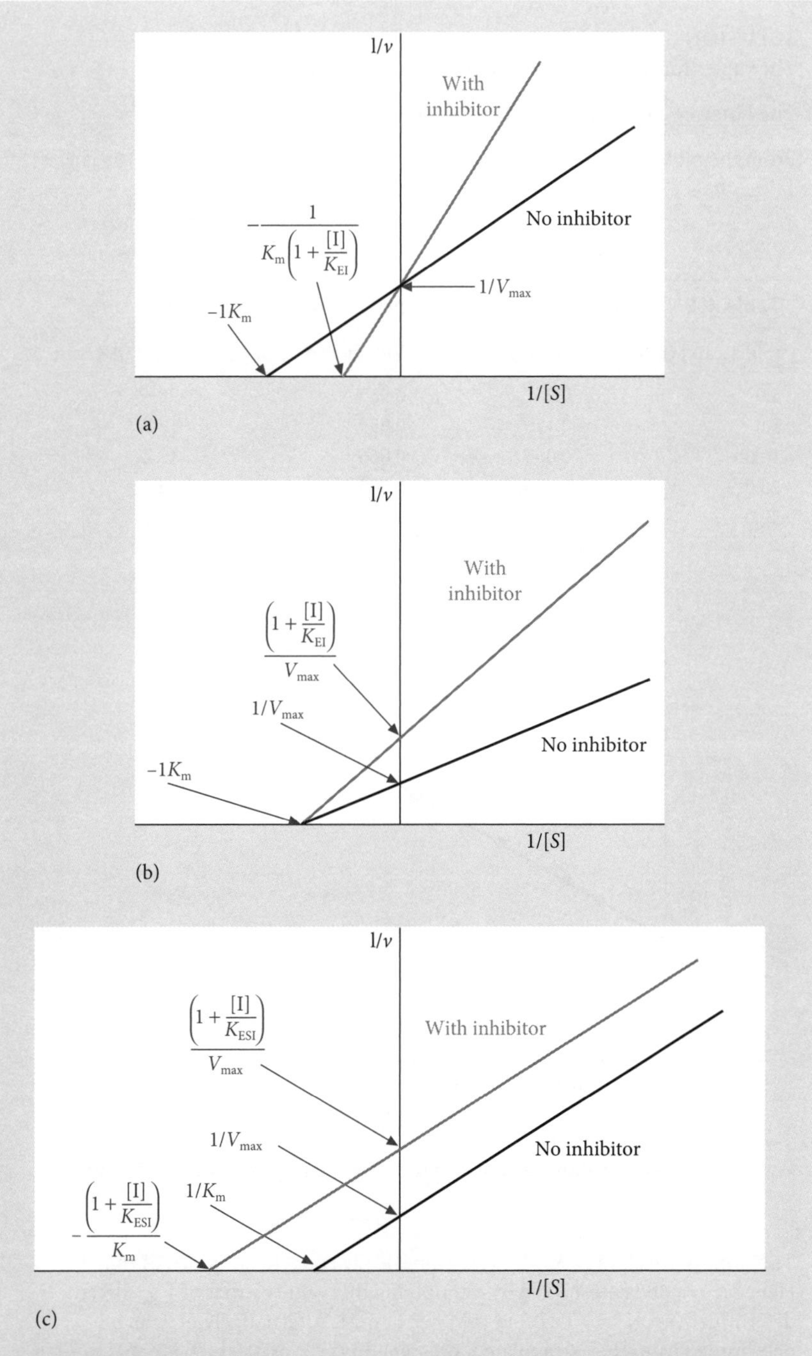

Fig. 4.16 The effects of different types of inhibitor on Lineweaver–Burk plots of enzyme kinetic data. (a), (b), and (c) show competitive, non-competitive, and uncompetitive inhibition, respectively.

SOLUTION

The values for the Lineweaver–Burk plot can be drawn up (Table 4.5):

The Lineweaver–Burk plot of these data is shown in Figure 4.17.

From the plot, the value of the intercept on the *y*-axis is 8.5 (ΔA min^{-1})$^{-1}$. This is $1/V_{max}$, hence $V_{max} = 1/8.5\ \Delta A\ \text{min}^{-1} = 0.118\ \Delta A\ \text{min}^{-1}$.

Table 4.5 Values of parameters for the Lineweaver–Burk plot

[Substrate] (mM)	1/[*S*] (mM^{-1})	Velocity (ΔA min^{-1})	1/[*v*] (ΔA min^{-1})$^{-1}$
4.0	0.25	0.035	28.6
8.0	0.125	0.054	18.5
12.0	0.083	0.066	15.2
20.0	0.050	0.080	12.5
30.0	0.033	0.090	11.1

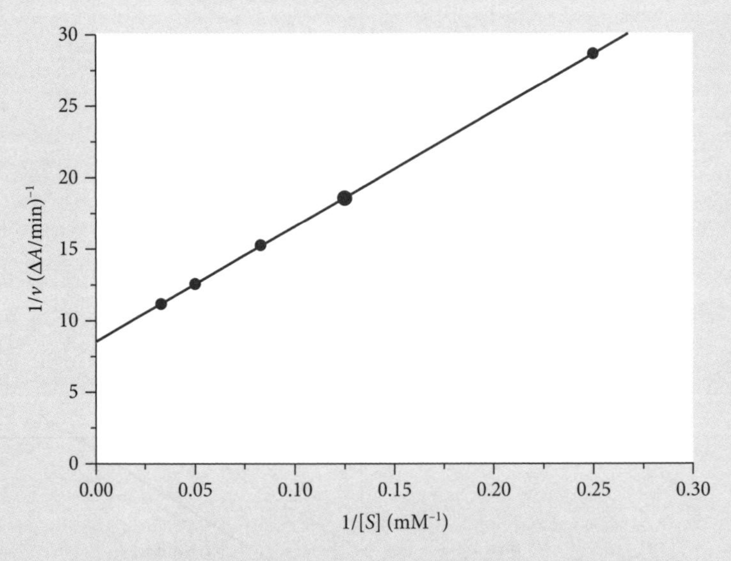

Fig 4.17 Lineweaver–Burk plot of enzyme kinetic data in the worked example.

The slope of this line ($= K_m/V_{max}$) is numerically equal to 80 mM/(ΔA min^{-1}). Hence K_m can be determined by multiplying this value by that of V_{max} already determined; i.e. $K_m = 80 \times 0.118$ mM $= 9.44$ mM. Alternatively, K_m can be determined from the extrapolated intercept on the *x*-axis, which is equal to $-1/K_m$. (Note that if the straight line is extended (extrapolated) beyond the $1/v$ axis as shown, the intercept is negative because the values of 1/[*S*] on this side of the origin are negative.) This allows the value of K_m to be calculated directly (in this case the value of the intercept is -0.106 mM^{-1}, hence $K_m = 1/0.106$ mM $= 9.44$ mM.

It is always good practice to assess whether your values of V_{max} *and* K_m *are reasonable or not in view of the original data provided. We can confirm that this is the case for the worked example above as follows:*

- The value of V_{max} appears to be a reasonable estimate of the limiting value towards which the actual values of v are tending as $[S]$ increases
- The value of K_m should represent the value of $[S]$ at which $v = \frac{1}{2}V_{max}$. In the present example, V_{max} is 0.118 ΔA min^{-1}, the value of $[S]$ when $v = 0.059$ ΔA min^{-1} (i.e. $\frac{1}{2}V_{max}$) lies between 8.0 and 12.0 mM. Thus, the value of K_m (9.44 mM) derived from the plot is reasonable.

WORKED EXAMPLE 

The rate of hydrolysis of a penicillin analogue catalysed by β-lactamase varied with substrate concentration as shown in Table 4.6. What are the values of K_m and V_{max} for this reaction?

Table 4.6 Variation of rate of the β-lactamase-catalysed reaction with substrate concentration

[Substrate] (μM)	Rate (nmol min^{-1})
1	0.22
2	0.38
3	0.50
5	0.68
10	0.90
30	1.16

STRATEGY

The various transformations of the Michaelis–Menten equation illustrated in Figs. 4.13–4.15 can be used to derive the parameters; alternatively, non-linear regression can be used to fit the data directly to the equation.

SOLUTION

All three plots give very good straight lines indicating that the experimental data conform well to the Michaelis–Menten equation. The values obtained using the Lineweaver–Burk, Eadie–Hofstee, and Hanes–Woolf plots are: V_{max} 1.36, 1.37, and 1.36 nmol min^{-1}, respectively, and K_m 5.26, 5.18, and 5.12 μM, respectively. The non-linear regression fitting gives V_{max} = 1.36 (standard error 0.01) nmol min^{-1} and K_m = 5.10 (standard error 0.06) μM. The very low standard errors (approximately 1% of the actual values of the parameters) confirm the goodness of the fit of the experimental data to the to the Michaelis–Menten equation. The values of the parameters can be seen to be reasonable by the criteria given above. An extension of this problem is given in the self-test exercise **ST 4.10** at the end of this section.

WORKED EXAMPLE

The guanidinium ion is known to act as an inhibitor of the trypsin-catalysed hydrolysis of a model substrate, *N*-benzoyl-L-arginine-4-nitroanilide (BAPNA). Analyse the data in Table 4.7 to determine the type of inhibition being shown and evaluate the appropriate dissociation constant for the enzyme–guanidinium complex.

STRATEGY

The Lineweaver–Burk plots for the data obtained in the presence and absence of inhibitor will indicate the type of inhibition observed (see Fig. 4.16).

Table 4.7 Inhibition of the trypsin-catalysed hydrolysis of BAPNA by guanidinium ions

[BAPNA] (mM)	Velocity (μmol min^{-1} mg^{-1})	
	−guanidinium	+guanidinium (3 mM)
0.3	3.8	2.0
0.6	5.8	3.5
1.0	7.2	4.8
1.5	8.2	5.9
2.0	8.9	6.8
2.5	9.3	7.4

SOLUTION

From any of the appropriate plots (e.g. Lineweaver–Burk) it can be seen that in the presence of guanidinium the V_{max} remains constant (11.5 μmol min^{-1} mg^{-1}), but that K_m is increased (from 0.6 to 1.4 mM). Thus, guanidinium acts as a competitive inhibitor; it should be noted that the arginine side chain of the substrate contains the guanidinium group. There is a 2.3-fold increase in K_m, so $(1 + ([I]/K_{EI})) = 2.3$. Putting $[I] = 3$ mM, this gives $K_{EI} = 2.3$ mM.

The action of proteases on nitroanilide substrates is described in Chapter 9, section 9.9.4. The structures of arginine and the guanidinium ion are given in Chapter 1, section 1.8.

WORKED EXAMPLE

The data in Table 4.8 were obtained for the binding of NAD^+ to the enzyme malate dehydrogenase (15 μM), which consists of two polypeptide chains. Analyse the data using the Scatchard equation (see Chapter 2, section 2.5.2) to obtain the number of binding sites and the K_d for the interaction.

Table 4.8 Binding of NAD^+ to malate dehydrogenase

$[NAD^+]_{free}$ (μM)	r
5	0.33
10	0.57
20	0.89
30	1.09
50	1.33
75	1.50
100	1.60

STRATEGY

The data must be put into the correct form for the Scatchard plot ($r/[NAD^+]_{free}$ vs r (Table 4.9). It is important to make sure that the correct units are noted.

Table 4.9 Values of parameters for the Scatchard plot

$[NAD^+]_{free}$ (μM)	r	$r/[NAD^+]_{free}$ $(\mu M)^{-1}$
5	0.33	0.067
10	0.57	0.057
20	0.89	0.045
30	1.09	0.036
50	1.33	0.027
75	1.50	0.020
100	1.60	0.016

SOLUTION

The Scatchard plot is shown in Fig. 4.18. The data points fall on a straight line of slope $-0.04\ \mu M^{-1}$. Thus $K_d = -(-1/0.04)\ \mu M = 25\ \mu M$. The intercept on the x-axis is equal to the n, the number of binding sites. There are thus two binding sites for NAD^+ on malate dehydrogenase, i.e. one per polypeptide chain. Since the plot is linear, these sites are equivalent and independent.

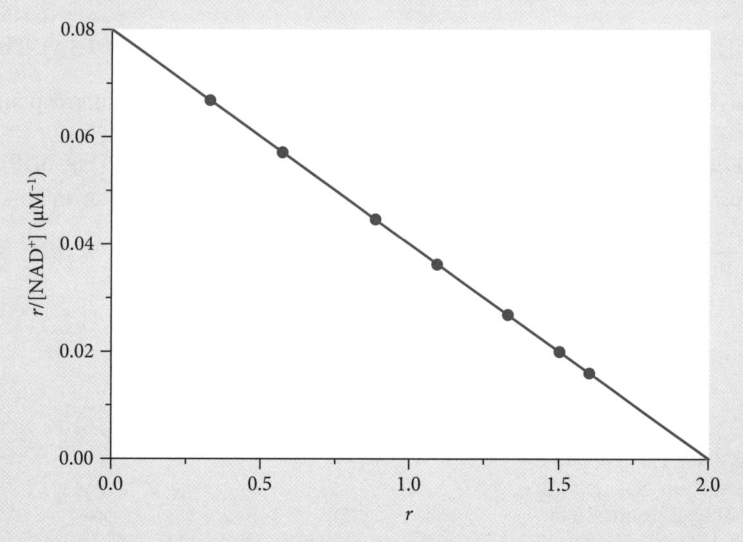

Fig. 4.18 Scatchard plot analysis of data for the binding of NAD^+ to the enzyme malate dehydrogenase in the worked example.

Note that it is essential to express the quantities derived from of the analysis of kinetic or binding data in the correct units. The units of K_m or K_d are those of concentration (M, mM, μM, etc. or in the case of a species of unknown molecular mass this might be expressed as % (w/v)). The units of V_{max} are those in which v is expressed (e.g. μmol min^{-1} or ΔA s^{-1}). The parameter r in the binding equation does not have units as it represents a ratio. The value of r when there are multiple binding sites should be correlated with other data such as the number of subunits in a multi-subunit protein (e.g. four binding sites for NADH in lactate dehydrogenase, a tetrameric protein).

If a ligand binds to multiple sites on a protein in a cooperative fashion, the binding can be conveniently analysed by the Hill equation (eqn. 4.25):

$$Y = \frac{[L]^h}{K + [L]^h} \qquad 4.25$$

Rearranging and taking logarithms of both sides, we obtain:

$$\log\left(\frac{Y}{1-Y}\right) = h \log [L] - \log K \qquad 4.27$$

This equation is of the form $y = mx + c$ (see Chapter 2, section 2.5.1). Thus, a plot of $\log\left(\frac{Y}{1-Y}\right)$ vs $\log [L]$ is a straight line of slope h, with the intercept on the $y = 0$ line equal to $(\log K)/h$. This plot is known as the Hill plot (Fig. 4.19). We can use a similar type of plot for analysing enzyme kinetic data; in this case $\log\left(\frac{v}{V_{max} - v}\right)$ is plotted against $\log [S]$.

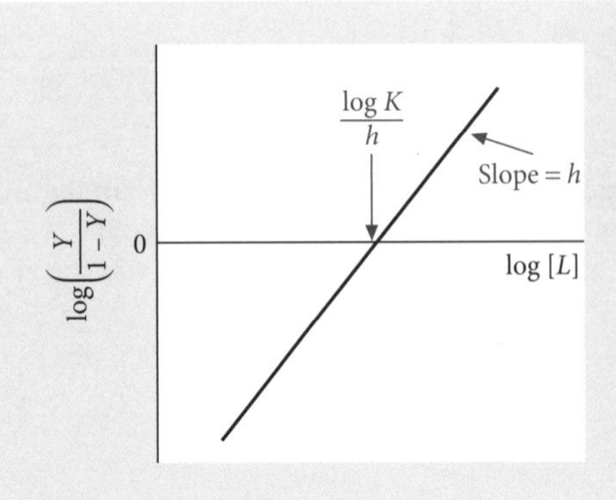

Fig. 4.19 The Hill plot for the analysis of cooperative binding of ligand (L) to a protein according to eqn. 4.27. From the slope of the line (h) and the value of $\log [L]$ at half saturation (i.e. when $Y = 0.5$ and hence $\log\left(\frac{Y}{1-Y}\right) = 0$), the value of K can be evaluated as shown.

SELF TEST ?

Check that you have mastered the key concepts at the start of the section by attempting the following questions.

ST 4.10 In a further experiment on the hydrolysis of the pencillin analogue catalysed by β-lactamase the following data were obtained. What are the values of K_m and V_{max} for the reaction?

[Substrate] (μM)	Rate (nmol min^{-1})
1	0.24
2	0.34
3	0.56
5	0.76
10	0.90
30	0.95

ST 4.11 Addition of an inhibitor (80 μM) led to no change in the V_{max} of an enzyme-catalysed reaction, but the K_m for the substrate was raised from 22 to 68 μM. What can you conclude about the interaction of the inhibitor with the enzyme?

ST 4.12 A Scatchard plot was used to analyse the binding of a ligand to a protein with four polypeptide chains. The term $r/[L]_{free}$ was plotted on the y-axis and r on the x-axis (r is the mol ligand bound per mol enzyme and $[L]_{free}$ is the concentration of free ligand). The plot showed a straight line of negative slope, with an intercept on the y-axis (i.e. when $x = 0$) of 0.058 μM^{-1}; the intercept on the x-axis (i.e. when $y = 0$) was 3.85. What can you conclude about the binding of the ligand to the protein?

Answers

ST 4.10 The experimental points do not fall on perfect straight lines in the various linear transformation plots. The best fit lines give the following parameters: Lineweaver–Burk, V_{max} 1.14 nmol min^{-1}, K_m 3.58 μM; Eadie–Hofstee, V_{max} 1.26 nmol min^{-1}, K_m 4.46 μM; Hanes–Woolf, V_{max} 1.12 nmol min^{-1}, K_m 3.29 μM. The values of K_m and V_{max} obtained by the different methods show considerable variation from each other, illustrating the problem of deciding what is the most appropriate straight line when the data do not conform exactly to the Michaelis–Menten equation. The difficulties are most likely to be caused by random experimental errors. The values obtained by non-linear regression are $V_{max} = 1.11$ (standard error 0.08) nmol min^{-1} and $K_m = 3.16$ (standard error 0.68) μM. The standard error of the estimate of K_m is very high (21.5% of the actual value), confirming the poor fit to the equation. In practice, standard errors of less than 10% (and ideally of less than 5%) of the actual values should be aimed for. It would be sensible in the present case to attempt to refine the experimental techniques and to collect further data to improve the reliability of the estimates.

ST 4.11 The inhibitor is competitive, from the change in K_m, K_{EI} can be calculated to be 38 μM.

ST 4.12 There are 3.85 binding sites per enzyme, i.e. very close to one per polypeptide chain; these are equivalent and independent. The slope of the Scatchard plot $= -\frac{0.058}{3.85}$ μM^{-1}, i.e. −0.0151 μM^{-1}. This is equal to $-\frac{1}{K_d}$, so that $K_d = 66$ μM.

Attempt Problems 4.8–4.11 at the end of the chapter.

4.5 Calculation of k_{cat} and k_{cat}/K_m for an enzyme-catalysed reaction

KEY CONCEPTS

- Calculating k_{cat} for an enzyme-catalysed reaction given appropriate data
- Calculating the ratio k_{cat}/K_m for an enzyme-catalysed reaction and understanding its importance as a measure of the efficiency of enzyme action

We can, of course, also calculate k_{cat} if we know the molar concentration of substrate converted per unit time and the molar concentration of enzyme present.

The parameter k_{cat} is a rate constant which denotes how rapidly (under saturating conditions), one molecule of enzyme can convert one molecule of substrate to product. In order to calculate k_{cat}, it is necessary to know V_{max} in terms of the number of moles substrate converted per unit time and the amount of enzyme present in terms of the number of moles. (The latter is calculated from the mass of enzyme used in the assay and the molecular mass of the enzyme.) The calculation of k_{cat} is illustrated in the worked example below.

✓ WORKED EXAMPLE

Dehydroquinase catalyses the conversion of 3-dehydroquinate to 3-dehydroshikimate and water. The V_{max} for the reaction catalysed by the type II enzyme from *Streptomyces coelicolor* is 0.011 μmol converted per min when 0.025-μg enzyme is added. The molecular mass of each subunit (i.e. each active site) is 16.5 kDa. Calculate the value of k_{cat} for this enzyme.

STRATEGY

This type of calculation involves a number of steps. It is important to set these out (with explanations) and to make sure that you keep track of the units.

Note the importance in the calculation of k_{cat} of specifying the correct molecular mass. In this case, the enzyme consists of 12 polypeptide chains (subunits) each of molecular mass 16.5 kDa. The value of k_{cat} obtained refers to the events at the catalytic site on each subunit.

SOLUTION

The V_{max} is 0.011 μmol min^{-1}, i.e. 0.011/60 μmol s^{-1} = 1.83×10^{-10} mol s^{-1} (183 pmol s^{-1}). The amount of enzyme present is 0.025 μg, which is equivalent to $0.025 \times 10^{-6}/16\,500$ mol = 1.52×10^{-12} mol (1.52 pmol). (Note 1 mol enzyme is 16 500 g.)Hence, k_{cat} = 183/1.52 pmol pmol^{-1} s^{-1} = 120 mol mol^{-1} s^{-1} (this could also be expressed as 120 s^{-1}). At a molecular level, the value of k_{cat} means that every 8.3 ms, a molecule of substrate (3-dehydroquinate) diffuses to and binds to the enzyme, is converted to products (3-dehydroshikimate and water) which then dissociate from the enzyme.

The values of k_{cat} generally range from about 10 s^{-1} for enzymes catalysing complex biosynthetic reactions to about 10^6 s^{-1} for enzymes such as catalase or carbonic anhydrase which catalyse chemically simple reactions (the breakdown of hydrogen peroxide to give water and oxygen, and the hydration of CO_2 to give carbonic acid, respectively).

The ratio k_{cat}/K_m is a useful parameter to characterize an enzyme. It can indicate the specificity of an enzyme towards a given substrate in cases where an enzyme can act on a number of possible substrates. The absolute value of the ratio can also indicate the catalytic efficiency of an enzyme. In the case of dehydroquinase in the worked example, the K_m for 3-dehydroquinate is 105 μM, so that the value of $k_{cat}/K_m = 1.14 \times 10^6\ M^{-1}\ s^{-1}$. However, for a number of enzymes, such as triosephosphate isomerase, fumarase and acetylcholinesterase, the value of k_{cat}/K_m is about $10^8\ M^{-1}\ s^{-1}$, which effectively represents the diffusion limit, i.e. the rate of the overall reaction is limited by the rate at which substrate can encounter the active site of the enzyme, with all subsequent steps (the chemical reaction, dissociation of products, etc.) occurring more rapidly. This phenomenon is sometimes referred as 'catalytic perfection'.

Although an enzyme might have achieved 'perfection' in terms of its catalytic efficiency, it may possibly still be 'improved' in terms of regulatory properties or stability in different physiological situations.

SELF TEST ?

Check that you have mastered the key concepts at the start of the section by attempting the following questions.

ST 4.13 For the reaction catalysed by fumarase (fumarate ⇌ malate + H_2O), the K_m for fumarate is 4.5 μM and the V_{max} is 45 000 mol fumarate mol^{-1} enzyme min^{-1}. Calculate k_{cat} (in units of s^{-1}) and the ratio k_{cat}/K_m (in units of $M^{-1}\ s^{-1}$) and comment on the value of this ratio.

ST 4.14 The enzyme β-lactamase catalyses the hydrolysis of penicillin and a number of its derivatives. In a 1-mL reaction volume to which 0.015 μg enzyme was added, the V_{max} was found to be 10.2 μM substrate hydrolysed per minute and the K_m was 5.2 μM. The molecular mass of the enzyme is 29.6 kDa. Calculate k_{cat} and k_{cat}/K_m for this enzyme-catalysed reaction.

Answers

ST 4.13 The values of k_{cat} and k_{cat}/K_m are 750 s^{-1} and $1.67 \times 10^8\ M^{-1}\ s^{-1}$, respectively; the latter value is close to the diffusion-controlled limit, i.e. to 'catalytic perfection'.

ST 4.14 The values of k_{cat} and k_{cat}/K_m are 335 s^{-1} and $6.44 \times 10^7\ M^{-1}\ s^{-1}$, respectively.

Attempt Problems 4.12 and 4.13 at the end of the chapter.

4.6 Radioactivity

KEY CONCEPTS

- Knowing the three types of radiation emitted by unstable nuclei
- Knowing the definitions of the units of radioactivity (Curie and Becquerel)
- Understanding the terms efficiency of counting and specific radioactivity and how they are used in calculations

The nucleus of an unstable isotope will decay to produce one of the following types of radiation, namely α (corresponding to He nuclei), β (corresponding to

Table 4.10 Properties of the radioactive isotopes most commonly used in studies of proteins

Isotope	Half-life	Maximum energy (eV)*
^{3}H	12.3 years	1.9×10^{4}
^{14}C	5730 years	1.56×10^{5}
^{31}P	14.3 days	1.71×10^{6}
^{35}S	87.4 days	1.67×10^{5}

*The electrons emitted will have a range of energies; the values shown are the maximum energies observed in each case. 1 eV (electron volt) is equivalent to 96.5 kJ mol^{-1}.

electrons), and γ (corresponding to high-energy photons). In some cases, more than one type will be produced e.g. ^{131}I decays giving both β- and γ-rays.

The much higher energy of the electrons emitted by ^{32}P means that it is very important to adopt stringent safety procedures when working with this isotope.

The principal isotopes used in biochemical research are β-emitters, although γ-emitting isotopes of iodine have a number of specialized applications, particularly in immunology. Incorporation of a radioactive isotope provides a very sensitive means of tracking a particular compound, especially in studies of enzyme-catalysed reactions or metabolism, and in measuring its binding to another molecule. Since radioactivity arises from the decay of individual nuclei, it is possible to measure very small amounts of a radioactive compound (in the pmol or fmol range). The isotopes most commonly employed in studies of proteins are ^{3}H, ^{14}C, ^{32}P, and ^{35}S, the properties of which are listed in Table 4.10. (There are no convenient radioactive isotopes of N or O available.)

The Curie is named in honour of Pierre and Marie Curie, who worked in France on the identification and purification of radioactive elements (principally polonium and radium). Marie was Polish by birth (Maria Sklodowska). The Becquerel is named in honour of Henri Becquerel, who discovered the phenomenon of radioactivity in a salt of uranium.

The amount of radioactivity in a substance is quoted in two types of units. The older, but still used, unit is the Curie (Ci), which is defined as 3.7×10^{10} disintegrations per second (dps). This corresponds to the activity of 1g pure radium. Because the Curie is such a large amount of radioactivity, we typically use much smaller quantities (mCi or μCi) in experiments. The other unit is the Becquerel (Bq), which corresponds to 1 dps; this is, conversely, such a small amount of radioactivity that we would generally be using kBq or MBq in experiments.

A radioactive compound will be supplied with a stated specific radioactivity, e.g. ^{3}H-alanine might be supplied at 50 Ci mol^{-1} (this would correspond to 1850 GBq mol^{-1}). For experimental purposes, this would usually be mixed with a large excess of unlabelled alanine to give a specific radioactivity of, for example 0.5 Ci mol^{-1}. *Knowledge of the specific radioactivity of a sample containing a radioactive compound allows us to convert the observed radioactivity into the amount (mol) of that compound present.*

Radioactivity is usually measured by liquid scintillation counting. The sample is dissolved or suspended in a suitable solvent containing one or more fluorescent compounds (known as fluors). Emission of an electron by the radioactive nucleus will excite the fluor (this can occur via excitation of a solvent molecule). Emission of radiation (scintillations) by the fluor can be detected by a photomultiplier. The efficiency of detection of the radioactive decay depends on a number of factors including the physical state of the radioactive material and the energy of the electrons emitted. In the case of ^{3}H-labelled compounds the efficiency is fairly low (typically 20–30%), but is much higher for ^{14}C-labelled, and especially for ^{32}P-labelled compounds. *We must measure the efficiency of counting to be able to convert the observed counts per second (cps) to disintegrations per second (dps)*

Radioactivity is a very sensitive tool because events could, in theory at least, be detected at very nearly the scale of individual atoms and molecules. This makes it ideally suited to the measurements of very small amounts of material such as may occur in biological systems. In addition, the introduction of radioactive isotopes does not affect the chemical behaviour of atoms significantly (although there can be effects on the rates of some reactions), so that they serve as ideal probes and tracers for molecules in complex systems. In this book, we shall confine our attention to the use of radioactive isotopes for assays of enzyme activity and for the monitoring of binding processes.

In practice, radioactivity should only be used as a probe with a small proportion of the relevant molecules containing the radioactive isotope. When an isotope decays it can generate a new element with different chemical characteristics, which may lead to damage to the molecule concerned. Thus, ^{32}P and ^{35}S decay by β-emission to generate ^{32}S and ^{35}Cl, respectively. The radioactive compound supplied would generally be mixed with a large molar excess of the non-radioactive compound to obtain the working solution.

WORKED EXAMPLE 

A sample of ATP labelled in the γ-phosphate position with ^{32}P was purchased and mixed with a solution of unlabelled ATP to give a specific radioactivity of 5 Ci mol^{-1}; this material was then stored in the freezer. After 43 days an experiment was carried out in which this solution was used to monitor the incorporation of phosphate into a protein of molecular mass 60 kDa. After 1 mL of the protein solution (1.3 mg mL^{-1}) had been incubated with the radioactive ATP and necessary factors for reaction, the mixture was treated with trichloroacetic acid to stop the reaction and precipitate the protein. The radioactivity incorporated was 25 200 counts per minute (cpm), and the efficiency of counting was 90%. What was the extent of incorporation of phosphate into the protein?

STRATEGY

Given the relatively short half-life of the ^{32}P isotope, it is necessary to calculate the specific radioactivity on the day of the experiment. Calculation of the extent of incorporation requires knowledge of the amounts (numbers of moles) of protein and of radioactivity incorporated. The working should be clearly set out and explained.

SOLUTION

The ATP sample was used 43 days after purchase; reference to Table 3.1 shows that this time corresponds to 3 half-lives of the ^{32}P isotope. Thus, on the day of the experiment, the specific radioactivity will be $0.5 \times 0.5 \times 0.5$ (i.e. 0.125) times the original, namely 0.625 Ci mol^{-1}, or $0.625 \times 3.7 \times 10^{10}$ dps mol^{-1} $= 2.313 \times 10^{10}$ dps mol^{-1}.
The radioactivity incorporated was 25 200 cpm or 25 200/60 cps = 420 cps. Since the efficiency was 90%, this would correspond to $420 \times 100/90$ dps = 467 dps
Using the value for the specific radioactivity we can calculate that this would correspond to $467/(2.313 \times 10^{10})$ mol = 20.2 nmol phosphate incorporated. The amount of protein taken was 1.2 mg; this corresponds to $1.2 \times 10^{-3}/60\,000$ mol protein = 20 nmol protein. Thus, the incorporation of phosphate amounts to 20.2/20 = 1.01 mol mol^{-1} protein.

WORKED EXAMPLE

The half-life of ^{14}C is 5730 years. What is the maximum specific radioactivity (Bq mol^{-1}) of a sample of [^{14}C] glucose, which has the molecular formula $C_6H_{12}O_6$? (Avogadro's number is 6.02×10^{23} mol^{-1}.)

STRATEGY

From the half-life, the rate constant can be calculated. When multiplied by Avogadro's number, this will give the initial rate of radioactive decay for 1 mol of ^{14}C atoms; this value has to be multiplied by 6 to obtain the rate for 1 mol glucose, which has 6 carbon atoms.

SOLUTION

The $t_{1/2}$ (5730 years) corresponds to 1.81×10^{11} s; thus the rate constant $= 0.693/(1.81 \times 10^{11})$ s^{-1}, i.e. 3.835×10^{-12} s^{-1}. The initial rate of decay for 1mol ^{14}C atoms $= 3.835 \times 10^{-12} \times 6.02 \times 10^{23}$ dps (Bq) $= 2.31 \times 10^{12}$ Bq. For glucose, the rate would be 1.39×10^{13} Bq mol^{-1} or 13.9 TBq mol^{-1} (T = tera = 10^{12}).

SELF TEST

Check that you have mastered the key concepts at the start of the section by attempting the following questions.

ST 4.15 The radioactivity in a sample of [^3H] leucine was measured as 8650 cpm; the efficiency of counting was 22%. The specific radioactivity of the sample was 1.43 GBq mol^{-1}. Calculate the dps of the sample and the number of moles of leucine present.

ST 4.16 What is the maximum radioactivity of 1 mol of ^{32}P atoms, given that the half-life is 14.3 days and Avogadro's number is 6.02×10^{23} mol^{-1}.

Answers

ST 4.15 There are 655 dps in the sample and hence 4.58×10^{-7} mol, i.e. 0.458 μmol present.

ST 4.16 The maximum radioactivity is 3.77×10^{17} Bq mol^{-1}.

Attempt Problems 4.14–4.16 at the end of the chapter.

4.7 Problems

Full solutions to odd-numbered problems are available to all in the student section of the Online Resource Centre at www.oxfordtextbooks.co.uk/orc/price/. Full solutions to even-numbered problems are available to lecturers only in the lecturer section of the Online Resource Centre.

4.1 For the formation of the complex between Mg^{2+} ions and the chelating agent $EDTA^{4-}$ at 20°C, $\Delta G^0 = -48.7$ kJ mol^{-1} and $\Delta H^0 = 13.1$ kJ mol^{-1}. Calculate the ΔS^0 for the formation of the $MgEDTA^{2-}$ complex, and comment on your answer.

4.2 For the hydrolysis of ATP (ATP + $H_2O \rightleftharpoons$ ADP + P_i), $\Delta G^{0\prime}_{310} = -31.0$ kJ mol^{-1}. In resting muscle at 310 K, the concentrations of ATP, ADP and P_i are 10, 3, and 1 mM, respectively. Is the ATP hydrolysis reaction at equilibrium in resting muscle? If not, what is the value of ΔG under these conditions?

4.3 The average daily energy requirements of an adult male human are 10 000 kJ.

Using the value obtained in Problem 4.2 for the ΔG corresponding to ATP hydrolysis and synthesis under cellular conditions, calculate the mass of ATP turned over per day. Assume that 1 mol of ATP corresponds to 550 g.

4.4 Glycogen phosphorylase catalyses the following reaction in the mobilization of glycogen reserves:

$$(\text{glycogen})_n + P_i \rightleftharpoons (\text{glycogen})_{n-1} + \text{glucose-1-phosphate}$$

for which $\Delta G^{0\prime} = 3.05$ kJ mol^{-1} at 37°C. Assuming that the concentrations of P_i and glucose-1-phosphate are equal, does the equilibrium lie in favour of glycogen synthesis or degradation at 310 K? In muscle the concentrations of P_i and glucose-1-phosphate are 10 mM and 30 μM, respectively. Does glycogen phosphorylase catalyse the net synthesis or degradation of glycogen in muscle?

4.5 The radioactivity of a sample of ATP labelled at the γ-phosphate with the isotope ^{32}P was measured at different times with the following results.

Time (days)	0	5	10	20	30	50
Radioactivity (cpm)	53 450	41 950	32 920	20 270	12 490	4740

Show that the decay of the isotope is a first-order process and determine the rate constant and half-life for this process.

4.6 When the enzyme creatine kinase is incubated with iodoacetamide, there is a complete loss of activity due to reaction of a single cysteine side chain per polypeptide chain of the enzyme. The rate of the reaction of enzyme (5 μM polypeptide chains) with different concentrations of iodoacetamide was studied by determining the remaining activity of samples withdrawn at stated times.

	Activity remaining (%)			
	0.25 mM iodoacetamide	0.50 mM iodoacetamide	0.75 mM iodoacetamide	1.0 mM iodoacetamide
Time (min)				
0	100	100	100	100
1	95	92	88	84
2	92	84	77	71
4	84	71	60	51
8	71	51	36	26
14	55	30	17	9
20	43	18	8	3

Show that the reactions obey pseudo-first order kinetics and derive the pseudo-first order rate constants in each case and the second-order rate constant for the reaction of creatine kinase with iodoacetamide under these conditions.

4.7 Some typical K_d values for interactions of biochemical interest are avidin–biotin 10^{-15} M, antigen–antibody 10^{-10} to 10^{-9} M, and enzyme–substrate 10^{-6} to 10^{-4} M. Evaluate the ΔG^0_{310} values corresponding to these dissociation constants.

4.8 What would be the effects of competitive, non-competitive and uncompetitive inhibitors on the Eadie–Hofstee and Hanes–Woolf enzyme kinetic plots?

4.9 Chymotrypsin catalyses the hydrolysis of the ester substrate *N*-acetyl-L-tyrosine ethylester (ATEE). The reaction is inhibited by indole. The following data were obtained in the absence and presence of 1.3 mM indole. Determine the type of inhibition observed and the dissociation constant for the interaction of enzyme with the inhibitor.

[ATEE] (mM)	Velocity (μmol min^{-1} mg^{-1})	
	Without indole	With indole
2.0	9.8	3.9
4.0	14.4	5.8
6.0	17.1	6.8
10.0	20.0	8.0
15.0	21.9	8.8
20.0	23.0	9.2

4.10 Dehydroquinase catalyses the conversion of 3-dehydroquinate to 3-dehydroshikimate and water. The type II enzyme from *Helicobacter pylori* is inhibited by citrate. From the following data obtained in the absence and presence of 2 mM citrate, determine the type of inhibition observed and the dissociation constant for the interaction of enzyme with the inhibitor.

[3-dehydroquinate] (μM)	Velocity (μmol min^{-1} mg^{-1})	
	Without citrate	With citrate
50	4.4	3.0
100	6.9	5.0
150	8.5	6.5
250	10.3	8.4
500	12.4	10.9
750	13.3	12.1

4.11 Benzamidine is known to bind to the active site of trypsin, leading to inhibition of the enzyme. The binding was studied under conditions when the concentration of enzyme (3.5 μM) was very small by comparison with the concentrations of benzamidine added so that essentially all the ligand can be considered free. Using an appropriate plot, analyse the following data to determine the K_d for the trypsin–benzamidine interaction.

[Benzamidine] (mM)	Y (fractional saturation)
0.1	0.29
0.2	0.44
0.5	0.67
0.8	0.76
1.2	0.83
1.6	0.87

4.12 The following data were obtained during a study of the papain-catalysed hydrolysis of a model 4-nitroanilide substrate. The reaction was monitored by the increase in absorbance at 405 nm as the 4-nitroaniline product was formed. (The molar absorption coefficient for 4-nitroaniline at 405 nm is 9500 M^{-1} cm^{-1}; the substrate does not absorb at this wavelength.)

[Substrate] (mM)	Velocity (ΔA_{405} min^{-1})
0.10	0.029
0.25	0.054
0.50	0.076
0.75	0.089
1.00	0.096
1.50	0.105

The assay volume was 3 mL in a cuvette of 1-cm pathlength. The amount of papain added was 0.8 μg and the molecular mass of the enzyme is 23 kDa. Using an appropriate plot, determine the K_m for the substrate and the V_{max} for the reaction in terms of ΔA_{405} min^{-1}. Calculate the specific activity and k_{cat} for the papain-catalysed reaction.

4.13 The V_{max} for the type II dehydroquinase from *Mycobacterium tuberculosis* is 0.0134 μmol min^{-1} when 1.7 μg enzyme is added to the assay mixture. What is the k_{cat} for the enzyme? (The molecular mass of each subunit of the enzyme is 18 kDa). If the K_m for 3-dehydroquinate is 25 μM, what is the k_{cat}/K_m ratio?

4.14 The binding of ^{14}C-labelled L-phenylalanine to the enzyme pyruvate kinase was studied. After equilibrium had been achieved, the radioactivity bound to 2 mL of a solution of enzyme (0.85 mg mL^{-1}) was found to be 6250 cpm. The specific radioactivity of the ^{14}C-L-phenylalanine was 0.15 Ci mol^{-1}, and the efficiency of counting was 70%. The enzyme has a molecular mass of 240 kDa and contains four identical subunits. What is the stoichiometry of binding of L-phenylalanine to pyruvate kinase?

4.15 The amino acid methionine has the molecular formula $C_5H_{11}NO_2S$. A sample of [^{35}S] methionine was stated to have a specific radioactivity of 1.9×10^{16} Bq mol^{-1}. What per cent of the *S* atoms in the sample are present as ^{35}S? The half-life of ^{35}S is 87.4 days and Avogadro's number is 6.02×10^{23} mol^{-1}.

4.16 Ornithine decarboxylase catalyses the reaction: ornithine $\rightleftharpoons$ 1,4-diaminobutane (putrescine) + CO_2. The reaction can be monitored by measuring the release of $^{14}CO_2$ from ornithine which is labelled with ^{14}C at the carbon atom of the carboxyl group. In a reaction mixture of volume 0.4 mL, the concentration of [^{14}C]-ornithine was 10.8 μM. The specific radioactivity of the [^{14}C]-ornithine was 5.8 Ci mol^{-1}. After 10 min, the reaction was stopped and the radioactivity released as $^{14}CO_2$ was found to be 14 580 cpm; the efficiency of counting was 75%. What was the rate of the reaction expressed as nmol product formed per minute and what proportion of the original substrate present had been converted to products after 10 min?

References for Chapter 4

Price, N.C. and Stevens, L. (1999) *Fundamentals of Enzymology*, 3rd edn. Oxford University Press, Oxford, 478 pp.

Price, N.C., Dwek, R.A., Ratcliffe, R.G., and Wormald, M.R. (2001) *Principles and Problems in Physical Chemistry for Biochemists*, 3rd edn. Oxford University Press, Oxford, 401 pp.

Appendix

Appendix 4.1 Binding of ligands to macromolecules: saturation curves

Consider the binding of ligand L to a protein P which has a single binding site:

$$\mathrm{P} + \mathrm{L} \rightleftharpoons \mathrm{PL}$$

For which $K_d = \dfrac{[P] \times [L]}{[PL]}$, hence $[PL] = \dfrac{[P] \times [L]}{K_d}$

At a given concentration of free ligand ($[L]$), the fractional saturation of the binding sites (Y) equals the concentration of bound ligand divided by the total concentration of protein.

Hence $Y = \dfrac{[PL]}{[P] + [PL]}$

$$Y = \frac{\dfrac{[P] \times [L]}{K_d}}{[P] + \dfrac{[P] \times [L]}{K_d}}$$

Dividing each term in the numerator and denominator by $[P]$:

$$Y = \frac{\dfrac{[L]}{K_d}}{1 + \dfrac{[L]}{K_d}}$$

Multiplying each term in the numerator and denominator by K_d

$$Y = \frac{[L]}{K_d + [L]} \qquad 4.22$$

When there are n binding sites for ligand on the protein, the fractional saturation equation is analogous to eqn. 4.22. It is expressed in the following terms:

$$r = \frac{n \times [L]}{K_d + [L]} \quad 4.23$$

where r is the average number of molecules of ligand bound per molecule of protein, n the number of binding sites, $[L]$ is the concentration of free ligand, and K_d is the dissociation constant.

It is, however, considerably more complex to derive this equation since it is necessary to take into account the statistical nature of the relationship between the successive dissociation constants for binding of ligand. (This is discussed in Chapter 6 of the book by Price and Stevens (1999)).

In the case of enzyme kinetics we can consider the fractional saturation of the active sites of the enzyme by the substrate S. One way of calculating the fractional saturation is to make the assumption that in the mechanism $E + S \rightleftharpoons ES \rightarrow E + P$ the breakdown of ES to yield E + P does not perturb the $E + S \rightleftharpoons ES$ equilibrium significantly. In this case, the fractional saturation can be written as:

$$Y = \frac{[S]}{K_d + [S]}$$

In this case the K_d is denoted by the K_m (Michaelis constant), so:

$$Y = \frac{[S]}{K_m + [S]}$$

Complete saturation of the active sites corresponds to the limiting rate (V_{max}), so the actual rate (v) at a substrate concentration ($[S]$) is given by $\frac{v}{V_{max}}$, which corresponds to the fractional saturation, i.e. Y.

$$\text{Hence } \frac{v}{V_{max}} = \frac{[S]}{K_m + [S]}, \text{ so}$$

$$v = \frac{V_{max} \times [S]}{K_m + [S]} \quad 4.26$$

We can also derive an equation of this form by making the steady state assumption, namely that the concentration of the intermediate (ES) remains constant, i.e. the rate at which it is formed equals the rate at which it is broken down. In this case, K_m represents a function of various rate constants of the individual steps in the reaction. Only in the limiting case (i.e. that the rate of ES breakdown to give E + P equals zero) does K_m become equal to the dissociation constant of ES to give E + S. (For further details see Chapter 4 of Price and Stevens (1999).)

B

Goals and methods

The important properties of proteins and how to explore them

5

5.1 Introduction: the context for studies and data analysis

KEY CONCEPT

- Appreciating the complexity of cellular systems in terms of the numbers of distinct protein species present

'There's no big mystery to being an enzymologist. All you have to have is a razor blade and a liver'. Gordon Tomkins to Julius Axelrod, circa 1950.

The characterization of proteins, including enzymatic proteins, requires a good understanding of their important properties and the approaches employed to explore them. This chapter, underpinned by the conceptual toolkit in Chapter 1, aims to enhance your understanding of the goals sought and methods used in characterizing proteins. The simplicity of the approach suggested by Gordon Tomkins fails to convey the challenges associated with applying the key concepts and tools, outlined in Chapters 2–4, to the characterization of proteins within complex biological systems. A typical eukaryotic genome may well have in excess of 10 000 genes which, in turn, can encode probably over 10 times as many distinct proteins, resulting from differential RNA processing and post-translational modification. To establish the role of the several hundred thousand proteins would require the structural and functional characterization of each one. The enormity of this task is compounded by the diverse nature of protein function and structure. In section 5.2, we shall consider how to establish the function and structure of a protein and how its activity may be regulated. In section 5.3, the range of assays used to monitor the biological activity of proteins is outlined; such assays underpin protein characterization and purification. The classical approach to studying proteins requires their purification from their native source (section 5.4.1) using bespoke purification procedures for individual proteins. This task has been simplified greatly with the advent of protein expression in recombinant systems (section 5.4.2). In addition, the overexpression of recombinant proteins has enhanced the structural characterization of proteins, as reflected in the large

As examples, the genomes of the eukaryotic species budding yeast *Saccharomyces cerevisiae*, nematode worm *Caenorhabditis elegans* and human contain about 6000, 12 000 and 23 000 genes, respectively.

As described in section 5.8, the PDB is the database for all three-dimensional structures of proteins solved to atomic resolution by X-ray crystallography or nuclear magnetic resonance (NMR). As of July 2008, there were approximately 47 000 structures in the PDB, of which about 85% had been solved by X-ray crystallography. It should be noted that many structures in the PDB refer to the same protein in different forms or complexes and that relatively few structures of membrane proteins have been determined.

increase in the number of structures deposited in the Protein Data Bank (PDB) in recent years. Section 5.5 will present a brief outline of the methods employed to determine the structures of proteins.

Having established the structure and function of a protein, it is important to understand how it is regulated within the cellular environment and the types of interactions in which it is involved. Typically, this is achieved by monitoring the effects of a number of physical and chemical variables on protein activity, as described in sections 5.6 and 5.7. In section 5.8, we shall consider the use of bioinformatics in exploring the properties of proteins and, finally, experimental design will be outlined in section 5.9.

This chapter should lead to a sound understanding of the goals and methods employed to separate, identify, and characterize proteins that will enable data acquisition and handling in the specific examples outlined in subsequent chapters.

5.2 The key questions about a protein

KEY CONCEPTS

- Being aware of the range of functions of proteins
- Understanding the levels of protein structure
- Identifying appropriate methods to explore protein structure and protein interactions

All proteins share the common structural feature of being composed of amino acids which are linked by peptide bonds (Chapter 1, sections 1.1–1.3); however, it is the *sequence* of amino acids within a given protein that dictates its unique function and structure. The characterization of a novel protein requires an appreciation of the diversity of protein function and structure. It is also important to appreciate that biological systems are not static entities; they respond to environmental, developmental and metabolic signals with concomitant changes in the structure and function of proteins associated with these processes. Finally, while it is convenient to study proteins in isolation (section 5.4), it must be remembered that they almost always occur within complex cellular environments, interacting with other proteins, metabolites and cellular structures. Thus, to achieve complete characterization, we need to establish how proteins interact with other molecules under physiologically relevant conditions.

5.2.1 What is the function of the protein?

Proteins fulfil a diverse range of roles within the cell which can be categorized into the following general groups:

Catalytic proteins: Within a typical cell, around 60% of all proteins are catalytic proteins, known as enzymes. Enzymes enhance the rates of specific reactions and are subject to regulation by interaction with other molecules (known as effectors) and covalent modification (section 5.6). The activity of an enzyme is also affected by changes in pH, temperature and substrate concentration.

The word enzyme was coined by Kühne in 1878 and is derived from the Greek meaning 'in yeast', since many of the early studies of catalysis were performed using extracts of yeast.

An extreme example of the catalytic power of enzymes is provided by the antioxidant enzyme catalase that catalyses the following reaction:

$$H_2O_2 \rightleftharpoons H_2O + \tfrac{1}{2}O_2$$

One molecule of catalase from the yeast *Pichia pastoris* breaks down 850 000 molecules of hydrogen peroxide per second, that is it is at least a million-fold more effective than the chemical catalyst, iron. This catalytic power arises from the ability of enzymes to lower the free energy barrier ($\Delta G^{\ddagger}$) of a given reaction (see Chapter 4, section 4.2). The 'energy profile' of a typical reaction (Chapter 4, Fig. 4.1) indicates that an energy barrier limits the rate of conversion of reactants to products. The transition state of a reaction is represented by the highest point of this energy profile. Enzymes speed up reactions by offering an alternative reaction pathway with a lower energy barrier, i.e. lower the energy of activation ($\Delta G^{\ddagger}$).

Catalase contains iron in the form of a haem group in its active site. The environment of this iron, which is generated by the structure of the protein, is the key to the catalytic power of the enzyme.

Binding proteins: Like enzymes, binding proteins are highly specific, interacting only with target binding partners, which may take the form of anything from a small molecule to a complex macromolecular structure; these partners are collectively known as ligands. However, unlike enzymes, binding proteins usually display no catalytic activity towards their binding partner. Antibodies are good examples of binding proteins, with each antibody typically presenting two binding sites for antigens (Fig. 5.1). Antibodies play a central role in the adaptive immune response of vertebrates by recognizing and binding antigens presented on the surface of viruses, bacteria and other infectious agents.

The thermodynamic and kinetic aspects of protein–ligands interactions are discussed in Chapter 4, sections 4.2 and 4.3.

The multiple binding sites on antibodies can generate high affinities for antigens and lead to large cross-linked species which are insoluble, thus aiding their destruction.

Transport proteins: Transport proteins bind ligands in one location and release them at a target location. In the case of the well-characterized transport protein haemoglobin, whilst blood is circulating around the lungs, haemoglobin within red blood cells is responsible for binding oxygen; oxygen is then released as blood circulates at locations around the body, in which the oxygen levels are low. The efficiency of oxygen binding and release arises from the interaction between the binding sites, or cooperativity, within this tetrameric protein (i.e. a protein consisting of four subunits or polypeptide chains; see Chapter 4, section 4.3.2).

Structural proteins: Subcellular, cellular and multicellular structures are maintained by a range of structural proteins. This group of proteins constitutes the most abundant proteins within biological systems and includes proteins such as collagen, keratin and the cytoskeletal proteins (e.g. actin and spectrin). Actin (see Fig. 5.2), is the most abundant protein within the cytoplasm of eukaryotic cells, providing a scaffold to maintain and remodel cellular structures.

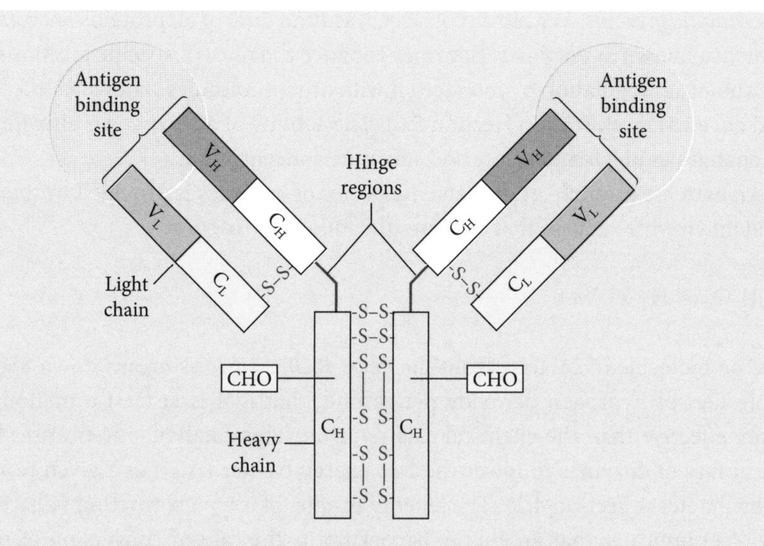

Fig. 5.1 Representation of general structure of an antibody molecule with the antigen binding sites located at the N-termini of the heavy (V_H) and light (V_L) chains. The constant regions (C_H and C_L) are linked by disulphide bonds (S–) and contain a carbohydrate modification (CHO).

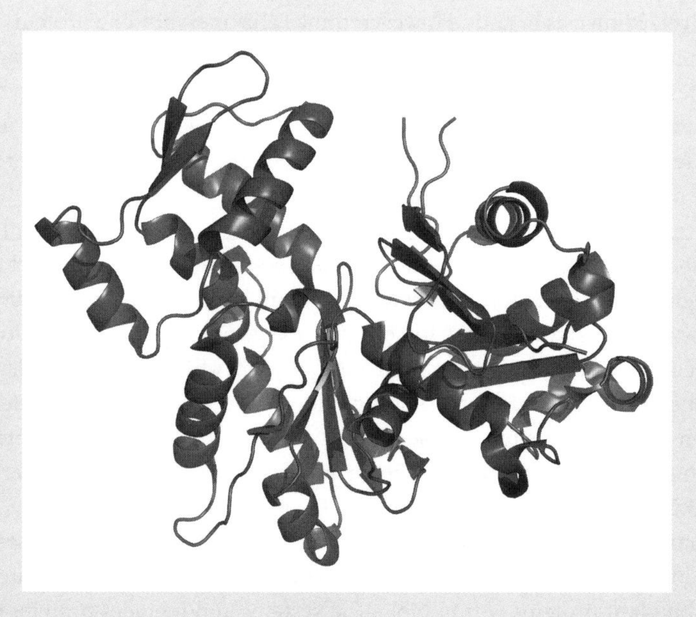

Fig. 5.2 Structure of a monomeric form of actin (Protein Data Bank code 2A5X). This structural protein usually occurs as part of multimeric structures, such as actin filaments.

Signalling proteins: Communication between cells relies on the production of signal molecules in one cell type that are detected by receptors (proteins) located on the surface of a second cell type or target cell; the interaction between any given signal and its receptor leads to a cellular response, a process termed 'signal transduction'. Cellular signal molecules can take the form of small molecules (such as adrenalin, also known as epinephrine) or macromolecules. Insulin, a protein produced in the β cells of the pancreas is one such signal molecule which is detected by insulin receptors located in the cell membrane of many cell types. Insulin is produced in response to high blood glucose levels and the subsequent binding of the hormone to insulin receptors leads to a reduction in blood glucose.

The terms adrenalin(e), which is used in the UK and Europe, and epinephrine (used in the USA) are both derived from the location of the secretory adrenal gland that is adjacent to the kidneys. The roots are: Latin, *ad* (against), *ren* (kidney) or Greek, *epi* (close to), *nephros* (kidney).

Motor proteins: Movement within biological systems is a process that is accompanied by the utilization of ATP. A number of motor proteins, including dynein and myosin, can harvest the energy released by ATP hydrolysis to generate movement. Within muscle structures, ATP hydrolysis drives the movement of myosin relative to actin filaments resulting in muscle contraction.

It has been estimated that the actin–myosin system in skeletal muscle can be up to about 60% efficient in converting the chemical energy of the fuel (ATP hydrolysis) into mechanical work. This is more efficient than power stations powered by coal or gas, which can achieve efficiencies in the range 30–40%.

Storage proteins: Storage proteins fulfil an essential role within the cell by storing minerals or essentially acting as a source of amino acid nutrients, poised for release in response to an appropriate metabolic or developmental signal. Seed storage proteins are released and degraded on seed germination to provide essential nutrients for the developing seedling. A number of storage proteins, such as the iron storage protein, ferritin, sequester ligands, which may prove toxic to the cell. Iron would tend towards its toxic ferric state within biological systems; however, this essential mineral is stored safely within proteins such as ferritin until it is required for processes such as haem synthesis.

Under physiological conditions, iron can exist in the ferrous (Fe^{2+}) or ferric (Fe^{3+}) state. The ferric state forms large complexes with anions and hydroxide ions, which are highly insoluble and toxic. The role of ferritin is to sequester iron in the ferric state, complexed with phosphate and hydroxide ions, until it is required for processes such as haem biosynthesis. Ferritin is composed of 24 identical subunits which associate to form a hollow spherical structure. Each 24-mer can accommodate up to 4500 iron ions within this hollow structure (see Fig. 5.3).

In all cases, the function of each of these protein groups is dependent on the structure of the protein, i.e. form fits function: catalytic proteins have residues in and around the active site, which present an environment to promote specific chemical reactions and binding proteins present specific binding sites to allow recognition and binding of target molecules.

Increasingly, the function of a novel protein is determined by establishing its amino acid sequence (either directly using amino acid sequencing or indirectly by translating the sequence of the gene encoding the novel protein) and then conducting a homology search of the ever-expanding sequence databases (section 5.8.5). Confirmation of the probable function requires the purification of the protein and an assay to test its function, e.g. an enzyme assay to measure the activity of the putative catalytic protein. Assays can also be used to test the possible influence of environmental and metabolic effectors on the activity of the protein. A complete understanding of protein function requires solving (or prediction) of its structure.

Fig. 5.3 24-mer structure of the iron storage protein ferritin (Protein Data Bank code 1FHA). 24 identical subunits pack together to form a hollow shell that can accommodate up to 4500 iron ions in the ferric (Fe^{3+}) state.

5.2.2 What is the structure of the protein?

Over the past 20 years, our understanding of protein structure has been greatly enhanced by the near exponential growth in the number of structures that have been solved (see Fig. 5.4). A number of factors have contributed to this growth, including: the ability to overexpress many proteins (which has provided the quantity and quality of material required for structural studies), enhanced computing capabilities and developments in the biophysical techniques employed to determine protein structure. Complete structural characterization of a protein requires:

- *Determination of the amino acid sequence*: This may be deduced from the nucleotide sequence of a gene encoding a particular protein or from direct amino acid sequencing (Chapter 8, section 8.3). The amino acid sequence of a protein can be used to generate a wealth of structural information, including the

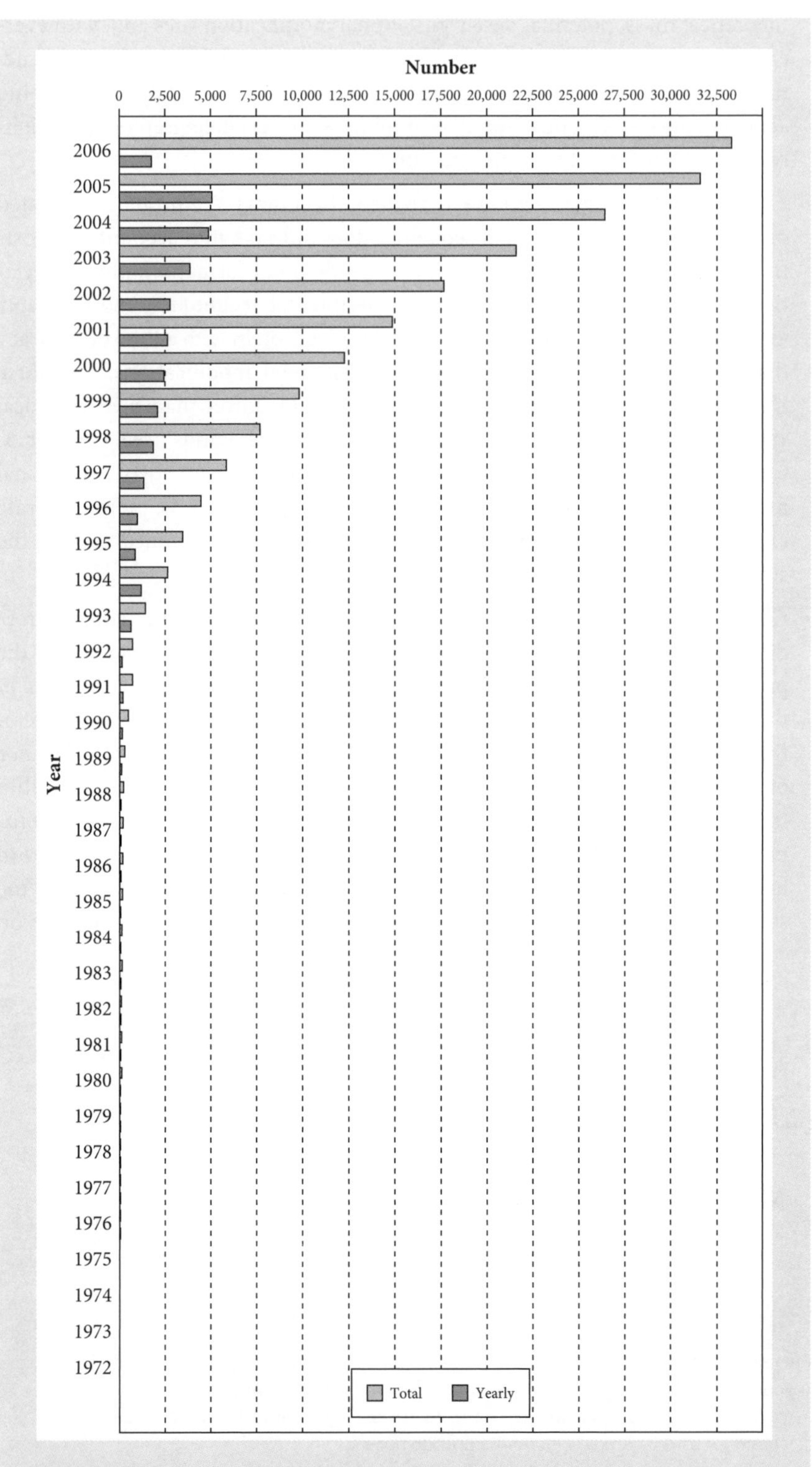

Fig. 5.4 Growth of protein structures solved over the past 35 years.

Many of the DNA-derived protein sequences in the databases correspond to proteins of unknown function or proteins which may not be found in cells, certainly under normal growth conditions.

theoretical mass, potential post-translational modification sites and, with ever-expanding sequence databases and analysis tools, the ability to identify structural and functional motifs (see Chapter 1, section 1.3.2). In addition, the amino acid sequence is essential for determining X-ray crystallographic and high-resolution NMR spectroscopic structures.

- *Experimentally determined mass*: The theoretical mass of a protein, calculated on the basis of the amino acid sequence, does not take into account any post-translational modifications, such as disulphide bond formation, phosphorylation, glycosylation, and proteolysis. Experimental techniques, most notably mass spectrometry, can determine the mass of a protein with a level of accuracy that permits characterization of post-translational modifications, e.g. horse lysozyme has an experimental mass which is 8 Da lower than the theoretical value, indicating the formation of four disulphide bonds (see Chapter 1, section 1.3.2.5). In the case of proteins with more complex post-translational modifications, e.g. the hormone protein, insulin (see Fig. 5.5), more elaborate experiments are required to relate the experimentally determined mass to the structure of the post-translationally modified protein.
- *Characterization of secondary and tertiary structure*: Proteins adopt three-dimensional structures that are dictated by the amino acid sequence of the protein (see Chapter 1, sections 1.4 and 1.5). An inspection of the tens of thousands of structures which have been solved to date reveals a vast range of three-dimensional structures, from the simplicity of defensin to the elegance of ATP synthase (Fig. 5.6), which enable proteins to fulfil specific roles within the cell. Despite the variety of three-dimensional structures, a number of common principles determine the structures adopted by proteins: residues with hydrophobic amino acid side chains tend to be buried in the interior of the structure, whereas residues with hydrophilic side chains tend to be exposed on

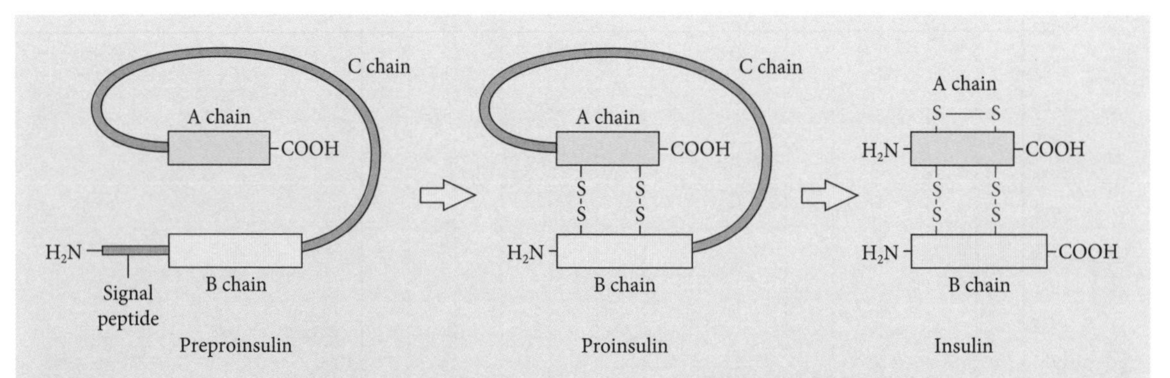

Fig. 5.5 Maturation of insulin from preproinsulin. The 24 amino acid signal peptide is initially cleaved from the N-terminus of preproinsulin to generate proinsulin. Two disulphide bonds form within proinsulin, which is then proteolytically cleaved to remove a stretch of 51 amino acids (C-chain) to generate the mature form of insulin, which contains two polypeptide chains (A and B) linked by two disulphide bonds.

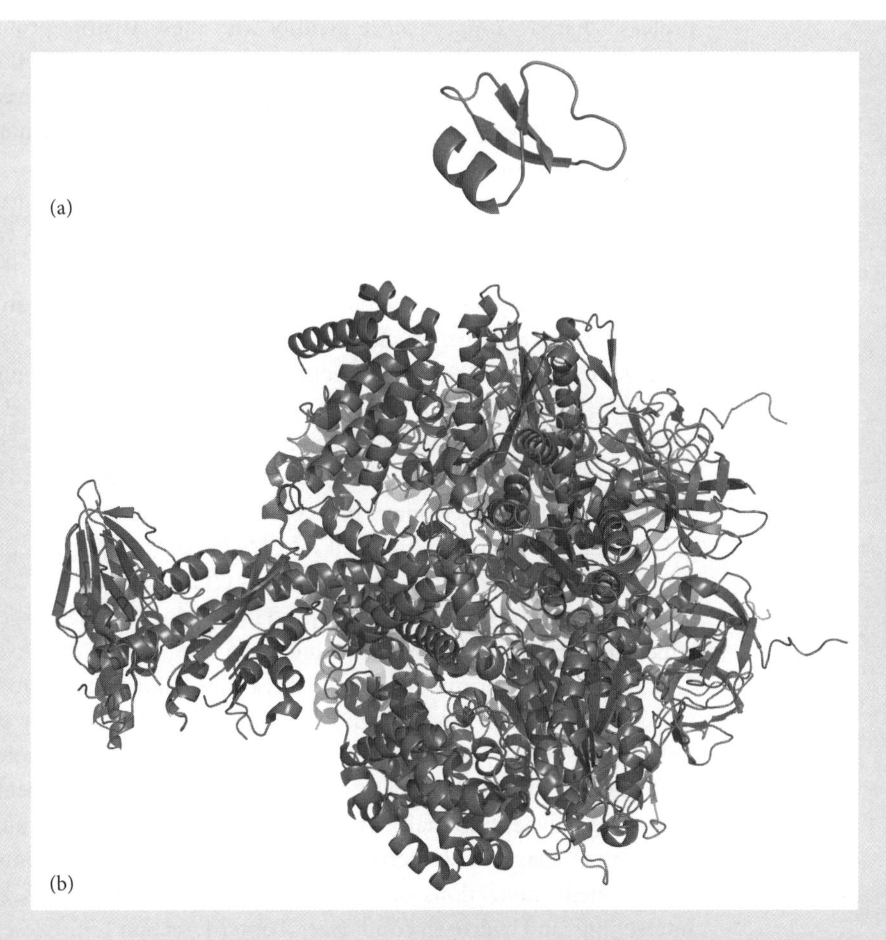

Fig. 5.6 Structures of (a) human β-defensin-1 and (b) ATP synthase. Defensin (Protein Data Bank code 1IJU) is a small antimicrobial protein with a relatively simple structure, whereas ATP synthase (Protein Data Bank code 1QO1) is an elegant machine that converts the energy associated with a proton motive force to generate ATP.

the surface; they adopt close-packed structures and contain elements of secondary structure as a result of main chain polar groups forming hydrogen bonds (see Chapter 1, section 1.5.1). The predominant forms of secondary structure are α-helices and β-sheets. The secondary structure of a protein can be inferred theoretically from its amino acid sequence or it can be measured experimentally.

Theoretical methods employed to determine the secondary structure include homology modelling and prediction methods. To determine the secondary structure of an unknown protein using homology modelling requires a similar protein (>25% sequence identity) of known structure which can serve as a model to establish the structure of the unknown protein (Sander and Schneider, 1994). As an example, prior to solving the structure of HIV protease, the structures of aspartic proteases from a number of sources had been determined. As HIV protease shared >25% sequence identity with these aspartic proteases, it was possible to use homology modelling-based methods to predict the secondary structure of HIV protease. Indeed, this predicted structure proved pivotal in the design of drugs which inhibited HIV protease, which in turn inhibited the replication cycle of HIV.

The preferences of amino acids for types of secondary structure are mentioned in Chapter 1, section 1.4.4.

In the absence of a similar protein, prediction methods can be used to predict the secondary structure of the unknown protein. To date, two types of prediction methods are used: statistical analysis of the sequence of the unknown protein to calculate the likelihood of a given amino acid to occur in a particular type of secondary structure element (Chou and Fasman, 1974) or techniques such as PHD (Rost, 1996), which generate multiple sequence alignments (with proteins of lower levels of identity). This alignment is submitted to a neural network system to predict the secondary structure of the unknown protein. With accuracies of up to 72%, such prediction methods are a useful tool to complement experimental estimates of secondary structure content.

CD is based on the difference in absorption of right and left circularly polarized components of plane-polarized light and can be used to detect chirality (optical activity) in molecules. The regular secondary structures of proteins such as α-helix and β-sheet are chiral.

One technique which has proved valuable in estimating the secondary structure content of proteins is circular dichroism (CD) (Kelly *et al.*, 2005). Regular secondary structure elements within proteins produce characteristic CD spectra which arise from absorption at 190 and 220 nm by peptide bonds (see Fig. 5.7). Ultimately, the determination of the three-dimensional structure of a protein by X-ray crystallography or high-resolution NMR spectroscopy provides not only a measure of the secondary structure content, but also molecular detail of the length and spatial arrangement of secondary structure elements within the folded polypeptide chain. The overall three-dimensional arrangement of a polypeptide chain, i.e. its tertiary structure, is maintained by multiple weak, non-covalent interactions such as electrostatic, van der Waals', hydrophobic interactions, and hydrogen bonds (see Chapter 1, section 1.5). These interactions are also involved in maintaining subunit–subunit contacts, i.e. the quaternary structure of proteins.

- *Quaternary structure*: In general, larger proteins (typically >50 kDa) tend to exist as multiple subunits giving rise to the level of structure known as the quaternary structure. The quaternary structure can range in complexity from two identical subunits, e.g. ribulose-bisphosphate carboxylase (Rubisco: EC 4.1.1.39) from photosynthetic bacteria (2 × 55 kDa) to multiple non-identical subunits, e.g. Rubisco from plants and algae which exists as hexadecamer with eight large subunits and eight small subunits (8 × 55 kDa and 8 × 15 kDa). Characterization of multi-subunit proteins requires determination of the overall molecular mass of the protein, identification of the types of subunits and the molecular mass of each type, calculation of the number of each subunit type within the protein and the structural arrangement of the subunits. The molecular mass of multi-subunit proteins must be determined under non-denaturing conditions to maintain subunit–subunit interactions, using techniques such as gel filtration chromatography and ultracentrifugation (Chapter 8, section 8.2). Analysis by SDS-PAGE

Rubisco catalyses the first dark reaction of photosynthesis (i.e. the addition of CO_2 to the five-carbon sugar ribulose bisphosphate). It is a relatively inefficient enzyme in catalytic terms and is thought to be the most abundant protein on earth.

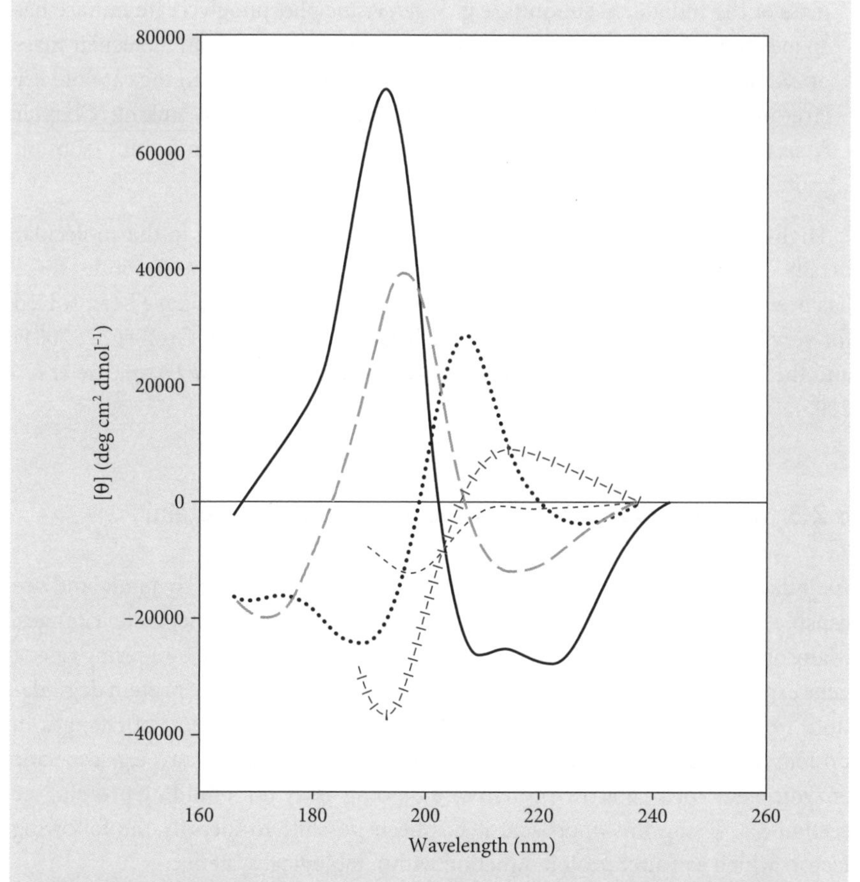

Fig. 5.7 Far-UV CD spectra of various types of protein secondary structure. Solid line, α-helix; long dashed line, anti-parallel β-sheet; dotted line, type I β-turn; cross dashed line, extended 3_1-helix or poly (Pro) II helix; short dashed line, irregular structure.

(see Chapter 8, section 8.2.1) can reveal the molecular mass of individual subunits and the number of subunit types within a multi-subunit protein: if only one species appears on an SDS-PAGE gel, it usually indicates that only one type of subunit exists within the protein; two species can suggest two subunit types, and so on. It is worth noting that there may be instances when non-identical subunits may have similar mobilities on SDS-PAGE (i.e. subunits of similar size but different amino acid sequence appear as one band) and further analysis, such as mass spectroscopy or amino acid sequencing, will be required to confirm whether all subunits are identical or not. It is often possible to determine the number of subunits within a multi-subunit protein knowing the molecular mass of the multi-subunit protein, the number of subunit types and the molecular

It should be noted that SDS-PAGE is not a high-resolution technique and it is doubtful if two proteins which had subunits of similar molecular masses (within 2%) would be efficiently separated by this technique. Mass spectrometry (see Chapter 8, section 8.2.5) would be useful for resolving proteins of similar masses.

mass of the individual subunits, e.g. *S. cerevisiae* phosphoglycerate mutase has an overall molecular mass of 110 kDa, one subunit type with an molecular mass of 27.5 kDa, which suggests it exists as a tetramer. However, in the case of very large proteins with multiple subunits, techniques such as cross-linking (Chapter 8, section 8.5) can determine the number of subunits within a multi-subunit protein.

High-resolution X-ray crystallography is required to provide the molecular details of subunit arrangements within multi-subunit proteins. Recently, there has been a marked increase in the number of structures which have been solved for very large multi-subunit proteins, such as the proteasome (Groll *et al.*, 2001) and the heterodecameric RNA polymerase II from *S. cerevisiae* (Armache *et al.*, 2005).

5.2.3 What factors might affect the function of the protein?

The turnover rates of different proteins in eukaryotic cells vary enormously, with some proteins having a very short half-life of only a few minutes, whereas others exist within the cell for weeks. The half-life of a protein depends on its function, e.g. proteins involved in the regulation of cell division and transcription have a short half-life (minutes), whereas central metabolic enzymes have a long half-life (several days).

As indicated at the start of section 5.2, biological systems are dynamic and are capable of responding to environmental, developmental and metabolic changes. Many of these changes affect the amount of protein present (e.g. by altering rate of gene expression (transcription and translation) and/or the rate of protein degradation) or by altering the activity of that protein. Characterization of changes in protein function, in response to a factor, requires a biological assay, e.g. a specific enzyme assay for a catalytic protein or a binding assay for a binding protein, see section 5.3. Using this approach, it has been possible to identify the following factors which can alter protein function within biological systems:

Non-covalent binding of other molecules ranging from small molecules to regulatory protein subunits and specific macromolecules, collectively known as *effectors*. It is possible to characterize these interactions *in vitro* using binding assays (see Chapter 6, section 6.4) coupled with appropriate analysis of the saturation curves (see Chapter 4, section 4.3).

Availability of ligands, including substrates for enzymes or cofactors, which can vary greatly within the cell. This effect is most obvious when varying the substrate concentration at values close to the Michaelis constant (K_m), which can lead to sizeable changes in activity (see Chapter 4, section 4.3.3 and Fig. 4.8).

For example, antibodies are available commercially that can distinguish between phosphorylated side chains of serine and tyrosine in proteins.

Reversible covalent modification, involving the addition and removal of specific chemical groups, can have a dramatic effect on the property of a protein. Many of these modifications are listed in Chapter 1, section 1.3.2.5 and can be readily identified by mass spectrometry or binding of antibodies which recognize specific post-translational modification groups such as the phosphoryl group.

Irreversible covalent changes, including the targeted proteolytic cleavage of inactive precursors to generate functionally active proteins, can be characterized using SDS-PAGE, size exclusion chromatography and mass spectrometry.

5.2.4 How does the protein interact with other molecules?

By considering the complexity of the cellular environment, together with the concentration of macromolecules within the cell (a typical prokaryotic cell has a protein concentration >200 mg mL^{-1}, and even higher concentrations are found in erythrocytes (>300 mg mL^{-1}) and the mitochondrial matrix (>500 mg mL^{-1})), it soon becomes clear that proteins do not exist in isolation. Proteins interact with many different types of molecules within the cell via non-covalent interactions such as hydrogen bonds, ionic, van der Waals', and hydrophobic interactions (see Chapter 1, section 1.7). As noted in Chapter 4, section 4.3 and in section 5.2.1 the types of molecules with which proteins interact include other proteins, nucleic acids, lipids, low molecular mass molecules, and substrates (for enzymes), and are collectively referred to as *ligands*.

To address how proteins interact with ligands, we should aim to determine:

- the three-dimensional structure of the protein–ligand complex
- the site on the protein which interacts with the ligand and the molecular details of the interaction
- the number of ligands interacting with the protein (stoichiometry)
- the strength of the interaction and the rate constants involved.

Binding studies, or enzyme kinetic studies for catalytic proteins (as described in Chapter 4, section 4.3), provide information relating to the stoichiometry, rate constants, and the strength of the interaction, whereas structural studies, such as site-directed mutagenesis and side chain modifications (see Chapter 9, section 9.10), indicate the amino acids that are important for protein–ligand interactions.

Assays for biological activity

KEY CONCEPT

- Appreciating the range of assays for biological activity measurements and their limitations

A specific assay is required for purification of a protein and to address key questions, relating its structure and function. In general, an assay is used during protein purification to gauge the success of the purification protocol, i.e. enhancement of the specific activity and maintenance of a high yield of biologically active protein (Chapter 3, section 3.9). Assays are also used to assess the effect of factors which may influence the biological activity of a protein, providing information relating to the function and structure of the protein. A good assay is one which is quick, simple to conduct, highly specific for the protein of interest, and relatively inexpensive. Before looking at some specific examples of different assay types, we must consider the care which must be exercised when interpreting assay data in general:

Is the activity measurement due solely to the presence of the protein of interest? There may be other factors present which may contribute to the activity measurement. A series of control assay measurements in the absence of biologically active protein will indicate whether the assay components make a contribution to the measured response, e.g. an increase in absorbance arising from non-enzyme-catalysed substrate degradation in an enzyme assay, or non-specific binding partners in immunoblots and ELISAs. The quality of assay components is critical to the success of the assay. A good example of this is provided by the enzyme-coupled assay for the glycolytic enzyme, phosphoglycerate mutase (PGAM): 3-phosphoglycerate is converted to 2-phosphoglycerate by PGAM and then 2-phosphoglycerate is then converted to phosphoenolpyruvate (PEP) by the coupling enzyme enolase (see Fig. 5.8). The assay is started by the addition of PGAM, with the resultant formation of phosphoenolpyruvate (PEP) monitored by measuring the increase in absorbance at 240 nm. During a study of yeast PGAM, it was noted that prior to the addition of this enzyme to the assay, some unexpectedly high-activity measurements were obtained. Further analysis revealed that the commercial preparation of enolase (purified from horse liver) was contaminated with PGAM.

In a coupled assay, the coupling enzyme(s) is (are) added in a large excess, perhaps 50-fold more than the enzyme being assayed. Any coupling enzymes should therefore be extremely pure.

Is the activity measurement proportional to the amount of protein present? If twice as much protein is added to the assay, is a doubling of the activity observed? If the answer is no, this may reflect the fact that some component of the assay is limiting the activity measurements.

Such limiting factors could include an insufficiency of a coupling enzyme, or that the reaction occurs too quickly for the detection system to give an accurate measurement of the rate.

The most convenient type of assay is the continuous assay in which a response is measured directly, e.g. change in absorbance, fluorescence or pH, following the addition of protein and the response can be recorded continuously throughout the course of the reaction. Less convenient, but no less useful, are discontinuous assays that involve reactions which are initiated by the addition of a protein and then samples are removed at specific time intervals. These samples are subsequently quenched (i.e. biological activity is stopped) and analysed.

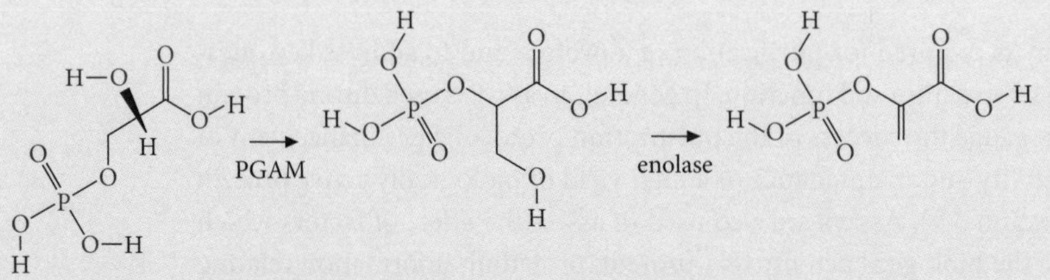

PGAM → enolase →

Fig. 5.8 Phosphoglycerate mutase (PGAM) assay contains 3-phosphoglycerate and the coupling enzyme enolase. The reaction is initiated by the addition of PGAM, which converts 3-phosphoglycerate to 2-phosphoglycerate and then enolase converts this to phosphoenol pyruvate (PEP), which is detected at 240 nm.

5.3.1 Catalytic proteins (enzymes)

The aim of assaying an enzyme is to measure the rate of product formation or substrate utilization. This typically relies on a difference in the spectroscopic properties between the substrate and the product. For example, in the following reaction catalysed by alcohol dehydrogenase (EC 1.1.1.1):

$$\text{ethanol} + \text{NAD}^+ \rightarrow \text{acetaldehyde (ethanal)} + \text{NADH} + \text{H}^+$$

the reduced form of nicotinamide dinucleotide coenzyme (NADH) absorbs radiation at 340 nm, however, the oxidized form (NAD^+) does not, allowing the direct continuous monitoring of product formation. Not all enzyme-catalysed reactions have natural reactants and products with suitable spectroscopic properties. In such cases, synthetic chromogenic substrates may prove useful, e.g. the enzyme 4-nitrobenzyl esterase, which is used in the synthesis of the antibiotic, Loracarbef, catalyses a reaction which produces little spectroscopic change. However, a nitrophenyl derivative of the substrate generates the product 4-nitrophenol, which is readily detected at 400 nm.

Loracarbef is s cephalosporin-derived antibiotic. During the synthesis of loracarbef, carboxyl groups are protected with 4-nitrobenzyl alcohol, which is subsequently removed by the enzyme 4-nitrobenzyl esterase.

Alternatively, the reaction of interest can be coupled to a second reaction, which will produce a spectroscopic change, e.g. the assay for hexokinase (EC 2.7.1.1) involves the addition of the coupling enzyme glucose-6-phosphate dehydrogenase (EC 1.1.1.49).

Hexokinase:

$$\text{ATP} + \text{D-glucose} \rightarrow \text{ADP} + \text{D-glucose 6-phosphate}$$

Glucose-6-phosphate dehydrogenase:

$$\text{D-glucose 6-phosphate} + \text{NADP}^+ \rightarrow \text{D-glucono-1,5-lactone 6-phosphate} + \text{NADPH} + \text{H}^+$$

Whilst the reaction catalysed by hexokinase produces no spectroscopic change, glucose-6-phosphate dehydrogenase generates NADPH which absorbs at 340 nm, providing a indirect continuous assay for hexokinase activity.

Alternative detection methods, such as change in pH, may provide a means of monitoring enzyme activity.

5.3.2 Binding proteins

Protein binding assays can be used to both identify ligands and to characterize the nature of the protein–ligand interaction. Binding assays can be subdivided into two categories: those which are based on biophysical changes in the protein or ligand upon protein–ligand complex formation and those which employ direct quantitation of free and bound ligand. Biophysical changes can be monitored using a range of spectroscopic methods, such as absorbance, fluorescence, CD and NMR. These techniques rely on significant changes in the spectra of the unbound

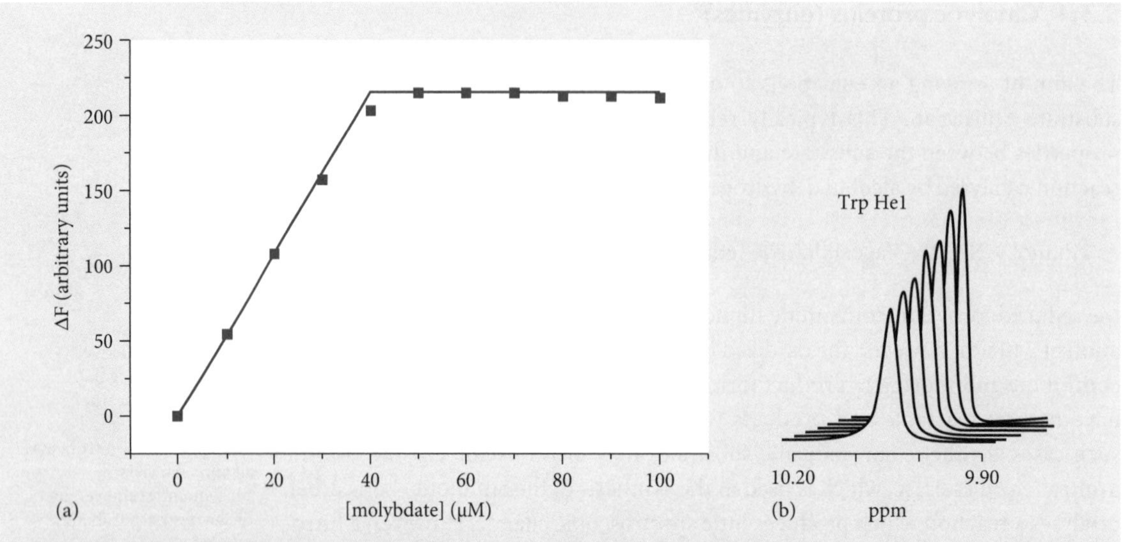

Fig. 5.9 Changes in the spectral properties of proteins on the addition of a ligand. (a) Change in fluorescence. Binding of molybdate ions to the molybdate-sensing protein ModE from *E. coli*. Aliquots of ligand were added to a solution of protein (40 μM) and changes in the fluorescence at 350 nm were monitored. Saturation occurs at a concentration of 40 μM ligand, showing that there is one binding site per polypeptide chain. Binding of ligand leads to an approximately 50% quenching in fluorescence (Boxer *et al.*, 2004). (b) Chemical shift in NMR spectrum. This stack plot shows the spectral effects on addition of increasing amounts of Cu(II) on the tryptophan residue within a prion protein peptide (PHGGGWGQ). The $CuSO_4$ was added in aliquots of 0.0033 mole-equivalents up to 0.02 mole-equivalents (Viles et al., 1999).

The chromatin-associated protein Hbsu from *Bacillus subtilis* binds DNA in a sequence-independent manner and is important in bacterial nucleoid formation. Wild-type Hbsu does not contain any Trp residues and by introducing a Trp at position 47 it was possible to generate a mutant form of the protein, which was indistinguishable from the wild type while providing a spectroscopic means for determining dissociation constants (Groch *et al.*, 1992).

protein or ligand and the complexed state. In titration studies, where the degree of spectral change is assumed to be directly proportional to the ligand concentration, it is possible to determine the degree of binding (see Fig. 5.9). In the absence of a suitable intrinsic spectroscopic change, it is possible to design chemically modified versions of the protein or ligand to facilitate a simple binding assay. A variety of means can be used to introduce spectroscopic labels into proteins. For example, site-directed mutagenesis can be employed to substitute a non-fluorescent amino acid by the fluorescent tryptophan (Trp) at a selected position in the protein.

An alternative approach to introducing fluorophores involves *in vitro* chemical modification of amino acids; a fluorescent group can be introduced by reaction of a suitable reagent with Cys side chains, for example the introduction of the fluorescent probe IAEDANS to Trp repressor protein mutants to characterize tryptophan and DNA binding (Chou and Matthews, 1989). Recent advances in bacterial and yeast expression systems allow efficient site-specific introduction of unnatural amino acids *in vivo*; use of engineered tRNA and aminoacyl tRNA synthase within these systems permits the introduction of a range of unnatural amino acids, including fluorophores (Magliery, 2005). Site-specific, *in vivo*, incorporation of unnatural fluorescent amino acids has been used to modify green fluorescent protein (GFP) at residue 66. The mutant GFPs were successfully overexpressed and purified and were found to have unique spectral properties (Wang *et al.*, 2003).

Another possibility is to introduce the Trp analogue 7-azaTrp in place of Trp in an expressed protein by growth of the host organism on a medium containing this amino acid, with the biosynthetic pathway for Trp inhibited. 7-azaTrp has spectroscopic properties which can be readily distinguished from those of Trp.

Replacement of all tryptophan residues in λ bacteriophage lysozyme with 7-aza Trp has been used to probe the structure and function of this enzyme. In addition, the 7-aza Trp-modified version of lysozyme facilitated its successful crystallization under the microgravity conditions on space shuttle flights (Evrard *et al.*, 1998).

In the case of proteins that undergo significant conformational changes as a result of ligand binding, it may be possible to monitor these changes by determining the sedimentation coefficients using ultracentrifugation.

Traditional direct quantitation methods rely on partitioning techniques, such as equilibrium dialysis and membrane filtration, in which the protein and bound ligand are separated from free ligand. A typical dialysis binding assay would involve placing the protein within a dialysis membrane, which is then placed in a solution of ligand. At equilibrium, the concentration of free ligand will be the same inside and outside the dialysis membrane. The concentration of bound ligand can either be calculated from the difference between the free ligand concentration at the start of dialysis and the free ligand concentration at equilibrium, or from the measured concentrations of ligand on the protein side of the membrane (free ligand plus bound ligand) and on the other side of the membrane (free ligand). Similarly, the ability of membrane filtration to retain protein and protein complexed with ligand, but not free ligand, can be exploited to detect ligand binding. This simple technique requires a means of measuring the amount of ligand retained by the filter, i.e. complexed with protein.

Binding studies using equilibrium dialysis or membrane filtration are greatly facilitated if the ligand has some convenient spectroscopic property or is radioactively labelled.

More recently, the use of solid phase techniques, such as surface plasmon resonance (SPR), have been employed to detect and characterize protein–ligand interactions. This technique is based on the detection of an increase in mass resulting from protein–ligand complex formation. A typical assay would involve immobilization of the protein on the surface of a sensor, followed by introduction of a ligand. Protein–ligand interactions, resulting in an increase in mass, give rise to an increase in signal. Likewise, dissociation of a ligand from immobilized protein results in a decrease in mass, producing a decrease in signal (see Fig. 5.10). Thus, solid phase techniques are proving useful in identifying potential ligands and in determining the kinetic and affinity properties in binding assays (see Chapter 10, section 10.7).

Essentially, the SPR technique measures the rate constant for the association (complex formation) and dissociation (complex breakdown) steps. As described in Chapter 4, section 4.3, the equilibrium constant can be derived from the ratio of these rate constants.

5.3.3 Transport proteins

Transport protein assays are designed to measure the rate of transport of a ligand from one location to another, e.g. transport of glucose into erythrocytes by an integral membrane glucose transporter. While binding assays (see section 5.3.2) on purified transport proteins provide information relating to the stoichiometry and affinity of the protein–ligand interaction, they do not necessarily provide a measure of transport. *In vivo* and *in vitro* transport assays can involve relatively complex systems, such as whole cells or proteoliposomes, and require some mechanism to monitor the initial ligand concentrations in one location and ligand concentration in its final destination. One of the most direct methods employed to

There is a whole family of glucose transporter (GLUT) proteins which share some common structural features, but have different tissue distributions and affinities for glucose. For example, GLUT2 has a low affinity for glucose and is found in the liver and pancreas. GLUT4 has a higher affinity for glucose and is stimulated by insulin; it plays a particularly important role in glucose uptake by muscle and adipose tissue.

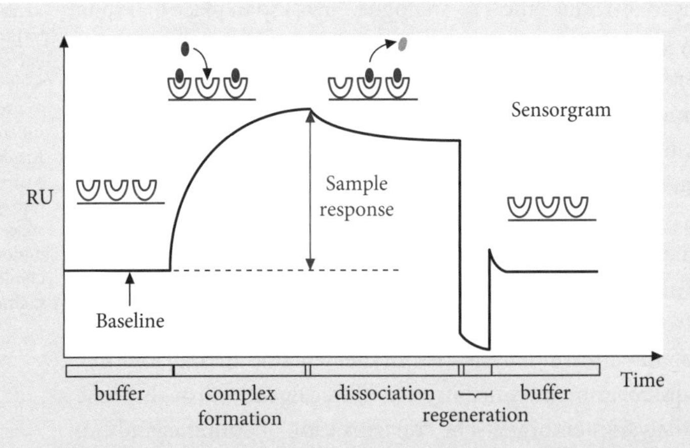

Fig. 5.10 A typical surface plasmon resonance (SPR) sensogram. A baseline signal is produced by the continuous flow of buffer over the protein immobilized on the surface of a sensor chip. Introduction of ligand into the flow of buffer may result in an association of the ligand with the immobilized protein and this is detected by an increase in signal. Subsequent removal of the ligand and a return to the continuous flow of buffer alone will generate a decrease in signal which reflects the dissociation of the ligand from the immobilized protein. The rate constants for the association and dissociation steps are obtained by curve-fitting procedures. SPR occurs when light is reflected off thin metal films. SPR has been exploited to detect protein–ligand interactions by immobilizing a protein on the surface of a thin metal film (sensor chip) and subsequently adding ligands to produce a change in mass (protein plus ligand), which in turn alters the surface of the metal film. Surface changes produce a change in the angle of the reflected light, i.e. a change in the SPR signal. RU, response units.

measure the activity of transport proteins involves the use of isotopically labelled ligands. In a typical assay, a known amount of isotopically labelled ligand is added to the system and incubated for a fixed period of time, after which the reaction is stopped by introducing an inhibitor or by rapid isolation of the final destination of the ligand, e.g. isolation of cells or proteoliposomes using centrifugation or size exclusion chromatography. Functional studies of transport proteins require assays to be conducted in the presence of varying amounts of ligand and in the presence or absence of effectors (see section 5.2.3).

5.3.4 Other types of proteins

This section will describe how to test the biological function of proteins which cannot be measured using the more conventional assays outlined in the previous sections. GFP, which occurs naturally in the jellyfish *Aequorea victoria* and has been used as a reporter molecule in many prokaryotic and eukaryotic systems, is assayed based on its ability to fluoresce at 510 nm, following excitation at 395 nm. A more unusual assay has been developed for the taste-modifying glycoprotein, miraculin. Miraculin, which is isolated from the red berries of the West African shrub, *Richadella dulcifica*, has the unusual property of making sour tastes seem sweet. The sweet-inducing activity of miraculin is measured by administering a small amount of miraculin to subjects, followed by sour citric acid solutions; the subjects then assign an apparent sweetness value to the citric acid solutions (Theerasilp and Kurihara, 1988).

One final example of less conventional assays is that used to monitor the effects of cytokines. Cytokines are a family of proteins, secreted primarily by leukocytes, which allow cell–cell communication. Cytokines are often assayed by monitoring the effects they have on cell cultures, such as cell proliferation, differentiation, and stimulation of immune functions.

The cytokines are a diverse group of signalling proteins that include interferons, several interleukins, and a range of growth factors. Cytokines are secreted by many different cell types which then bind to specific receptor proteins located on the surface target cells, resulting in a biological response. Each cytokine acting on specific target cells produces a specific response that may be cell differentiation, growth, or tissue development.

In the case of proteins with biological functions which cannot be measured easily using direct assays, the advent of heterologous expression systems has presented a convenient means of monitoring the purification of such proteins, circumventing the need for an assay. Overexpressed proteins can account for a substantial proportion of the total cell protein (see section 5.4.2) and as a result can be readily identified by measurements of molecular mass using SDS-PAGE; the most abundant species with the correct molecular mass (theoretical or known) can be identified in crude extracts and in samples throughout the purification stages (see Fig. 5.11). When the protein with the correct mass is purified, its identity should be confirmed by mass spectrometry and by partial amino acid sequencing or peptide mass fingerprinting (Chapter 8, section 8.3).

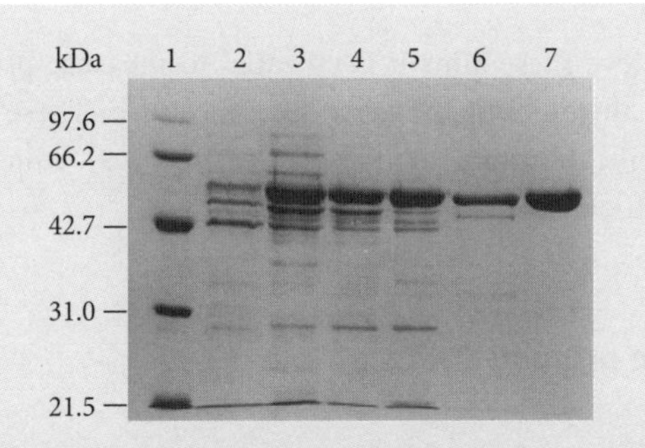

Fig. 5.11 SDS-PAGE analysis of tomato leucine aminopeptidase (LAP-A) overexpression and purification. LAP-A was overexpressed in *E. coli* and purified using a four-step procedure, during which LAP-A could be followed as the most intense band, measuring 55 kDa on SDS-PAGE analysis (from Gu *et al.*, 1999) Lane 1, protein molecular weight markers; lane 2, *E. coli* lysate prior to IPTG induction; lane 3, *E. coli* lysate following IPTG induction; lane 4, heat denaturation; lane 5, ammonium sulphate precipitation; lane 6, MonoQ chromatography; lane 7, hydrophobic interaction chromatography.

5.4 Purification of proteins

KEY CONCEPTS

- Knowing the objectives of protein purification
- Being aware of the experimental considerations required to meet these objectives

As outlined in the previous sections, structural and functional studies can yield information about the important properties of proteins. This information and its interpretation are simplified greatly by studying purified proteins, i.e. structural and functional properties derived from experiments involving 'protein X' can be attributed to 'protein X' alone and not to some other molecular species. The aims of any protein purification scheme are:

To retain maximum biological activity: The success of structural and functional studies hinges on the purified protein behaving in a manner similar to its behaviour within the cell, i.e adopting its native structure and being fully functional.

To ensure the protein is indeed pure: When characterizing a protein it is essential that the properties measured can be attributed to the protein of interest and are not due to the presence of contaminating macromolecules.

To maximize the amount of protein recovered: An efficient purification scheme will recover as much biologically active, pure protein as possible from the source material.

In order to achieve these aims it is essential to measure protein content and biological activity throughout the isolation procedure. These data can be used to construct a purification table (Chapter 3, section 3.9), which will give a clear indication of the efficiency of the procedure.

5.4.1 Wild-type proteins

The isolation of proteins from their natural source exploits their heterogeneous biochemical properties. A range of purification techniques have been developed to isolate individual proteins according to their unique set of biochemical characteristics, encompassing molecular mass, charge, stability, hydrophobicity, solubility, and specific ligand binding sites. Most isolation procedures consist of a number of steps, each employing a different purification technique, although it is often preferable to minimize the number of steps to ensure retention of the maximum quantity of biologically active protein.

The task of devising a new purification procedure can be made easier by gathering as much biochemical information relating to the protein of interest as possible. Predicted properties such as molecular mass, pI, hydrophobicity, post-translational modifications, and putative ligands can be deduced from the amino acid sequence of the protein or a homologue (section 5.8). Alongside experimental information relating to the purification of similar proteins, an isolation strategy can be readily developed. Careful selection of the starting material will improve the chance of devising a successful strategy; ideally, the starting material should be readily available and rich in the protein of interest (Chapter 7, section 7.1). The starting material, together with subsequent purification steps, influences cell lysate preparation. This requires a balance between optimal release of active protein from the staring material with providing conditions conducive to subsequent purification steps. As a result it is important to optimize the following factors in the cell lysate preparation: pH, ionic strength, temperature, solubility, protease inhibitors, reducing agents, and ligands to confer stability/activity.

Many of the preliminary steps in protein isolation procedures involve precipitation methods, which employ precipitants such as ammonium sulphate, polyethylene glycol, or ethanol to separate proteins by solubility. Precipitants weaken the forces between the protein and the aqueous solvent that keep the protein in solution. Precipitation induced by changes in pH and temperature can also be employed to separate proteins. Most subsequent separation techniques in an isolation procedure employ chromatographic methods, which separate proteins according to size, charge, hydrophobicity, and ligand binding (Chapter 7, section 7.2). Most procedures require a few chromatographic steps and in some favourable cases it may be possible to isolate the protein in one simple chromatographic step. A successful isolation procedure will produce material with a high specific activity and purity, suggested by SDS-PAGE analysis or mass spectrometry.

An example of a single-step purification is provided by the purification of fructose-bisphosphate aldolase from rabbit muscle. The protein precipitating between 50% and 52% saturation ammonium sulphate is essentially pure enzyme. (Note that at 25°C, a saturated solution of ammonium sulphate is about 4 M). Use of ammonium sulphate precipitation in protein purification is detailed in Chapter 7, section 7.2.1.

5.4.2 Recombinant proteins

Protein purification has been simplified greatly by the advent of recombinant DNA technologies that enable high levels of protein expression. Following the development of DNA cloning in bacterial systems in the mid-1970s, it soon became possible to express proteins, encoded by cloned DNA, in heterologous systems. One of the earliest successes was the overexpression of mouse dihydrofolate reductase in *Escherichia coli* (Chang *et al.*, 1978). A large number of prokaryotic and eukaryotic overexpression systems are now available, including *E. coli*, *Aspergillus nidulans*, *S. cerevisiae*, *P. pastoris*, insect cells, mammalian cells, and transgenic plants and animals. In addition to providing large quantities of material, these systems have enabled the expression of site-directed mutants and the introduction of purification tags.

In most cases the overexpressed protein accounts for between 1% and 10% of the total cell protein in the expression system. In extreme cases, this proportion can be in excess of 25%, e.g. the therapeutically important proteins human insulin and interleukin-2 are produced in *E. coli* at levels in excess of 25% total cell protein.

To achieve maximum expression of biologically active protein, it is important to consider a number of factors, including the natural source of the protein, whether it is post-translationally modified, the level of expression required, whether it is secreted, its subcellular location, its toxicity to the expression host, and whether a purification tag is required. As a general rule, optimal expression systems closely mimic the natural production of the protein, e.g. if producing a eukaryotic, post-translationally modified, secreted protein, overexpression may be achieved in an *S. cerevisiae* or *P. pastoris* system in which the gene encoding the protein is linked to an appropriate signal sequence that directs the protein out of the cell. In many cases, the use of these signal sequences can simplify the purification of the expressed proteins from the growth medium, without the need to make cell extracts.

In some cases, attempts to overexpress proteins in *E. coli* can result in the formation of inclusion bodies which are composed of partially folded, overexpressed protein. Although inclusion bodies present a rich source of pure and stable protein, which can be readily isolated by centrifugation, recovery of folded, fully functional protein from this partially folded state can prove challenging. Typically recovery involves solublization of the inclusion body protein using a denaturant and then removing the denaturatant slowly (by dialysis, dilution or filtration) to promote refolding of the protein.

E. coli expression systems are by far the most commonly used, by virtue of their well-characterized molecular biology and the high levels of overexpression which can be achieved. Although mostly limited to producing intracellular proteins which are not post-translationally modified, the vast range of *E. coli* expression vectors, which can produce proteins fused to purification tags, makes *E. coli* the system of choice. Engineering a tagged protein to facilitate purification requires adding DNA encoding the tag to either the 5′ or 3′ end of the gene encoding the protein of interest to generate recombinant protein with a tag at the N- or C-terminus (see Fig. 5.12). Purification tags can take the form of enzymatically active

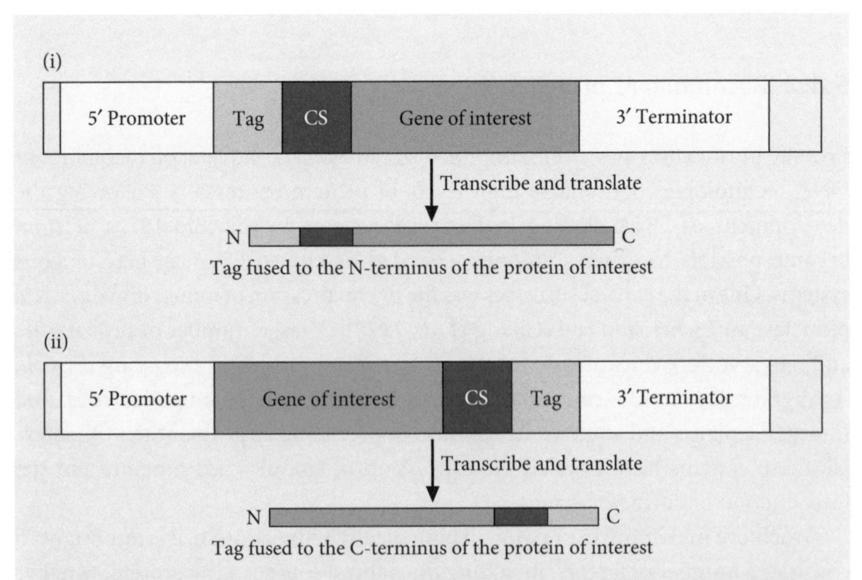

Fig. 5.12 Portion of expression plasmid with tag sequence at 5′ (i) or 3′ (ii) end of the gene producing N- or C- terminally tagged protein, respectively. In some cases, as shown here, a stretch of amino acids containing a target cleavage sequence (CS) (shown in black) is included to allow selective removal of the tag.

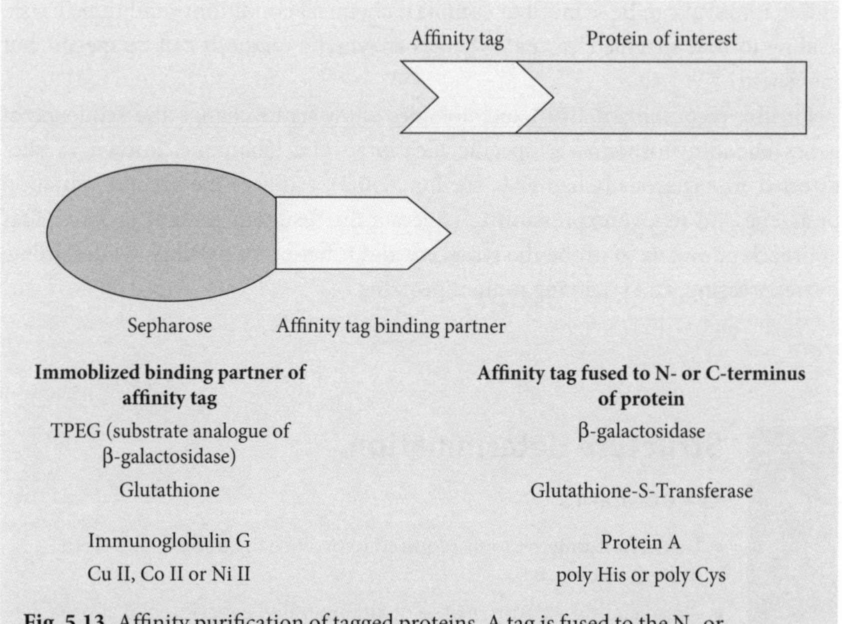

Fig. 5.13 Affinity purification of tagged proteins. A tag is fused to the N- or C-terminus of the protein of interest to facilitate purification, which relies on a specific interaction between the affinity tag and the immobilized binding partner of the affinity tag.

fusion proteins such as β-galactosidase and glutathione-*S*-transferase (Chapter 7, section 7.3.4), affinity proteins (e.g. protein A, exploiting its affinity for IgG) or metal-chelating affinity tags (see Fig. 5.13). Metal affinity tags, which include poly-histidine and poly-cysteine tags, have a high affinity for divalent metal ions such as copper, nickel, and cobalt. This property forms the basis of immobilized metal affinity chromatography: divalent metal ions are immobilized on a chelating chromatography medium (e.g. NTA (nitrilotriacetic acid)-agarose) and selectively retain proteins fused to poly-histidine or poly-cysteine affinity tags. Subsequent elution, for example with imidazole solutions in the case of His-tags, can produce large quantities of pure tagged protein in one chromatographic step (Chapter 7, section 7.3.8). Removal of purification tags may be necessary to restore function to the protein of interest or to enhance its solubility. Removal of a tag can be achieved by chemical or, more commonly, enzymatic methods. A number of examples of cleavage target sites and associated cleavage agents are given below.

Cleavage site	Cleavage agent
Asp–Asp–Asp–Asp–Lys↓–X	Enteropeptidase
Leu–Val–Pro–Arg↓–X	Thrombin
Ile–(Glu/Asp)–Gly–Arg↓X	Factor Xa
Met↓–X	Cyanogen bromide
Asn–Gly	Hydroxylamine

Tag removal can be somewhat complex: chemical conditions tend to be harsh, leading to non-specific cleavage, whereas enzymatic methods can be specific but inefficient.

Finally, recombinant DNA technologies allow us to change the sequence of genes encoding proteins at specific locations. The technique, known as site-directed mutagenesis (Chapter 9, section 9.10.3), requires the specific mutation of a gene and its overexpression to generate the resultant mutant protein. This approach allows us to probe the structure and function of proteins by designing, overexpressing, and purifying mutant proteins.

5.5 Structure determination

KEY CONCEPTS

- Understanding the tools required to explore the different levels of protein structure
- Appreciating the importance of an integrated experimental approach to provide a more complete picture of protein structure

Prior to the early 1980s, many protein crystal structures were studied without the availability of their amino acid sequences. Nevertheless, in each case it was usually possible to produce an atomic interpretation of the X-ray diffraction pattern and to trace the path of the polypeptide chain, although the later availability of the sequence allowed detailed interpretation of the diffraction data to be undertaken. Many of these proteins, e.g. α-chymotrypsin, lactate dehydrogenase, subtilisin, and myoglobin, were studied because they could be readily purified from their natural source and formed crystals which gave rise to clear, interpretable, X-ray diffraction patterns.

The ultimate aim of protein structure determination is to gather and interpret data to reveal the three-dimensional structure of the protein at an atomic level. Whilst X-ray crystallography and high-resolution NMR spectroscopy (for proteins of molecular mass <30 kDa) generate structural data, additional experimental information relating to the primary, secondary, tertiary, and quaternary structure plus post-translational modifications, is required to interpret these data and solve the structure.

The apparent molecular mass, which is indicative of the number of amino acids within individual subunits, can be estimated by SDS-PAGE, gel filtration, ultracentrifugation, and mass spectrometry. The order of the amino acids within a polypepticle chain can be determined directly or indirectly. The direct approach would use Edman degradation or tandem mass spectrometry (Chapter 8, Section 8.3) to degrade sequentially peptides derived from the protein (see Fig. 5.14). The indirect approach would require translation of the gene encoding the protein. A comparison of the theoretical molecular mass calculated from the primary structure with the experimentally determined mass can provide evidence of post-translational modification. The nature of the post-translational modification can be determined by SDS-PAGE combined with specific removal of modifications, immunoblotting using antibodies raised to specific post-translational modifications, and mass spectrometry to analyse the mass of peptides with modifications (Chapter 8, section 8.2).

A number of techniques, including chemical modification of surface exposed residues, fluorescence, CD, and NMR spectroscopy, can indicate the nature of the secondary and tertiary structure of a protein (Chapter 8, section 8.4). The availability of secondary and tertiary structure prediction tools (section 5.2.2) together

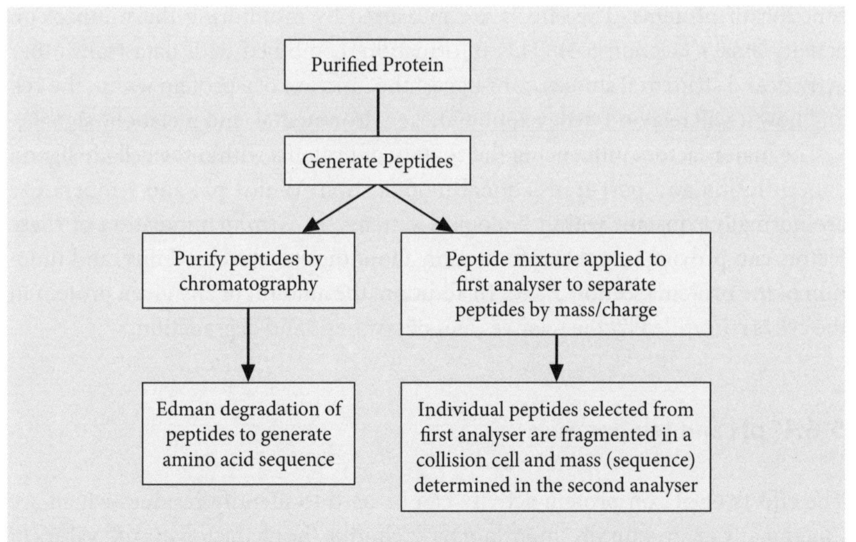

Fig. 5.14 Ouline of scheme for the 'direct' determination of the primary structure of a protein using Edman degradation or tandem mass spectrometry.

with the primary structure can be employed to complement experimentally derived data.

Determination of the quaternary structure of a protein requires combining information relating the molecular mass and number of types of individual subunits to the apparent mass of the native multi-subunit protein, estimated by gel filtration or ultracentrifugation. Complex multi-subunit proteins may require further characterization using techniques, such as cross-linking (Chapter 8, section 8.5), to establish the number of subunits present.

Combining all of this experimental evidence permits the accurate interpretation of data generated by X-ray crystallography or NMR spectroscopy, and eventual structure determination. The advent of bioinformatics has enabled the determination of theoretical protein structures, using homology searches and prediction methods (sections 5.2.2 and 5.8). Again, this approach can be used to complement experimentally derived data.

5.6 Factors affecting the activity of proteins

KEY CONCEPTS

- Defining the major factors which influence the activity of proteins
- Understanding how to monitor their effects on protein structure and function

Studying the effects of factors such as post-translational modification, ligands, pH, and temperature on protein activity can provide insight into the structure and

Note that while we can draw some conclusions about how the protein may function within the cell, the *in vitro* assay conditions will be very different from *in vivo* conditions (e.g. high protein concentration, low concentration of ligands, protein complex formation, compartmentalization).

function of proteins. The effects are measured by monitoring their impact on activity assays (section 5.3). This information, combined with data from other activity and structural studies, can suggest the function of a protein within the cell and how it will respond to developmental, environmental, and metabolic signals.

The major factors influencing the activity of proteins within the cell are ligand concentration and post-translational modification. Whilst pH and temperature are normally constant within biological systems, *in vitro* manipulation of these factors can provide valuable information about the structure, stability, and function of the protein (section 5.2.3). In addition, the amount of any given protein in the cell is controlled by the relative rates of synthesis and degradation.

5.6.1 pH and temperature

The effects of pH are not only important in studies of activity, also they are an essential consideration in developing successful purification strategies in which isolectric focusing (Chapter 6, section 6.2.3) or ion-exchange chromatography (Chapter 7, section 7.2.2) are employed.

The effects of pH on protein activity can be used to identify residues which are functionally or structurally important by exploiting the characteristic pK_a values of amino acid side chains (see Chapter 1, section 1.2.3.2). A typical activity response to changes in pH is shown in Fig. 5.15. Under extreme pH conditions, the lack of activity is due to protein denaturation. pH conditions which elicit the highest level of activity (the pH optimum) promote side chain side ionization states which are essential for optimal activity. Identification of important residues requires activity measurements at pH values close to the pH optimum. Resultant changes in the ionization state of key residues produce changes in activity which can be used to estimate pK_a values. In some cases, pK_a values can be assigned to particular types of amino acids. In the example shown in Fig. 5.15, the measured pK_a values of 4.0 and 10.5 are indicative of glutamic acid and lysine, respectively.

Care must be exercised when interpreting apparent pK_a values as the pK_a of side chains of amino acids within proteins can be very different from the pK_a of free

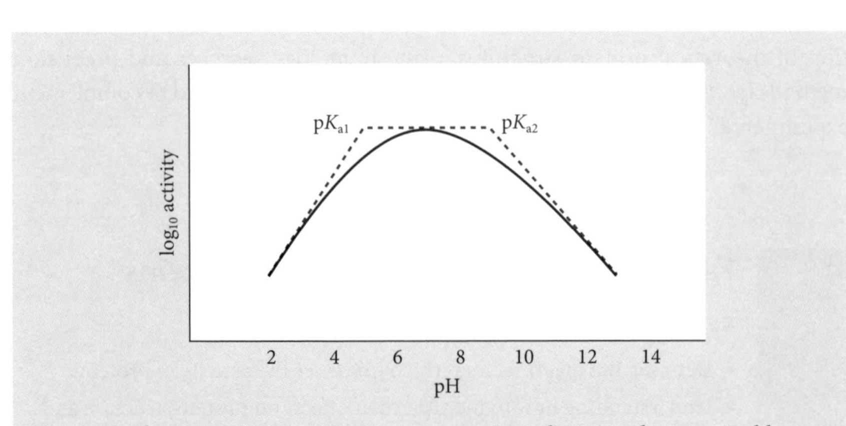

Fig. 5.15 Change in protein activity at varying pH conditions with two ionizable groups involved.

amino acid side chains. It is also important to ensure that pH changes are not having an effect on other assay components. It is therefore good practice to confirm pK_a values derived from activity measurements using other techniques such as spectroscopic titration studies, chemical modification, and site-directed mutagenesis.

In the case of enzyme-catalysed reactions involving H^+ as a reactant or product, a change in pH will influence the position of equilibrium. The reaction catalysed by creatine kinase (E.C. 2.7.3.2) provides a good example of this (note that in most reactions ATP and ADP occur as complexes with Mg):

$$\text{MgADP}^{-} + \text{phosphocreatine}^{2-} + \text{H}^{+} \rightleftharpoons \text{MgATP}^{2-} + \text{creatine}$$

Creatine kinase maintains ATP levels within muscle cells during exercise. In the resting state, the pH within the muscle cell is around 7.0 but during prolonged exercise, involving anaerobic metabolism and the production of lactate, the pH can drop to about 6.3. As a result, a shift in equilibrium towards the right of this reaction occurs, to try to maintain the levels of ATP. Beyond a certain point, however, the build up of acidic substances will alter the activity of key proteins in muscle and lead to muscle fatigue.

By monitoring protein activity over a range of temperature conditions, it is possible to characterize structural changes which can be used to infer the functional properties of proteins. A typical response in protein activity is shown in Fig. 5.16. Activity measurements at lower temperatures indicate that as temperature increases, so the activity increases (see the Arrhenius equation, Chapter 4, section 4.2.3). Most mammalian enzymes tend to show maximal activity in the range 40–45°C, known as the optimum temperature; at temperatures above this range there is a rapid loss of activity resulting from protein denaturation.

The optimum temperature for an enzyme can depend on various factors such as the length of time the enzyme is incubated at the temperature used for assay. Biochemists now prefer to use the term 'apparent optimum temperature'.

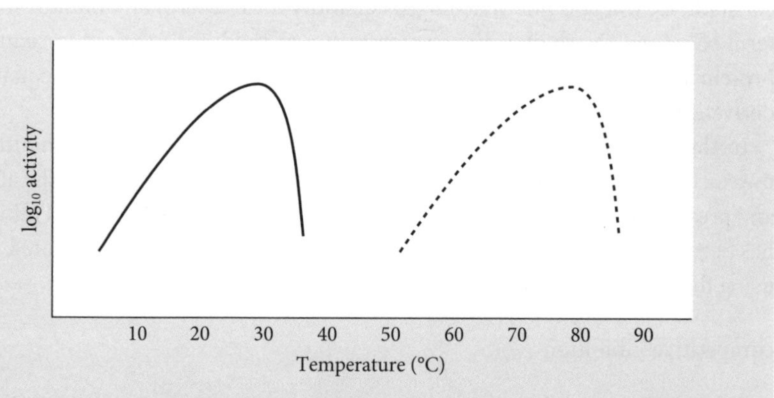

Fig. 5.16 Effect of temperature on protein activity. Most proteins have a temperature optimum of 30–40°C (solid line) whereas thermophilic proteins are much more stable, exhibiting optimum temperature values of >70°C (dotted line).

Thermophilic organisms are categorized according to their growth temperatures: the optimal growth temperature for thermophiles is >50°C, for extreme thermophiles >65°C (e.g. *Thermus aquaticus*, the source of *Taq* polymerase), and for hyperthermophiles >90°C (e.g. *Pyrococcus furiosus*). Mesophilic organisms have growth temperature optima in the range 20–37°C.

Further information can be gleaned by monitoring activity changes in response to temperature changes under varying conditions, such as the presence of ligands and post-translational modifications. These types of experiments are important in characterizing the influence of ligands and post-translational modifications on protein activity. Generally, we can correlate temperature-induced activity changes with temperature-induced structural changes, using a number of methods such as spectroscopic techniques, chemical modification, and site-directed mutagenesis.

Proteins from thermophilic organisms respond to temperature changes in a similar way; however, in these cases, the optimum temperature is much higher. Although thermophilic proteins are generally structurally similar to their mesophilic counterparts, they have adopted a range of strategies to enhance their thermostability including additional electrostatic interactions, helix dipoles, helix capping, and shorter surface loops (Cowan, 1995).

5.6.2 Inhibitor and activator molecules

Inhibitors and activators are important effectors of protein activity within the cell. By measuring the influence of these effectors on protein activity *in vitro*, it is possible to establish how proteins are regulated in a cellular environment. The effect of inhibitor and activator molecules on protein activity assays (section 5.3) combined with saturation curve analyses (Chapter 4, section 4.3) can be used to calculate protein:effector stoichiometry, the strength of the interaction, and their influence on the rate constants of individual steps in the process.

Typical inhibition studies involve measuring protein activity in the presence and absence of inhibitor. The effect of the inhibitor can be quantified by calculating K_i, the binding constant of the inhibitor to the protein: a small K_i, value indicates a high-affinity inhibitor, whereas a large K_i, value suggests a low-affinity inhibitor.

Biochemists usually quantify the effects of inhibitors in terms of an inhibitor constant, K_i, whereas pharmacologists quantify the effects of an inhibitor with the term IC_{50} (or $I_{0.5}$), which is the concentration of inhibitor required to reduce the protein activity by 50%. This is related to the K_i by the Cheng–Prusoff equation, a derivation of which is given in the appendix to this chapter.

In the case of enzymes, it is possible to determine the type of inhibition by measuring the reaction rate at a variety of substrate concentrations in the absence and presence of an inhibitor (Chapter 4, section 4.3.4). Once the type of inhibition has been identified it is possible to calculate K_i (in this case usually denoted as K_{EI}) using the following equations:

competitive inhibition

$$K_i = \frac{[I]}{\left(\frac{K_m^I}{K_m} - 1\right)} \tag{5.1}$$

non-competitive inhibition

$$K_i = \frac{[I]}{\left(\frac{V_{\text{max}}}{V^I_{\text{max}}} - 1\right)} \quad 5.2$$

where K_m and V_{max} are the constants in the absence of inhibitor, whereas K^I_m and V^I_{max} are the constants in the presence of the inhibitor and these are determined as outlined in Chapter 4, section 4.4. A more accurate method of determining K_i involves extending this approach to look at the effect of several concentrations of inhibitor to generate a series of Lineweaver–Burk plots (Engel, 1981).

The activation of proteins by effector molecules can be characterized by conducting protein activity assays in the presence and absence of activator. Changes in saturation curves are reflected in the rate constants that can be used to measure the affinity of the activator for the protein (K_A) and the type of activation, e.g. allosteric activation, in which the binding of the activator promotes cooperative substrate/ligand binding, analogous to oxygen binding to haemoglobin outlined in Chapter 4, section 4.3.2.

Repeating these experiments with modified versions of effectors can improve our understanding of protein/effector specificity and provide structural information concerning the nature of the effector binding site (see Fig. 5.17).

Over the last decade, generating libraries of modified protein ligands by a technique known as combinatorial chemistry has emerged as a powerful tool in protein characterization and drug discovery.

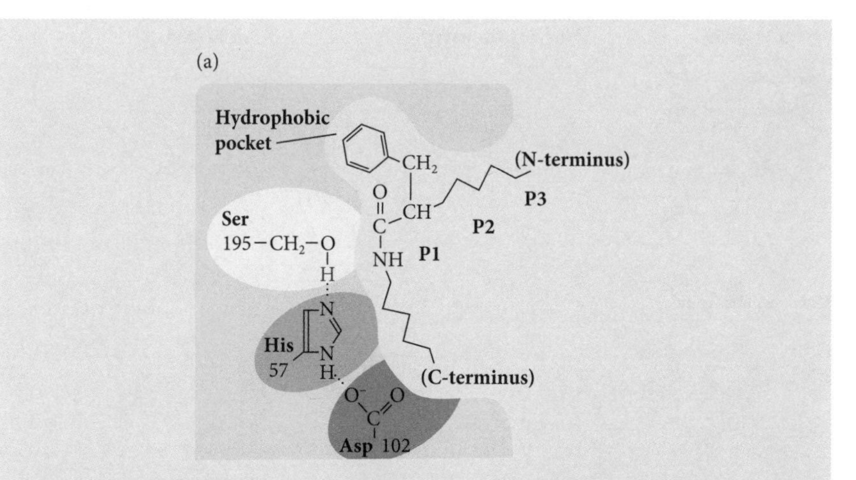

Fig. 5.17 Mapping binding sites using modified ligands. (a) In the 1960s the active sites of serine proteases (e.g. chymotrypsin) were mapped using substrate mimics of varying length and composition. By varying substrate length (P1, P2, P3, etc.) and composition, it was noted that substrates with large hydrophobic residues at P1 were good substrates for chymotrypsin. Structure studies subsequently revealed that the substrate specificity observed relates to the shape and hydrophobic nature of the active site.

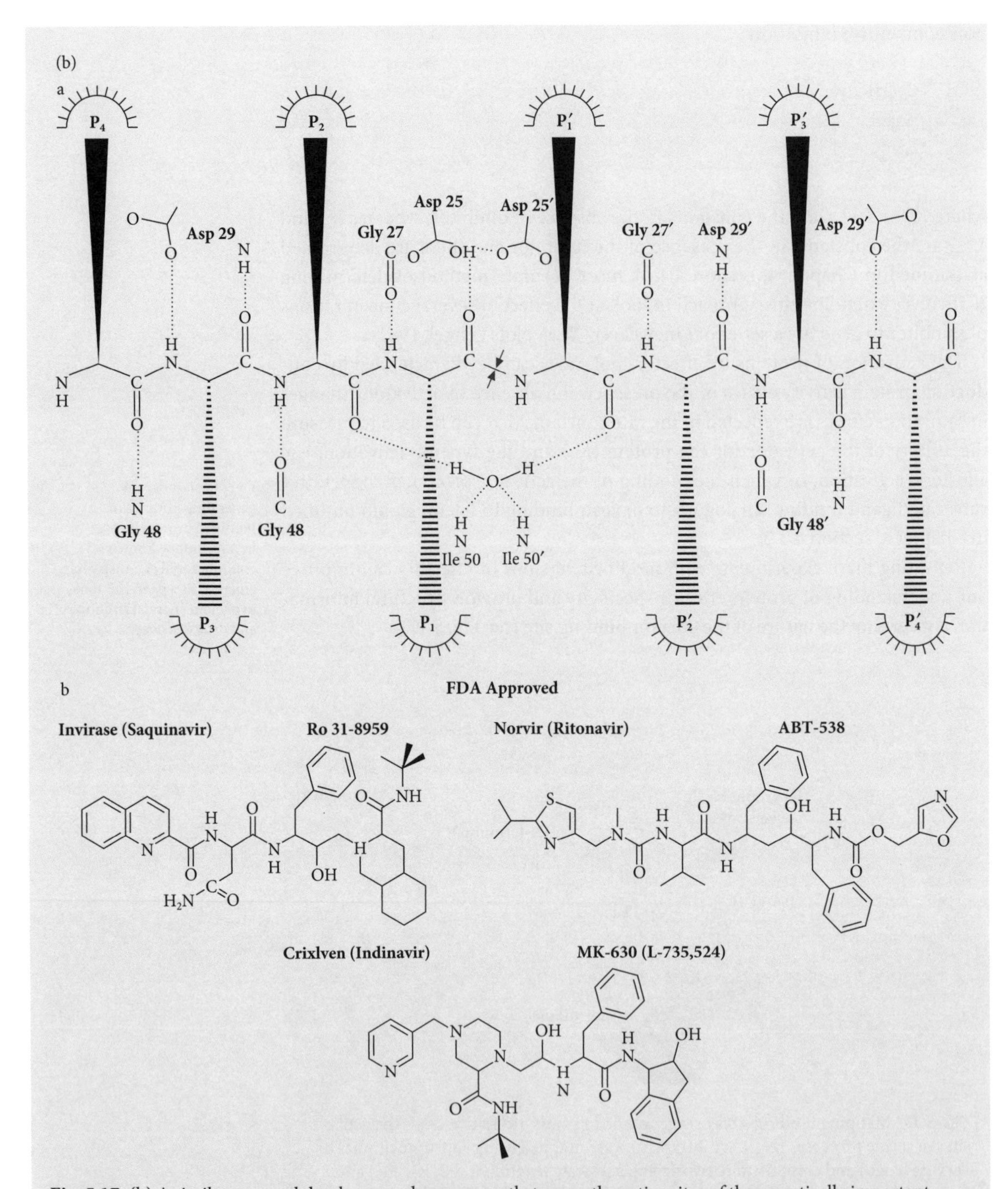

Fig. 5.17 (b) A similar approach has been used more recently to map the active sites of therapeutically important proteases and subsequently to inhibit them. A good example of this is provided by the protease inhibitors designed for the effective treatment for HIV infection. a representation of HIV protease complexed with putative substrate, b number of effective HIV protease inhibitors which mimic the substrate (from Wlodawer and Vondrasek, (1998)).

In many cases, activity changes induced by the presence of inhibitors and activators are associated with measurable changes in protein structure. Effector-induced structural changes can be monitored using spectroscopic techniques, such as absorbance, fluorescence, CD, and NMR, and can provide valuable information which complements measurements of effector-induced activity changes.

5.6.3 Post-translational modifications

Post-translational modification is one of major effectors of protein activity within the cell, indeed, its importance has been re-examined recently following the completion of numerous genome sequencing projects, in particular those of higher organisms. These projects have revealed that the complexity of higher organisms is due to the differential RNA processing and post-translational modification of gene products. Thus, the human genome is thought to encode about 23 000 gene products, but these are thought to give rise to several hundred thousand distinct proteins. There are many forms of post-translational modification (see Table 5.1) which can influence protein activity. All of these covalent modifications are the result of enzyme-catalysed reactions which confer rapid and amplified changes in protein activity in response to small cellular signals.

Protein activity assays alongside saturation curve analyses (Chapter 4, section 4.3) can be used to characterize the effects of post-translational modifications. The exploration of these effects requires uniform protein preparations, both with and without post-translational modification. This can be checked using SDS-PAGE, Western blotting with modification-specific antibodies, and/or mass spectrometry. Developing standard activity assays of post-translationally modified protein to include activity measurements in the presence of inhibitors/activators and the effects of temperature can help determine how the protein is regulated *in vivo*.

Changes in protein activity resulting from post-translational modification may be accompanied by structural changes. Secondary and tertiary structural changes can be monitored using spectroscopic techniques, whereas changes in the

Table 5.1 Examples of post-translational modifications

Amino acid modifications	
Modifications	**Example**
Cysteine: disulphide bond formation	Lysozyme
Lysine biotinylation	Acetyl CoA carboxylase
Serine phosphorylation	Glycogen phosphorylase
Threonine phosphorylation	Cyclin-dependent kinase
Tyrosine phosphorylation	Cortactin
Addition of prosthetic groups—thiamine diphosphate	Pyruvate dehydrogenase
Proteolytic processing	Chymotrypsin
Protein targetting (signal sequences)	Penicillin acylase

quaternary structure can be determined using gel filtration, ultracentrifugation, and cross-linking studies (Chapter 8, section 8.2).

A good example of a protein which undergoes post-translational modification resulting in activity and structural changes is the enzyme glycogen phosphorylase (E.C. 2.4.1.1). Phosphorylase exists in two interconvertible forms: phosphorylase *a*, which is phosphorylated at Ser 14 by a specific kinase, and phosphorylase *b*, which is not phosphorylated. In the absence of AMP, phosphorylase *b* is inactive, whereas phosphorylase *a* actively catalyses the cleavage of glycogen to produce glucose-1-phosphate:

$$(1,4\text{–}\alpha\text{-}\mathrm{D}\text{-glucosyl})_n + P_i \rightleftharpoons (1,4\text{–}\alpha\text{-}\mathrm{D}\text{-glucosyl})_{n-1} + \alpha\text{-}\mathrm{D}\text{-glucose 1-phosphate}$$

Conversion of phosphorylase *b* to phosphorylase *a* results in tertiary and quaternary structural changes which are associated with the regulatory properties of both forms of this key enzyme.

5.7 Interactions with other macromolecules

KEY CONCEPTS

- Appreciating the importance of protein–protein interactions *in vivo*
- Understanding the need to use appropriate experimental conditions to study these interactions

Whilst it is convenient to study isolated proteins *in vitro*, this does not reflect how they function *in vivo*. Within the cell, proteins are part of a highly organized system involving subcellular structures and high concentrations of macromolecules, including other proteins, nucleic acids, and lipids. Recent developments in techniques used to study protein interactions (section 5.3.2) are beginning to reveal the complexity of these interactions, many of which are weak and transient, but nonetheless important for the expression of protein function. The best-characterized protein–protein, protein–nucleic acid, or protein–lipid complexes are those which are most stable and have survived the harsh cell disruption procedures used during complex isolation.

The study of the interaction with other molecules can involve the identification of molecules which interact with the protein of interest or characterization of the effect of these molecules on protein activity. Typically, the identification of novel effectors involves the immobilization of the protein of interest which then serves as a bait to bind potential effectors. A great deal of care must be exercised when establishing the conditions which will promote and maintain protein–effector interactions as these interactions are highly sensitive to pH, ionic strength and the presence of other effectors. Failure to optimize conditions can result in false positive results, i.e. certain conditions can promote non-specific interactions, or produce false negative data in which interactions are not identified due to weakened binding. The influence of 'other molecules' on protein activity can be determined using activity assays (section 5.3) and saturation curve analyses (Chapter 4, section 4.3). The binding of 'other molecules' may be accompanied by structural changes which can be monitored as outlined in the previous section, providing structural data to complement activity measurements.

5.8 Use of bioinformatics

KEY CONCEPTS

- Appreciating the range of databases and bioinformatic tools available to assist protein characterization
- Understanding the theoretical basis of the tools used to calculate properties of a protein from its sequence

The study of proteins is no longer an exclusively laboratory-based activity pursued by biochemists; instead it is possible to use computer-based methods to examine proteins in detail. The field of bioinformatics has harnessed the exponential growth in the amount of information relating to nucleotide sequences, protein sequences, and biomolecular structures. As a result, all of this information has been organized into web-based databases which can be accessed and analysed using a range of computer programs. In this section, we shall consider some of these databases and how they can be employed to explore protein structure and function. In addition, we shall look at a range of programs which can be used to analyse database information to assist protein characterization.

5.8.1 Web resources and databases

Nucleotide sequences, amino acid sequences, and protein structures are collected within a number of web-based databases. The aim of each of these databases is not only to collect information but also to present it in an annotated form to help biologists understand better the significance of the data. More recently, there has been an effort to integrate data sets (e.g. linking individual nucleotide/protein sequences to related three-dimensional structures, metabolic pathway databases, enzyme databases, disease databases, organism-specific databases, two-dimensional gel databases, and associated references), allowing researchers to characterize more fully the structural and functional properties of proteins.

The major databases for RNA and DNA sequences are EMBL, Genbank, and DDBJ.

The International Nucleotide Sequence Database Collaboration (http://www.insdc.org/) combines the efforts of European Molecular Biology Lab (EMBL) in Hinxton, UK, GenBank in Bethesda, Maryland, USA and DNA Data Bank of Japan (DDBJ) in Mishima, Japan.

Sequencing technologies have enabled the completion of increasing numbers of genome-sequencing projects, which in turn has maintained a very large growth of these databases, with the EMBL Nucleotide Sequence Database containing 130 million sequence entries, comprising in excess of 2×10^{11} nucleotides in 2008. Search tools permit retrieval of relevant database entries which can be further analysed using a suite of molecular biology tools (e.g. restriction site identification, sequence comparisons, and translation to identify open reading frames) or cross-referenced with other databases.

One of the main protein sequence databases is SwissProt (http://www.ebi.ac.uk/swissprot/), which contains sequences which have been generated either directly by amino acid sequencing or indirectly from translating open reading frames of genes within TrEMBL (a database containing translated coding regions of EMBL/GenBank/DDBJ nucleotide databases). The SwissProt database is linked with about 50 other databases, allowing extensive characterization of the protein of interest.

An increasing number of whole genome-specific databases have become available over the past decade. Each genome is annotated to identify individual genes and these are cross-referenced with protein sequence and three-dimensional structure databases. The genomes characterized to date reflect their importance either as biological models (e.g. *E. coli, C. elegans, S. cerevisiae*) or as commercial organisms (e.g. rice, maize, chicken, salmon) or as disease-related organisms (human, *Helicobacter pylori, Anopheles gambiae*). Fig. 5.18 displays the NCBI websites for the *C. elegans* and the *A. gambiae* genomes.

The bacterium *H. pylori*, although isolated over 100 years ago, has only recently been shown to cause stomach or duodenal ulcers.

A. gambiae, the mosquito, carries the malaria parasite and is the main vector of malaria in Africa, where it is estimated that a million children die each year from this disease. Characterization of the genomes of such disease agents will enhance the development of disease prevention, detection, and treatment.

The PDB is currently maintained by the Research Collaboratory for Structural Bioinformatics (RSCB), involving Rutgers (NJ), San Diego (CA) and Madison (WI), and is located at http://www.rcsb.org/pdb. Linked sites are maintained by the European Bioinformatics Institute in Cambridge, UK (http://www.ebi.ac.uk/msd/) and in Osaka, Japan (PDBj; see http://www.pdbj.org).

All three-dimensional biomolecular structures are deposited in the PDB database. A search of the PDB database using the name of a protein or its PDB entry code, will provide an image of the protein structure together with information about the protein and its source, reference to the paper describing structure determination, the amino acid sequence, ligands present in the structure, secondary structure composition, and the atomic coordinates. The atomic coordinates can be downloaded from the PDB site and analysed in detail using molecular viewer software such as Rasmol (http://www.openrasmol.org) or its derivative Protein Explorer (http://www.umass.edu/microbio/rasmol). Such tools facilitate an exploration of protein structure, including ligand binding sites, catalytic residues, key structural residues, and protein–protein interactions.

5.8.2 Sequence analysis

Protein sequences, derived from experimental data or database entries, can generate a wealth of information that can assist protein characterization. Empirically derived information will be considered in this section; sequence comparisons

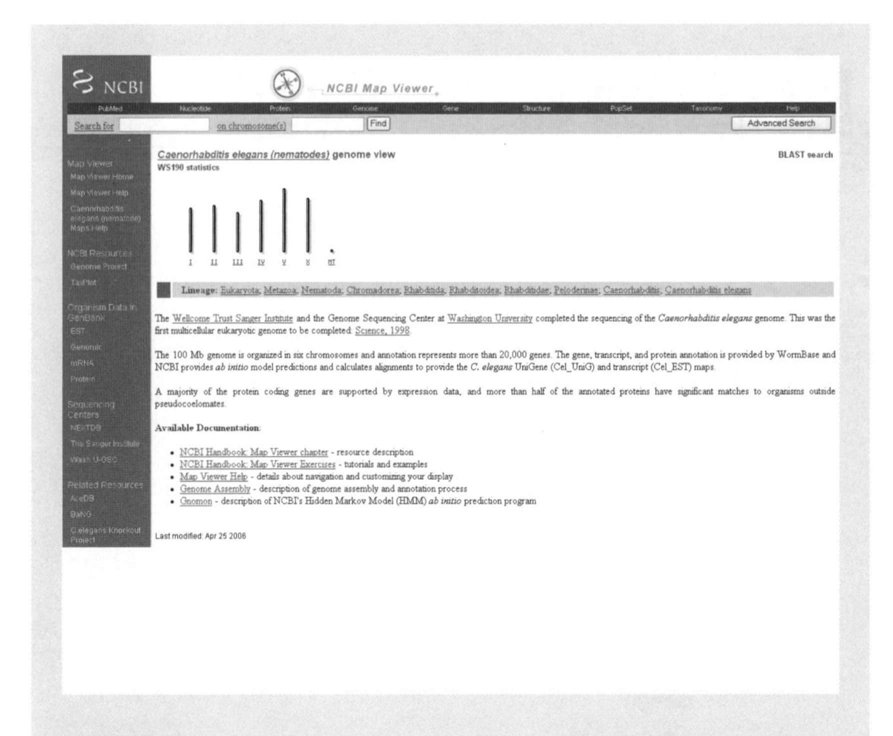

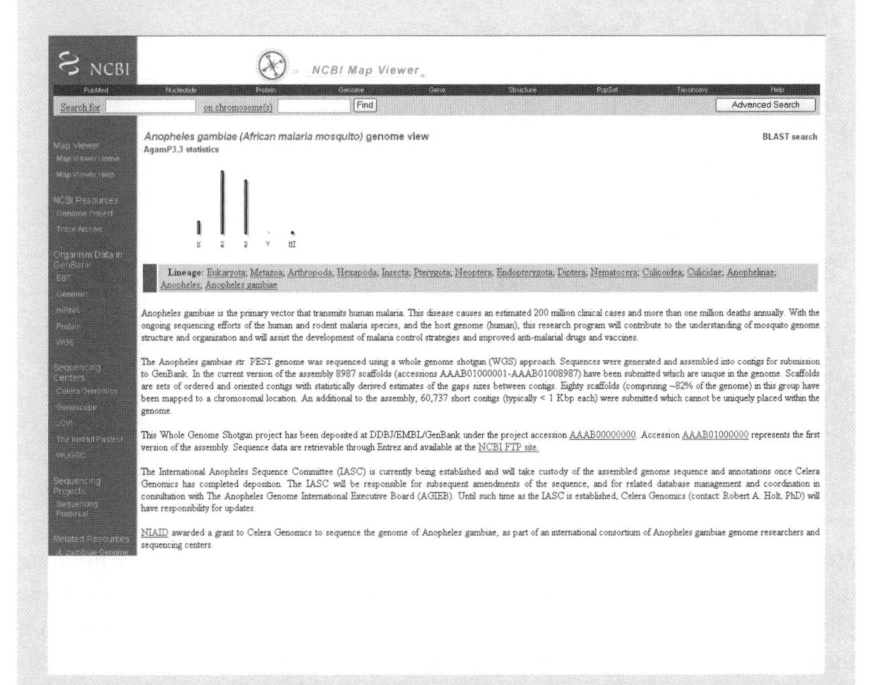

Fig. 5.18 NCBI website entries for the complete genomes of *C. elegans* and *A. gambiae*. All genome sequencing projects, complete and incomplete, can be accessed via http://www.genomesonline.org/.

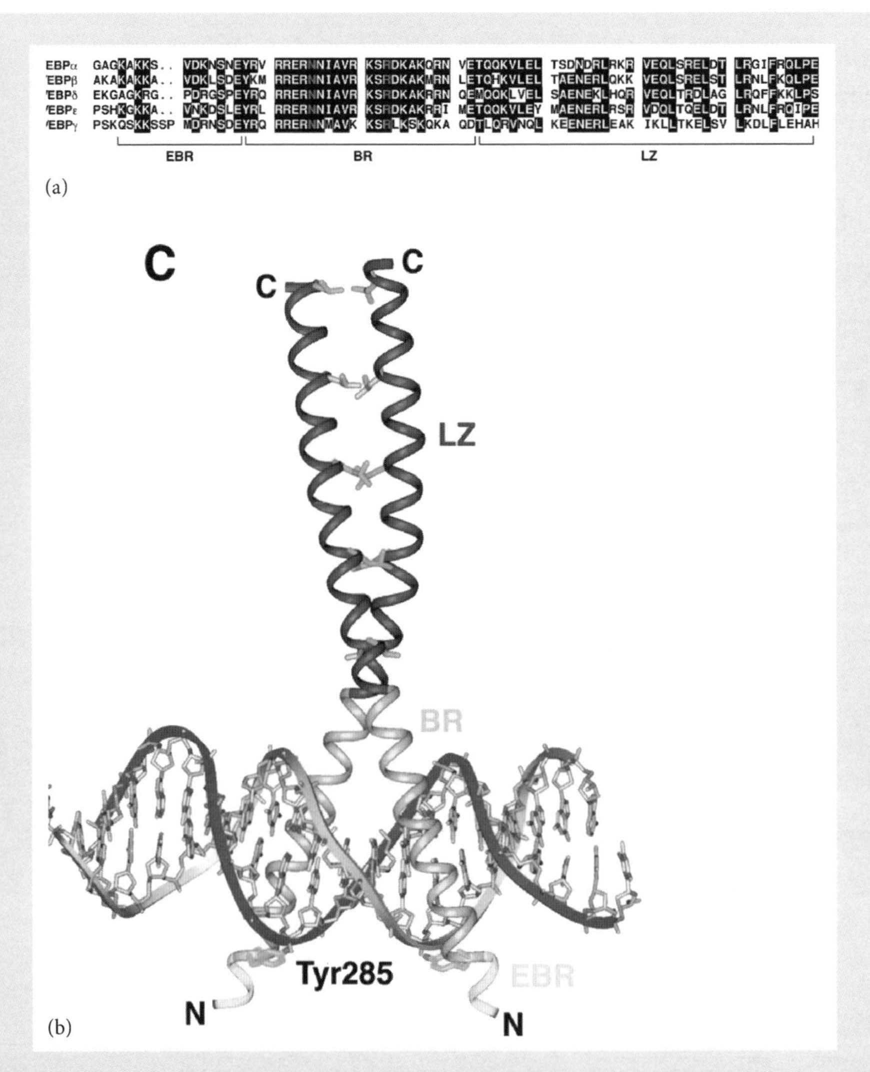

Fig. 5.19 DNA binding protein/DNA complex. (a) Alignment of leucine zipper type DNA binding proteins, where LZ is the leucine zipper region, BR is the basic region, which is rich in arginine and lysine, and interacts with DNA, and EBR is the extended basic region. (b) Overall structure of the binding protein/DNA complex (from Miller *et al.*, 2003).

The ExPASy (Expert Protein Analysis System) web address is http://www.expasy.org/.

are considered in section 5.8.3 and properties predicted from sequences are outlined in section 5.8.5. A number of software packages which calculate the physico-chemical properties of proteins from their sequences are available on the web, e.g. ProtParam, available on the ExPASy server. ProtParam can calculate a number of parameters, including amino acid composition, molecular mass, pI (isoelectric point), and absorption coefficient, from complete sequences or selected stretches of sequence.

The amino acid composition derived from a protein sequence, or selected stretch of sequence, will indicate the percentage of amino acids with acidic, basic, and hydrophobic side chains. This information alone can provide clues to the function of a protein e.g. DNA-binding proteins have stretches which are rich in basic amino acids (to neutralize the negative charges on the sugar–phosphate backbone of the DNA (see Fig. 5.19)) and proteins that are active in acidic environments have a high ratio of acidic:basic amino acids, such as the protease pepsin with ~11:1 ratio of acidic:basic amino acids.

The amino acid composition can also assist experimental structural characterization such as determining the molecular mass using ultracentrifugation (Chapter 8, section 8.2.4) which requires the partial specific volume of a protein; this can be calculated from the amino acid composition data.

ProtParam and similar packages calculate the theoretical mass of a protein from the sum of the masses of the amino acids in the protein sequence (Chapter 1, section 1.3.2.1). The molecular mass of a protein is an essential property in protein characterization experiments as it allows the conversion of concentration values (e.g. mg mL^{-1}) to units of molarity. Molarity units are an absolute necessity in ligand binding studies (how many molecules of ligand bound per molecule protein?), enzyme kinetics (how many molecules of product generated per enzyme molecule per second?), and chemical modification studies (how many side chains are modified per molecule of protein?). In addition, the molecular mass gives an indication of the relative mobility of the protein on polyacrylamide gel electrophoresis and gel filtration chromatography. Differences between theoretically and experimentally derived mass values can provide evidence of post-translational modification, which can be characterized using mass spectrometry (Chapter 8, section 8.2.5) or predicted using the tools outlined in section 5.8.5.

The theoretical isoelectric point (pI) is calculated from a protein sequence using the number of charged groups and their average pK_a values (Bjellqvist *et al.*, 1993), see Chapter 1, section 1.3.2.2. This calculation is based on the protein existing in a denatured state in which the pK_a values of the side chains correspond to those of small amino acid model compounds. However, the experimental values are derived from studies of native proteins where the pK_a values of side chains would be expected to vary markedly depending on their environment. Despite these reservations, predicted pI values are a useful tool in devising electrophoresis or ion-exchange chromatography experimental conditions.

It is possible to compute the absorption coefficient (Chapter 1, section 1.3.2.3, and Chapter 3, section 3.6) of a protein knowing the amino acid sequence and the molar absorption coefficients (E) of tyrosine, tryptophan, and disulphide bonds at A_{280} using the following equation:

This equation is taken from Pace *et al.* (1995). It applies to those proteins known to contain disulphide bonds; if the protein is intracellular and therefore does not contain disulphide bonds, the term in N_{Cys} is omitted.

$$E_{Prot} = N_{Tyr} \times E_{Tyr} + N_{Trp} \times E_{Trp} + N_{Cys} \times E_{Cys} \quad 5.3$$

where N_X is the number of amino acid X per polypeptide chain. The values of E_{Tyr}, E_{Trp} and E_{Cys} are 1490, 5500 and 62.5 M^{-1} cm^{-1}, respectively. Table 5.2 presents a

Table 5.2 Sample calculations of the absorption coefficients for a range of proteins.

Protein	Molecular mass (Da)	N_{Tyr}	N_{Trp}	N_{Cys}	E_{Prot} (calc) (1 M)*	E_{Prot} (calc) (1 mg mL^{-1})*	E_{Prot} (exp) (1 mg mL^{-1})
Insulin (bovine)	5 734	4	0	6	5 960 (6 335)	1.04 (1.10)	0.97
Lysozyme (hen)	14 314	3	6	8	37 470 (37 970)	2.62 (2.65)	2.63
Chymotrypsinogen (bovine)	25 666	4	8	10	49 960 (50 585)	1.95 (1.97)	1.98
Phosphoglycerate kinase (yeast)	44 607	7	2	1	21 430 (21 430)	0.48 (0.48)	0.49
Pyruvate kinase (rabbit muscle)	57 917	9	3	9	29 910 (30 410)	0.52 (0.53)	0.54
Serum albumin (bovine)	66 296	20	2	35	40 800 (42 925)	0.62 (0.65)	0.66

*The numbers in brackets assume that all (or the maximum number) of Cys residues form disulphide bonds. Disulphide bonds would be expected in extracellular proteins such as insulin, lysozyme, chymotrypsinogen, and serum albumin, but not in intracellular proteins such as phosphoglycerate kinase or pyruvate kinase.

number of examples of proteins with calculated absorption coefficients, as well as the experimentally measured values.

Division of E_{Prot} by the molecular mass of the polypeptide chain of the protein (in Da) gives the absorption coefficient for a 1 mg mL^{-1} solution of the protein.

The calculation of E_{Prot} refers to a protein in its native state and is based on average values of the absorption coefficients for the amino acids in native proteins. There are other equations which allow the E_{Prot} of the denatured protein to be calculated with reasonable accuracy; however, the value obtained must then be adjusted by an experimentally determined factor to give the correct value for the native protein. These calculations will not be reliable if the protein contains prosthetic groups which absorb in the near UV portion of the spectrum.

The absorption coefficient at A_{280} is essential for simple and accurate estimates of the concentration of protein present in most purification and characterization techniques (Chapter 6, section 6.1.5). In the case of proteins that contain very few tyrosine and tryptophan residues, e.g. collagen, it may be possible to estimate the concentration of protein present using absorbance measurements at A_{205} due to the peptide bond absorbance (Chapter 6, section 6.1.6).

The term 'homologous' in terms of sequence comparisons implies sequences are descended from a common ancestor. As the name implies, 'similarity' is simply a measure of how similar sequences are, irrespective of the evolutionary route, i.e. could arise from divergence, convergence, gene duplication, or a combination of these.

5.8.3 Sequence comparisons

Sequence comparisons (Chapter 1, section 1.3.2.7) have proved increasingly informative in protein characterization as a result of the growth of sequence databases. Whilst pairwise alignments provide a measure of the similarity of two sequences, the information arising from multiple sequence alignments, using more than two sequences from a protein family, can be used to:

Suggest the function of unknown proteins: The rationale of this approach is to relate the sequence of an unknown protein to the multiple alignment of a family of similar known proteins. This approach assumes that sequence features common to the multiple alignment and the unknown protein will confer similar function and structure. The growth of sequence databases alongside improved software packages has greatly improved the accuracy of this approach. Most packages use sequence and structure information to generate multiple alignments of families of known proteins, which can be used to identify unknown protein function, e.g. PSI-BLAST, Gapped BLAST and PSI-BLAST (Altschul *et al.*, 1997). The significance of the similarity of an unknown protein to multiple alignments identified by programs such as BLAST is given a score. This score provides a measure of the probability that similarities between sequences have occurred by chance. In the case of PSI-BLAST an *E*-score is given, ranging from 0 to the number of sequences in the database: *E*-scores less than 0.2 are indicative of homology and values greater than 1.0 indicate that the similarities are just as likely to have arisen by chance. If we consider the serine proteases, we find that the similarity between trypsin (EC 3.4.21.4) and chymotrypsin (EC 3.4.21.1) produces an *E*-score of $\sim 1 \times 10^{-40}$, whereas the similarity between trypsin and subtilisin (EC 3.4.21.62) produces an *E*-score >1.0. Structural studies have established that trypsin and chymotrypsin are closely related and are both members of the trypsin-like superfamily (section 5.8.4). However, while trypsin and subtilisin have similar catalytic residues (the charge relay residues Asp, His and Glu, see Chapter 9, section 9.8.2), the non-catalytic residues are not similar. Structural studies have revealed that trypsin is composed mainly of β-sheet and is a member of the trypsin-like superfamily, whereas subtilisin is an α-helix/β-sheet class of protein and is a member of the subtilisin-like superfamily.

BLAST is an acronym for Basic Local Alignment Search Tool.

Identify functionally and/or structurally important residues: Multiple alignments of related protein sequences can be used to identify residues that are identical across all sequences. Absolute identity is often indicative of functionally and/or structurally important residues, e.g. residues essential for the catalytic mechanism of enzymes or residues essential for prosthetic group interactions. A good example of this is provided by haemocyanin, the oxygen transporter in many chelicerates and arthropods, which contains two copper-binding sites, copper A and copper B. Each copper is coordinated to three histidine residues which are highly conserved across all species (see Fig. 5.20), highlighting the importance of these residues.

As well as complete identity, comparisons of sequences can show that certain amino acids may be functionally conserved (e.g. always non-polar such as Leu, Ile, or Val), or may be freely variable (e.g. Lys, Asp, Gly, or Phe). Clearly freely variable amino acids are unlikely to play any crucial role in either the structure or the function of the protein concerned.

Improve structure prediction tools: It is possible to predict the three-dimensional structure of a protein by modelling its sequence onto a homologous structure of a known protein (parent). This approach has become more accurate with the availability of multiple structures of homologous proteins that reveal the highly conserved structural features of the family.

Detect and characterize evolutionary relationships: The classical approach to phylogeny, based on macroscopic observations of form and function, has been replaced with the comparison of molecular information, including protein sequences. Sequence alignments can be used to infer the evolutionary relationship

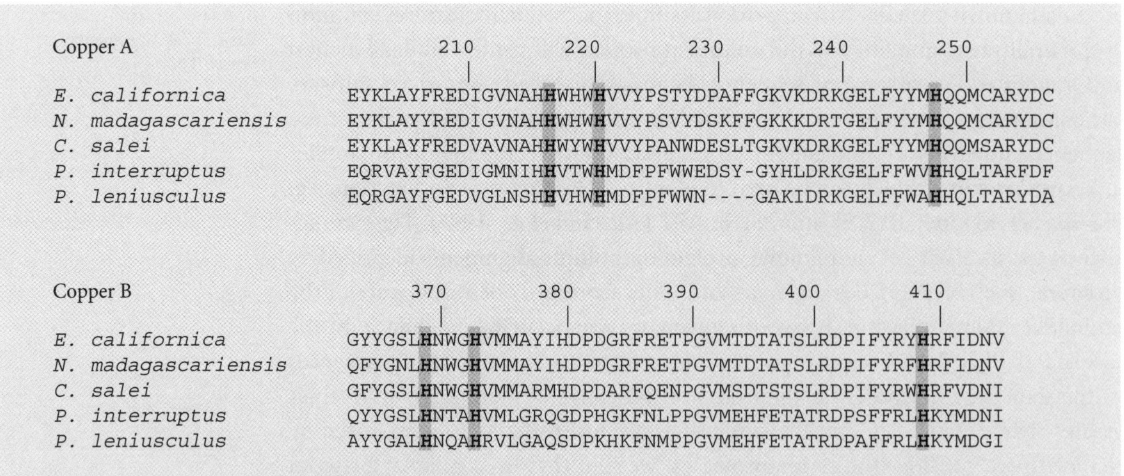

Fig. 5.20 Alignment of copper-binding sites Copper A and Copper B, of selected haemocyanin sequences from *Eurypelma californica* (American tarantula), *Nephila inaurata madagascariensis* (Red legged golden orb-web spider), *Cupiennius salei* (Wandering spider), *Panulirus interruptus* (California spiny lobster), and *Pacifastacus leniusculus* (Signal crayfish). Functionally important histidine residues which coordinate to copper are shown in bold.

between organisms (by comparing the same protein from different organisms) or between proteins (by comparing related proteins from the same organism). Protein sequences greater than 100 amino acids in length and which share 30% identity are assumed to be homologous, i.e. are members of the same family, whereas those sharing less than 15–20% identity are similar but are not necessarily derived from a common evolutionary ancestor. In such cases additional information such as three-dimensional structures may be needed to confirm homology (Rost, 1999).

5.8.4 Structure comparisons

Protein structure comparisons are powerful tools in establishing structure, function, and evolutionary relationships, particularly when comparing distantly related proteins with low levels of sequence identity. The development of databases which group proteins with related structures in a hierarchal manner, e.g. SCOP (Structural Classification of Proteins; http://scop.mrc-lmb.cam.ac.uk/scop/) and CATH (Class Architecture Topology Homologous superfamily; http://www.cathdb.info/latest/cath_info.html), has enabled the rapid identification of the close relatives of any particular protein. In order of decreasing hierarchy, the SCOP database organizes protein domains according to family (homologous with respect to structure inferring a common evolutionary ancestor), superfamily (similar structure/function suggesting a probable relationship between proteins), and fold (similar secondary structure arrangement but distantly related function and overall structure). The CATH database organizes protein domains into homologous superfamilies (sharing a common ancestor), topology (similar shape and connectivity of secondary structure), architecture (similar secondary structure arrangement), and class (similar secondary structure composition).

Having identified structures related to the protein of interest, it is then possible to use molecular visualization tools (section 5.8.1) to overlay related structures onto the protein of interest to characterize protein function, identify residues of structural and functional importance, and infer evolutionary relationships, in a manner analogous to sequence comparisons.

5.8.5 Predicting possible functions of proteins

The classical experimental approach to determining the function of a novel protein requires structure characterization (section 5.2.2), determination of the factors influencing activity (section 5.2.3), and identification of ligands (section 5.2.4). However, the work by Max Perutz in the 1960s on myoglobin and haemoglobin indicated that the three-dimensional structure of a protein, and hence its function, is determined by its amino acid sequence. The subsequent expansion of sequence and structure databases has enabled the development of tools to predict structure and function from amino acid sequences.

Max Perutz and John Kendrew were awarded the Nobel Prize in chemistry in 1962 for their studies on the structures of globular proteins, including haemoglobin and myoglobin. Perutz noted that the tertiary structures of haemoglobin and myoglobin subunits were similar and proposed that this was as a result of similar amino acid sequences. The subsequent determination of the amino acid sequences of these proteins confirmed they shared a substantial degree of identity (of the order of 30%) that conferred similar structures and functions.

In many cases, it will be possible to identify homologues of a novel protein using tools such as *PSI-BLAST*. Assuming the complete sequence of the novel protein shares a high degree of identity with a family of well-characterized homologous proteins and contains key functional residues, it will be possible to assign function. More functional information can be obtained using comparative tools to build a high-quality structure model of the novel protein, e.g. details of an active site can be used to infer substrate specificity.

Folded proteins often consist of a number of structural units known as domain folds, with each fold often associated with a particular function. In the absence of a family of well-characterized homologues, fold recognition methods can be used to match the novel sequence to known folding patterns. This process involves aligning the query sequence to a number of similar sequences of known fold, e.g. nucleotide binding fold, globin-like fold, or immunoglobulin fold (see Chapter, section 1.5.2). Once aligned, approximate models of the novel protein are constructed and evaluated to identify the most accurate model.

Secondary structure prediction methods are outlined in this chapter, section 5.2.2 and in Chapter 8, section 8.4.

In addition to suggesting function from the predicted structure, there are a number of bioinformatic tools that can predict the subcellular location of proteins from their amino acid sequences. Transmembrane helices and their topology can be predicted with a high degree of accuracy, based on sequence length and composition. For example, the predicted transmembrane helices of bovine rhodopsin are in good agreement with the experimentally determined structure of this seven transmembrane helix protein (bovine rhodopsin PDB entry 1F88). Likewise there are a number of methods that can identify N-terminal pre-sequences which determine organelle destination, e.g. mitochondrial transfer peptides (25–45 residues), nuclear location signals (4–6 residues), signal peptides for extracellular proteins (20–30 residues), and peroxisomes target signals (either the C-terminal signal Ser–Lys–Leu as in catalase or a 30 residue N-terminal signal sequence, as occurs in thiolase). Examples of other functional motifs are outlined in Chapter 1, section 1.3.2.6.

NetPhos http://www.cbs.dtu.dk/services/NetPhos/

NetNGlyc http://www.cbs.dtu.dk/services/NetNGlyc/

CASTp http://sts.bioengr.uic.edu/castp/

PONDR http://www.pondr.com

Factors affecting protein activity, including post-translational modifications and ligand interactions, can also be predicted using bioinformatic tools. Post-translational modification sequence motifs (see Chapter 1, section 1.3.2.6) can be identified by a number of pattern recognition programs such as NetPhos to predict serine, threonine and tyrosine phosphorylation sites in eukaryotic systems and NetNGlyc to predict N-glycosylation sites in human proteins. Analysis of the surface of protein structures, using programs such as CASTp, can identify and characterize accessible pockets such as ligand binding sites and active sites. As a final example, regions of proteins that do not adopt stable three-dimensional structure even under native conditions can be predicted using PONDR (Prediction of Natural Disordered Regions); in many proteins such regions can play important roles in adjusting the specificity of ligand recognition.

Prediction of protein structure, factors influencing activity, and ligand binding sites using bioinformatic tools can reveal the possible function of a protein and provide valuable information to complement classical experimental approaches.

5.9 Experimental design

KEY CONCEPTS

- Understanding the process of successful experimental design
- Appreciating the need to comply with relevant safety and ethical requirements

A sound understanding of the goals and methods employed to characterize proteins can be best put into practice alongside good experimental design. Whether you are aiming to study the function, structure, or the nature of effectors of a particular protein, it is important to consider some essential, if somewhat obvious, aspects of experimental design. A thorough literature review coupled with use of a range of bioinformatic tools (e.g. sequence data of a similar protein, the calculated molecular mass, isoelectric point, and absorption coefficient at 280 nm) will yield a wealth of information to assist experimental design. In this section, we shall consider the design of two types of experiments to demonstrate how to obtain relevant and specific information to address key questions being asked about a protein.

Example 1

Suppose that we wish to characterize the quaternary structure of a signal-transducing G-protein and explore the binding of a GTP analogue to it. Prior to characterization studies, it is essential to test the purity and structural and functional integrity of the G-protein. SDS-PAGE and mass spectrometry can be employed to check that possible contaminating proteins are absent. SDS-PAGE combined with specific activity measurements (GTPase assay) will provide a good indication of the structural and functional integrity of the protein. To maintain this integrity in later experimental procedures, it is essential to determine the optimal conditions for storage and activity measurements of the G-protein. This will require exploring a range of possible storage and assay conditions; however, guidance may be available from previous studies on similar systems.

All protein characterization experiments hinge on two important methods: an accurate method to determine the concentration of the protein and a specific assay to determine the activity of the protein. As each protein concentration determination method has its limitations, it would be worthwhile determining the concentration of the sample of G-protein using a number of different methods (see Chapter 6, section 6.1) to ensure the reliability of the estimate. A reliable assay of activity requires high-quality reagents and a number of control measurements to validate the procedure. In the case of a G-protein, a simple colourimetric assay involving malachite green to measure phosphate release is used to monitor GTPase activity. This assay includes a suitable buffer, salts, GTP, malachite green, and G-protein. Omission of each assay component in turn will indicate the specificity of the assay. It is also important to determine whether the assay measurements are proportional to the amount of G-protein added, i.e. doubling the amount of G-protein added to the assay should generate a two-fold increase in activity. If this is not the case, it may be that some assay component is present in limiting quantities and the concentrations of the various components may need to be adjusted to overcome the problem.

A number of experimental approaches will be required to determine the quaternary structure of the G-protein including gel filtration chromatography (Chapter 8, section 8.2.3), SDS-PAGE (Chapter 8, sections 8.2.1 and 8.2.2), and mass spectrometry (Chapter 8, section 8.2.5). All three methods require a range of suitable standard proteins, of known molecular mass, to calibrate the procedure. Analysis of the G-protein using each of these methods may require a buffer exchange step; mass spectrometry and SDS-PAGE are sensitive to high salt concentrations. On the other hand, poor gel filtration data may be obtained if the ionic strength of the buffer is too low, since non-specific interactions between protein and chromatographic media can occur. It is also important to consider the amount of protein analysed using each of these techniques, as too much or too little may compromise the quality of the data.

To explore the effects of a GTP analogue on the G-protein, activity assays and binding assays should be conducted. Prior to gathering data in the presence of the G-protein, it will be important to assess the impact of the presence of the GTP analogue on the activity assay and binding assay, e.g. is it soluble under the assay conditions, will it cause a change in pH or a change in absorption? Preliminary measurements in the presence of the GTP analogue will be required to establish the correct concentration range to use. Ideally, we should be aiming to cover the concentration range from 0.1 to 10 times the K_d for the GTP analogue in the binding assays. It is worthwhile exploring more than one binding assay method to confirm our findings, e.g. a direct binding measurement using equilibrium dialysis and a spectroscopic titration method to monitor ligand-induced structural changes.

Throughout all of these approaches, it is good practice to test the outcome of empirical calculations using experimentally derived data. For example, dilution of a stock solution of G-protein should result in a suitable A_{280} value or introduction of a calculated amount of G-protein to an assay should generate a predictable activity measurement.

Example 2

Use of the artificial substrate MUNANA leads to a much more convenient and sensitive assay for the enzyme compared with the natural substrates (removal of sialic acid groups from oligosaccharides, glycoproteins and glycolipids).

Suppose that we wish to obtain kinetic parameters for the enzyme neuraminidase (EC 3.2.1.18) from the virus H5N1 (the causative agent of bird flu) acting on the fluorogenic substrate 2′ (4-methylumbelliferyl)-α-D-*N*-acetylneuraminic acid (MUNANA). Neuraminidase is crucial for the degradation of glycan structures at the surfaces of viral-infected cells and its action is required for the release of viral particles. We also wish to study the effects of the inhibitor 5-*N*-acetyl-3-(1-ethylpropyl)-1-cyclohexene-1-carboxylic acid (Tamiflu) (Fig. 5.21), on neuraminidase activity.

As in the previous example, preliminary experiments should be undertaken to establish the purity and quality of the neuraminidase, determine optimal storage conditions, and test the ability of the assay to provide reliable and accurate results. To obtain data that will establish the kinetic parameters of neuraminidase, the concentration of the substrate MUNANA should ideally be varied over the range from 0.1 to 10 times the K_m. The success of this approach requires that the MUNANA remains soluble in the assay over this concentration range and does

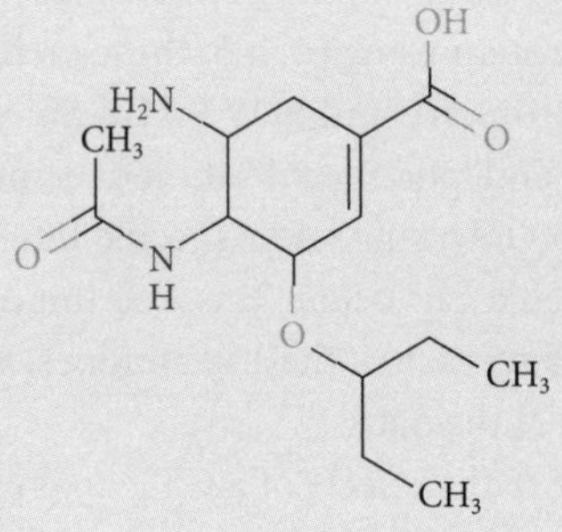

Fig. 5.21 Structure of the neuraminidase inhibitor, 5-N-acetyl-3-(1-ethylpropyl)-1-cyclohexene-1-carboxylic acid (Tamiflu).

not alter the properties of the assay (e.g. change pH or spectroscopic properties). Repeating this range of activity measurements in the presence of increasing concentrations of Tamiflu will provide a measure of the type and strength of neuraminidase inhibition. A series of preliminary experiments will be required to establish the concentration ranges of substrate and inhibitor required and to check that Tamiflu, in the absence of neuraminidase, has no effect on the assay procedure.

Two final considerations in the design of any protein characterization experiment are safety and ethical issues. All experiments should meet the safety standards required by national health and safety regulatory bodies and institutional safety guidelines. Likewise, experiments involving animals or human subjects (or materials derived from them) are required to adhere to national ethical legislation and institutional requirements.

References for Chapter 5

Altschul, S.F., Madden, T.L., Schaffer, A.A., *et al.* (1997) *Nucl. Acids Res.* **25**, 3389–402.

Armache, K.-J., Mitterweger, S., Meinhart, A., and Cramer, P. (2005) *J. Biol. Chem.* **280**, 7131–34.

Bjellqvist, B., Hughes, G.J., Pasquali, C., *et al.* (1993) *Electrophoresis* **14**, 1023–31.

Boxer, D.H., Zhang, H., Gourley, D.G., *et al.* (2004) *Org. Biomol. Chem.* **2**, 2829–37.

Chang, A.C.Y, Nunberg, J.H., Kaufman, R.J., *et al.* (1978) *Nature* **275,** 617–24.

Cheng, Y. and Prusoff, W.H. (1973) *Biochem. Pharmacol.* **22**, 3099–108.

Chou, P.Y. and Fasman, G.D. (1974) *Biochemistry* **13**, 222–45.

Chou, W.Y. and Matthews, K. S. (1989) *J Biol Chem* **264**, 18314–19.

Cowan, D.A. (1995) *Essays Biochem.* **29**, 193–207.

Engel, P.C. (1981) *Enzyme Kinetics*, 2nd edition, Chapman & Hall, London.

Evrard, C., Fastrez, J., and Declercq, J.-P. (1998) *J. Mol. Biol.* **276**, 151–64.

Groch, N., Schindelin, H., Scholtz, A. S., *et al.* (1992) *Eur. J. Biochem.* **207**, 677–85.

Groll, M., Koguchi, Y., Huber, R. and Kohno, J. (2001) *J. Mol. Biol.* **311**, 543–48.

Gu, Y.Q., Holzer, F.M. and Walling, L.L. (1999) *Eur. J. Biochem.* **263**, 726–35.

Kelly, S.M., Jess, T.J., and Price, N.C. (2005) *Biochim. Biophys. Acta* **1751**, 119–39.

Magliery, T.J. (2005) *Med. Chem. Rev.* **2**, 303–23.

Miller, M. Shuman, J.D., Sebastian, T., *et al.* (2003) *J. Biol. Chem.*, **278**, 15178–84.

Pace, C.N., Vajdos, F., Fee, L., *et al.* (1995) *Protein Sci.* **11**, 2411–23.

Rost, B. (1996) *Methods Enzymol.* **266**, 525–39.

Rost, B. (1999) *Protein Eng.* **12**, 85–94.

Sander, C. and Schneider, R. (1994) *Nucl. Acids Res.* **22**, 3597–99.

Theerasilp, S. and Kurihara, Y. (1988) *J. Biol. Chem.* **263**, 11536–39.

Viles, J.H., Cohen, F.E., Prusiner, S.B., *et al.* (1999) *Proc. Natl. Acad. Sci. USA* **96**, 2042–47.

Wang, L., Xie, J., Deniz, A.A. and Schultz, P.G. (2003) *J. Org. Chem.* **68**, 174–76.

Wlodawer, A. and Vondrasek, J. (1998) *Annu. Rev. Biophys. Biomol. Struct.*, 27, 249–84.

Appendix

Appendix 5.1 Derivation of the Cheng–Prusoff equation (Cheng and Prusoff, 1973)

According to the Michaelis–Menten equation (Chapter 4, eqn. 4.26), the rate of a reaction (v_1) at a substrate concentration $[S]$ is:

$$v_1 = \frac{V_{max} \times [S]}{K_m + [S]}$$

In the presence of a competitive inhibitor at a concentration $[I]$ with an inhibitor constant K_{EI}, the K_m is increased by a factor $\left(1 + \frac{[I]}{K_{EI}}\right)$ (Chapter 4, section 4.3.4), so the new rate (v_2) is given by:

$$v_2 = \frac{V_{max}[S]}{K_m\left(1 + \frac{[I]}{K_{EI}}\right) + [S]} \qquad 5.4$$

If $[I] = IC_{50}$, $v_2 = 0.5 \times v_1$, i.e. $v_1 = 2 \times v_2$

$$\frac{V_{max} \times [S]}{K_m + [S]} = \frac{2 \times V_{max} \times [S]}{K_m\left(1 + \frac{IC_{50}}{K_{EI}}\right) + [S]}$$

1) Dividing both sides by $V_{max} \times [S]$, and then cross-multiplying we obtain:

2) $2 \times K_m + 2 \times [S] = K_m + \frac{K_m \times IC_{50}}{K_{EI}} + [S]$

3) Collecting terms in K_m and $[S]$, we obtain:

4) $K_m + [S] = \frac{K_m \times IC_{50}}{K_{EI}}$

5) Rearranging we obtain:

6) $K_{EI} = \dfrac{K_m \times IC_{50}}{K_m + [S]}$

7) Dividing the top and bottom terms on the right-hand side by K_m, we obtain:

8) $$K_{EI} = \frac{IC_{50}}{1 + \frac{[S]}{K_m}} \qquad 5.5$$

which is the Cheng–Prusoff equation.

C

Data analysis in practice

Analytical methods

6

In this chapter, we shall focus on the analytical methods used to study proteins. We do not aim to provide an exhaustive list of all the methods needed to answer all of the key questions about protein structure and function (Chapter 5, section 5.2). Instead, we have focused on the most commonly used analytical techniques, namely determination of protein concentration, electrophoresis, enzyme assays, and binding assays. The techniques described in this chapter can be readily used in undergraduate practical classes to establish some of the key characteristics of a protein, including purity and concentration, relative molecular mass, enzymatic activity, and binding properties. For each of the protein determination methods, there is a short introduction describing its theoretical basis and a section outlining experimental design considerations. In addition, for the first two of these methods, specimen results are provided to allow data analysis with a view to enhancing the reader's understanding of these techniques and their application to characterizing proteins. The remaining techniques are discussed in greater detail in later chapters and references to these are given.

6.1 Determination of protein concentration

KEY CONCEPTS

- Knowing the key features of the various protein concentration determination methods
- Selecting an appropriate protein concentration determination method for a given protein sample

The determination of the amount of protein present in a solution is a widely used procedure in biochemistry. It is important to know the amount of protein at different stages in a purification procedure so that the specific activity (Chapter 3, section 3.8) of the protein being purified can be calculated. This will allow a purification table (Chapter 3, section 3.9), charting the success of the purification procedure, to be constructed. If we have a purified protein, knowledge of the amount of protein and its molecular mass will allow the number of moles of protein to be calculated and hence its properties expressed in molecular terms (e.g. the calculation of k_{cat} for an enzyme (Chapter 4, section 4.5) or the number of binding sites for a particular ligand (Chapter 4, sections 4.3 and 4.4).

When dealing with complex protein mixtures such as cell lysates and intermediate steps in a purification scheme, the methods generally used are comparative, i.e. they are based on the use of a reference, or standard, protein. Some of the more commonly used comparative methods are described including dye binding, bicinchoninic acid (BCA), Lowry and Biuret methods (sections 6.1.1–6.1.4). Details of the reagents are given in the Appendix to this chapter.

For sections 6.1.1–6.1.4 there is a standard pattern to the protein assays:

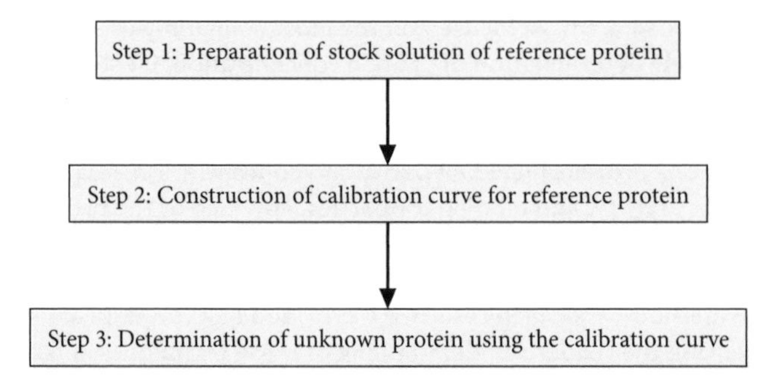

The most commonly used reference protein is bovine serum albumin (BSA), so Step 1 is common to all the methods. In any particular case, the precise details of Steps 2 and 3 will depend on various factors, including the sensitivity of the assay method.

When dealing with a purified protein, the most commonly used methods are spectrophotometric and are very often based on measurements of A_{280} (section 6.1.5). An alternative spectrophotometric method is based on measurements of A_{205} in conjunction with A_{280} (section 6.1.6).

In the last part of this section (section 6.1.7), some of the experimental aspects that must be considered when selecting and designing a protein determination method are outlined.

6.1.1 Dye binding method

Background

The absorbance spectrum of the dye Coomassie Brilliant Blue G-250 (Fig. 6.1) undergoes a measurable change when it binds to proteins. The protonated form of Coomassie Brilliant Blue G-250 dye has an orange/brown appearance (λ_{max} 465 nm); however, on binding to a protein the dye becomes deprotonated and has an intense blue appearance (λ_{max} between 595 and 620 nm) and is detected at 595 nm in the method outlined below. The unprotonated form of the dye results from

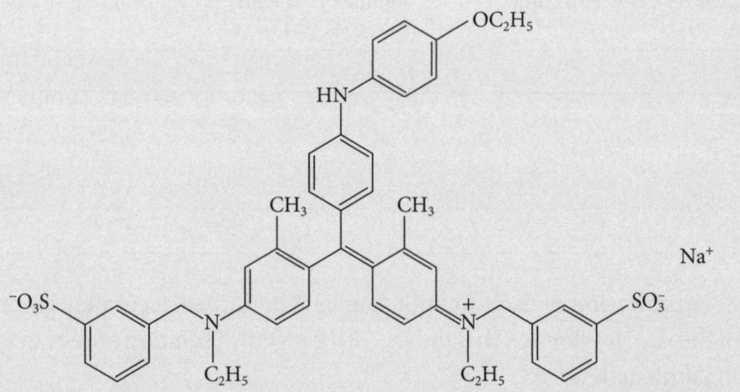

Fig. 6.1 The chemical structure of Coomassie Brilliant Blue G-250, the reagent used in the dye binding method to determine protein concentration. Coomassie Brilliant Blue G-250 forms a dye–protein complex which exhibits an absorbance maximum of between 595 and 620 nm.

the dye-anion interacting with side chain NH_3^+ groups. The NH_3^+ group content varies from protein to protein and NH_3^+ groups have varying reactivity towards Coomassie Brilliant Blue G-250. As a result, the degree of colour change varies from protein to protein.

Experimental aspects

Dye binding is a highly sensitive method of protein estimation that is suitable over the range 1–25 μg protein. Due to variability in protein–dye interaction, the standard curve should be constructed using, if at all possible, a protein of similar composition to the protein(s) to be determined. The method outlined is that described by Read and Northcote (1981).

A variation of this protocol for determination of smaller amounts of protein (1–5 μg) has been described in which a microtitre plate is used with smaller total volumes (210 μL). However, since the plate reader may not be readily available in an undergraduate laboratory, we shall only give details here of the 'conventional' larger-scale method.

Step 1: Preparation of the reference protein solution

A stock solution of BSA is prepared by weighing out 25-mg BSA and dissolving this in 5-mL water. At this stage, it is important to agitate the container *gently* to avoid excessive frothing (and hence denaturation of the protein).

The stock solution of BSA is diluted to give a series of BSA solutions of increasing concentrations, each with a final volume of 1 mL. You should complete the table below by calculating the volumes of the protein solution and water required.

You may find it useful to refer to Chapter 3, section 3.5.2 to check on how to dilute stock solutions.

Required BSA concentration ($mg\ mL^{-1}$)	Volume of 5 $mg\ mL^{-1}$ BSA stock (mL)	Volume of water (mL)
0		
0.2		
0.4		
0.6		
0.8		
1.0		

It is important to use the correct type of cuvette when measuring the absorbance at 280 nm. Glass cuvettes and certain plastic cuvettes (e.g. polystyrene) absorb strongly below 300 nm. Quartz cuvettes, which do not absorb significantly above 200 nm, are expensive. It is good practice to check the absorbance at the chosen wavelength with water or buffer in the cuvette.

The concentration of each solution can be determined accurately by measurement of the A_{280}. It is known that the A_{280} of a 1 $mg\ mL^{-1}$ solution of BSA in a cuvette of 1-cm pathlength is 0.66.

Question **Why is it important to measure the concentration of the solution of BSA by spectrophotometry, rather than just relying on the amount of protein weighed out?**

Answer **The sample of BSA (a lyophilized (freeze dried) powder) may contain impurities such as buffer salts or adsorbed moisture. Hence, the material weighed out is likely to be <100% protein. Determination of the concentration by measuring the A_{280} also provides a check on the weighing and pipetting procedures.**

Specimen A_{280} values (relative to the first solution with no BSA) for a typical experiment are shown in the table below. The actual BSA concentrations can be calculated using the known absorption coefficient (0.66 for a 1 $mg\ mL^{-1}$ solution in a cuvette of 1-cm pathlength).

Nominal BSA concentration ($mg\ mL^{-1}$)	A_{280}	Actual BSA concentration calculated from A_{280} ($mg\ mL^{-1}$)
0	0	0
0.2	0.120	0.182
0.4	0.238	0.361
0.6	0.365	0.553
0.8	0.472	0.715
1.0	0.602	0.912

A plot of A_{280} vs the volume of BSA stock added shows a very good straight line (see Fig. 6.2), confirming that the Beer–Lambert law (Chapter 3, section 3.6) applies over this range. The slope of this plot shows that the original stock solution of BSA is in fact 4.54 $mg\ mL^{-1}$, rather than the nominal 5 $mg\ mL^{-1}$ that was made up by weight.

Step 2: Construction of calibration curve for the reference protein

To obtain solutions over the required range of protein, 25 μL of each of the BSA solutions prepared in step 1 is added to 0.975 mL water and then 1.5 mL Coomassie

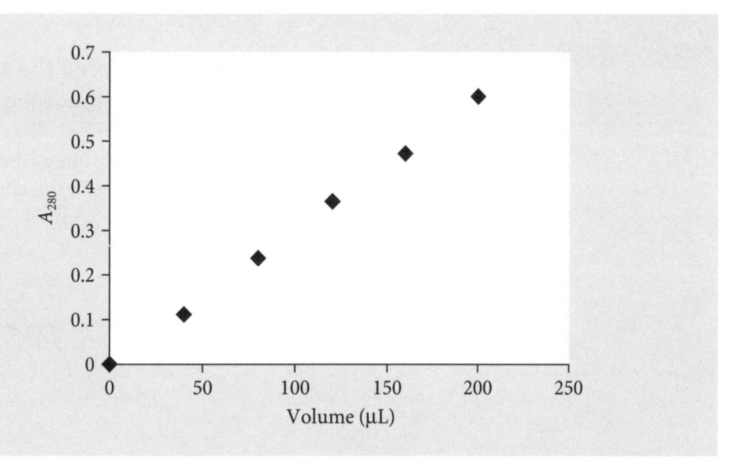

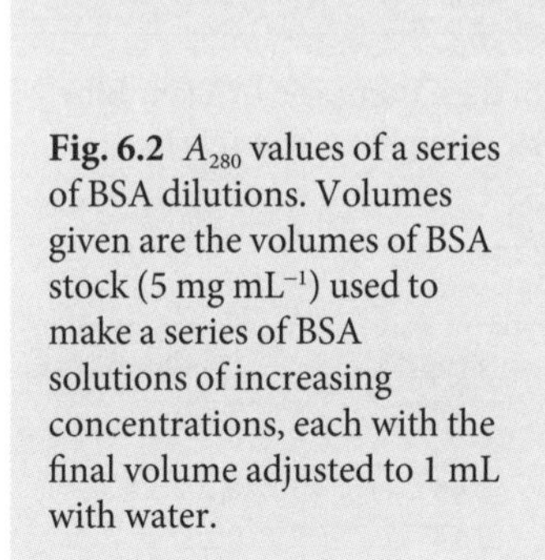

Fig. 6.2 A_{280} values of a series of BSA dilutions. Volumes given are the volumes of BSA stock (5 mg mL^{-1}) used to make a series of BSA solutions of increasing concentrations, each with the final volume adjusted to 1 mL with water.

Brilliant Blue reagent (see Appendix 6.1) is added to each diluted protein solution. After incubation, the A_{595} of each tube is read relative to the first solution, which contains no protein.

As the Coomassie Brilliant Blue dye binds to the surface of glass and quartz cuvettes, requiring harsh cleaning methods to remove the dye, it is preferable to use disposable polystyrene cuvettes.

To present your data as a standard curve you will have to calculate the amount of protein used in each assay. For example, if your nominal 0.2 mg mL^{-1} BSA solution was in fact 0.182 mg mL^{-1}, the amount of protein taken for the calibration curve is $0.182 \times (25/1000)$ mg, i.e. 0.00455 mg or 4.55 µg. You should be able to calculate the amounts of protein in the other solutions and complete the entries in the second column of the table. Plot the A_{595} values against the amount of protein used in each assay.

BSA concentration calculated from A_{280} (mg mL^{-1})	Amount of protein used in each assay (µg)	A_{595}
0	0	0
0.182	4.55	0.209
0.361		0.382
0.553		0.537
0.715		0.683
0.912		0.810

The plot of A_{595} vs protein is not a straight line but tends to level off somewhat at higher amounts of protein. Thus, it is important to ensure that readings of the unknown protein fall within the range over which the standard curve is linear.

Step 3: Determination of unknown protein using the calibration curve

The experiment is usually performed to determine the protein present in complex mixtures. However, it is instructive to explore the variation in response between a number of purified proteins. Suitable proteins to use might be hen egg white lysozyme, bovine trypsin, and hen ovalbumin To calculate the amount of protein used in each assay, you will need the A_{280} values for 1 mg mL^{-1} solutions of these proteins in a cuvette of 1-cm pathlength; these are 2.63, 1.57, and 0.69, respectively.

SELF TEST

Check that you have mastered the key concepts at the start of this section by attempting the following question.

ST 6.1 What can you conclude about the response of the Coomassie Brilliant Blue reagent to the different proteins in the table below? How might you be able to account for your findings?

Protein used in each assay (μg)	A_{595}			
	BSA	Lysozyme	Trypsin	Ovalbumin
0	0	0	0	0
5	0.23	0.22	0.08	0.12
10	0.42	0.43	0.14	0.22
15	0.59	0.60	0.19	0.30
20	0.75	0.76	0.24	0.38
25	0.89	0.90	0.29	0.45

Answer

ST 6.1 The responses of BSA and lysozyme are similar; that of trypsin is only about 33% that of BSA, and that of ovalbumin about 50% of BSA. There is a general correlation between the amount of dye bound to each protein and the number of Lys + Arg side chains present, but other factors, including the number of aromatic amino acid side chains, appear to be important.

6.1.2 Bicinchoninic acid method

Background

Under alkaline conditions, Cu^{2+} present in the BCA reagent (see Appendix 6.1) complexes with peptide bonds and becomes reduced to Cu^{+}. Cu^{+} subsequently interacts with BCA to form a purple BCA–Cu^{+} complex that can be detected at 562 nm.

Experimental aspects

The BCA method is a highly sensitive method for determining protein concentrations, with sensitivity limits of 0.2–50 μg. The BCA method also has the advantages of being a one-step procedure and employing a reagent which is stable for up to 1 week. Care should be taken to ensure that glucose, NH_4^+, and EDTA are absent from the procedure as these compounds interfere with the process of colour development.

Step 1: Preparation of the reference protein solution

A stock solution of BSA (5 mg mL^{-1}) should be prepared as outlined in section 6.1.1 and then diluted two-fold with water to give a nominal concentration of 2.5 mg mL^{-1}.

You should complete the table below to show how this stock solution of BSA is further diluted to give a series of BSA solutions of increasing concentrations, each with a final volume of 1 mL.

Required BSA concentration (mg mL^{-1})	Volume of 2.5 mg mL^{-1} BSA stock (mL)	Volume of water (mL)
0	0	1.0
0.5		
1.0		
1.5		
2.0		
2.5		

The required BSA concentrations (0.5 mg mL^{-1} up to 2.5 mg mL^{-1}) must be diluted by a factor of 10 to allow accurate recording of A_{280} values (see Chapter 3, section 3.6). Specimen A_{280} values (relative to the first solution containing no BSA) for a typical experiment are shown in the table below.

Nominal BSA concentration (mg mL^{-1})	A_{280} of 1:10 dilution of BSA solution*	A_{280} of BSA solution	Actual BSA concentration calculated from A_{280} (mg mL^{-1})
0	0	0	0
0.5	0.035	0.35	0.53
1.0	0.068	0.68	
1.5	0.110	1.10	
2.0	0.141	1.41	
2.5	0.176	1.76	

*1:10 dilution of each BSA solution is prepared by pipetting 100 μL of each serial dilution into a quartz cuvette containing 900 μL of water and reading the A_{280} relative to the solution with no BSA as a blank.

Given that a 1 mg mL^{-1} solution of BSA should have an A_{280} of 0.66 in a cuvette of 1-cm pathlength, you can use the measured absorbance to calculate the accurate protein concentration value for each of the BSA solutions in this dilution series and thus complete the right-hand column of the table.

Step 2: Construction of the calibration curve for the reference protein

To obtain solutions over the required range of protein:

1) 980 μL of BCA reagent is added to 20 μL of each of the protein solutions prepared in step 1.

2) After incubation at 60°C for 1 h, the solutions are cooled to room temperature.

3) The absorbance at 562 nm is read against a blank sample which contains no BSA.

The incubation step at 60°C gives the BCA method greater sensitivity and less protein-to-protein variation than using lower temperatures.

The amount of protein used in each assay is calculated knowing:

A. The concentrations you have calculated in completing the previous table.

B. 20 μL of protein sample is added to 980-μL reagent.

For example, 20 μL of the 0.53 mg mL^{-1} protein solution contains (20/1000) × 530 μg = 10.6 μg. You should be able to calculate the amounts in the other solutions and complete the table below.

BSA concentration calculated from A_{280} (mg mL^{-1})	Amount of protein in assay (μg)	A_{562}
0	0	0
0.53	10.6	0.58
		1.17
		1.88
		2.29
		2.66

The A_{562} values are then plotted against the amount of protein used in each assay.

The plot of A_{562} vs protein is a reasonable straight line particularly at A_{562} values up to about 1. It would be advisable to remeasure the solutions with absorbance values greater than about 1.5 after an appropriate degree of dilution (e.g. two- or three-fold), since the accuracy of many spectrophotometers is limited at such high absorbance values.

The absorbance readings above 1.5 in the table were obtained after suitable dilution of the solutions.

The table below indicates that, compared with the Coomassie Brilliant Blue method in section 6.1.1, there is less protein-to-protein variation using the BCA method.

Protein used in each assay (μg)	A_{562}		
	BSA	Chymotrypsin	Ovalbumin
0	0	0	0
10	0.57	0.71	0.52
20	1.15	1.42	1.05
30	1.75	2.29	1.68
40	2.28	2.79	2.05
50	2.79	3.24	2.38

Step 3: Determination of unknown protein using the calibration curve

See Step 3 in section 6.1.1 for guidance about experiments to be performed.

SELF TEST ?

Check that you have mastered the key concepts at the start of this section by attempting the following question.

ST 6.2 Using the calibration curve constructed in step 2 of the BCA method, calculate the amount of protein in a diluted cell extract that generates an A_{562} of 1.45. Given that the cell extract was diluted 25-fold and that we have 12 mL of undiluted cell extract, calculate the total amount of protein in the extract.

Answer

ST 6.2 A reading of 1.45 corresponds to ~26 μg protein in 20 μL diluted cell extract. The concentration of the diluted extract will be 26/20 μg μL^{-1} = 1.3 μg μL^{-1} = 1.3 mg mL^{-1}. With a dilution factor of 25, the undiluted cell extract will be $25 \times 1.3 = 32.5$ mg mL^{-1} and with a total volume of 12 mL of cell extract, the total amount of protein in the extract will be $12 \times 32.5 = 390$ mg.

6.1.3 Lowry method

Background

The Lowry method (or Folin–Ciocalteu method) relies on the chelation of Cu^{2+} ions by peptide bonds and their subsequent reduction to form Cu^{+}. The presence of Cu^{+}, together with the side chains of tryptophan, tyrosine, cysteine, and histidine, facilitates the reduction of phosphomolybdic–tungstic acid, resulting in the production of a characteristic blue colour that can be quantified by measuring the absorbance at 750 nm.

Experimental aspects

The Lowry method can be used to detect protein in the range 5–100 μg; however, it is sensitive to the presence of a range of compounds, many of which are used routinely in protein isolation, e.g. zwitterionic buffers, EDTA, dithiothreitol, 2-mercaptoethanol, NH_4^+, tyrosine, tryptophan, cysteine, and cystine. This problem can be minimized by including interfering compounds in the control reaction, with no protein present, and then correcting the sample readings. Alternatively, attempts could be made to selectively remove the interfering compound, for example by dialysis of the sample.

The Lowry reagent interacts with peptide bonds as well as the side chains of tryptophan, tyrosine, cysteine, cystine, and histidine. It is therefore important to construct a standard curve using a protein similar to the one of interest, as proteins of varying composition and size may give rise to a different colour response.

Step 1: Preparation of the reference protein solution

A stock solution of BSA (5 mg mL^{-1}) should be prepared as outlined in section 6.1.1 and then diluted 20-fold with water to give a nominal concentration of 0.25 mg mL^{-1} (250 µg mL^{-1}).

You should complete the table below to show how this stock solution of BSA is further diluted to give a series of BSA solutions of increasing concentrations, each with a final volume of 1 mL. Specimen A_{280} values of these solutions are shown.

Nominal BSA concentration (µg mL^{-1})	Volume of 100 µg mL^{-1} BSA stock (mL)	Volume of water (mL)	A_{280}	BSA concentration calculated from A_{280} (µg mL^{-1})
0	0	1.0	0	0
50			0.032	48
100			0.061	
150			0.087	
200			0.118	
250			0.148	

Given that a 1 mg mL^{-1} solution of BSA should have an A_{280} of 0.66 in a cuvette of 1-cm pathlength, you can use this absorbance to calculate the accurate protein concentration value of each of the BSA solutions in this dilution series and thus complete the right-hand column of the table.

Step 2: Construction of the calibration curve for the reference protein

The reactions for the Lowry calibration curve require incubation of 400 µL of each diluted protein sample with 400 µL Reagent B (see Appendix 6.1) at room temperature for 10 min. 200 µL Reagent C (see Appendix 6.1) is then added, mixed, and incubated at room temperature for 30 min. The absorbance at 750 nm is recorded against the first sample, which contains no BSA.

To present your data in the form of a standard curve you will have to calculate the amount of protein used in each assay. For example, if your nominal 50 µg mL^{-1} BSA solution gives an A_{280} reading of 0.032, then the BSA concentration calculated using the absorption coefficient (1 mg mL^{-1} has $A_{280} = 0.66$) is 48 µg mL^{-1}. As 400 µL (0.4 mL) of protein sample is used in each Lowry assay, the amount used in the case of the 48 µg mL^{-1} BSA solution is $0.4 \times 48 = 19.2$ µg protein.

BSA concentration calculated from A_{280} (µg mL^{-1})	Amount of protein used in each assay (µg)	A_{750}
0	0	0
48	19.2	0.33
		0.63
		0.95
		1.27
		1.61

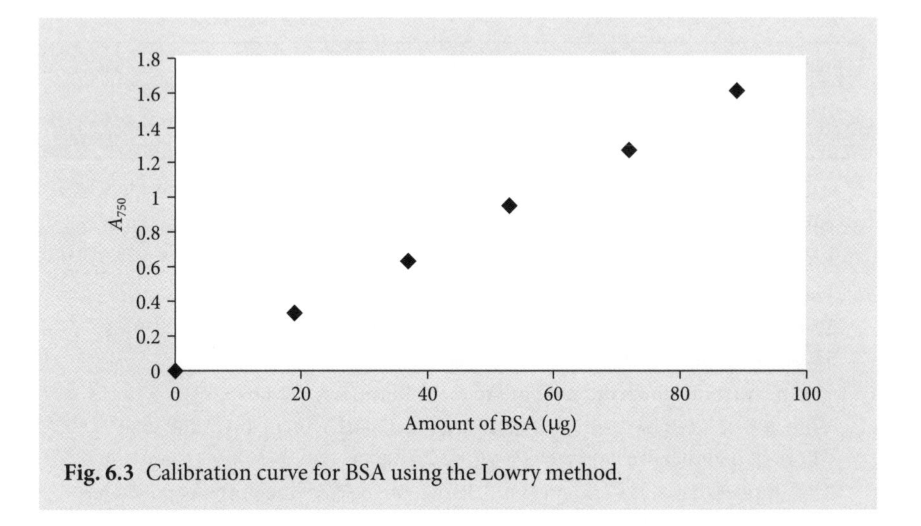

Fig. 6.3 Calibration curve for BSA using the Lowry method.

A plot of A_{750} values against the amount of protein used in each assay should be constructed (Fig. 6.3).

The table below indicates that compared with the Coomassie Brilliant Blue method, section 6.1.1, there is less protein-to-protein variation when using the Lowry method.

Protein used in each assay (μg)	A_{750}		
	BSA	Chymotrypsin	Myoglobin
0	0	0	0
20	0.34	0.38	0.29
40	0.67	0.74	0.56
60	0.99	1.12	0.85
80	1.33	1.47	1.14
100	1.64	1.87	1.44

Step 3: Determination of an unknown protein using the calibration curve

See Step 3 in section 6.1.1 for guidance about experiments to be performed.

SELF TEST ?

Check that you have mastered the key concepts at the start of this section by attempting the following question.

ST 6.3 During an affinity chromatography procedure, the amount of protein applied to the affinity column and the amount of protein that did not bind to the column were determined using the Lowry method. Using the calibration curve in Fig. 6.3, calculate the amount of protein applied to the column and the amount which failed to bind given the following information:

Step	Total volume of extract (mL)	A_{750}
Applied to column	15	1.3
Fraction which failed to bind	24	0.66

Answer

ST 6.3 From the calibration curve, Fig. 6.3, we can determine the amount of protein present in each sample.

For the extract applied to the column, $A_{750} = 1.3 = 73$ μg

With 400 μL used per assay, we have 73 μg in 400 μL = 182 μg mL^{-1} and with 15 mL of extract, we have 2.73 mg protein.

For the fraction which did not bind to the column, $A_{750} = 0.66 = 39$ μg.

With 400 μL used per assay, we have 39 μg in 400 μL = 98 μg mL^{-1} and with 24 mL of material, we have 24×98 μg = 2.35 mg protein.

This implies that 2.73 – 2.35 mg, i.e. 0.38 mg, or 13.9% of the protein applied has bound to the affinity column.

6.1.4 Biuret method

Background

The peptide bonds of proteins react with Cu^{2+} ions present in Biuret reagent (see Appendix 6.1) under alkaline conditions to produce a purple complex that can be detected by measuring A_{540} (see Fig. 6.4).

The uniform occurrence (i.e. frequency on a w/w basis) of peptide bonds in virtually all proteins results in minimal variation in the complex formation between different proteins. However, as noted below, the method is very much less sensitive than most other methods for protein determination.

Experimental aspects

The Biuret method is sensitive over the range 0.5–5 mg. While most standard reagents do not interfere with this assay, ammonium ions should be avoided.

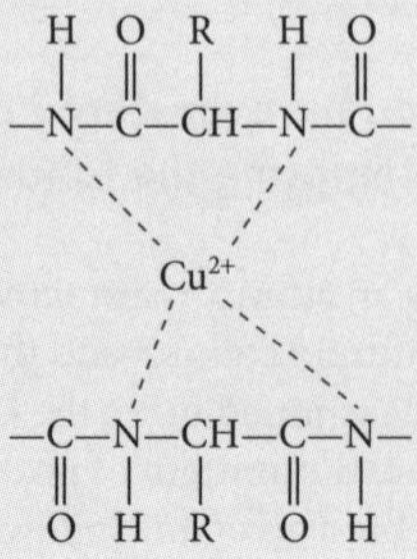

```
 H  O  R     H  O
 |  ‖  |     |  ‖
—N—C—CH—N—C—
        Cu2+
—C—N—CH—C—N—
 ‖  |  |     ‖  |
 O  H  R     O  H
```

Fig. 6.4 Complex formed between copper and adjacent peptide bonds in the Biuret reaction. This complex is purple and can be detected by its absorbance at 540nm.

Step 1: Preparation of the reference protein solution

An initial stock solution of BSA is prepared by dissolving 100 mg BSA in 5 mL water. At this stage, it is important to agitate the container *gently* to avoid excessive frothing (and hence denaturation of the protein).

This stock solution (20 mg mL^{-1}) is subsequently diluted to give a series of BSA solutions of increasing concentrations, each with a final volume of 1 mL, by completing the table below.

Required BSA concentration (mg mL^{-1})	Volume of 20 mg mL^{-1} BSA stock (μL)	Volume of water (μL)
0		
4		
8		
12		
16		
20		

The required BSA concentrations (4 mg mL^{-1} up to 16 mg mL^{-1}) are considerably higher than those in the methods in sections 6.1.1–6.1.3 and as a result must be diluted by a factor of 25 to allow accurate recording of A_{280} values (see Chapter 3, section 3.6). Specimen A_{280} values (relative to the first solution containing no BSA) for a typical experiment are shown in the table below. The actual BSA concentrations can be calculated using the absorption coefficient at 280 nm (0.66 for a 1 mg mL^{-1} BSA solution).

Nominal BSA concentration (mg mL^{-1})	A_{280} of 1:25 dilution of BSA solution*	A_{280} of BSA solution	Actual BSA concentration calculated from A_{280} (mg mL^{-1})
0	0	0	0
4	0.11	2.75	4.17
8	0.21	5.25	7.95
12	0.33	8.25	12.50
16	0.45	11.25	17.05
20	0.56	14.00	21.21

*1:25 dilution of each BSA solution prepared by pipetting 40 μL of each diluted solution into a quartz cuvette containing 960 μL of water and reading the A_{280} relative to a blank solution with no BSA.

The slope of a plot of A_{280} values against the volume of stock solution of BSA used to construct the dilution series indicates that the actual concentration of the stock solution of BSA is 20.8 mg mL^{-1} rather than the nominal 20 mg mL^{-1}, which was made up by weight.

Step 2: Construction of calibration curve for the reference protein

Biuret reagent is added to each of the BSA solutions in the dilution series by mixing 300 μL of each diluted protein standard with 1.4 mL Biuret reagent. The

Biuret–protein mix is incubated at room temperature for 10–15 min, and then the absorbance at 540 nm is measured using a spectrophotometer.

Presentation of your data as a standard curve requires that you calculate the amount of protein used in each assay. For example, for the BSA solution with a concentration of 4.17 mg mL^{-1}, and with 300 μL used in the assay, the amount used = (300/1000) × 4.17 mg = 1.25 mg.

BSA concentration calculated from A_{280} (mg mL^{-1})	Amount of protein used in each assay (mg)	A_{540}
0	0	0
4.17	1.25	0.16
7.95		0.30
12.50		0.51
17.05		0.62
21.21		0.82

A plot of the A_{540} values against the amount of protein used in each assay should be constructed.

This experiment can be extended by testing the response of the Biuret reagent to other proteins as indicated for the Coomassie Brilliant Blue method (end of section 6.1.1). The data in the table below indicate that there is very limited protein-to-protein variability when using the Biuret method.

Protein used in each assay (mg)	A_{540}		
	BSA	Ovalbumin	Trypsin
0	0	0	0
1.5	0.19	0.20	0.21
3.0	0.37	0.39	0.41
4.5	0.56	0.59	0.62
6.0	0.75	0.79	0.86
7.5	0.97	1.02	1.07

Step 3: Determination of an unknown protein using the calibration curve

See Step 3 in section 6.1.1 for guidance about experiments to be performed.

? SELF TEST

Check that you have mastered the key concepts at the start of this section by attempting the following question.

ST 6.4 A protein sample was concentrated using an ultrafiltration centrifugal device (Chapter 7, section 7.2.4). The Biuret method was used to determine the concentration of the filtrate and of the concentrate. Using the standard curve from Step 2, calculate the concentration of protein and the total amount of protein in the filtrate and the concentrate, given the following information:

Sample	Volume (mL)	A_{540}
Concentrate	2.1	0.4
Filtrate	5.8	0.06

Answer

ST 6.4 Using the A_{540} from the Biuret assay and the standard curve from step 2, we can calculate the amount of protein used in each assay.

Concentrate

A_{540} is 0.4 = 3.1 mg protein in 300 μL. This is equivalent to (1000/300) × 3.1 mg mL^{-1} = 10.33 mg mL^{-1}. With a volume of 2.1 mL, we have 10.33 × 2.1 mg = 21.7 mg protein in the concentrate.

Filtrate

A_{540} is 0.06 = 0.45 mg protein in 300 μL. This is equivalent to (1000/300) × 0.45 mg mL^{-1} = 1.5 mg mL^{-1}. With a volume of 5.8 mL, we have 1.5 × 5.8 mg = 8.7 mg protein in the filtrate.

6.1.5 Absorbance at 280 nm

Background

Proteins absorb radiation in the near UV (260–320 nm) due to their tyrosine, tryptophan, and to a lesser extent cystine, content. As a result, it is common practice to detect and to estimate the concentration of proteins using the absorbance at 280 nm. Conversion of an A_{280} value to an accurate protein concentration requires knowledge of the molar absorption coefficient (ε), which can be determined from the amino acid composition. The molar absorption coefficient values of tyrosine, tryptophan, and cysteine are used to calculate the molar absorption coefficient (ε) of a protein using the following eqn. 6.1 (modified from Gill and von Hippel, 1989). A related approach has been described in Chapter 5, section 5.8.2.

$$\varepsilon \text{ of the protein } (280\text{ nm: M}^{-1}\text{ cm}^{-1}) = (\text{number of Tyr} \times 1280) + (\text{number of Trp} \times 5690) + (\text{number of Cys} \times 60) \quad \textbf{6.1}$$

The internal redox environment of cells is generally sufficiently reducing that the cysteine side chains of proteins are maintained in the reduced state. Proteins that are destined for secretion from the cell undergo oxidation (resulting in the formation of disulphide bonds) in specialized compartments such as the lumen of the endoplasmic reticulum or the periplasm where there is a more oxidizing environment.

This equation assumes that all the cysteine side chains are involved in the formation of disulphide bonds (cystines). If none of them are (as would be the case for intracellular proteins), eqn. 6.1 becomes eqn. 6.2:

$$\varepsilon \text{ of the protein } (280\text{ nm: M}^{-1}\text{ cm}^{-1}) = (\text{number of Tyr} \times 1280) + (\text{number of Trp} \times 5690) \quad \textbf{6.2}$$

These calculations provide the molar absorption coefficient of a denatured (unfolded) protein in 6.0 M guanidinium chloride, 20 mM phosphate buffer, pH 6.5. However, A_{280} measurements of solutions of native (i.e. properly folded) proteins in a range of buffers generally give similar values (within ±10% of this value), for example, T4 lysozyme has an A_{280} folded: A_{280} unfolded ratio of 0.967.

The equation of Pace *et al.* (1995) (see Chapter 5, section 5.8.2) uses somewhat different values to calculate the absorption coefficient of the native (folded) protein (eqn. 6.3):

$$\varepsilon \text{ of the protein (280 nm: M}^{-1}\text{ cm}^{-1}) = (\text{number of Tyr} \times 1490) + (\text{number of Trp} \times 5500) + (\text{number of Cys} \times 62.5) \quad \textbf{6.3}$$

The absorbance values for each of the contributing amino acids used by Pace *et al.* (1995) were chosen to give the best overall fit for a wide variety of well-characterized proteins.

(assuming all cysteine residues in the protein are present as cystine; if none are, the last term is again neglected).

To calculate the absorbance of a 1 mg mL^{-1} protein solution, it is necessary to remember that a 1 M solution of protein, of molecular mass X Da, is X mg mL^{-1} (see Chapter 3, section 3.4). We therefore divide the molar absorption coefficient by the molecular mass in Da to obtain the absorbance of a 1 mg mL^{-1} protein solution:

$$A_{280}\text{ (1 mg mL}^{-1}) = \varepsilon \text{ of the protein}/X$$

The A_{280} value for a 1 mg mL^{-1} solution varies from protein to protein; hen egg white lysozyme has a value of 2.63 primarily due to its high tryptophan content, whilst BSA has a value of 0.66.

It should be noted that the values of A_{280} for 1 mg mL^{-1} protein, calculated according to eqns. 6.1 or 6.2, are available from analytical tools such as Protparam within the Expasy system (http://us.expasy.org/tools/protparam.html) for any given protein stored in Swiss-Prot or TrEMBL databases or for any user-entered sequence (see Chapter 5, section 5.8.2).

Experimental aspects

An accurate A_{280} measurement of a protein solution requires the absorbance value to fall between 0.1 and 1.0 (see Chapter 3, section 3.6). Dilution of the protein solution may be necessary to meet this requirement. A suitable buffer should be selected for the dilution steps to minimize precipitation of the protein. The buffer should also be of sufficient ionic strength to prevent protein adsorbing to the surface of the cuvette, typically a buffer strength >10 mM should minimize adsorption. Quartz cuvettes, or suitable plastic cuvettes which do not absorb significantly above 260 nm, are required for A_{280} measurements.

Preparation of reference protein solutions

This protocol describes how to use A_{280} values together with molar absorption coefficients to make an accurate estimate of protein concentrations. Two proteins of differing molar absorption coefficients have been selected to highlight the effect of proteins of differing amino acid composition on A_{280} values.

Protein 1: Bovine serum albumin

A 5 mg mL^{-1} stock solution of BSA is made by weighing out 50 mg BSA and dissolving, with gentle mixing, in 10 mL 50 mM sodium phosphate buffer, pH 7.0.

The 5 mg mL^{-1} stock solution is diluted to give a series of BSA solutions of increasing concentration, each with a final volume of 1 mL. Complete the second and third columns of the table below.

Nominal BSA concentration (mg mL^{-1})	Volume of 5 mg mL^{-1} BSA stock (mL)	Volume of water (mL)	A_{280}	BSA concentration calculated from A_{280} (mg mL^{-1})
0			0	
0.2			0.12	
0.4			0.25	
0.6			0.38	
0.8			0.51	
1.0			0.64	

WORKED EXAMPLE 

Use the A_{280} values of the series of BSA solutions described above to determine an accurate protein concentration value, given that a 1 mg mL^{-1} solution of BSA should have an A_{280} of 0.66.

SOLUTION

The plot of A_{280} vs concentration of protein is very nearly linear, see Fig. 6.5; from the slope of the best straight line it can be calculated that the concentration of the original BSA stock solution is $5 \times 0.95 = 4.75$ mg mL^{-1}.

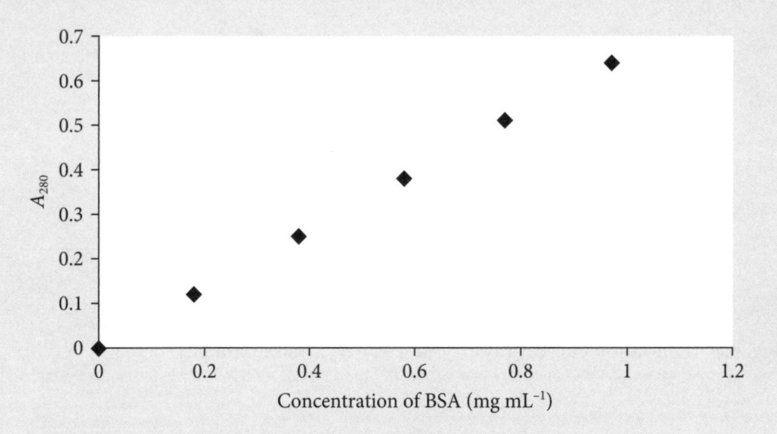

Fig. 6.5 Calibration curve for BSA using A_{280} measurements.

Protein 2: Hen egg white lysozyme

A 1.0 mg mL^{-1} stock solution of hen egg white lysozyme is made by weighing out 10 mg of hen egg white lysozyme and dissolving, with gentle mixing, in 10 mL 50 mM sodium phosphate buffer, pH 7.0.

The 1.0 mg mL^{-1} stock solution is diluted to give a series of hen egg white lysozyme solutions of increasing concentration, each with a final volume of 1 mL. Complete the second and third columns of the table below.

Nominal lysozyme concentration (mg mL^{-1})	Volume of 1 mg mL^{-1} lysozyme stock (mL)	Volume of water (mL)	A_{280}	Lysozyme concentration calculated from A_{280} (mg mL^{-1})
0			0	
0.05			0.15	
0.10			0.31	
0.15			0.46	
0.20			0.62	
0.25			0.77	

✓ WORKED EXAMPLE

Use the A_{280} values of the series of lysozyme solutions described above to determine an accurate protein concentration value, given that a 1 mg mL^{-1} solution of hen egg white lysozyme should have an A_{280} of 2.63.

SOLUTION

The plot of A_{280} vs concentration of protein is very nearly linear, see Fig. 6.6; from the slope of the best straight line it can be calculated that the concentration of the original lysozyme stock solution is 1.18 mg mL^{-1}.

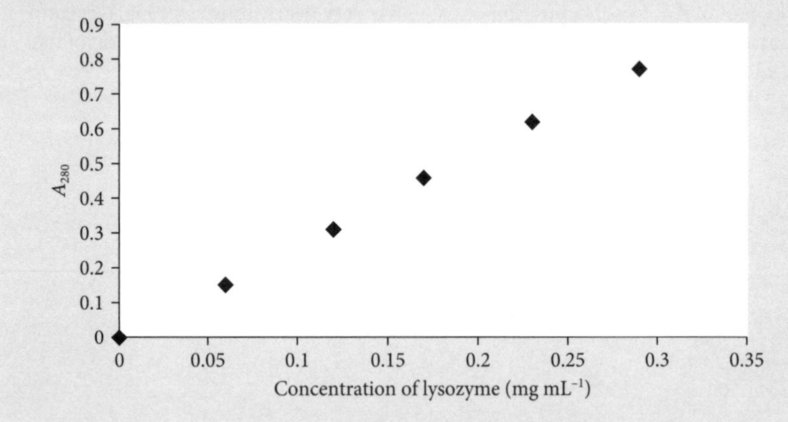

Fig. 6.6 Calibration curve for lysozyme using A_{280} measurements.

6.1.6 Absorbance in the far-UV

Background

Peptide bonds within proteins absorb radiation in the far UV (i.e. below 240 nm). Since such bonds occur in regular proportions in most proteins, it is possible to use absorbance measurements in this region (205 nm) to estimate the concentration of a protein. This method is often employed for proteins and peptides, which are low in tyrosine and tryptophan content, and consequently are not readily amenable to estimation by measurements of absorbance at 280 nm (section 6.1.5). This is also the method of choice if there is significant nucleic acid contamination as nucleic acids absorb very little radiation at 205 nm. Proteins absorb radiation at 205 nm to a much greater extent than radiation at 280 nm and as a result considerably less material is required for protein determination using A_{205} measurements (1–100 μg mL^{-1}) than A_{280} measurements (20–2000 μg mL^{-1}).

On the basis of absorption by peptide bonds, A_{205} data can be used to estimate the concentration of a protein solution using eqn. 6.4, which is based on the average absorption value for a variety of proteins:

$$\text{protein concentration (mg mL}^{-1}) = A_{205}/31 \qquad 6.4$$

An alternative, more reliable, approach developed by Scopes (1974) takes into account that the aromatic amino acids tyrosine and tryptophan contribute to the absorption at 205 nm. This is done by measuring A_{205} and A_{280} (usually at different dilutions of the protein solution) and then using the following equation (6.5) to calculate the A_{205} (1 mg mL^{-1}) value for that protein:

$$A_{205}\ (1\ \text{mg mL}^{-1}) = 27 + 120\ (A_{280}/A_{205}) \qquad 6.5$$

In this equation, 27 represents the absorbance of a 1 mg mL^{-1} solution of protein at 205 nm and the factor 120 is an empirical value (chosen from studies of a variety of standard proteins) to take into account the contribution of the aromatic amino acids at 205 nm.

Experimental aspects

In addition to the general considerations outlined in the Experimental procedure in section 6.1.5, A_{205} values should be corrected for absorbance by system components such as the buffer. As A_{205} measurements can detect very low concentrations of protein, this method is sensitive to losses of protein due to adsorption; this can be minimized by maintaining buffer concentrations >10 mM. Most of the spectrophotometers available in undergraduate laboratories give only poor performance at wavelengths as low as 205 nm, so only sample data are included in problem 6.1.

6.1.7 Choice of method for protein determination

The choice of which of the methods described in sections 6.1.1–6.1.6 to use in any particular case will depend on a number of factors.

If a pure protein is being measured, methods based on the A_{280} (section 6.1.5) or a combination of A_{280} and A_{205} (section 6.1.6) are probably the best and most convenient. It should be remembered, however, that successful application of spectrophotometric methods depends on there being no contaminating chromophores present; this should be established by recording an absorption spectrum over a suitable range (400–240 nm, or to 200 nm if the A_{205} is to be used). Measurements at 280 nm will be unreliable if there are significant levels of nucleic acids or oligonucleotide fragments present. Proteins generally show an A_{280}/A_{260} ratio of about 1.7; for nucleic acids the ratio is about 0.62. Hence significant contamination of proteins by nucleic acids can be easily detected by measuring this ratio.

Measurements at 205 nm will be unreliable if compounds which absorb in the far UV are present; these include some buffer components (e.g. those based on piperazine sulphonic acids such as HEPES or MOPS) and imidazole. Imidazole often presents a problem in the far UV since high concentrations are often employed to elute His-tagged proteins from immobilized metal-binding columns (see Chapter 7, section 7.3.8). Before measuring the absorbance in the far UV it is important to remove as much imidazole as possible from the protein by extensive dialysis, ultrafiltration, or gel filtration.

Light scattering usually follows an inverse power dependence on wavelength, λ; i.e. it is proportional to $1/\lambda^n$, where n is in the range 3–4 for most proteins. The greater scattering of light of lower wavelengths by dust particles in the atmosphere is the reason that the sky appears blue, rather than red.

It is also important to check that the sample does not contain significant amounts of aggregated protein since this will lead to light scattering that will increase the apparent absorbance values, especially at low wavelengths. The shape of the baseline of the absorption spectrum over the range from 400 to 310 nm will indicate whether scattering is a problem; a flat, or very nearly flat, baseline shows that scattering is not significant.

If it is not possible or desirable to use the spectrophotometric methods (for example because a mixture of proteins is being measured or the composition of a protein is unknown), one of the comparative methods listed in sections 6.1.1–6.1.4 should be used. Table 6.1 compares some of the key features of the various methods.

Table 6.1 The key features of comparative protein-determination methods

Comparative method	Sensitivity (μg)	Protein-to-protein variability	Assay complexity	Sensitivity to interference
Coomassie	1–25	Very high	Very simple	
BCA	0.2–50	Limited	Simple with 1 h incubation	NH_4^+, EDTA, glucose
Lowry	5–100	Medium/high	Simple with 10 and 30 min incubations	Zwitterionic buffers, EDTA, DTT, mercaptoethanol, NH_4^+, tyrosine, tryptophan
Biuret	500–5000	Minimal	Simple with 10-min incubation	Sucrose, Tris, glycerol, NH_4^+

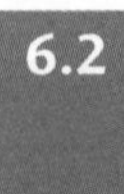

Electrophoresis methods

KEY CONCEPTS

- Appreciating the range of electrophoresis methods and their applications
- Knowing the variables to consider in designing an electrophoresis experiment

Electrophoresis techniques enable the separation of proteins according to their size, net charge or a combination of size and net charge. Such techniques can be performed on an analytical scale or preparative scale and are routinely employed in the characterization of proteins. Electrophoresis under denaturing conditions (section 6.2.1) is the standard method used to gauge the success of protein purification protocols (Chapter 7, section 7.1) and to indicate the molecular mass of a given protein. Electrophoresis under non-denaturing conditions (section 6.2.2) is used to analyse proteins or protein complexes while retaining their structure and, in some instances, their function. Non-denaturing conditions are a necessity for most activity gel stains and the isolation of intact protein complexes. Isoelectric focusing (IEF) (section 6.2.3) separates proteins according to their net charge and is useful as a protein purification tool, as a means of assessing the purity of protein preparations and as the first stage in two-dimensional gel electrophoresis (section 6.2.4). At the end of this section, we shall look at the use of immunoelectrophoresis (section 6.2.5) in the separation, detection, and quantitation of proteins which cross-react with specific antibodies.

6.2.1 One-dimensional SDS-PAGE (denaturing PAGE)

The simplicity and efficiency of sodium dodecyl sulphate polyacrylamide gel electrophoresis (SDS-PAGE) have made this technique routine in protein characterization. SDS-PAGE is used most often for the estimation of protein subunit molecular mass and the assessment of the purity of protein preparations. While it might be possible to scale up the technique as a protein purification step, SDS-PAGE is used more routinely on an analytical scale. SDS-PAGE, as developed by Laemmli (1970), is the most widely used electrophoretic system in protein analysis. This is a discontinuous system, i.e. involves the separation of proteins in two gels of differing acrylamide concentrations, namely a stacking gel and a separating gel (see Fig. 6.7).

Laemmli (1970) Cleavage of Structural Proteins during the Assembly of the Head of Bacteriophage T4 *Nature* **227**, pp. 680–5, describes the development of 'an improved gel electrophoresis' method. This paper remains the most cited article in the journal *Nature*.

It is difficult to remove SDS from the protein–SDS complexes and thus allow the protein to regain its original folded structure and hence biological activity.

Theory

In general, SDS-PAGE separates proteins according to their size. SDS-PAGE involves the application of an electric field across a polyacrylamide gel that results

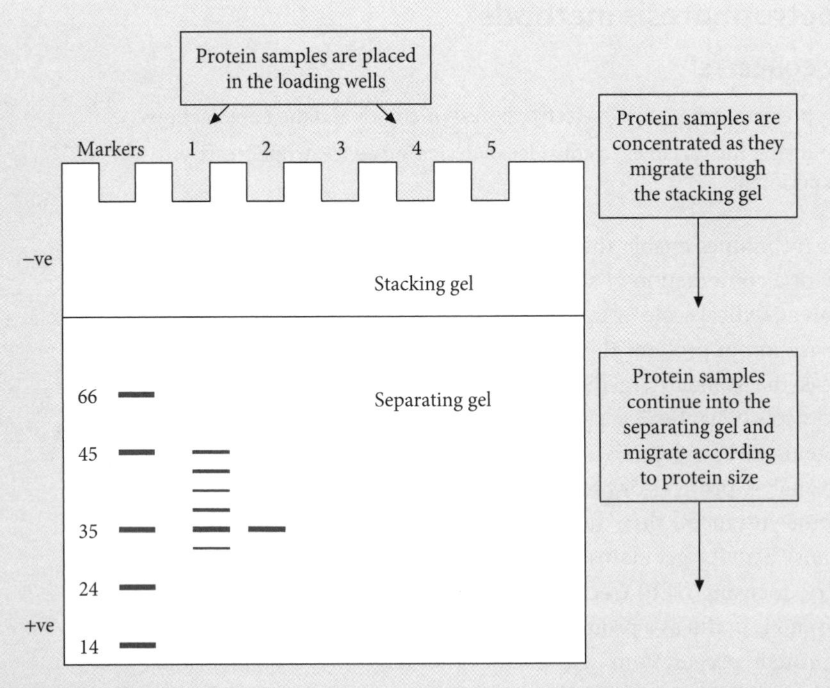

Fig. 6.7 Schematic representation of SDS-PAGE. SDS-treated proteins are loaded into the wells at the top of stacking gel and migrate towards the positive electrode such that the sample is concentrated into a compact band. The concentrated protein samples continue to migrate into the separating gel. Once within the separating gel, proteins migrate according to their size, with smaller proteins travelling through the acrylamide gel more quickly. In this gel, the lane labelled 'Markers' contains molecular mass markers (molecular mass indicated in kDa) and lanes 1 and 2 contain protein samples. Lane 1 contains a complex mixture of proteins spanning 30–45 kDa, whereas lane 2 appears to contain a single protein species of 35 kDa.

The stoichiometry of binding means that, on average, one dodecylsulphate group is bound to very nearly two amino acid units in the polypeptide chain.

in the migration of SDS-treated proteins towards the anode. The SDS treatment of proteins confers a constant negative charge: size ratio (for most proteins 1.4 mg of SDS binds to each mg of protein) and the structure of the protein–SDS complex becomes a uniform rod shape, i.e. both charge and dimension depend only on molecular mass.

As a result, SDS-treated proteins move through the gel according to their size only, with larger proteins being retarded and moving more slowly through the polyacrylamide gel (see Fig. 6.7). Proteins are visualized at the end of the electrophoresis process by using a suitable detection procedure, e.g. by staining with Coomassie Brilliant Blue or silver.

The predictable nature of SDS-treated proteins in SDS-PAGE permits the comparison of the mobility of an unknown protein, relative to the mobility of SDS-treated calibration proteins (termed 'marker' proteins), to estimate the molecular mass of the unknown protein. In SDS-PAGE, the mobility of a protein is found to depend in a linear fashion on the log (molecular mass) of the protein; thus we can

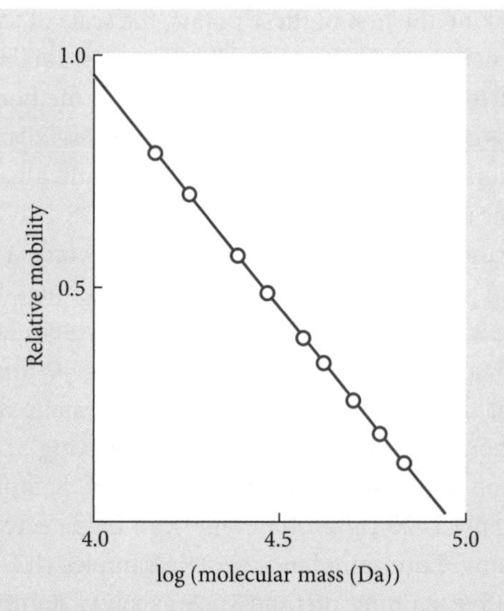

Fig. 6.8 Sodium dodecylsulphate polyacrylamide gel electrophoresis of proteins. Ten per cent polyacrylamide gels were used and mobilities are expressed relative to bromophenol blue. The proteins used (in order of increasing subunit molecular mass) were: lysozyme, myoglobin, trypsin, carbonic anhydrase, glyceraldehyde-3-phosphate dehydrogenase, fructose-bisphosphate aldolase, fumarase, catalase, and bovine serum albumin.

construct a calibration plot of mobility against log (molecular mass) of the marker proteins (see Fig. 6.8). We can then determine the molecular mass of an unknown protein by measuring the distance it travelled on the gel and referring to the calibration plot.

The molecular mass measured by SDS-PAGE is the mass of the subunit (or polypeptide chain), since virtually all oligomeric proteins are dissociated into their constituent subunits by SDS. A heterooligomeric protein, i.e. one consisting of different subunits, will show multiple bands on SDS-PAGE. For example, G proteins would show three bands corresponding to the α, β, and γ subunits, which are typically of molecular masses 40, 35, and 8 kDa, respectively.

Experimental design

There are a number of variables that must be considered when analysing a protein by SDS-PAGE. These include:

- scale of analysis
- complexity of the sample
- suitable detection method
- percentage of polyacrylamide gel required to resolve the protein of interest
- type of marker proteins required
- reducing or non-reducing conditions

Taking the first of these points, the scale of the analysis will dictate the amount of protein required for SDS-PAGE. Small and large format gels are available, with small format gels (8 cm × 7 cm) being the method of choice for the rapid separation of less complex samples. Larger format gels (20 cm × 16 cm) are selected when high resolution is required, e.g. resolving an individual protein from a complex mixture.

The amount of protein analysed by SDS-PAGE will vary with the complexity of the sample. A protein sample containing only a few proteins will typically require ~1 μg protein/lane for a small format Coomassie Brilliant Blue-stained gel. More complex samples will require ~20 μg protein/lane for a small format Coomassie Brilliant Blue-stained gel. In both cases, 10-fold less protein should be analysed when using the more sensitive silver stain detection method. The detection method used to visualize proteins following SDS-PAGE is an important consideration when designing the experiment. Samples with medium-to-high protein content (1–20 μg protein/lane) can be detected using Coomassie Brilliant Blue staining. Low abundance protein samples (0.1–1 μg protein/lane) will require a sensitive staining method such as silver staining. A more sensitive method for detecting and quantitating specific proteins that have been separated by SDS-PAGE is immunoblotting. Immunoblotting involves the transfer of proteins separated by PAGE from a gel onto a membrane, usually made of nitrocellulose or polyvinylidene fluoride (PVDF), which is subsequently probed with a specific antibody, termed the primary antibody. Typically, antibody–antigen interactions are detected by a secondary antibody which is conjugated to an enzyme. Addition of enzyme substrate produces either a colourimetric or fluorescent product, indicating the location of secondary antibody bound to a specific primary antibody–antigen complex. This is a highly sensitive and specific detection method.

The effective separation range of polyacrylamide gels varies with the pore size of the gel. Gels with larger pores are used to separate larger proteins, whereas gels with smaller pore sizes are used to resolve smaller proteins. The pore size of the polyacrylamide gels varies with the concentration of acrylamide and the cross-linker, bis-acrylamide, in the gel mixture, with increasing [acrylamide] and [bis-acrylamide] giving rise to smaller pore sizes. Most protein analyses by SDS-PAGE use gels in the range 5–15% acrylamide, with the effective separation range of the most commonly used gels shown in Table 6.2.

It is standard practice to include a lane of marker proteins that permit the estimation of the molecular mass of the protein of interest. It is important to select a suitable range of markers that span mobilities greater than and less than the

Table 6.2 Effective separation range of uniform polyacrylamide gels

% Acrylamide (w/v)*	5	10	12	15
Separation range (kDa)	57–212	20–80	12–60	10–43

The range over which proteins will migrate in a linear fashion with respect to the log (molecular mass) of the protein is given for each gel type.

*Acrylamide solution contains acrylamide:bis-acrylamide in the ratio 30:0.8. The percentage acrylamide present in the polyacrylamide gel is expressed as the total concentration of the acrylamide monomer present.

protein of interest to ensure an accurate estimate of the molecular mass. It is also important to ensure that the mobility of the standard proteins behave in a linear fashion relative to log (molecular mass) in the selected gel. Standard markers typically span the range 10–100 kDa, whereas high molecular mass markers are more useful in the range 100–500 kDa. Modified markers are also available for specific detection methods such as silver staining or immunoblotting.

When analysing a protein by SDS-PAGE, the possible presence of disulphide bonds must be considered. Treatment of proteins with a reducing agent, such as 2-mercaptoethanol, will disrupt disulphide bonds within or between subunits. Thus, in standard molecular mass estimations or purity analyses, a reducing agent is included during SDS treatment to ensure that cysteine side chains are maintained in the reduced state to generate linearized monomers. Analysis of protein–protein interactions involving disulphide bonds should be conducted in the presence and in the absence of reducing agents to characterize the nature of the interaction in more detail.

6.2.2 Non-denaturing gel electrophoresis (Ferguson plots, activity stains)

Non-denaturing gel electrophoresis is used to analyse proteins or protein complexes while retaining their structure and, in some instances, their function. Under non-denaturing gel electrophoresis conditions, proteins are separated by their size, shape, and charge. This contrasts with denaturing gel electrophoresis conditions (section 6.2.1), which promote protein denaturation and the denatured proteins are separated according to size only.

Theory

Non-denaturing gels separate proteins according to their size, shape, and charge. As a result, non-denaturing PAGE can be used to determine the molecular mass, the charge or subunit composition of a protein. The overall procedure for non-denaturing gel electrophoresis is very similar to SDS-PAGE, with the exception that the proteins are not treated with SDS, heat, or reducing agents during the procedure. As a result the protein structure is retained, including the oligomeric structure of non-covalently linked subunits, and (in many cases) biological activity is also retained. The retention of oligomeric structures in non-denaturing gels allows the characterization of the quaternary structure of proteins. The retention of biological activity allows the detection and quantitation of specific proteins using appropriate activity stains.

Although proteins are separated by various factors (size, shape, and charge) under non-denaturing gel conditions, it is nevertheless possible to estimate the molecular mass of a protein using Ferguson plots. Initially, the relative mobility of calibration proteins in a series of polyacrylamide gels, with different acrylamide

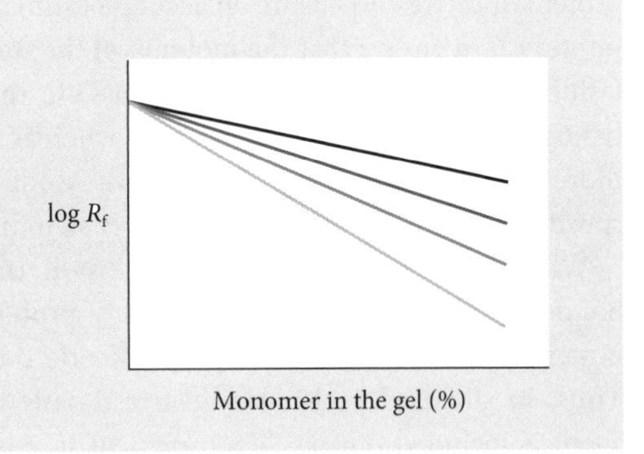

Fig. 6.9 Ferguson plot showing four standard proteins with the same charge but different mass values. Determination of the mass of a protein of unknown mass requires comparison of its behaviour with the standard proteins and plotting the slopes of the lines against the molecular mass.

concentrations, is recorded. A plot of log (relative mobility) against gel concentration will generate a slope for each calibration protein that can then be plotted against molecular mass to create a Ferguson plot. Comparison of the slope generated by an unknown protein with the calibration proteins will provide an estimate of its molecular mass (see Fig. 6.9).

This approach was used to estimate the native molecular mass of a novel intracellular acid phosphatase (EC 3.1.3.2), from tomato (Bozzo *et al.*, 2004). Electrophoresis of the native enzyme in a 7% non-denaturing gel yielded a single band, indicative of a single protein species. Subsequent non-denaturing gel electrophoresis of the native acid phosphatase was conducted with different acrylamide concentrations. The calibration proteins used in each gel were: bovine milk α-lactalbumin (14.2 kDa), bovine erythrocyte carbonic anhydrase (29 kDa), chicken ovalbumin (45 kDa), BSA (monomer, 66 kDa; dimer, 132 kDa) and jack bean urease (trimer, 272 kDa; hexamer, 545 kDa). Ferguson plot analysis indicated that this novel acid phosphatase has a molecular mass of 142 kDa.

Experimental design

The design considerations for a non-denaturing gel experiment are similar to those outlined for SDS-PAGE. One specific consideration is the type of calibration proteins selected for non-denaturing PAGE. Calibration proteins with similar charge values but different mass properties should be used when determining the mass of a protein, whereas calibration proteins with similar mass values but different charge properties should be used when determining the charge of a protein.

6.2.3 Isoelectric focusing

The overall charge carried by a protein depends on its composition, conformation, presence of prosthetic groups, and the pH of the surrounding environment. The specific pH at which a protein carries a charge of zero is known as the pI value (see

Table 6.3 pI values of selected proteins

Protein	pI value
Albumin (hen)	4.6
Albumin (bovine, BSA)	4.9
Alcohol dehydrogenase (bakers yeast)	6.6
Haemoglobin (human)	7.1
Cytochrome *c* (horse)	10.6
Histone (bovine)	10.8
Lysozyme (hen)	11.0

Chapter 1, section 1.3.2.2). Typically, the pI values of proteins are within the range pH 3–10, with most proteins having a pI value <7.0: proteins rich in acidic groups have pI values closer to 3, proteins rich in basic residues have pI values closer to 10 and proteins with a similar amount of acidic and basic residues have pI values near pH 7. Examples of proteins and their associated pI values are given in Table 6.3.

Isoelectric focusing (IEF) is a high-resolution technique which separates proteins according to their pI values. Indeed, proteins that are structurally very similar, and have pI values that differ by as little as 0.01 pH units, can be resolved by IEF. In IEF, proteins are subjected to electrophoresis across a pH gradient: proteins carrying a net positive charge migrate towards the cathode and lose or gain protons (if below or above pI, respectively) until they reach the pH value in the gradient that matches their pI. Likewise, proteins carrying a net negative charge migrate towards the anode and lose or gain protons (if above or below pI, respectively) until they reach the pH value in the gradient which matches their pI. Once a protein has migrated to the position where the pH matches its pI value, it will be concentrated or 'focused'. Thus, a mixture of proteins applied to IEF will separate according to their characteristic pI values and, following suitable detection methods, will appear as a series of well-defined bands across the pH gradient.

Experimental design

IEF can be conducted on an analytical scale (typically 10 μg) or on a preparative scale (typically 1–10 mg). Until recently, most analytical scale IEF was performed using polyacrylamide gel electrophoresis, with the incorporation of synthetic carrier ampholytes into the polyacrylamide gel mix, to generate the pH gradient. More recently, pre-formed polyacrylamide gels containing immobilized pH gradients have become the method of choice due to their commercial availability. On a preparative scale, use of the Rotofor cell (a system which fractionates proteins by IEF in free solution) has proved an effective IEF technique in the separation and recovery of protein in the milligram to gram scale.

Ampholytes are mixtures of polyaminoacids with different charge properties and hence different pI values.

The success of both analytical and preparative IEF relies on the ability to resolve the proteins in the sample. Good resolution requires optimization of both the pH gradient range and the size of the electric field applied across the selected pH gradient. Typically, a broad range pH gradient is used in preliminary experiments and the behaviour of the protein of interest directs the selection of narrower range pH gradients in subsequent experiments to enhance the resolution, e.g. start with a pH 4–10 gradient and if the protein of interest focuses at ~pH 5, progress to focus the protein using a gradient from pH 4–6.

Sample preparation also requires optimization in both analytical and preparative IEF. Care must be exercised to ensure that the proteins in the sample are soluble and are free from salts that will interfere with the focusing process. Protein solubilization may require the presence of urea or detergents (non-ionic or zwitterionic) to maintain a continuous pH gradient and to generate well-focused protein bands.

6.2.4 Two-dimensional gel electrophoresis

Theory

Two-dimensional gel electrophoresis (two-dimensional PAGE) is used to separate proteins in two dimensions:

1) IEF (section 6.2.3) is used in the first dimension to separate proteins according to their pI value. At the end of the first dimension, proteins are focused, in bands, along the length of an IEF gel/strip. The gel or strip is then applied to the top of an SDS-PAGE gel.
2) SDS-PAGE (section 6.2.1) is used in the second dimension to separate proteins according to their molecular mass. At the end of the second dimension, the gel is stained using a suitable method to visualize the separated proteins which appear as a series of spots on the gel (see Fig. 6.10).

This technique was first developed in the mid-1970s (O'Farrell, 1975); however, it was not until the mid-1990s that this technique became more widely used. Two developments were responsible for this new-found popularity:

1) the commercial availability of pre-formed polyacrylamide gels containing immobilized pH gradients for the IEF step in the first dimension. Prior to the mid-1990s, researchers prepared their own immobilized pH gradients, a task which was technically challenging and which generated non-standardized gels.
2) improved mass spectrometry which enabled the rapid identification of proteins separated by two-dimensional PAGE. Peptide mass fingerprinting and MS/MS sequencing (Chapter 8, sections 8.3.2 and 8.3.4) could be conducted on protein spots isolated from 2D-PAGE gels.

The resolution of two-dimensional PAGE is such that complicated mixtures of proteins such as cell lysates containing many thousands of different proteins can

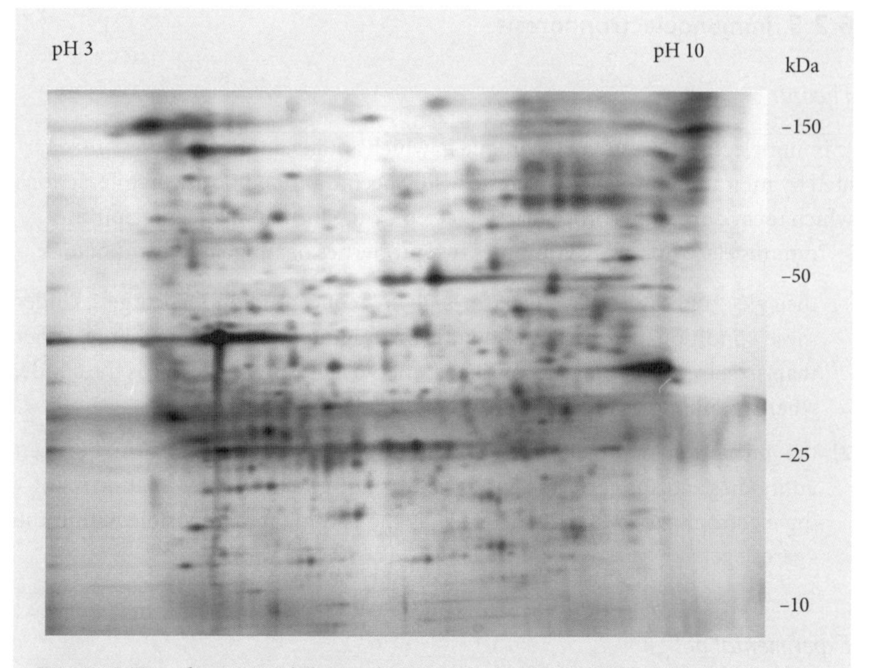

Fig. 6.10 Two-dimensional electrophoresis gel of a *Dictyostelium* cell lysate. Courtesy of D. Lamont, University of Dundee.

be analysed. In addition, it is also possible to characterize subtle post-translational modifications (Chapter 8, section 8.1.4) and to quantitate each protein within complex mixtures. As a result, two-dimensional PAGE is used to identify changes in protein expression patterns in response to hormonal, environmental, and developmental signals.

Experimental design

As with IEF experiments, each new two-dimensional PAGE experiment requires optimization of a number of variables including:

- solublization of the proteins
- loading of the optimal amount of sample to allow separation and detection of proteins which are abundant as well as those which are less abundant
- selection of a suitable pH range for the first dimension
- selection of an appropriate acrylamide gel for optimal separation of the proteins of interest in the second dimension

6.2.5 Immunoelectrophoresis

Theory

Immunoelectrophoresis is a technique in which antigens and antibodies are separated by migration through an electric field. Specific antibody–antigen interactions which form during this process are detected by the formation of a precipitin.

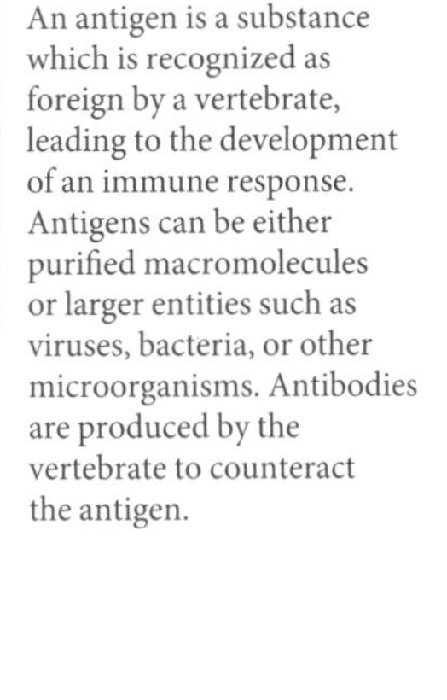

An antigen is a substance which is recognized as foreign by a vertebrate, leading to the development of an immune response. Antigens can be either purified macromolecules or larger entities such as viruses, bacteria, or other microorganisms. Antibodies are produced by the vertebrate to counteract the antigen.

Immunoelectrophoresis combines two properties of antigens and antibodies:

1) their electrophoretic movement through low-concentration agarose gels. Under these conditions antibodies and antigens migrate according to charge rather than size. In an electric field, antigens will typically migrate towards the anode, whereas antibodies move towards the cathode.
2) the formation of precipitates (precipitin) when complexes form between antibodies and antigens. Specific antibody–antigen complex formation following electrophoresis is easily detected as the precipitates are visible within the agarose gel.

Experimental design

There are a number of immunoelectrophoretic techniques that exploit the properties outlined above. The most common techniques, however, are cross-over

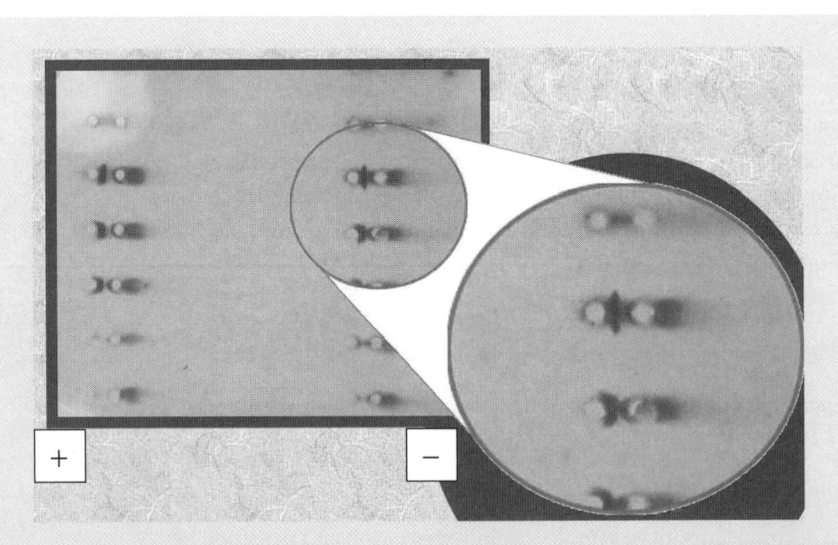

Fig. 6.11 Cross-over immunoelectrophoresis (CIEP) gel. For each experiment, two wells are created in the agarose gel. Antigen is placed in the well next to the cathode and the antibody is placed in the well next to the anode. During electrophoresis the antigen migrates towards the anode and the antibody migrates towards the cathode. A specific antibody–antigen interaction is characterized by the formation of a visible precipitate (precipitin). Some of the applications of this technique include the detection of blood proteins at crime scenes and the detection of viral proteins. Image from http://www.paleoresearch.com/services/pra.html.

electrophoresis, quantitative immunoelectrophoresis, and two-dimensional immunoelectrophoresis. In cross-over electrophoresis, antibody and antigen are loaded into separate wells of an agarose gel. Application of an electric field across the agarose gel results in the migration of the antigen to the anode and the antibody to the cathode. A visible precipitate will indicate whether a specific antibody–antigen interaction has occurred (see Fig. 6.11).

In quantitative immunoelectrophoresis (or rocket immunoelectrophoresis), antibody is added to the agarose during gel preparation. Once the gel is set, antigen is applied to a well and during electrophoresis the antigen migrates from the well into the gel towards the anode while the antibody moves towards the cathode. Specific antigen–antibody interactions will result in the formation of a tall sharp, 'rocket-shaped' precipitin. The height of the precipitin is proportional to the amount of antigen present in each well.

WORKED EXAMPLE 

Using quantitative immunoelectrophoresis, calculate the amount of *Candida albicans* antigen present in a patient sample using the calibration data below.

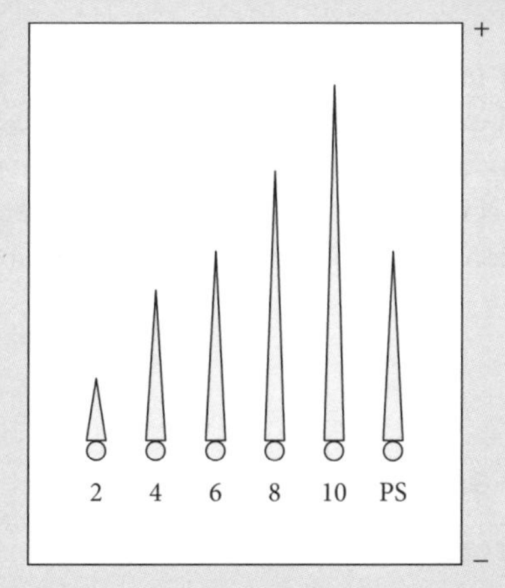

Quantitative (or rocket) immunoelectrophoresis assay was conducted in an agarose gel impregnated with anti-*C. albicans* antibody. The wells contained increasing amounts of *C. albicans* antigen (2, 4, 6, 8, and 10 μg). Following electrophoresis, precipitin rockets formed. A patient's sample (PS) was included in this assay.

STRATEGY

Use the calibration data to construct a calibration curve of the height of the precipitin rockets against the amount of antigen present in each well (Fig. 6.12). To calculate the amount of *C. albicans* antigen in the patient's sample, measure the height of the precipitin rocket and use the calibration curve to estimate the amount of *C. albicans* antigen present.

SOLUTION

Using a calibration curve of the height of the precipitin rockets against the amount of antigen present (Fig. 6.12), an estimated 5.6 μg of *C. albicans* antigen is present in the patient's sample.

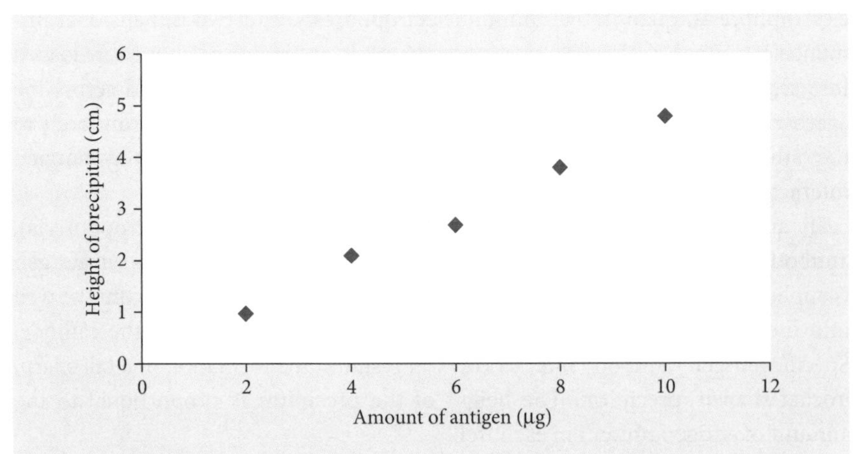

Fig. 6.12 Quantitative immunoelectrophoresis calibration curve showing the height of the precipitin rockets plotted against the amount of *C. albicans* antigen present in each well.

In two-dimensional immunoelectrophoresis, antigens are initially separated by non-denaturing gel electrophoresis (section 6.2.2) and the subsequent gel is applied to the top of an agarose gel impregnated with an antibody. Application of an electric field will result in the migration of the antigens into the agarose gel with precipitin formation signalling a specific antigen–antibody interaction.

6.3 Enzyme assays

KEY CONCEPTS

- Selecting the appropriate method to assay enzyme activity
- Making sure that valid kinetic data are obtained

An enzyme assay is a test of the amount of catalytic activity present in a sample of enzyme. In this section, we shall discuss some of the main types of assays which can be performed and the key points to consider to ensure that reliable data are obtained. This topic has been mentioned in outline in Chapter 5, section 5.3.1, and will be discussed further in Chapter 9, section 9.2, where the design of assays for particular types of enzymes will be described.

Apart from detecting the presence of a given enzyme in a sample, enzyme assays are the means by which we can obtain quantitative information about the catalytic activity, including parameters such as the Michaelis constant (K_m) and maximum (or limiting) velocity (V_{max} or in terms of mol mol^{-1} s^{-1}, k_{cat}) (see Chapter 4, section 4.3.3) and the catalytic efficiency k_{cat}/K_m (see Chapter 4, section 4.5 and Chapter 9, section 9.3.5). The variation of these parameters with, for example pH and temperature can provide some information about the catalytic mechanism of the enzyme (see Chapter 9, section 9.7).

By studying the activity under a variety of conditions we may be able to evaluate how the enzyme works under conditions close to those operating in vivo, allowing insights into the way that the activity may be regulated under various metabolic circumstances. This information may help in the design of specific inhibitors, which could have therapeutic potential for the treatment of disease states where the activity of an enzyme has been shown to be too high, leading to the accumulation of an undesirable product. Examples of enzyme inhibitors which have proved to be valuable drugs include (a) HMG-CoA reductase inhibitors, such as simvastatin and related compounds, to treat high blood cholesterol levels, (b) angiotensin-converting enzyme (ACE) inhibitors, such as captopril and its derivatives, to treat hypertension, and (c) xanthine oxidase inhibitors, such as allopurinol, to treat gout.

The normal approach to studying enzyme activity is to use the so-called 'steady-state conditions' under which the concentrations of enzyme–substrate complexes remain essentially constant (see Chapter 4, section 4.3.3). In this type of experiment, the enzyme is present at a very small molar concentration, typically 0.1% or less of the molar concentration of the substrate. Under these conditions the equations used for analysis of the data are reasonably straightforward and the data are collected over the timescale of minutes, allowing conventional means of mixing and recording to be employed. In Chapter 9, section 9.9.4, we shall describe the use of rapid reaction techniques to examine events in the so-called presteady-state period; under these conditions the molar concentrations of enzyme and substrate are comparable.

It should be noted that in the cell, the molar concentrations of enzyme and substrate can be comparable. For example, the concentration of fructose bis-phosphate aldolase in muscle is 66 μM, whereas the concentrations of fructose bis-phosphate is 65 μM (Price and Stevens, 1999). Hence extrapolation of results from steady-state kinetics to the in vivo situation can be difficult.

To set up an enzyme assay we need to define the substrate(s) and product(s) of the reaction and determine which property we can exploit to distinguish between them so that the progress of the reaction can be monitored. In principle, any enzyme can be assayed using the following process:

- Mixing enzyme and substrate(s) at a defined time
- Stopping the reaction after a given period, e.g. by adding trichloroacetic acid (which will precipitate and inactivate the enzyme)
- Analysing the mixture to ascertain the extent of conversion of substrate to product.

For low molecular mass substrates and products of moderate polarity, reverse phase high performance liquid chromatography (RP-HPLC) would be a very appropriate analytical method, since it is rapid, sensitive, and highly reproducible.

An example of the use of RP-HPLC to monitor a reaction is the assay of chorismate synthase from *Escherichia coli*. Using this technique, the product (chorismate) could be readily separated from (substrate) 5-enolpyruvylshikimate 3-phosphate (White *et al.*, 1987).

Some enzymes that catalyse reactions consisting of multiple steps can be conveniently assayed by focusing on one of the steps. For example, the hydrolysis of

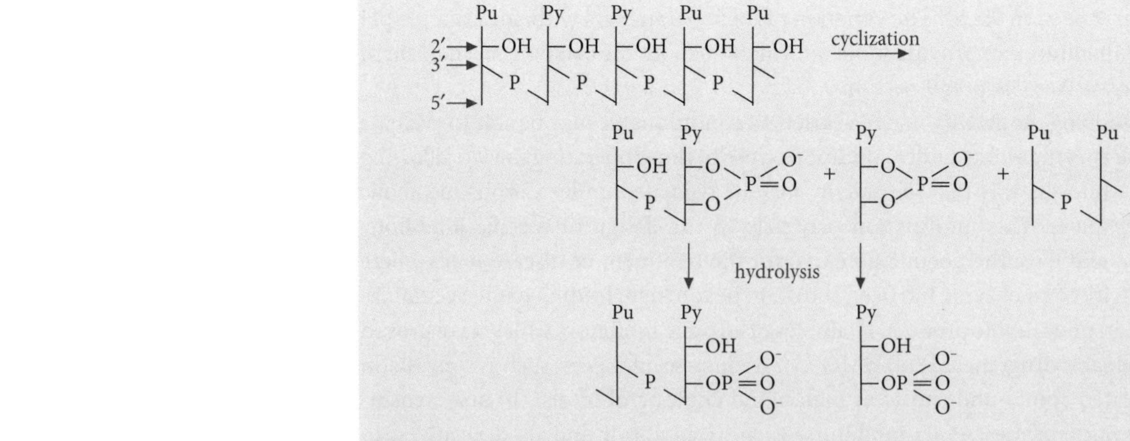

Fig. 6.13 The mechanism of the RNase-catalysed hydrolysis of RNA. The ribose unit of each nucleotide is represented by a vertical line and the 3′–5′ phosphodiester bond between successive nucleotide units is represented by –P–. Pu and Py denote purine and pyrimidine bases, respectively. The reaction involves formation of a 2′,3′-cyclic phosphate ester intermediate, which is then hydrolysed to give the final 3′-phosphate product.

RNA by pancreatic ribonuclease consists of two steps, namely the formation and breakdown of a pyrimidine 2′,3′-cyclic phosphate intermediate (Fig. 6.13). If such an intermediate (e.g. 2′,3′-cyclic CMP) is used as a substrate for ribonuclease, there is a small but measurable change in absorbance at 284 nm on conversion to the final product, 3′-CMP. This can be used as the basis of a continuous assay (section 6.3.1) for the enzyme (Crook *et al.*, 1960).

A second example of this approach is provided by aminoacyl–tRNA ligases (synthetases) that catalyse the overall reaction:

amino acid + tRNA + ATP ⟶ aminoacyl-tRNA + AMP + pyrophosphate.

The overall reaction is difficult to assay, but we can take advantage of the fact that it occurs in two steps, the first of which involves the activation of the amino acid:

amino acid + ATP ⟶ aminoacyl-AMP + pyrophosphate

This reaction is conveniently assayed in a discontinuous fashion (section 6.3.2) by observing the amount of incorporation of ^{32}P from [^{32}P]-labelled pyrophosphate into ATP in the presence of the appropriate amino acid and enzyme.

6.3.1 Continuous assays

By far the most convenient assays are those where the conversion of substrate to product can be continuously observed. If there is a spectroscopic signal which changes (such as absorbance or fluorescence) as a result of the reaction this can provide a very convenient means of monitoring the activity.

One of the best-known types of continuous assay involves dehydrogenases; these are enzymes that catalyse redox reactions in which one substrate is oxidized and the second one reduced. For most (but, by no means all) dehydrogenases the redox system used is the $NAD(P)^+/NAD(P)H + H^+$ couple, for example

Alcohol dehydrogenase

ethanol + $NAD^+ \rightleftharpoons$ ethanal + NADH (+ H^+)

Glucose-6-phosphate dehydrogenase

D-glucose-6-phosphate + $NADP^+ \rightarrow$ D-glucono-1,5-lactone-6-phosphate + NADPH + H^+

Since NAD(P)H absorbs radiation at 340 nm but $NAD(P)^+$ does not, we have a convenient way of monitoring the reaction. The molar absorption coefficient for NAD(P)H at 340 nm is 6220 M^{-1} cm^{-1} and this value allows us to express the absorbance changes in terms of changes in the concentration of NADH, and hence in terms of changes in the amount of NADH (see Chapter 3, section 3.8). In section 6.3.4, we will show how we can use such calculations to show that assays of dehydrogenases are being performed under experimentally appropriate conditions.

In some cases, even if the natural substrate does not show a convenient spectrophotometric change on conversion to product, we may be able to use a non-natural substrate which does show a change as the reaction proceeds.

This approach is widely used in the assay of hydrolases, which catalyse hydrolysis reactions. Examples include 2-nitrophenylgalactoside, 4-nitrophenylphosphate, and *N*-benzoylarginine-4-nitroanilide, which are used for the assay of β-galactosidase, alkaline phosphatase, and trypsin, respectively (Table 6.4).

On hydrolysis, these substrates form 2-nitrophenol, 4-nitrophenol, and 4-nitroaniline, respectively, all of which show strong absorbance at the blue end of the visible spectrum (around 400 nm); the corresponding substrates show little or no absorbance at these wavelengths.

2-nitrophenylgalactoside + $H_2O \rightarrow$ 2-nitrophenol + galactose

4-nitrophenylphosphate + $H_2O \rightarrow$ 4-nitrophenol + phosphate

N-benzoylarginine-4-nitroanilide + $H_2O \rightarrow$ *N*-benzoylarginine + 4-nitroaniline

Other types of non-natural substrates are based on derivatives of coumarin or umbelliferone; these yield intensely fluorescent derivatives on hydrolysis.

Table 6.4 Assays of hydrolases using chromogenic substrates

Hydrolase	Substrate providing convenient spectral change	Reaction product	Wavelength at which product can be observed
β-Galactosidase	2-Nitrophenylgalactoside 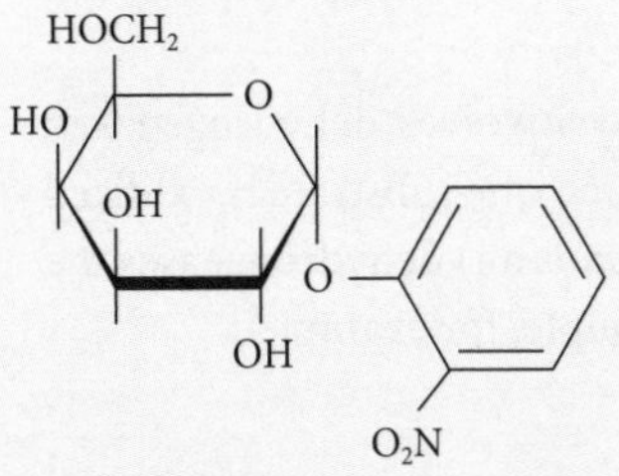	2-Nitrophenol 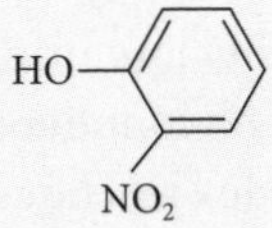	400 nm*
Alkaline phosphatase	4-Nitrophenylphosphate 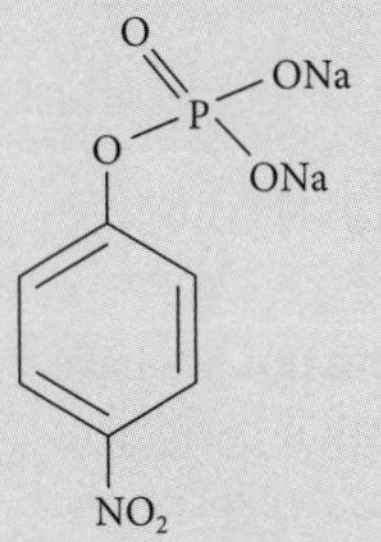	4-Nitrophenol 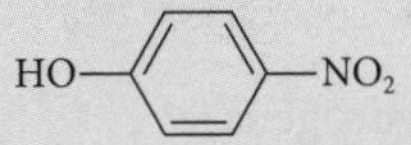	400 nm*
Trypsin	*N*-Benzoylarginine-4-nitroanilide 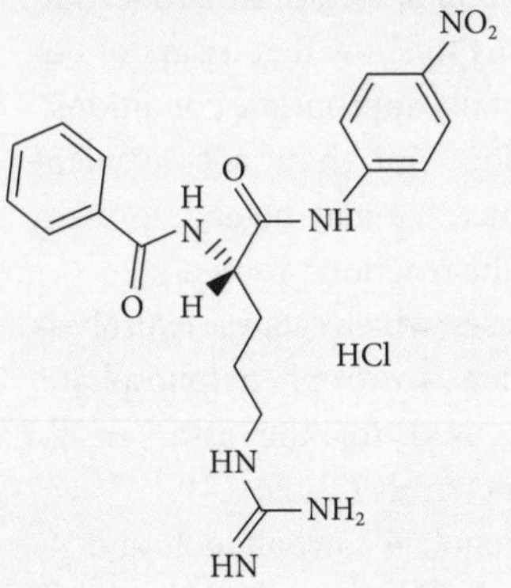	4-Nitroaniline 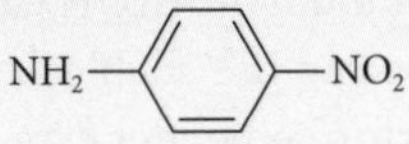	405 nm

*The wavelength refers to the absorption by the anionic form of the product.

In other cases, it may be possible to monitor a reaction continuously by devising a suitable coupled assay system; this approach is discussed in section 6.3.3.

6.3.2 Discontinuous assays

Discontinuous assays are assays of the 'stop and sample' type (see section 6.3) and are used where there is no convenient spectroscopic or other property to monitor in a continuous fashion. After taking a sample from the reaction mixture, it is important to inhibit the enzyme as rapidly as possible before analysis is carried out to determine the extent of the reaction, so that a true picture of the rate of the reaction can be obtained. Common ways of inactivating enzymes include rapid changes in pH or rapid heating. Another widely used method is the addition of trichloroacetic acid that denatures and precipitates almost all proteins. Whatever method is used, it is important to establish that (a) the enzyme is in fact essentially instantaneously inactivated, and (b) that the conditions used do not lead to any breakdown of the substrate or product being measured.

Discontinuous assays are frequently used when the enzyme acts on a macromolecular substrate. For example, to assay general levels of proteolytic activity (rather than a specific protease) it is common to use the milk protein casein as a substrate. After taking a sample and adding trichloroacetic acid to precipitate not only the enzyme, but also any remaining undigested casein, the extent of hydrolysis is measured by the absorbance at 280 nm of the supernatant after addition of trichloroacetic acid, followed by centrifugation. The absorbance arises from the small peptides produced by digestion of the casein, since these are not precipitated by the trichloroacetic acid.

The application of this type of procedure to assay enzymes such as DNA polymerase, which catalyse the synthesis of macromolecules, is discussed in Chapter 9, section 9.2.4. Here the synthesis is usually monitored by the incorporation of a radioactively labelled precursor into the macromolecule (e.g. [^{3}H]-labelled thymidine into DNA).

Stop and sample assays would also be appropriate when the modification of a macromolecule is being monitored. For example, we might wish to monitor the action of a kinase, which is known to phosphorylate a target protein in a signalling pathway at a specific amino acid (for example at a particular serine side chain). In this case, we could examine the incorporation of ^{32}P from γ-^{32}P-labelled ATP into the target protein or use mass spectrometry or a specific immunoassay to quantify the extent of modification of the target protein.

6.3.3 Coupled assays

When the enzyme-catalysed reaction of interest does not lead to any convenient spectroscopic signals, it is often possible to couple this reaction to a second one in which such a change occurs. In Chapter 5, section 5.3.1, we saw how the hexokinase(HK)-catalysed reaction:

$$\text{D-glucose} + \text{ATP} \rightarrow \text{D-glucose-6-phosphate} + \text{ADP}$$

could be coupled to that catalysed by glucose-6-phosphate dehydrogenase:

$$\text{D-glucose-6-phosphate} + \text{NADP}^+ \rightarrow \text{D-glucono-1,5-lactone-6-phosphate} + \text{NADPH} + \text{H}^+$$

The second reaction proceeds with an increase in absorbance at 340 nm due to the formation of NADPH. So long as sufficient coupling substrate ($NADP^+$) and coupling enzyme (G6PDH) are added, the D-glucose-6-phosphate formed in the HK reaction is essentially instantly oxidized in the G6PDH reaction.

If we do not add a sufficient excess of the coupling enzyme(s) and substrate(s), we may well be measuring the rate of the coupling reaction, rather than that of the reaction of interest.

In performing coupled assays it is important to add a considerable excess of coupling enzyme (usually at least 50-fold greater in terms of units of enzyme activity) compared with the enzyme being assayed. In addition, it would be important to add the coupling substrate at a concentration well above its K_m value for the coupling enzyme. It is essential that the coupling enzyme and substrate are highly purified; in particular, the coupling enzyme should be free of any detectable amount of the enzyme being assayed (see Chapter 5, section 5.3).

6.3.3.1 *Assays of reactions involving ATP*

The hydrolysis of ATP (by the action of an ATPase such as that associated with the muscle protein myosin) is monitored by a stop and sample procedure in which the liberated inorganic phosphate could be estimated by its reaction with molybdate to give the blue phosphomolybdate complex. Alternatively the reaction could be assayed in a continuous manner by a coupled reaction involving the glyceraldehyde-3-phosphate dehydrogenase-catalysed reaction:

$$\text{glyceraldehyde-3-phosphate} + NAD^+ + P_i \rightarrow \text{1,3-bis-phosphoglycerate} + NADH + H^+$$

P_i is the abbreviation used to denote inorganic phosphate.

The formation of NADH gives a convenient increase in absorbance at 340 nm.

A second possibility would be to couple the formation of ADP in the ATPase reaction to the pyruvate kinase reaction:

$$ADP + PEP \rightarrow ATP + \text{pyruvate}$$

and then estimate the pyruvate formed using the lactate dehydrogenase reaction:

$$\text{pyruvate} + NADH + H^+ \rightarrow \text{lactate} + NAD^+$$

Pyruvate kinase is an unusual kinase in that the equilibrium of the reaction lies very much to the side of ATP formation; this reflects the high phosphate transfer potential (i.e. high negative free energy of hydrolysis) of the substrate PEP (phosphoenolpyruvate).

This second approach requires the use of two coupling enzymes (PK and LDH) and two coupling substrates (PEP and NADH). All the coupling substrates and enzymes must be highly purified and added in appropriate excess to ensure that the coupling steps are not rate limiting. The double coupled assay approach involving pyruvate kinase and lactate dehydrogenase is widely used for the assay of kinases (enzymes which catalyse the transfer of a phosphoryl group from ATP to an acceptor molecule, resulting in the formation of ADP.)

6.3.4 How to ensure that valid enzyme kinetic data are obtained

A number of points should be checked to make sure that an assay procedure has been set up in an experimentally meaningful fashion. This may require some detailed preliminary work, but it is an important prerequisite for obtaining valid results.

Firstly, it is important to use buffer components and substrates of high purity (analytical grade). Use of less pure reagents may lead to the introduction of potential inhibitors, for example heavy metals, which could damage enzymes by reaction with important cysteine side chains.

Secondly, a control experiment should be performed in the absence of enzyme to check that there is no significant blank reaction, or to ascertain the extent of such a reaction so it can be used to correct the rate observed when enzyme is added. The control reaction also provides a useful check on the stability of the recording device (e.g. spectrophotometer) over the time period of the assay.

Thirdly, it is important to ensure that experimental variables such as temperature and pH are controlled by suitable thermostat and buffer systems, respectively. The buffer system should be chosen to provide adequate buffering capacity at the chosen pH, especially if there is an uptake or release of H^+ ions during the reaction, or stock solutions of charged substrates are to be made up at high concentrations. It is also important that the buffer should not interfere with the reaction. For example, high concentrations of phosphate buffers should not be used if an enzyme is dependent on divalent metal ions for activity, since the metal ions could be complexed by the phosphate ions present.

Fourthly, it should be checked that the enzyme is stable under the conditions used for assay. This is particularly important when prolonged incubations are used for assays, or when the assay is performed under conditions such as elevated temperatures or extremes of pH where its three-dimensional structure and hence activity may be compromised. Under such conditions, it may be advisable to make the incubation time for the assay as short as possible.

Fifthly, the initial rate should be measured. This will minimize complications caused by any of the following factors:

- depletion of substrate (there will be little change in the concentration of substrate from the initial value)
- product inhibition (there will be little build up of product, so the degree of any possible inhibition will be small)
- occurrence of the reverse reaction (only small amounts of product are formed, so the rate of the reverse reaction will be minimal).

In order to measure the initial rate, it is important to ensure that data recording starts as soon as possible after mixing and that the amount of enzyme added is not so large as to mean that a significant portion of the reaction has occurred during this 'dead time'.

Finally, it should be checked that the rate of the reaction is constant over the time period used to measure the rate, and that this rate is proportional to the amount of enzyme added (this can be checked by adding half of the amount of enzyme; the observed rate should be halved). This latter requirement provides a very important check on the validity of coupled assay systems. For example, consider a coupled assay system in which the amount of the enzyme of interest added to the assay is halved and the observed rate decreases by 1.7-fold (to 59% of the initial value). Such a result would indicate that insufficient coupling enzyme or substrate has been added (or alternatively that the rate of the reaction catalysed by the enzyme of interest is too great, and the amount of that enzyme should be decreased).

A simple calculation will illustrate how the appropriate amount of enzyme to be added to an assay can be judged (see the Worked example below).

✓ WORKED EXAMPLE

Consider the reaction catalysed by lactate dehydrogenase (assayed in the direction pyruvate + NADH + $H^+ \rightarrow$ lactate + NAD^+). Under certain conditions the specific activity of the enzyme is 500 μmol min^{-1} mg^{-1} enzyme. The molar absorption coefficient of NADH at 340 nm is 6220 M^{-1} cm^{-1}; NAD^+ does not absorb at this wavelength. Assays are conducted in a total volume of 3 mL in a cuvette of 1-cm pathlength. How much enzyme should be added to obtain a rate of decrease of A_{340} of 0.1 min^{-1} (this is a convenient rate to record in a conventional spectrophotometer)?

STRATEGY

The absorbance change min^{-1} is used to derive the change in concentration (and hence change in number of moles of NADH) min^{-1}. This can be divided by the specific activity to derive the amount of enzyme to be added. It is essential to express quantities in the correct units during the calculation.

SOLUTION

A decrease in A_{340} of 0.1 min^{-1} corresponds to a change of concentration of 0.1/6220 M min^{-1}, i.e. 16.1 μM min^{-1}. In 3 mL reaction volume, this corresponds to 0.482 μmol min^{-1}. Using the value of the specific activity, this means that 0.482/500 mg enzyme should be added, i.e. 0.96 μg enzyme.

If, for example the initial concentration of NADH in the lactate dehydrogenase assay were 150 μM, the initial A_{340} would be about 0.93. On addition of a large excess of lactate dehydrogenase, this would decrease very quickly to a final value (typically 0.05 or less) and then stay effectively constant.

A mistake commonly made when performing enzyme assays is to add considerably too much enzyme to the assay mixture. In the example of lactate dehydrogenase discussed in the worked examples, if 10 μg enzyme were added, the reaction would occur so quickly that the reaction would be essentially completed by the time that the cuvette could be replaced in the spectrophotometer after mixing the enzyme and substrates. There would then appear to be a very low or zero rate of change in A_{340} with time and it might be tempting to conclude that the sample shows no activity. In fact the opposite is the case; there is too much activity as indicated by the very rapid decrease in A_{340} before the sample can be monitored. By contrast if the amount of enzyme added were far too low (for example 10 ng added to a reaction volume of 3 mL) the rate would be too small to measure accurately (the decrease in A_{340} would be approximately 0.001 min^{-1}).

It can also be difficult to make reliable measurements of the rate at low substrate concentrations. For example, if the rate of the lactate dehydrogenase reaction is to be studied at a [NADH] of 5 μM, the total change in A_{340} (assuming complete conversion to NAD^+) would be 0.031. In order to observe this rate accurately, it is important to add the appropriate amount of enzyme so that a constant rate can be observed for long enough (typically 1 min) to allow the rate to be measured accurately. This would require a number of preliminary experiments to be run in which the various amounts of enzyme are added; if too much enzyme is added, the reaction will be essentially over too quickly, if too little is added, the rate will be too slow to measure accurately.

SELF TEST ?

Check that you have mastered the key concepts at the start of this section by attempting the following questions.

ST 6.5 How would you assay the activity of a cAMP-dependent protein kinase that catalyses the transfer of the γ-phosphoryl group from ATP to the enzyme phosphofructokinase?

ST 6.6 Caspases are cysteine proteinases, which play key roles in processes such as apoptosis (programmed cell death) and inflammation. The enzymes cleave polypeptide chains on the C-terminal side of aspartate residues. How would you design assays for such enzymes?

ST 6.7 If in the case of the lactate dehydrogenase-catalysed reaction in the worked example in section 6.3.4, the reaction volume is 1 mL, in a cuvette of pathlength 1 cm, how much enzyme should be added to obtain the same rate (decrease in $A_{340}\ \text{min}^{-1} = 0.1$)?

Answers

ST 6.5 One possibility is to incubate phosphofrucotkinase with radioactively labelled ATP (with the γ-phosphoryl ^{32}P labelled) and stop the reaction at stated times after addition of enzyme. The products can be separated into high molecular mass and low molecular mass fractions by gel filtration and incorporation of radioactivity into the high molecular mass material measured. Alternatively, if it is not feasible to use radioactivity, the incorporation of the phosphoryl group could be monitored by mass spectrometry; there is an increase of 80 Da in the molecular mass of the protein.

ST 6.6 The most convenient substrates would be low molecular mass compounds such as 4-nitroanilides (see section 6.3.1). These would contain an aspartate residue whose carboxyl group forms an amide with 4-nitroaniline. There are at least 10 caspases with differing specificities, so it would be necessary to include additional amino acids on the N-terminal side of the aspartate to design a substrate for a particular caspase. (For example, caspase 3 activity can be assayed by using the model substrate *N*-acetyl-Asp–Glu–Val–Asp-4-nitroanilide).

ST 6.7 The approach used is the same as outlined in the worked example in section 6.3.4; however, the number of moles of NADH consumed is reduced by three-fold because of the smaller volume of the assay mixture. Hence a three-fold lower amount of enzyme should be added. The amount of enzyme to be added is 0.32 μg.

6.4 Binding assays

KEY CONCEPTS

- Selecting the appropriate method to measure a protein–ligand interaction
- Making sure that valid binding data are obtained

We would generally use the term 'small molecule' to refer to a molecule of mass ≤500 Da.

Many proteins function by binding (usually in a specific fashion) other molecules; these can include proteins, nucleic acids, carbohydrates, or small molecules, and are collectively termed ligands. In this section, we will discuss some of the principal ways in which protein–ligand interactions can be studied in a quantitative manner. To obtain a complete thermodynamic description of the binding, we would wish to answer the following questions:

- How many binding sites for the ligand are there on the protein?
- What is the strength of the protein–ligand interaction?
- If there are multiple ligand binding sites, are they independent or are there cooperative interactions between them?
- Is the strength of the protein–ligand interaction modified by other molecules?

In conjunction with these measurements we would also like to obtain structural data which will highlight the key types of interactions made between the protein and the ligand; these will involve high-resolution structural techniques such as X-ray crystallography or nuclear magnetic resonance (NMR) and are beyond the scope of this book.

Some of the key points have been already mentioned in Chapter 5, section 5.3.2, and will be discussed in more detail in Chapter 10.

In essence, the methods for measuring binding can be divided into two principal categories: those in which the protein–ligand complex is physically separated from either free protein or free ligand (section 6.4.1); and those in which some property is used to distinguish between the complex and the free interacting partners (section 6.4.2). A third, indirect, approach involves the so-called 'competition assays' in which we study the effect of the ligand of interest on a binding process we can more easily monitor (section 6.4.3). The aim of these approaches is to determine the concentrations of the protein–ligand complex and the free protein and free ligand. Division of the concentration of bound ligand by the concentration of total protein will give the stoichiometry of binding, a key parameter in the analysis of binding data (see Chapter 4, section 4.4). For example, if under certain conditions, the protein concentration in a solution were 12 μM and the concentration of bound ligand were 20 μM, the stoichiometry of binding would be $20/12 = 1.67$ mol ligand bound per mol protein. If it were known from other data that the protein had two binding sites for the ligand, the degree of saturation would be $20/(2 \times 12) = 0.83$.

Section 6.4.4 will briefly describe the principal ways in which binding data are analysed, and section 6.4.5 will give general guidelines on how reliable binding data can be obtained.

6.4.1 Methods involving separation of the protein–ligand complex

In this section, we will outline three methods in which the bound ligand (complexed with protein) is separated from free ligand on the basis of size.

6.4.1.1 Equilibrium dialysis

A dialysis membrane (usually cellophane) acts as a sieve with holes large enough to allow the passage of 'globular' molecules up to about 20 kDa in mass. This limit can be changed by various mechanical or chemical treatments of the membrane. Protein is added to the compartment on one side of the membrane, and a solution of the ligand is added to the other compartment. The ligand (assumed to be of mass much lower than 20 kDa) will diffuse freely across the membrane until equilibrium is achieved i.e. the ligand concentration in the non-protein compartment shows no further tendency to change. When equilibrium is achieved, the concentrations of protein and ligand in the protein compartment are measured, as well as the ligand concentration in the non-protein compartment.

The principle of equilibrium dialysis and the way that data are analysed are illustrated in Fig. 6.14. In the figure, a dialysis bag is used to contain the protein, but it is also possible to work on a smaller scale, for example by using small devices consisting of two perspex blocks each of which is machined out to provide a chamber of ca. 0.2-mL capacity. Each chamber can be filled or emptied via a suitably drilled channel. When the blocks are bolted together, a sheet of dialysis membrane is used to separate the two compartments.

Equilibrium dialysis is a useful technique to study the binding of a low molecular mass ligand to a protein provided that (a) the protein is larger than about 20 kDa (otherwise it would cross the membrane to a certain extent) and is available in the quantities required, and (b) there is a convenient way of estimating the ligand concentrations in the two compartments. An example of the use of

Radioactivity can provide a very convenient way of estimating ligand concentrations, since the measured radioactivity does not depend on whether the ligand is free or bound to protein.

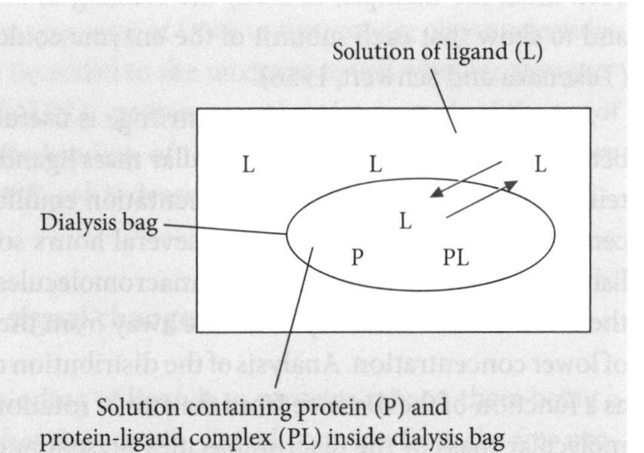

Fig. 6.14 Principles of equilibrium dialysis. At equilibrium $[L]_{free}$ is the same inside and outside the bag. Outside the bag $[L] = [L]_{free}$; inside the bag $[L] = [L]_{free} + [L]_{bound}$. Hence $[L]_{bound}$ can be established by difference. Division by the total concentration of protein gives the saturation of binding sites by the ligand.

between the ligand and an appropriate structural or spectroscopic feature of the protein itself or be a secondary effect resulting from a change in conformation of the protein on complex formation. However, any signal change that can be conveniently measured will monitor the extent of complex formation, irrespective of the origin of the change in precise structural terms. Some of the more common types of signal change are described in the sections below.

6.4.2.1 Changes in absorbance

A chromophore is a molecule or part of a molecule that can absorb radiation. For example, the phenol moiety in the amino acid tyrosine absorbs radiation at around 280 nm and would be termed a chromophore. In any protein, the peptide bonds absorb strongly below 230 nm and would also be termed chromophores.

The absorption spectrum of a protein or a ligand will alter to a small extent when the polarity of its environment changes, reflecting the different degrees of interaction between a chromophore and its surroundings. Changes in the environment of the protein or the ligand or both could occur on complex formation; however, the changes in absorbance are usually relatively small and considerable experimental care is needed to obtain reliable data. However, useful information can be obtained in certain cases, including the binding of the inhibitor 2′-CMP to pancreatic ribonuclease, where there is a 25% change in the absorbance of the inhibitor at 262 nm on binding (Anderson *et al.*, 1968), and the binding of NADH to lactate dehydrogenase, which is accompanied by a 15% change in absorbance of NADH at 340 nm (Chance and Neilands, 1952). The technique is not generally useful for examining protein–protein interactions because both partners will usually contribute to the absorbance in the 280 nm region and changes on complex formation are likely to be small. This use of absorbance changes to monitor binding is described in greater detail in Chapter 10, section 10.3.1.

6.4.2.2 Changes in fluorescence

Fluorescence is the emission of radiation from the lowest vibrational energy level of the excited electronic state of a molecule as it decays to the ground state (Fig. 6.16).

The fluorophore in GFP is formed by an oxidative cyclization reaction of the sequence of amino acids Ser–Tyr–Gly (numbers 65–67 in the sequence of the protein); see Chapter 7, section 7.3.7.

Generally a fluorescent molecule or group (known as a fluorophore) is aromatic or highly conjugated. As far as proteins are concerned, fluorescence is primarily due to the side chains of Tyr and Trp (the latter usually being much more highly fluorescent than the former); in addition some proteins contain other fluorophores, such as green fluorescent protein (GFP) from the jellyfish *Aequorea victoria*. Certain ligands also exhibit fluorescence under physiological conditions, e.g. NADH (but not NAD^+) and flavins such as FAD or FMN. In Chapter 5, section 5.3.2, it was mentioned how fluorescent labels could be introduced into proteins. There are also fluorescent ligands that can bind to certain types of sites in proteins. For example, 1-anilino-8-naphthalene sulphonate (ANS) binds to non-polar sites in proteins, resulting in considerable enhancement and a shift of fluorescence maximum to shorter wavelengths. ANS has been used to explore the haem binding sites in myoglobin and haemoglobin, and the fatty acid binding sites in serum albumin, for example.

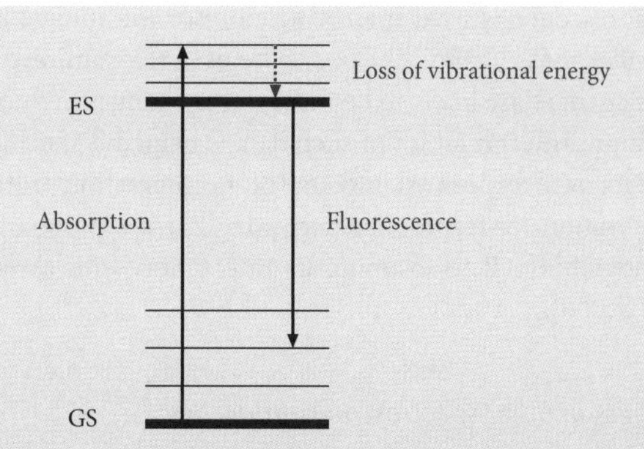

Fig. 6.16 The processes of absorption and fluorescence. The energies of the lowest vibrational levels of the ground (GS) and excited (ES) electronic states of a fluorophore are shown by heavy horizontal lines. The higher vibrational levels of each state are shown by the thin horizontal lines. Absorption of radiation (upward vertical arrow) occurs from the lowest vibrational level of the ground state to a higher vibrational level of the excited state. There is then a loss (dotted line) of vibrational energy to give the lowest vibrational level of the excited state. Fluorescence is the emission of radiation from this state to one of the higher vibrational levels of the ground state. In the case of proteins, fluorophores would include the side chains of tryptophan and tyrosine and various cofactors or ligands which bind to proteins such as flavins or NADH.

Fluorescence is generally much more sensitive to the environment than is absorbance, and is therefore more likely to show measurable changes on complex formation. For that reason, fluorescence is generally a more useful method for studying protein–ligand binding. As an example the binding of NADH to lactate dehydrogenase leads to only a 15% change in the absorbance of NADH in the 340 nm region, whereas there is a more than 5-fold enhancement in the NADH fluorescence at 440 nm. Furthermore, the fluorescence of lactate dehydrogenase is quenched by more than 5-fold on binding NADH. (Chance and Neilands, 1952; Velick, 1958). The binding of molybdate to the ModE protein from *E. coli* is accompanied by a 50% quenching of protein fluorescence, but there is less than 5% change in absorbance of the protein at 280 nm (Anderson *et al.*, 1997). Examples of the use of this approach are described in Chapter 10, section 10.3.2.

The greater sensitivity of fluorescence compared with absorbance to changes in the environment can be understood as follows. In the excited electronic state, the distribution of electrons is such that there is a greater charge separation (dipole) than in the ground state. Thus, the excited state can be stabilized to a greater extent in a polar environment compared with a non-polar environment than is possible in the ground state. This means that there is a greater range of energies (and hence wavelengths) in different environments for fluorescence than for absorbance. It should be remembered that during these electronic transitions the positions of the much heavier nuclei remain essentially fixed (the Franck–Condon principle).

7-Azatryptophan can be incorporated into overexpressed proteins in place of tryptophan. Its absorption and emission maxima are shifted to longer wavelengths (about 30 nm in each case) compared with the values for tryptophan.

Although fluorescence is used mainly to monitor the interaction of proteins with low molecular mass ligands, it can also be used to examine protein–protein interactions, since there are likely to be measurable changes in fluorescence upon complex formation. Interpretation of such data is helped if specific fluorophores can be attached to, or incorporated into, one of the interacting proteins. An example is the incorporation of a tryptophan derivative, 7-azatryptophan, into a defined part of immunoglobulin E to examine its interaction with a receptor protein (Harwood *et al.*, 2006).

6.4.2.3 Changes in other spectroscopic properties

A number of other spectroscopic properties of proteins or ligands have been used to monitor complex formation (Harding and Chowdhry, 2001). We will briefly describe two of these.

Circular dichroism

Circular dichroism (CD) depends on the differential absorption of the left- and right-circularly polarized components of plan polarized light by a chiral chromophore (see Chapter 5, section 5.2.2). CD signals arising from either the protein or the ligand could be used in these experiments. Changes in the CD spectrum of the protein would arise from structural changes in the protein on complex formation and are usually relatively minor. By contrast, there can be significant changes in the CD signal in the region where the ligand absorbs. This is because in many cases the free ligand is not chiral (i.e. it shows no CD signal), but acquires chirality (and hence shows a CD signal) when bound to the protein (which is chiral). CD has been widely used to monitor the binding of drugs such as diazepam to serum albumin (Drake, 2001). This example is described in greater detail in Chapter 10, section 10.3.3.

Nuclear magnetic resonance

Nuclear magnetic resonance (NMR) is based on the behaviour of certain nuclei in molecules (principally ^{1}H, but also by isotopic substitution ^{13}C and ^{15}N) in an applied magnetic field. The spectroscopic properties of each nucleus are sensitive to its chemical environment, so the technique can be used to monitor protein–ligand interactions and characterize them in thermodynamic terms. In addition, NMR can provide a great deal of structural information about the amino acid residues in the protein that interact with the ligand. The details are beyond the scope of this book, but a good introductory account is given by Lian (2001).

6.4.2.4 Other methods for studying ligand binding

A number of other more specialized methods to study complex formation are available. Changes in enthalpy on complex formation can be measured using the technique of isothermal titration calorimetry (see Chapter 10, section 10.6), and changes in refractive index can be measured using the technique of surface plasmon resonance (see Chapter 10, section 10.7).

6.4.3 Competition assays

In some cases, a protein–ligand interaction can be most easily studied by measuring the effect of the ligand of interest on a binding process we can more readily monitor; these indirect assays are termed competition-type assays. For example, the binding of an inhibitor of an enzyme-catalysed reaction might be difficult to study directly, especially if there is no convenient spectroscopic change on formation of the complex. However, the effect of the inhibitor on the catalysed reaction can be easily monitored, and the results analysed as described in Chapter 4, sections 4.3.4 and 4.4, to give the value for the dissociation constant of the enzyme–inhibitor complex.

This type of approach can be used for other types of biological activity. For example, the interaction of a receptor protein with a new candidate drug molecule could be monitored by observing the effect of that drug on the binding of the natural ligand for the receptor (see Chapter 5, section 5.6.2). Competition assays are discussed further in Chapter 10, section 10.8.

The key point as far as the effect of enzyme inhibitors is concerned is that the relevant kinetic parameter (K_m or V_{max}) is changed by a factor $(1 + ([I]/K_{EI}))$, where K_{EI} is the dissociation constant of the enzyme–inhibitor complex. Hence K_{EI} can be calculated knowing the magnitude of this change and the concentration of inhibitor present.

WORKED EXAMPLE 

The interaction of adrenaline with a certain β-adrenergic receptor has a K_d of 5 μM. Propranolol acts as an antagonist by competing with adrenaline for binding to the receptor. In the presence of 10-nM propranolol, the K_d for adrenaline is 15.9 μM. What is the K_d for the receptor–propanolol interaction?

STRATEGY

The factor by which K_d increases is given by $(1 + ([I]/K_I))$. Knowing this factor and the concentration of I added, K_I can be calculated.

SOLUTION

In the presence of propranolol, K_d increases by 3.18-fold. Hence $([I]/K_I) = 2.18$. Since $[I] = 10$ nM, $K_I = 4.6$ nM.

The very strong interaction between propranolol and β-adrenergic receptors underpins its clinical use as a β-blocker in the treatment of conditions such as hypertension, angina, cardiac arrhythmia, etc.

6.4.4 Analysis of binding data

Binding data should be analysed to give answers to the following questions:

- What is the stoichiometry of the interaction (i.e. how many ligand binding sites are there on the protein)?
- What is the strength of the interaction (i.e. what is the dissociation constant of the complex and hence the free energy of the interaction)?
- In the case of multiple ligand binding sites, what is the extent of any interactions between these sites (i.e. what is the degree of cooperativity)?

These points have been already been described in detail in Chapter 4, sections 4.3 and 4.4. However, for the present purposes, it is sufficient to note that the experimental data should allow the concentrations of free ligand, free protein, and the protein–ligand complex in a mixture to be determined. These data are then analysed using either non-linear regression or linear transformation approaches to give the best fit to an assumed theoretical model. One of the most widely used equations to analyse binding of a ligand to multiple sites in a protein is the Scatchard equation (see Chapter 4, section 4.4.):

$$r = \frac{n \times [L]}{K_d + [L]}$$

where r is the average number of molecules of ligand bound per molecule of protein, n is the number of binding sites, $[L]$ is the concentration of free ligand, and K_d is the dissociation constant.

This can be rearranged using the approach outlined in Chapter 2, section 2.5.2 to give:

$$\frac{r}{[L]} = \frac{n}{K_d} - \frac{r}{K_d}$$

A plot of $r/[L]$ vs r will give a straight line of slope $-1/K_d$, and an intercept on the x-axis of n. (The intercept on the y-axis is n/K_d.) In Chapter 10, we shall describe several examples of the use of Scatchard plots to analyse ligand binding data.

6.4.5 How to obtain reliable binding data

The key to obtaining data that can be analysed to give reliable estimates of the binding parameters is careful planning of experiments. There will be significant changes in the extent of dissociation of the protein–ligand complex when the concentrations of the interacting components are varied in the range around the value of the dissociation constant. It may well be necessary to perform some preliminary experiments to establish the appropriate concentration range over which to carry out titrations. Ideally, it would be desirable to cover the saturation curve from 10% to 90%, but this may not always be practicable and coverage from 25% to 75% saturation is more usual.

If the binding of a ligand to a protein is very tight, it may not be possible to detect significant dissociation of the protein–ligand complex. For example, the binding of molybdate to the ModE protein from *E. coli* can be conveniently studied by the quenching of protein fluorescence that occurs on binding. When 20 μM ModE dimer was titrated with increasing concentrations of molybdate, the change in fluorescence is essentially linear with [molybdate] up to very nearly the point of saturation of the binding sites (Fig. 6.17). This shows that, under these conditions,

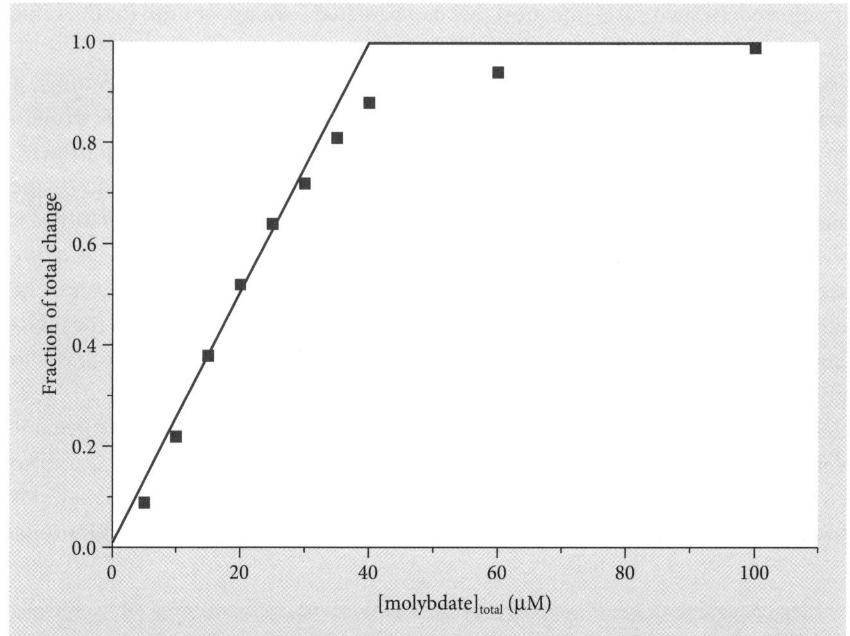

Fig. 6.17 Titration of ModE with molybdate monitored by fluorescence quenching. The data points are those obtained when molybdate is added to 20 μM ModE dimer. The straight line drawn through the points at lower molybdate concentrations is extrapolated to full site occupancy. The data over the entire range of molybdate concentrations can be fitted very closely assuming two binding sites per ModE dimer and a K_d of 0.8 μM.

essentially all the molybdate added is bound to the protein, i.e. there is little tendency of the complex to dissociate (Anderson *et al.*, 1997).

From extrapolation of the linear part of the titration curve to the point corresponding to complete saturation, the stoichiometry of binding can be readily deduced. The limiting concentration of bound molybdate is 40 μM and the concentration of ModE dimer is 20 μM. Thus, there are two binding sites per dimer, i.e. one per polypeptide chain. However, since the extent of dissociation of the complex is very small, estimates of the concentrations of free protein and free ligand under these conditions are unreliable and hence the dissociation constant cannot be determined accurately. In order to increase the extent of dissociation of the complex, it is necessary to lower the concentration of protein. Indeed, when the titration was repeated at a much lower concentration of ModE (1.5 μM), a typical hyperbolic saturation curve was obtained. Analysis of this curve showed that that the dissociation constant for the ModE–molybdate complex was 0.8 μM. In this case, the fluorescence measurements could be readily made over a wide range of protein concentrations, by using different excitation wavelengths. However, variation of protein concentration over a wide range may not always be possible; this would be the case, for example if absorbance changes were being monitored. In such cases, the most that can be stated is an upper limit for the value of K_d (e.g. $K_d < 1\ \mu M$).

If binding of a ligand to a protein is weak (K_d greater than about 0.5 mM), it is usually very difficult to obtain a sufficiently high concentration of the protein so that the complex represents a significant fraction of the total protein present. In such cases it is usually not possible to obtain an accurate estimate of the stoichiometry, but from titrations carried out at lower protein concentrations, it should be possible to estimate K_d. It is always true that the greater the range of the saturation curve that can be covered, the better the estimate of K_d that can be obtained; it is dangerous to extrapolate from less than 50% saturation of the binding sites. Where the binding of ligand is weak, the best that may be possible is to give a lower estimate of K_d (e.g. $K_d > 10mM$).

In summary, the key experimental points to observe about collecting reliable data are: (a) to use appropriate concentration ranges for protein and ligand to cover an adequate range of the saturation curve, and (b) to use detection methods to monitor the formation of complex that are appropriate to these concentrations.

? SELF TEST

Check that you have mastered the key concepts at the start of this section by attempting the following questions.

ST 6.8 In an equilibrium dialysis experiment, how would you establish that equilibrium had been achieved?

ST 6.9 How would you investigate the interaction between lactate dehydrogenase and its substrate pyruvate? Note that pyruvate has no convenient absorption band and does not fluoresce.

Answers

ST 6.8 This is most easily established by taking samples at different times from the non-protein compartment and determining the concentration of the free ligand. When this concentration shows no further tendency to change, equilibrium has been achieved.

ST 6.9 The equilibrium dialysis approach could be employed, using radioactively labelled (^{3}H or ^{14}C) pyruvate to monitor the ligand concentration. Other possibilities include isothermal titration calorimetry and investigating whether there is a change in the spectroscopic signals (absorbance, fluorescence, etc.) of the enzyme on addition of ligand. As described in Chapter 9, section 9.5, lactate dehydrogenase follows an ordered reaction mechanism. In fact, pyruvate does not bind to the enzyme until the substrate NADH has already bound.

6.5 Problems

Full solutions to odd-numbered problems are available to all in the student section of the Online Resource Centre at www.oxfordtextbooks.co.uk/orc/price/. Full solutions to even-numbered problems are available to lecturers only in the lecturer section of the Online Resource Centre.

6.1 Solutions of four proteins, BSA, hen egg white lysozyme, bovine trypsin, and hen ovalbumin, were made up by weight at 2 mg mL^{-1} in 50 mM sodium phosphate buffer, pH 7.0. The solutions were each diluted 10-fold into buffer and the A_{280} values measured. These solutions were then diluted a further 10-fold (i.e. 100-fold in total) into buffer and the A_{205} values determined. The results are shown in the table below.

Protein	A_{280} (10-fold dilution)	A_{205} (100-fold dilution)
BSA	0.122	0.548
Lysozyme	0.539	0.726
Trypsin	0.259	0.538
Ovalbumin	0.132	0.573

Use these data to calculate the concentrations of the original stock solutions of the proteins using (a) the A_{205} values and eqn. 6.4, (b) the A_{205} and the A_{280} values and eqn. 6.5, and (c) the A_{280} values and absorption coefficients at this wavelength given i.e. 0.66, 2.63, 1.57 and 0.69 for 1 mg mL^{-1} solutions of BSA, lysozyme, trypsin and ovalbumin respectively.

6.2 A His-tagged protein kinase was overexpressed in *E. coli* and purified by immobilized metal affinity chromatography. SDS-PAGE analysis of the purification procedure is shown below.

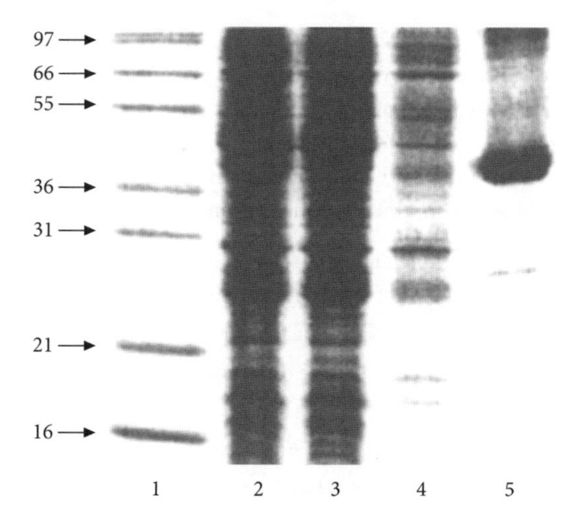

Lane	Sample
1	Molecular mass markers (masses given in kDa)
2	*E. coli* cell lysate
3	Breakthrough material that failed to bind to Ni-nitrilotriacetic acid matrix
4	Wash material that failed to bind to the Ni-nitrilotriacetic acid matrix
5	Imidazole eluate, containing His-tagged protein kinase

Comment on the degree of complexity of the samples analysed and calculate the molecular mass of the His-tagged protein kinase.

6.3 The following two-dimensional PAGE gel was obtained from tissue culture fibroblasts, which had been treated with a steroid hormone to determine which proteins were upregulated following hormone action. From a comparison of two-dimensional gels prepared from control and hormone-treated cells, a number of proteins whose expression had increased in the presence of the hormone were identified. These are indicated with the circle and the arrow.

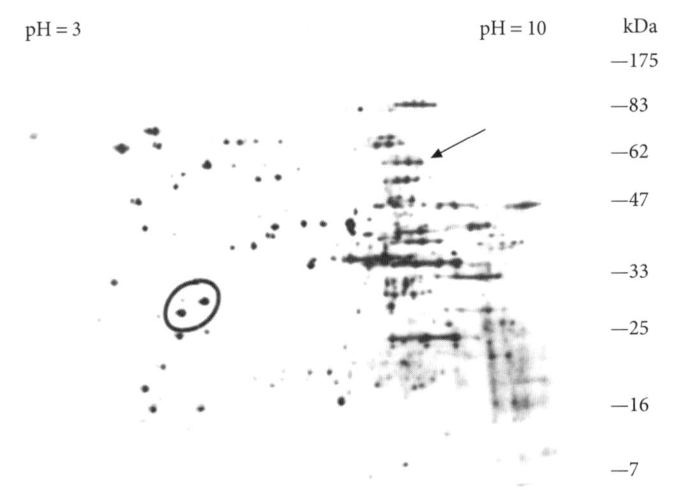

It was found that the circled proteins were integral membrane proteins. Further, analysis using Western blots and peptide mapping indicated that they were identical except that the smaller protein was missing approximately 20 amino acids.

The proteins indicated by the arrow were cytosolic and subsequent analysis showed that they contained identical amino acid sequences.

1) Determine the molecular masses of the two proteins enclosed in the circle and the protein band (three spots of identical molecular mass) indicated by the arrow.
2) Explain the significance of the difference in molecular masses of the circled proteins.
3) What might account for the differences in isoelectric points of the three proteins indicated by the arrow?

6.4 How would you assay alanine aminotransferase that catalyses the reaction below? There is no convenient absorbance change on reaction.

L-alanine + 2-oxoglutarate $\rightleftharpoons$ pyruvate + L-glutamate

6.5 The binding of the substrate (3-dehydroquinate) to the inactive Arg 23 → Lys mutant of type II dehydroquinase from *Streptomyces coelicolor* was studied by isothermal titration calorimetry (Krell *et al.*, 1996). At 25°C, the values of K_a and ΔH^0 for the association reaction are 910 M^{-1} and −25 kJ mol^{-1}, respectively. What are the values of K_d, ΔG^0 and ΔS^0 for the dissociation of the complex under these conditions?

6.6 What methods might be suitable to investigate the interaction between an antibody (molecular mass 150 kDa) and an antigen which is a protein of molecular mass 50 kDa?

References for Chapter 6

Anderson, D.G., Hammes, G.G., and Walz, F.G., Jr. (1968) *Biochemistry* 7, 1637–45.

Anderson, L.A., Palmer, T., Price, N.C., Bornemann, S., Boxer, D.H., and Pau, R.N. (1997) *Eur. J. Biochem.* **246**, 119–26.

Bozzo, G.G., Raghothama, K.G., and Plaxton, W.C. (2004) *Biochem. J.* **377**, 419–28.

Chance, B. and Neilands, J.B. (1952) *J. Biol. Chem.* **199**, 383–387.

Crook, E.M., Mathias, A.P., and Rabin, B.R. (1960) *Biochem. J.* **74**, 234–38.

Drake, A.F. (2001) Circular dichroism. In: *Protein-Ligand Interactions: Structure and Spectroscopy. A Practical Approach*, Harding, S.E. and Chowdhry, B.Z., eds. Oxford University Press, Oxford, pp. 123–67.

Gill, S.C. and von Hippel, P.H. (1989) *Anal. Biochem.* **182**, 319–26.

Harding, S.E. and Chowdhry, B.Z. (eds) (2001) *Protein–Ligand Interactions: Structure and Spectroscopy. A Practical Approach.* Oxford University Press, Oxford.

Harwood, N.E., Price, N.C., and McDonnell, J.M. (2006) *FEBS Lett.* **580**, 2129–34.

Krell, T., Horsburgh, M.J., Cooper, A., Kelly, S.M., and Coggins, J.R. (1996) *J. Biol Chem.* **271**, 24492–97.

Laemmli, U.K. (1970) *Nature* **227**, 680–85.

Lian, L.-Y. (2001) NMR studies of protein–ligand interactions. In: *Protein–Ligand Interactions: Structure and Spectroscopy. A Practical Approach*, Harding, S.E. and Chowdhry, B.Z., eds. Oxford University Press, Oxford, pp. 383–405.

O'Farrell, P.H.J. (1975) *J. Biol. Chem.* **250**, 4007–21.

Pace, C.N., Vajdos, F., Fee, L., Grimsley, G., and Gray, T. (1995) *Protein Sci.* **11**, 2411–23.

Price, N.C. and Stevens, L. (1999) *Fundamentals of Enzymology*, 3rd edn. Oxford University Press, Oxford, Chapter 8.

Read, S.M. and Northcote, D.H. (1981). *Anal. Biochem.* **116**, 53–64.

Schachman, H.K. and Edelstein, S.I. (1973) *Methods Enzymol.* **27**, 3–58.

Scopes, R.K. (1974) *Anal. Biochem.* **59**, 277–82.

Stevens, L. (1992) Buffers and the determination of protein concentrations. In: *Enzyme Assays: A Practical Approach*, Eisenthal, R. and Danson, M.J., eds. Oxford University Press, Oxford, pp. 317–35.

Stoscheck, C.M. (1990) *Methods Enzymol.*, **182**, 50–68.

Takenaka, Y. and Schwert, G.W. (1956) *J. Biol. Chem.* **223**, 157–70.

van Holde, K.E., Johnson, W.C., and Ho, P.S. (1998) *Principles of Physical Biochemistry.* Prentice-Hall, Englewood Cliffs, NJ, Chapter 13, pp. 544–54.

Velick, S.F. (1958) *J. Biol. Chem.* **233**, 1455–67.

White, P.J., Mousdale, D.M., and Coggins, J.R. (1987) *Biochem. Soc. Trans.* **15**, 144–45.

Appendix

Appendix 6.1 Preparation of reagents for protein determination

Preparation of the Coomassie Blue reagent (see section 6.1.1)

The exact procedure for the preparation of the reagent depends on the source of the dye. The dye from Serva (Serva Blue) contains approximately five times as much Coomassie Brilliant Blue per gram as that from Eastman.

Initially, a stock dye solution of Coomassie Brilliant Blue is prepared and requires 0.923 g of powder (Eastman), or 1 g of powder (Serva), to be dissolved in 200 mL 88% phosphoric acid (extreme care must be exercised when handling phosphoric acid) and 100 mL 95% ethanol. This solution is stable indefinitely at room temperature.

The stock dye solution of Coomassie Brilliant Blue is subsequently diluted to produce the Coomassie Brilliant Blue reagent. In the case of the Eastman stock dye solution, 150 mL of the stock dye solution is diluted to 600 mL, with water. The Coomassie Brilliant Blue reagent is then filtered through Whatman No. 1 filter paper. The Serva stock dye solution is diluted by mixing 30 mL of the stock dye solution with 80 mL 88% phosphoric acid and 40 mL 95% ethanol and is subsequently diluted to 600 mL with water and filtered as above. The Coomassie Brilliant Blue reagent should be stored in a dark bottle and will remain stable at room temperature for 3 months.

Preparation of the BCA reagent (see section 6.1.2) (from Stoscheck, 1990)

Reagent A: 1 g sodium bicinchoninate (BCA), 2 g Na_2CO_3, 0.16 g sodium tartrate, 0.4 g NaOH, and 0.95 g $NaHCO_3$ in 100 mL water and adjusted to pH 11.25 with 10 M NaOH.

Reagent B: 0.4 g $CuSO_4.5H_2O$ in 10 mL water.

BCA Reagent: 100 volumes of reagent A and 2 volumes of reagent B are mixed. The BCA reagent should appear green and will remain stable for up to 1 week.

Preparation of the Lowry reagent (see section 6.1.3) (from Stevens, 1992)

Reagent A: 0.1% (w/v) $CuSO_4.5H_2O$, 0.2% (w/v) sodium or potassium tartrate, 10% (w/v) Na_2CO_3. The Na_2CO_3, and the $CuSO_4$/tartrate should be dissolved separately in just less than half the final volume. The Na_2CO_3 solution is then slowly added to the $CuSO_4$/tartrate solution, with stirring, and then the final volume is adjusted with water. At 10°C, Reagent A will store indefinitely but near 0°C, the carbonate tends to cyrstallize out from the solution.

Reagent B: Mix 1 volume of Reagent A with 2 volumes of 5% (w/v) SDS and 1 volume of 0.8-M NaOH. Reagent B is stable at room temperature for 2–3 weeks.

Reagent C: 1 volume of the Folin–Ciocalteu reagent is added to 5 volumes of water.

Preparation of the Biuret reagent (see section 6.1.4)

The Biuret reagent is prepared by dissolving 1.5 g $CuSO_4.5H_2O$ and 6.0 g sodium potassium tartrate in 500 mL water. Next, 300 mL 10% (w/v) NaOH is added and finally water is added to a final volume of 1 L. The reagent will keep indefinitely in a plastic container, if 1 g KI is also added to this solution.

Purification of proteins

7

7.1 Introduction: aims of purification procedures

KEY CONCEPTS

- Knowing the aims of protein purification
- Knowing how to assess the success of a purification procedure

To address the key questions concerning the structure and the function of a protein (Chapter 5, section 5.2) it is essential that the protein is pure, as judged by the presence of a single species on SDS-PAGE and mass spectrometry (Chapter 8, sections 8.2.1 and 8.2.5, respectively). As a result any properties measured can be attributed to the protein of interest. However, in 'scouting'-type experiments, e.g. testing the activity of an overexpressed recombinant protein or simple quantitative analysis of protein expression in response to a signal, only partial purification is required.

The overall aim of any protein purification method is to generate a high yield of pure protein while retaining maximum activity. It is standard practice to record both the protein content (Chapter 6, section 6.1) and the activity (Chapter 6, sections 6.3 and 6.4) at each stage of a protein purification procedure to give an indication of the efficiency of the purification method. Taken together, SDS-PAGE (Chapter 8, section 8.2.1) and specific activity measurements (Chapter 3, sections 3.8 and 3.9) provide an assessment of the purity of the protein at each stage of the purification.

This chapter starts with an explanation of the basis of separation methods and proceeds to describe a number of purification methods, encompassing a range of starting materials, experimental conditions, and chromatographic methods. For each example of purification, a brief description of the protein is provided along with an outline of the experimental procedures required to purify the protein. Self test questions are provided throughout the remainder of this chapter to test your understanding of the key concepts of protein purification. In addition, sample data are provided for some of the procedures to enhance both data-handling abilities and an appreciation of the significance of the data at each purification step.

Detailed purification procedures are provided on the Online Resource Centre. This resource enables teachers to offer a range of laboratory practical classes on protein purification and allows students to appreciate the important aspects of purification procedures such as source selection, the degree of purity required, the amount of protein recovered, and assay design.

Some general considerations for any protein purification include the following.

Cell extract preparation

The purification of any protein starts with the careful preparation of a cell extract. The choice of preparation method is dictated by the nature of the starting material and compatibility with subsequent procedures such as chromatographic procedures and assay measurements. For example, a cell extract prepared with a high ionic strength extraction buffer may not be compatible with a subsequent ion-exchange chromatography step as these conditions will hinder the binding of the protein of interest to the ion exchanger. However, a cell extract prepared in a high ionic strength buffer may be compatible with a subsequent hydrophobic interaction chromatography (HIC) step.

Starting material

The starting material is generally selected to contain an abundance of the protein of interest and to be readily available. The ability to overexpress proteins in heterologous systems has made most proteins readily available in relatively large quantities (see Chapter 5, section 5.4). Other factors influencing the choice of starting material may include the cellular location of the protein, which may require subcellular fractionation of the cell extract, and comparative studies, e.g. studying the same protein from evolutionarily distant organisms.

Intacellular vs extracellular proteins

Purification of extracellular proteins, such as hydrolytic enzymes secreted from microorganisms, does not require cell lysis. In the case of extracellular proteins, growth media or supernatants from cell-harvesting centrifugation provide starting material which may require an initial concentration step to facilitate detection of protein content and activity. The purification of most other proteins requires cell disruption to generate a cell extract and the method of disruption depends on the nature of the cells. Plant, yeast, and bacterial cells all have rigid cell walls that require enzymatic or harsh mechanical disruption. Mammalian cells lack a rigid cell wall and require gentler mechanical disruption.

Yield and activity

To retain maximum yield and activity of the protein during purification keeping the number of steps in a purification procedure to a minimum will reduce the loss of active protein. In addition, a range of experimental conditions require optimization: pH, temperature, protection of labile thiol groups, and minimization of

Table 7.1 Some protease inhibitors used to minimize unwanted proteolysis

Protease type	Cell type containing protease	Inhibitor	Inhibition type
Aspartic	Animal tissue, yeasts, fungi	Pepstatin	Reversible (tight binding)
Cysteine	Plant tissue	E64 PMSF	Irreversible Irreversible
Serine	Animal tissue, plant tissue, yeasts, fungi, bacteria	3,4-Dichloroisocoumarin PMSF	Reversible (tight binding) Irreversible
Metallo-	Animal tissue, yeasts, fungi, bacteria	1,10-Phenanthroline EDTA	Reversible Chelating agent

E64, L-*trans*-epoxysuccinyl-leucamide-(4-guanidino)-butane; PMSF, phenylmethanesulphonyl fluoride; EDTA, ethylenediaminetetraacetic acid.

proteolysis. Efficient cellular disruption at the outset of protein purification procedures not only results in the release of the protein of interest, but also the unwanted release of active proteases. The range of proteases produced by each cell type is unique. However, most proteases belong to a small number of protease types, such as cysteine, metallo-, and serine proteases, and there are several broad range specificity inhibitors that inactivate these protease types, see Table 7.1 (for further information see North and Beynon (2001)).

Some general precautions can be employed to minimize proteolysis such as maintaining low temperatures and selecting pH conditions above or below the optimum for protease activity. Most commonly, a cocktail of protease inhibitors is included in the extraction buffer to minimize unwanted proteolysis. However, it is important to consider the compatibility of inhibitors with subsequent purification and analytical steps. A good example of this is the use of chelating inhibitors of metallo-proteases that are not compatible with metal affinity chromatography as they prevent adsorption of the protein to the chromatography media.

SELF TEST ?

Readers familiar with protein purification should attempt Self test 7.1 to determine whether they need to study this chapter in detail.

ST 7.1 Design a possible purification procedure for:

1. a 50-kDa protein with a predicted pI value of 8.2 from pig liver cells
2. a GST-tagged, 20-kDa protein overexpressed in *Escherichia coli*
3. an extracellular 6-kDa fungal polypeptide

Answers

ST 7.1.1 A cell extract is prepared by homogenization. Following centrifugation, to remove cell debris, the cell extract is diluted into a buffer of low ionic strength, ~pH 7.0. Under these conditions, the protein of interest will be positively charged and bind to a cation exchanger, such as CM-Sepharose or Mono-S. An increase in the ionic strength of the chromatography buffer will result in the selective elution of the protein of interest. An additional gel filtration step may be required using chromatography media such as Sephacryl S-300.

ST 7.1.2 Following growth and induction of the *E. coli* strain, which is overexpressing the GST-tagged protein, the cells are lysed by treatment with lysozyme and/or sonication. Following centrifugation to remove cell debris, the cell extract is diluted and applied to a GST affinity chromatography column. Non-specifically bound proteins are eluted by applying excess binding buffer. The GST-tagged protein is selectively eluted by applying increasing concentrations of reduced glutathione.

ST 7.1.3 Cells are removed from the culture medium by centrifugation. The culture medium, containing the extracellular 6-kDa fungal polypeptide, is collected and concentrated using a filtration device (section 7.2.4). The concentrated extracellular proteins are then applied to a suitable gel filtration chromatography medium, such as Sephacryl S-100.

7.2 Basis of separation methods

KEY CONCEPTS

- Understanding the range of separation methods available
- Appreciating the limitations and compatibilities of separation methods

Each protein has a unique set of biochemical properties which can be exploited to facilitate its isolation from complex biological sources. The most common biochemical properties that are exploited include surface polarity, molecular mass, and specific binding sites. In this section, we shall consider the basis of separation used in a range of protein isolation methods.

7.2.1 Precipitation

Proteins remain in solution as a result of interactions with surrounding solvent molecules. Perturbation of protein–solvent interactions can promote protein–protein interactions, leading to precipitation, i.e. the formation of insoluble protein–protein aggregates. The most common methods employed to promote precipitation are altering pH, increasing the ionic strength (e.g. by the addition of ammonium sulphate), and changing solvent polarity (e.g. by the addition of water-miscible organic solvents such as acetone or ethanol). All of these methods can be used on a relatively large scale and so are useful in the early stages of protein purification. Large volumes of dilute cell extracts can be treated so that the protein of interest can be concentrated into a small volume while retaining most of the initial activity. Purification factors (Chapter 3, section 3.9) of up to about 10-fold can be achieved.

Altering pH: The ionization state of α-amino, carboxyl, and side chain groups of amino acids contribute to the overall charge carried by a particular protein (Chapter 1, section 1.2.3.2). By altering the surrounding pH, we can change the ionization state of these groups and the overall charge carried by the protein. The pH at which a protein carries no net charge, i.e. carries an equal number of positive and negative charges, is known as the isoelectric point (pI) (see Chapter 1, section 1.3.2.2). At the pI, repulsive electrostatic forces between protein molecules are minimal thus promoting precipitation. This technique can be employed to precipitate the protein of interest or to remove contaminating proteins.

Addition of ammonium sulphate: Typically, residues at the surface of a protein interact with water molecules to maintain solubility. Addition of high concentrations of salts, such as ammonium sulphate, is thought to reduce protein solubility by decreasing the concentration of water available for protein–solvent interactions. This leads to the exposure of hydrophobic surface residues which then interact with each other to promote precipitation. The nature of the residues on the surface of a protein dictates the concentration of ammonium sulphate required to precipitate a given protein. In a complex protein mixture, some proteins will precipitate at lower salt concentrations, whereas others will precipitate at higher salt concentrations. By carefully selecting the range of ammonium sulphate concentrations used, a degree of purification (up to about 10-fold) can be achieved.

Ammonium sulphate is widely used in the purification of proteins, since it is cheap, highly soluble in water (approximately 4 M at saturation at 25°C), and generally has a stabilizing effect on proteins. It should be noted that ammonium sulphate is a weak acid and if this is added to a weakly buffered solution, the pH of the solution may decrease. Addition of a base will restore the pH.

Typically, ammonium sulphate concentrations are described in terms of the percentage ammonium sulphate saturation (% saturation). An ammonium sulphate solution which is 100% saturated at 25°C contains the maximum amount of soluble ammonium sulphate, i.e. contains 767 g per litre. Using the table in Appendix, it is possible to calculate the amount of ammonium sulphate required to achieve a target % saturation value or to increase the % saturation from one value to another.

Changing the solvent polarity: Addition of solvents such as ethanol and acetone, promotes precipitation by decreasing the dielectric constant (Chapter 1, section 1.7.1) of a solution causing an increase in protein–protein electrostatic interactions. It should be noted that addition of organic solvents may compromise protein activity; however, the use of low temperatures will generally minimize this effect.

The mixing of solvents such as ethanol and water generates heat, as can be readily appreciated by observing the effect of addition of water to whisky in a tumbler, when the ethanol will evaporate and condense on the sides. The solvents should be cooled before addition.

? SELF TEST

Check you have mastered the key concepts at the start of this section by attempting the following question.

ST 7.2 Using the table in Appendix 7.1, calculate the quantity of ammonium sulphate required to increase the saturation of ammonium sulphate from an initial value of 40% to a final percentage of 60% for a cell extract with a volume of 55 mL.

Answer

ST 7.2 Using the table in Appendix 7.1, determine the amount of ammonium sulphate (in grams) to be added to 1 L to increase the % saturation from an initial 40% to a final value of 60%. Next, calculate the amount of ammonium sulphate to be added to 55 mL cell extract.

40% → 60% ammonium sulphate saturation = 132 g L^{-1}

55 mL cell extract would require 132 × 55/1000 = 7.26 g

7.2.2 Chromatography

The most commonly used matrix material is cross-linked agarose in products such as Sepharose and BioGel A. Other matrix materials include cross-linked dextrans (Sephadex) and cross-linked dextran and bis-acrylamide (SephAcryl).

Chromatography is a technique which can be used to isolate individual proteins or to analyse their structural properties. Chromatographic techniques separate proteins (mobile phase) according to their ability to adsorb to a matrix (stationary phase). The matrix usually consists of spherical, porous beads, which adsorb proteins and are packed into a column. A large number of matrix materials are commercially available. Typically, chromatography matrices are composed of beaded particles (50–400 μm in diameter) with open-pore structures, which increase the binding surface area and provide greater capacity to capture target proteins. In addition, chromatography matrices offer minimal binding of non-target proteins.

Most of these matrices are suitable for low-to-moderate pressure chromatography and operate under aqueous conditions for both preparative and analytical analyses. More robust matrices, typically silica particles of 5-μm diameter, are used for high-pressure liquid chromatography (HPLC). HPLC is used for small-scale analysis of organic compounds, including peptides, and often involves the use of organic solvents.

A typical chromatography procedure involves three steps (see Fig. 7.1):

1) *Adsorption*: Selection of a matrix with optimal adsorption properties and compatible buffer conditions will result in the adsorption of protein of interest to the column.

2) *Elution of non-adsorbed proteins*: Continued application of buffer which promotes the adsorption of protein of interest and elution of contaminating proteins which fail to adsorb to the column.

3) *Selective elution*: Application of a gradient of changing conditions (e.g. pH, ionic strength, presence of ligand) will result in the selective elution of the protein of interest (Fig. 7.2).

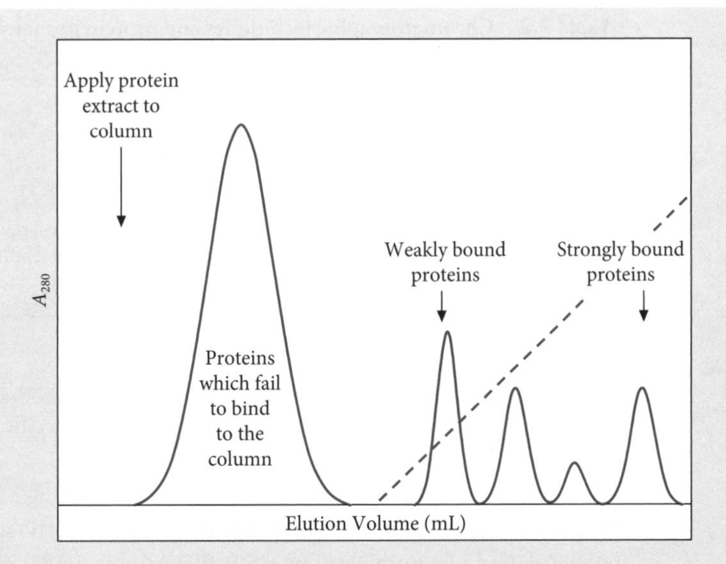

Fig. 7.1 Typical elution profile of a chromatographic procedure. Following application of the protein extract to the column, proteins which fail to interact with chromatographic media in the column emerge in the eluant. Proteins which do interact with the column are eluted by a change in elution conditions (dashed line). In the case of ion-exchange chromatography, this could be an increase in the ionic strength, or a change in the pH of the elution buffer. In the case of affinity chromatography, an increase in the presence of a ligand is typically used to elute proteins specifically. A decrease in the ionic strength of the eluant is used to elute proteins in hydrophobic interaction chromatography. Typically, the proteins which interact weakly with the column emerge from the column with a small change in the elution condition, whereas proteins that interact more strongly require a larger change in the elution conditions.

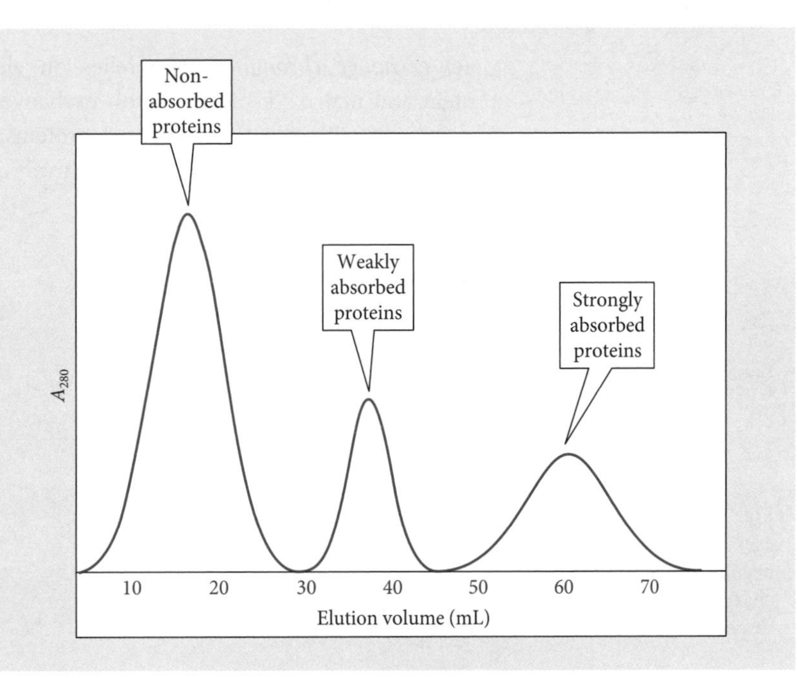

Fig. 7.2 Elution profile of a chromatographic procedure showing the elution of the non-absorbed proteins, weakly absorbed proteins and subsequent elution of strongly absorbed proteins.

Table 7.2 Chromatographic techniques and protein properties that form the basis of separation procedures

Chromatographic technique	Protein property
Ion exchange	Surface charge
Hydrophobic interaction	Surface hydrophobicity
Affinity	Specific binding sites, e.g. ligand binding, metal affinity
Gel filtration (size exclusion)	Size

Chromatography matrices with a range of specific properties offer the ability to purify proteins according to charge, hydrophobicity, specific binding sites or size (see Table 7.2).

The aim of a purification procedure is to isolate a single protein based on its specific properties. In practice, exploitation of one property, such as charge, results in the isolation of a number of proteins that exhibit similar characteristics, therefore most purification strategies employ a combination of chromatographic steps, exploiting differences in more than one protein property, for example ion exchange followed by gel filtration. This approach will reduce the likelihood of isolating contaminating proteins alongside the protein of interest as it is unlikely that contaminants and the protein of interest will share the same combination of specific properties. The resolution, binding capacity, and operational conditions differ from one chromatographic technique to another and must be considered in the design of a purification strategy.

Ion-exchange chromatography relies on electrostatic interactions between protein and matrix. Typically, anion exchange matrices are positively charged and interact with negatively charged proteins, whereas cation exchangers are negatively charged and interact with positively charged proteins (see Fig. 7.3 and Table 7.3).

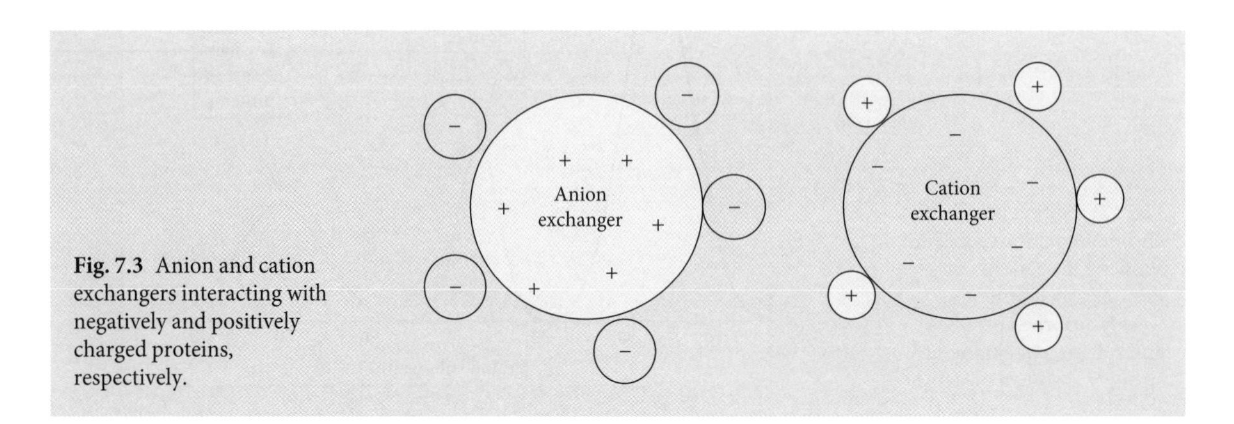

Fig. 7.3 Anion and cation exchangers interacting with negatively and positively charged proteins, respectively.

Table 7.3 Examples of ion-exchange groups used in ion-exchange chromatography

Anion exchanger	Anion-exchange group
Diethylaminoethyl (DEAE)	$-O-CH_2-CH_2-N^+H(CH_2CH_3)_2$
Quaternary ammonium (Q)	$-O-CH_2-CHOH-CH_2-O-CH_2-CHOH-CH_2-N^+(CH_3)_3$
Cation exchanger	**Cation-exchange group**
Carboxymethyl (CM)	$-O-CH_2-COO^-$
Methyl sulphonate (S)	$-O-CH_2-CHOH-CH_2-O-CH_2-CHOH-CH_2SO_3^-$

The adsorption phase of ion-exchange chromatography requires the protein to be applied:

- In the presence of a low ionic strength buffer
- At a pH value which will generate a net charge on the protein opposite to that carried by the ion exchanger

Selective elution from ion-exchange columns can be achieved by changing the pH, resulting in a change in the net charge carried by proteins and altering protein–matrix interactions. Alternatively, proteins can be eluted by increasing the ionic strength of the elution buffer, which introduces salt ions to compete with the protein in binding to the ion exchanger. Changes in pH or ionic strength can be introduced in a stepwise fashion or in a continuous gradient. The advantages of ion-exchange chromatography include the ability to load large volumes of dilute protein solutions onto ion-exchange columns and subsequent concentration of the target protein during adsorption and elution. In addition, ion-exchange chromatography can offer high adsorption capacity and excellent resolution properties.

Knowledge or prediction of the pI (Chapter 1, section 1.3.2.2 and Chapter 5, section 5.8.2) of the protein of interest will help to design the correct ion-exchange procedure. At pH values above the pI, a protein will carry a negative charge and adsorb to anion exchangers, whereas at pH values below the pI, a protein will carry a positive charge and adsorb to cation exchangers. The selected conditions are usually tested on a small scale to determine which matrix and pH combinations offer optimal adsorption, separation, and retention of activity.

In some cases, the adsorption phase of ion-exchange chromatography results in the adsorption of contaminating proteins only, whereas the protein of interest fails to bind and passes straight through the column giving rise to a certain degree of purification. However, this degree of purification is usually smaller than would be achieved by adsorption and elution procedures.

Hydrophobic interaction chromatography separates proteins according to their surface hydrophobicity properties. Under conditions of high ionic strength, hydrophobic protein surfaces will adsorb to hydrophobic matrices such as Phenyl Sepharose. Selective elution usually involves a decrease in the ionic strength of elution buffer, which decreases the strength of the hydrophobic interactions between the protein and the matrix. In most cases, proteins adsorb using buffers containing high concentrations of ammonium sulphate and are subsequently eluted by a reduction in the concentration of ammonium sulphate in either a stepwise or a continuous (gradient) fashion.

In some cases, the elution of tightly bound proteins may require the addition of an organic solvent such as ethylene glycol (1,2-ethanediol) or non-ionic detergents to the elution buffer.

Affinity chromatography exploits a specific reversible interaction between a protein and a ligand immobilized on a chromatography matrix. Protein–ligand

Table 7.4 Examples of immobilized ligands and target proteins used in affinity chromatography techniques with associated general adsorption and elution conditions

Immobilized ligand	Target protein	Adsorption conditions	Elution conditions
Protein A	IgG	Low ionic strength, neutral pH	High ionic strength, low pH
Mercapto-pyridine	IgM	High ionic strength	Low ionic strength
Nucleic acid/heparin	Nucleic acid binding proteins	Low ionic strength	High ionic strength
Cibacron Blue F3GA	Nucleotide (e.g. ATP or NAD^+) binding proteins	Low ionic strength	Low ionic strength plus appropriate nucleotide[a]
Glutathione	GST fusion proteins	Low ionic strength	Low ionic strength plus reduced glutathione
Cu^{2+}, Co^{2+}, Ni^{2+}, Zn^{2+}	His- or Cys-tagged fusion proteins, native proteins with histidine, cysteine, or tryptophan surface residues	High ionic strength	High ionic strength plus imidazole
Antigen	Antibody	[b]	[b]
Substrate mimic, inhibitor, cofactor	Enzyme	[b]	[b]

[a]Note that in some cases elution from Cibacron Blue can be brought about by high ionic strength alone, but this is probably not very selective.
[b]No general method as interaction types vary from protein to protein and require bespoke purification procedures.

interactions may involve electrostatic, hydrophobic, hydrogen bonding, or van der Waals' forces. The specificity of these interactions forms the basis of the selective adsorption of the protein of interest. Selective adsorption combined with selective protein elution, using compounds which will specifically bind to the protein or to the immobilized ligand, can generate a highly efficient purification step. It is also possible to elute adsorbed proteins by changing the pH or ionic strength. Optimal protein adsorption and elution conditions depend on the nature of the protein–ligand interactions (see Table 7.4).

Many of the ligands outlined in Table 7.4 are available commercially, coupled to chromatography matrices such as Sepharose. In these cases, the ligand has been coupled to the matrix to allow optimal binding of the target protein. Immobilization of a ligand on a chromatography matrix typically requires:

1) *A ligand* with a suitable chemically reactive group for attachment to the matrix. The mode of attachment must not compromise the reversible interaction between ligand and protein. In addition, favourable dissociation constants (Chapter 4, section 4.3) are required for the interaction between the immobilized ligand and target protein: it must be strong enough to promote adsorption (dissociation constant $<10^{-3}$ M) but weak enough (dissociation constant $>10^{-8}$ M) to allow efficient elution under conditions which will maintain protein stability.
2) *A matrix* with a suitable attachment site for the ligand. Typically, matrices are chemically activated to permit the coupling of the ligand. A number of

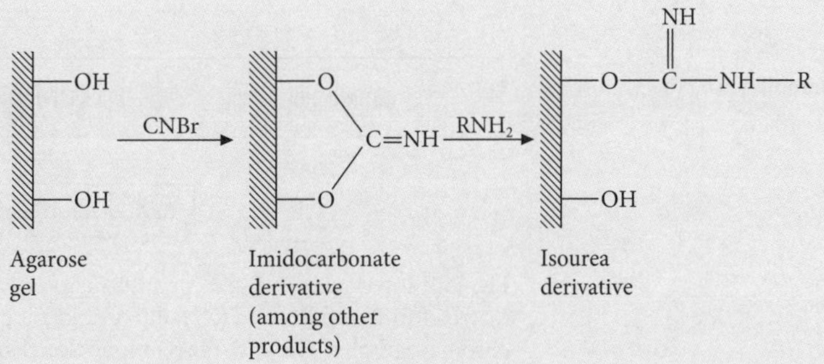

Fig. 7.4 Cyanogen bromide activation of agarose and immoblization of ligand.

Agarose gel —CNBr→ Imidocarbonate derivative (among other products) —RNH_2→ Isourea derivative

activation methods are available and depend on the nature of the matrix and the availability of compatible reactive groups on the ligand. One of the most common methods for immobilizing ligands involves cyanogen bromide activation of agarose to produce imidocarbonate derivatives, which react with amine groups to generate isourea linkages (see Fig. 7.4).

3) *A spacer arm* will be required in cases where direct coupling of the ligand to the matrix results in steric hindrance and subsequently the target protein will fail to bind to the immobilized ligand efficiently. The introduction of a spacer arm between the ligand and the matrix minimizes this steric effect and promotes optimal adsorption of the target protein to the immobilized ligand. (Note however that when a spacer arm of the $-(CH_2)_n$ type is used, there can be non-specific hydrophobic interactions between the spacer arm and the proteins).

The availability of a range of matrices with spacer arms of variable length and chemical groups for ligand attachment enables the development of tailored affinity chromatography media with specific ligands attached. Examples include the isolation of target enzymes using ligands which mimic enzyme substrates and the isolation of proteins using antibodies, raised against target proteins, as ligands (immunoaffinity chromatography). Immunoaffinity chromatography is used on a commercial scale to purify native enzymes and human plasma proteins, such as factor VII in the blood clotting cascade (Tomokiyo et al., 2003). It is possible to achieve purification factors of 2000-fold using this approach. In the case of factor VII, a monoclonal antibody directed against factor VII was attached to cyanogen bromide-activated Sepharose.

Immunoaffinity chromatography is also used to isolate recombinant proteins with purification tags attached (Chapter 5, section 5.4.2), which bind specific immobilized antibodies (Hearn and Acosta, 2001). Interleukin-2, an important signalling protein in immune responses, has been produced as a recombinant protein with an immunoaffinity purification tag attached. The purification tag used was a FLAG peptide (see Fig. 7.5), which enhanced the stability of interleukin-2 during expression and simplified its purification using immunoaffinity chromatography.

The FLAG peptide contains an enteropeptidase cleavage site (see Chapter 5, section 5.4.2) to facilitate its release from the fusion protein.

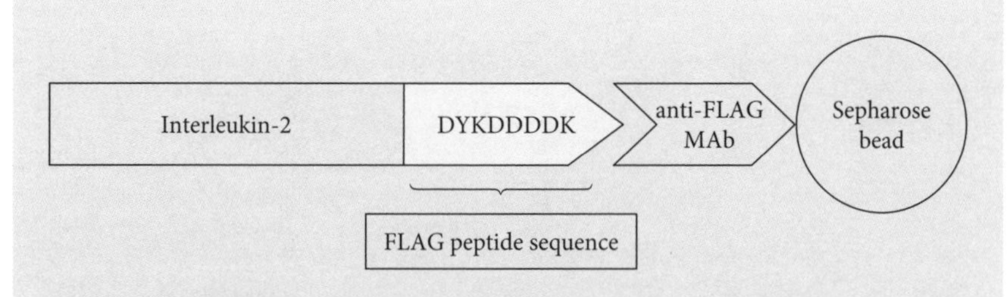

Fig. 7.5 Interleukin-2-FLAG peptide fusion protein generated to enhance interleukin-2 stability and to facilitate efficient purification by immunoaffinity chromatography. The anti-FLAG monoclonal antibody captures FLAG-tagged proteins that are subsequently eluted using a pulse of FLAG peptide which competes for the anti-FLAG monoclonal antibody binding site and releases the FLAG-tagged protein, in this case, interleukin-2.

Until the advent of recombinant DNA technology, the purification of individual proteins required the development and optimization of individual purification protocols. This task has been simplified greatly by the ability to fuse purification tags to proteins (see Chapter 5, section 5.4.2). Proteins with tags, such as glutathione-*S*-transferase (GST) or poly-histidine sequences, fused to the N- or C-terminus are readily purified using standard tag purification procedures. Two examples of affinity chromatography purification of such proteins are given later in this chapter: tagged with GST (section 7.3.4) or poly-histidine (section 7.3.8).

Gel filtration, unlike the chromatography methods outlined previously in this chapter, does not involve protein adsorption. Gel filtration (size exclusion) chromatography separates proteins according to their size on the basis that larger proteins do not enter the porous structure of matrix beads and travel through the column relatively quickly (see Fig. 7.6). Smaller proteins do enter the pores of matrix beads and as a result their movement through the column is relatively slow. The pore sizes can be adjusted by varying the degree of cross-linking of the matrix material; a range of such gel filtration matrices is available commercially (see Table 7.5).

To ensure that proteins are separated by size only during a gel filtration step, it is important to select an elution buffer of suitable pH and ionic strength to minimize unwanted interactions such as electrostatic and hydrophobic interactions. Typically, this involves a buffer of intermediate ionic strength and neutral pH.

Table 7.5 Fractionation range of selected examples of gel filtration matrices and their possible applications

Gel filtration matrix	Fractionation range (kDa)	Fractionation application
Sephacryl S-100	1–100	Small proteins and peptides
Superose 12	1–300	Small proteins and peptides
Sephacryl S-300	10–1500	Moderate-sized proteins
Superose 6	5–5000	Moderate-sized proteins
Sephacryl S-500	50–20 000	Large proteins, protein complexes

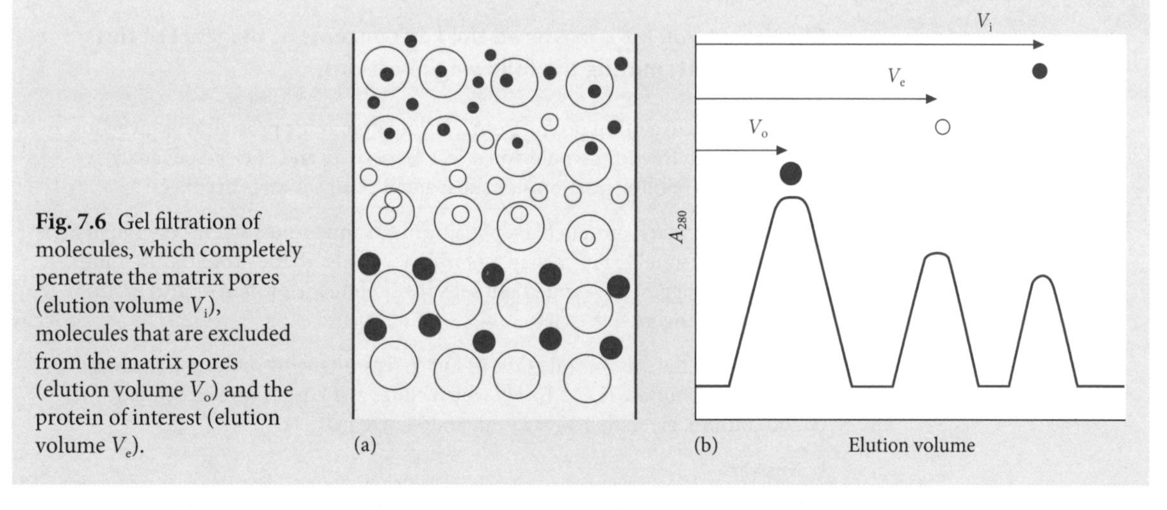

Fig. 7.6 Gel filtration of molecules, which completely penetrate the matrix pores (elution volume V_i), molecules that are excluded from the matrix pores (elution volume V_o) and the protein of interest (elution volume V_e).

Gel filtration can be used as a purification tool to isolate target proteins or as an analytical tool to determine the molecular mass of a protein. The basis of gel filtration as an analytical tool is the linear relationship between the logarithm of the molecular mass of a protein and its distribution coefficient, K_d:

$$K_d = (V_e - V_o)/(V_i - V_o) \quad \textbf{7.1}$$

where V_e is the elution volume of the protein of interest, i.e. the volume of eluant collected between the application of the sample to the column and the elution of the protein from the column, V_o is the void volume, i.e. the elution volume of a molecule which is excluded from the matrix pores (e.g. Dextran Blue, with a molecular mass of 2000 kDa, and V_i is the elution volume of a small molecule that penetrates the pores of the matrix, e.g. glucose (see Fig. 7.6).

Determination of the molecular mass of a protein by gel filtration requires calibration of the gel filtration column by measuring the K_d value of a range of proteins of known molecular mass (see Fig. 7.7).

In the case of small proteins, which penetrate the porous structure of the matrix, the V_e value will be close to the V_i value, $K_d = 1$. For very large proteins that fail to penetrate the porous structure of the matrix, the V_e value will be close to the V_o value, $K_d = 0$.

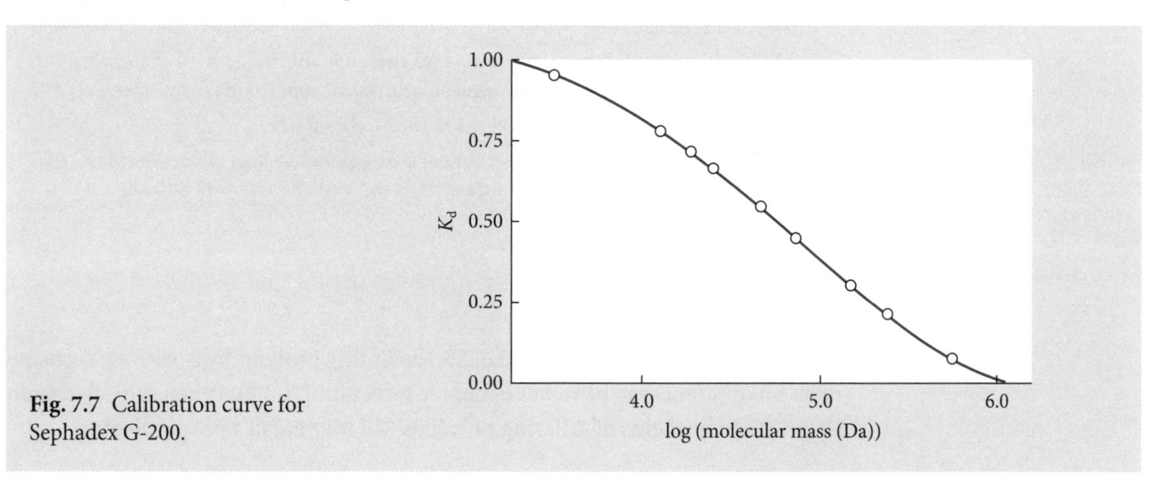

Fig. 7.7 Calibration curve for Sephadex G-200.

? SELF TEST

Check that you have mastered the key concepts at the start of this section by attempting the following questions.

ST 7.3 A typical elution profile from a chromatographic procedure is given in Fig. 7.2. Which fractions would you select for activity and SDS-PAGE analysis with a view to pooling fractions for subsequent purification procedures?

ST 7.4 Given that a protein has a pI value of 6.6 and is most stable between pH 7.5 and 8.5, select suitable ion exchangers which will adsorb the protein with minimal activity loss. Suggest how you might selectively elute the protein of interest from such ion exchangers.

ST 7.5 Using the calibration curve in Fig. 7.7, estimate the molecular mass of superoxide dismutase (EC 1.15.11) which elutes at 44 mL from this Sephadex G-200 column, given that $V_o = 23$ mL and $V_i = 58$ mL.

Answers

ST 7.3 Select fractions with the highest activity and the lowest protein content, i.e. highest specific activity.

Initially assay the fractions with high protein content (e.g. at elution volumes 20, 40, and 60 mL) and then continue to assay alternate fractions which occur before and after these fractions. Once activity is detected, continue to assay the neighbouring fractions to identify the activity peak. Only select the fractions with the highest specific activity. It is often the case that fractions at the start or the end of the activity peak will have a low activity: [protein] ratio, i.e. a low specific activity.

ST 7.4 To maintain the stability of the protein, it would be prudent to work between pH 7.5 and 8.5. Under these conditions, a protein with a pI value of 6.6 will carry a net negative charge and will bind to an anion exchanger such as diethylaminoethyl (DEAE) or quaternary ammonium (Q).

Typically, proteins are selectively eluted from ion exchangers by increasing the ionic strength of the buffer or by changing the pH of the buffer. A gradient of buffer change (e.g. apply 50 mM Tris HCl, pH 7.6 and then gradually increase the concentration of 50 mM Tris HCl, pH 7.6 containing 200 mM NaCl applied to the column) allows greater elution selectivity. By adjusting the duration or the composition of the gradient it may be possible to improve the resolution of the ion-exchange step.

ST 7.5 Use eqn. 7.1 to calculate K_d and then use this value with the calibration curve (Fig. 7.7), to estimate the molecular mass of superoxide dismutase.

$$K_d = (V_e - V_o)/(V_i - V_o) = (44 - 23)/(58 - 23) = 0.61$$

From the calibration curve a K_d value of 0.61 equates to log (molecular mass) of 4.50, therefore superoxide dismutase has a molecular mass = 32 000 Da.

7.2.3 Isoelectric focusing

As explained in Chapter 6, section 6.2.3, separating proteins by isoelectric focusing relies on differences in pI values (Chapter 1, section 1.3.2.2). When subjected to an electric field, proteins of differing pI values will migrate in a pH gradient.

pH gradients can be generated by applying an electric field to a mixture of ampholytes, i.e. polyamino acids carrying a range of charges, that is they have a range of pI values. Negatively charged ampholytes will migrate towards the anode until they encounter acid conditions, under which they will acquire protons. Positively charged ampholytes will migrate towards the cathode unitl they encounter alkaline conditions, under which protons dissociate (Fig. 7.8).

Once a given ampholyte encounters conditions where the local pH is equal to its pI value, it will stop migrating. Thus a collection of different ampholytes will generate a pH gradient in which the pH will increase from the anode to the cathode. This process can occur in solution, on a preparative scale, using a device such as the Rotofor IEF to purify proteins (Chapter 6, section 6.2.3). Alternatively, pH gradients can be generated on a solid phase matrix, using immobilized ampholytes arranged on an acrylamide matrix. Pre-prepared pH gradient strips with an acrylamide matrix are commercially available and are used on an analytical scale, e.g. as the first dimension in two-dimensional gels.

A protein carrying a positive charge will migrate towards the cathode and will lose or gain protons (if below or above pI, respectively) until it reaches the pH value in the gradient that matches its pI. A negatively charged protein will migrate towards the anode and will lose or gain protons (if below or above pI, respectively) until it reaches the pH value in the gradient which matches its pI.

7.2.4 Dialysis/ultrafiltration

Membrane-based techniques, such as dialysis and ultrafiltration, are commonly used during protein purification. The membranes employed in these techniques are available with a range of pore sizes, which allow the selective retention of proteins of a particular size. For example, dialysis (cellulose) membranes will retain

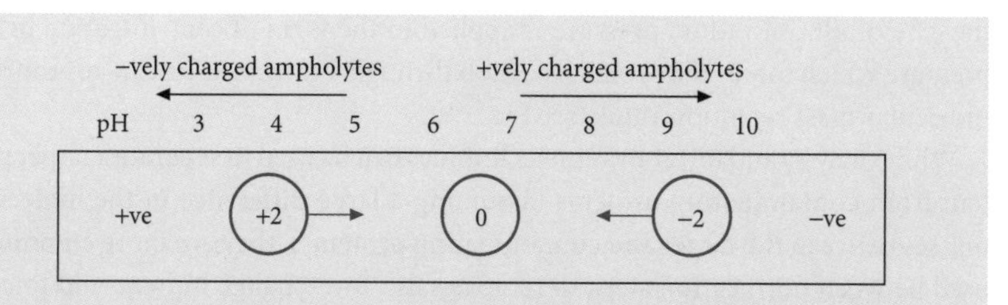

Fig. 7.8 Separation of proteins by isoelectric focusing. An electric field is applied to a mixture of ampholytes that migrate to form a pH gradient. Proteins applied to a pH gradient, in an electric field, will migrate to a pH where they carry no net charge. Circles represent proteins. Proteins with a net charge of zero will stop migrating. Proteins with a net charge of +2 will migrate towards the cathode until more alkaline conditions are encountered, protons lost, and the net charge becomes zero. Proteins with a net charge of −2 will migrate towards the anode until more acidic conditions are encountered, protons gained, and the net charge becomes zero.

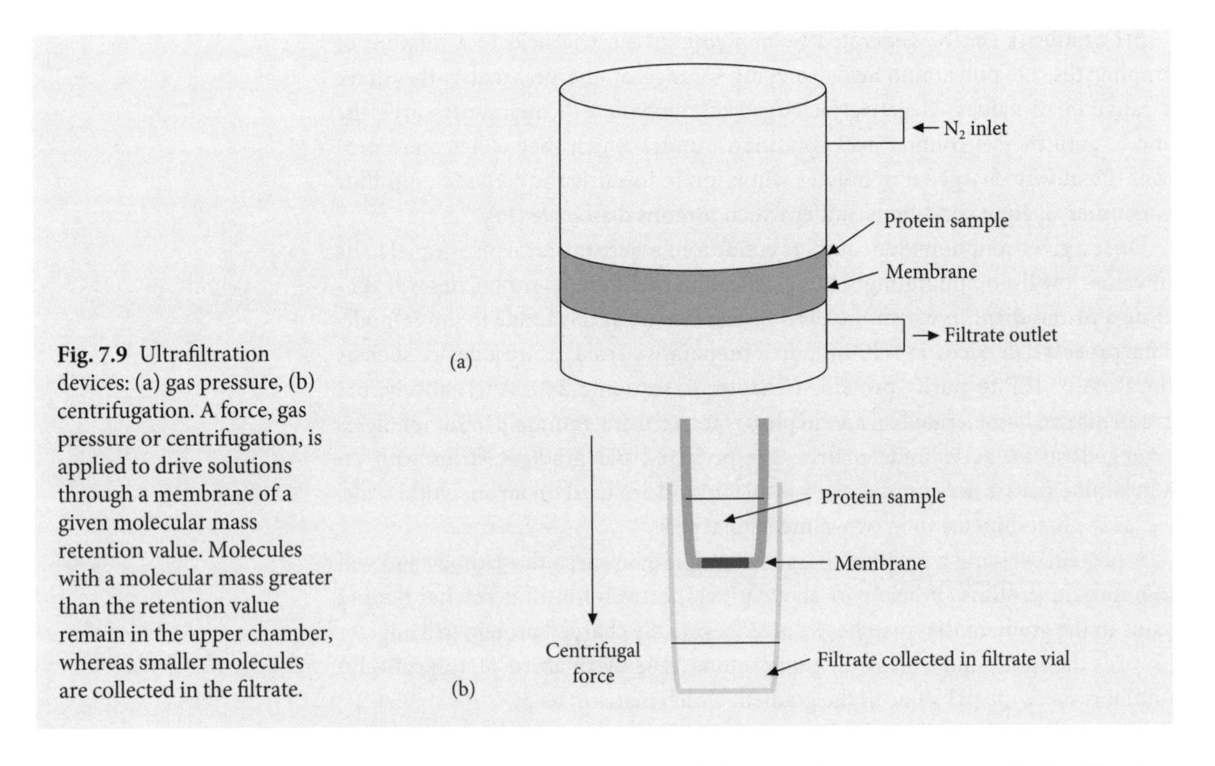

Fig. 7.9 Ultrafiltration devices: (a) gas pressure, (b) centrifugation. A force, gas pressure or centrifugation, is applied to drive solutions through a membrane of a given molecular mass retention value. Molecules with a molecular mass greater than the retention value remain in the upper chamber, whereas smaller molecules are collected in the filtrate.

proteins with a molecular mass of about 20 kDa or greater, and ultrafiltration devices can have a range of membranes that will retain proteins of molecular mass greater than 5, 10, 30, 50, or 100 kDa.

Both dialysis and ultrafiltration membranes act as molecular sieves allowing smaller molecules to pass through the membrane while larger molecules are retained. In a typical dialysis step, the protein is sealed within the dialysis membrane and placed in a large volume of buffer. Diffusion allows smaller molecules to move across the dialysis membrane to equalize the concentrations on the two sides. In the case of ultrafiltration, pressure is applied in the form of centrifugation or gas pressure which forces the protein solution through a membrane of an appropriate molecular mass retention value (see Fig. 7.9).

While dialysis and ultrafiltration techniques can be used to separate a target protein from contaminating proteins (assuming a large difference in the molecular masses between the target and contaminating proteins), they are more commonly used between purification steps to remove salts, to exchange buffers, and to concentrate dilute protein solutions.

7.2.5 Centrifugation

Centrifugation is a tool that is employed throughout most protein purification procedures. At the outset of many purification protocols, relatively low centrifugal fields (5000–50 000*g*) are used to harvest cells and remove cell debris.

Centrifugation is also the most efficient method of subcellular fractionation, allowing the isolation of subcellular components such as membrane fractions and intact organelles. This is particularly useful when studying a protein, which may be present in a number of locations within the cell and we want to isolate this protein from only one cellular location: high centrifugal fields can be used to collect soluble proteins and membrane fractions, whereas density gradient centrifugation can be used to isolate intact organelles and vesicles.

Protein precipitation (section 7.2.1) relies on the use of centrifugation for the quick and efficient separation of precipitated from non-precipitated material. Likewise, ultrafiltration centrifugal devices (section 7.2.4) require low centrifugal fields to drive solutions through selective membranes to exchange buffers, remove unwanted salts, and to concentrate protein samples.

7.3 Examples of protein purification

KEY CONCEPTS

- Understanding the range of purification methods
- Designing a possible purification method for a given protein

In this section, we shall look at the application of protein purification procedures to the purification of a number of specific protein examples. These examples have been selected to reflect the diversity of separation methods which can be employed. Initially, we shall look at the purification of four enzymes: lactate dehydrogenase (LDH), alcohol dehydrogenase (ADH), isocitrate dehydrogenase (ICDH), and glutathione-S-transferase (GST), which is used as an enzyme tag to facilitate the purification of recombinant proteins (Chapter 5, section 5.4.2). We shall then proceed to describe the purification of a carrier protein (myoglobin), a recognition protein (immunoglobulin (IgM)), a bioluminescent protein (green fluorescent protein (GFP)), and His-tagged proteins (see Chapter 5, section 5.4.2 for a description of the production of His-tagged proteins). Background information relating to each protein and purification protocol is presented in this chapter along with a guide to purification analysis and a sample data set. Solutions for these data sets are given at the end of the chapter. Details relating to materials, methods, and experimental procedure are given in the Online Resource Centre for this text to support teaching of this topic.

In all examples, at each purification step, it is essential to:

- Record the volume
  Estimation of the total protein content and total activity at each purification step requires knowledge of the volume of the extract at each step
- Determine the protein content
  A dye-binding method, as outlined in Chapter 6, section 6.1.1, is a convenient way to determine the protein content at each purification step

- Measure the activity
  Simple, specific assays are essential for measuring the activity of the protein of interest. Care must be exercised when interpreting data, in particular from complex extracts that occur in the early stages of purification, see Chapter 5, section 5.3 and Chapter 9, section 9.2. Activity measurements together with protein content measurements permit the calculation of the specific activity of the protein at each stage of the purification (Chapter 3, sections 3.8 and 3.9)
- Analyse by SDS-PAGE
  SDS-PAGE analysis provides a quick and convenient method to assess the purity of the protein at each stage of the purification (Chapter 8, section 8.2.1). For samples from the earlier stages of a purification procedure, ~20 μg of protein are required for a typical Coomassie Brilliant Blue-stained SDS-PAGE gel. Only ~2 μg of protein are required for samples from the later stages of the purification procedure, as these will be less complex and it should be easier to identify the protein of interest.

This information permits the assessment of the success of any given purification procedure.

It should be noted that the methods outlined for each protein purification procedure are not the only methods which could be employed to purify each protein. Other protocols may work equally well to produce a high yield of pure, active protein. While the design of each purification procedure was influenced by the general considerations outlined in section 7.1, the availability of suitable equipment and appropriate expertise must also be considered before embarking on any given procedure. In addition, the costs of materials and equipment may also impact on the protocol adopted.

7.3.1 Lactate dehydrogenase

Theory

LDH (EC 1.1.1.27) plays an important role in glycolysis under anaerobic conditions, catalysing the reduction of pyruvate to form lactate:

$$\text{pyruvate} + \text{NADH} + \text{H}^+ \rightleftharpoons \text{lactate} + \text{NAD}^+$$

Details are given in the Practical Activity 7.1, which is available for lecturers in the Online Resource Centre for this text to support teaching of this topic.

LDH is a cytosolic enzyme that has been isolated and characterized from a number of sources, including human muscle, pig heart, and chicken heart and muscle. Typically, LDH exists as a tetramer with an overall molecular mass of 140 kDa (see Fig. 7.10). Two types of subunit have been found in humans: the M-type subunit, which predominates in skeletal muscle, and the H-type subunit, which predominates in heart muscle. Combinations of these subunit forms give rise to five isoenzymes of LDH: M_4 (which predominates in skeletal muscle), M_3H, M_2H_2, MH_3, and H_4 (which predominates in heart muscle).

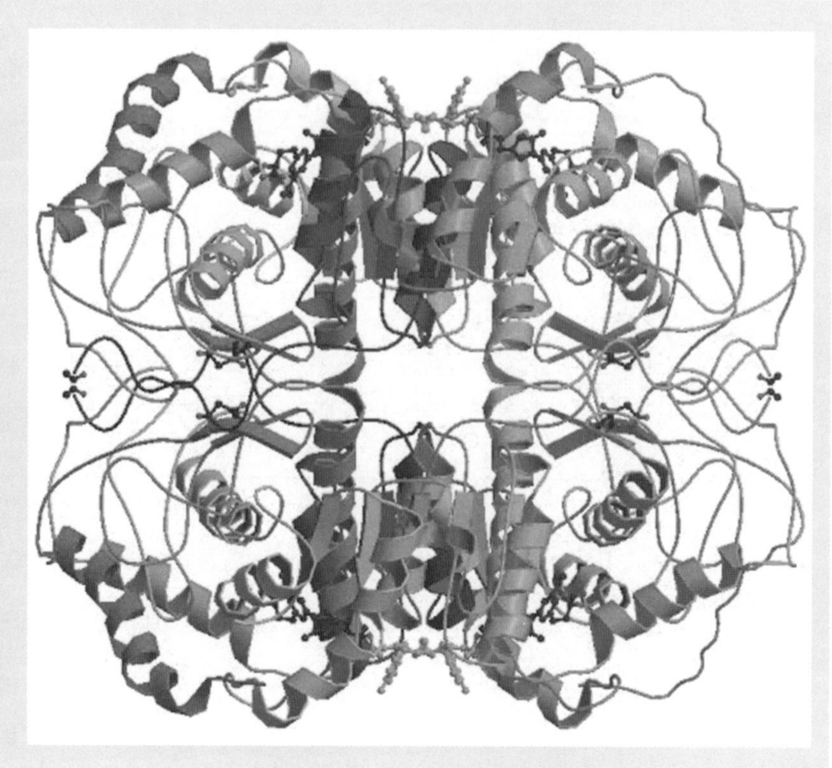

Fig. 7.10 Schematic representation of the tetrameric structure of lactate dehydrogenase from pig heart (PDB ID:5LDH).

Purification rationale

In this protocol, we describe the isolation of LDH from rabbit muscle. The abundance of LDH in muscle tissue of similar organisms is well known, therefore it is reasonable to assume that rabbit muscle will be a source rich in LDH. Following an ammonium sulphate precipitation step, LDH is purified using ion-exchange chromatography, with the selective elution of LDH from the ion-exchange column using the cofactor NADH. Ammonium sulphate fractionation only removes some contaminants from the cell extract and so a subsequent ion-exchange step is employed to purify the LDH further. Selective elution of LDH by NADH relies on the tight binding of NADH to LDH (see Chapter 9, section 9.5.3).

Once purified, the isoenzyme composition of the purified LDH can be analysed using non-denaturing electrophoresis combined with an activity stain. The small differences in the amino acid composition of each of the subunit forms gives rise to isoenzymes, with different charge properties, which can be separated by non-denaturing electrophoresis. In addition, the isoenzymes retain activity, which allows their rapid and specific detection following non-denaturing electrophoresis using an activity stain.

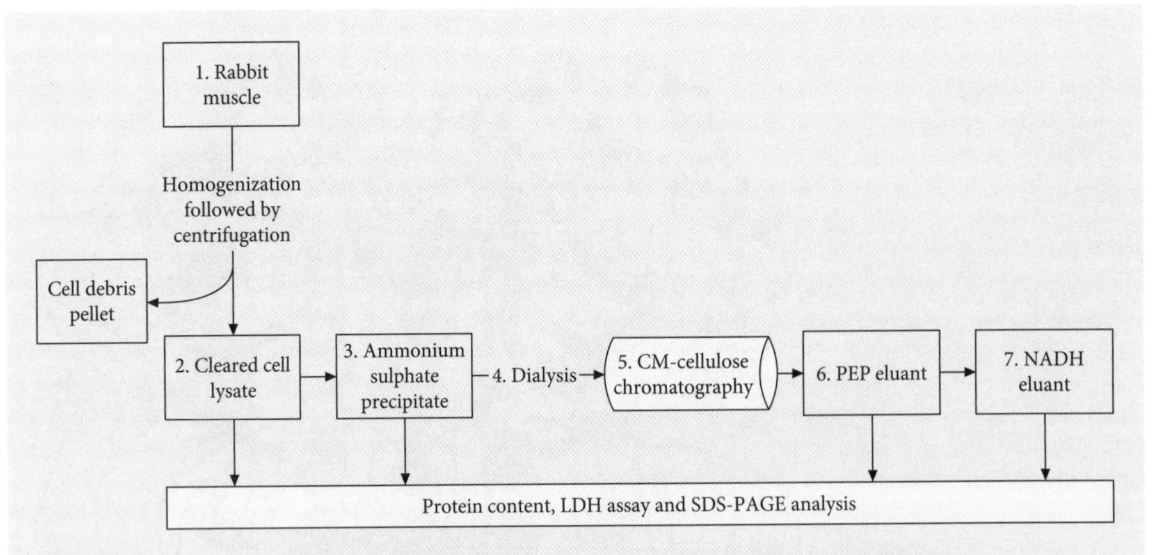

Fig. 7.11 Schematic representation of LDH purification. PEP, phosphoenol pyruvate; CM-cellulose, carboxymethyl-cellulose.

Experimental design

A two-step procedure for the purification of LDH from rabbit muscle is described: an initial ammonium sulphate fraction step followed by an ion-exchange chromatography step (see Fig. 7.11). This is adapted from the method of Scopes, 1977.

A number of muscle-derived LDH sequences from a range of organisms are available and we can use these sequences to predict their theoretical pI values, which range from 7 to 8.5 (pI prediction is outlined in Chapter 5, section 5.8.2). Assuming the sequence of rabbit muscle LDH is similar to these known sequences, we can infer a similar pI value for rabbit muscle LDH to design a suitable ion-exchange protocol.

Muscle-derived LDHs have a predicted pI value between 7 and 8.5, therefore pH conditions less than 7.0 are expected to promote the formation of positive charges on rabbit muscle LDH. This positively charged LDH will adsorb to a cation exchanger, in this case CM-cellulose.

Steps 1 and 2

Rabbit muscle is homogenized to disrupt the cells and release proteins. A centrifugation step removes cell debris.

Step 3

Ammonium sulphate fractionation is employed to remove some contaminants and enrich the LDH activity of the extract. Pyruvate kinase co-precipitates with LDH during the initial ammonium sulphate fractionation.

Steps 4 and 5

Dialysis step removes excess ammonium sulphate and enables proteins to bind to the ion-exchange column in the next purification step.

Step 6

Pyruvate kinase is eluted from CM-cellulose by applying the substrate phosphoenol pyruvate (PEP) thus minimizing contamination of the LDH preparation by pyruvate kinase.

Step 7

Selective elution of LDH is achieved by the addition of NADH and a small increase in the pH of the elution buffer. The substrate NADH binds to LDH tightly, as described in Chapter 9, section 9.5.3, and with an accompanying change in pH selectively promotes LDH desorption from the CM-cellulose column.

In addition to LDH activity measurements and SDS-PAGE analysis of the purified enzyme, the isoenzymes of LDH can be separated electrophoretically under non-denaturing conditions and detected using an activity stain (Chapter 6, section 6.2.2). A representation of the behaviour of muscle- and heart-derived LDH on non-denaturing electrophoresis is shown in Fig. 7.12.

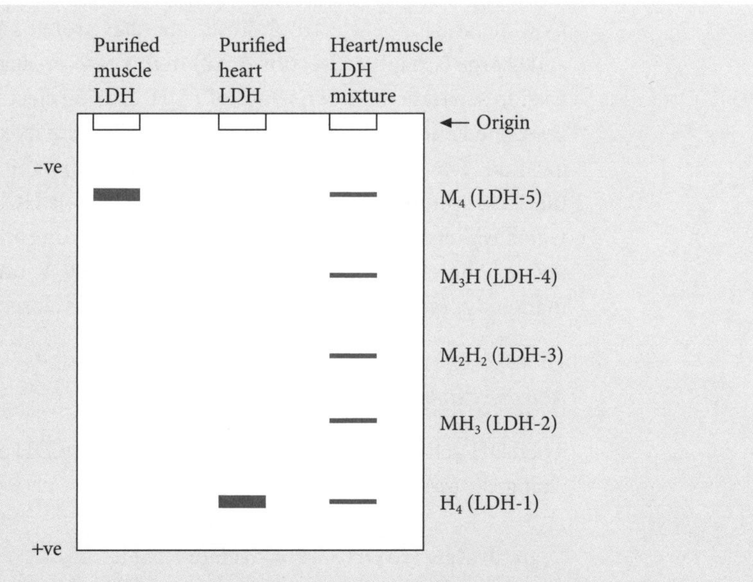

Fig. 7.12 Diagram of native gel LDH activity stain. Muscle LDH exists as tetramer of identical subunits (M_4) which appear as a single species on a non-denaturing gel. Similarly, heart LDH exists as tetramer of identical subunits (H_4) which appear as a single species on a non-denaturing gel. The differences in the size and charge of the M- and H-subunits allow the resolution of the M_4 and H_4 isoenzymes on a non-denautring gel. The lane on the right indicates the presence of five isoenzymes, which form following the heat treatment and subsequent cooling of an equimolar mixture of heart and muscle LDH. Heat treatment dissociates the subunits in the M_4 and H_4 species and subsequent cooling results in the formation of mixed tetramers.

Activity assay

The reaction catalysed by LDH is:

$$\text{pyruvate} + \text{NADH} + \text{H}^+ \rightleftharpoons \text{lactate} + \text{NAD}^+$$

At neutral pH the equilibrium favours the formation of lactate. The enzyme is normally assayed in this direction by monitoring the disappearance of NADH. NADH absorbs light at 340 nm but NAD^+ does not, so the assay simply involves measuring the rate of decrease of A_{340}.

The assay is typically conducted in 1 mL containing 50 mM potassium phosphate, pH 7.0, 0.15 mM NADH and 1 mM pyruvate, at 37°C. Prior to the addition of LDH extracts, the assay mixture should have an A_{340} value of 1.0. The assay is started by the addition of LDH extracts, diluted by 10 mM MES, pH 6.5 containing 100 μM EDTA such that the initial rate of decrease of A_{340} is less than 0.1 min^{-1} but greater than 0.01 min^{-1}. The absorbance change per minute should be calculated for the undiluted LDH extract and subsequently converted into a rate expressed as μmol per min, using the molar absorption coefficient (ε) for NADH at 340 nm, which is 6220 M^{-1} cm^{-1}.

Non-denaturing gels

Non-denaturing gel electrophoresis separates proteins according to both their size and charge (Chapter 6, section 6.2.2) In this case, an agarose non-denaturing gel is used to separate the isoenzymes of LDH. During electrophoresis enzyme activity is retained and can be detected using a specific activity stain to reveal the enzyme's location. The agarose gel is prepared at 1% (w/v) in 50 mM Tris HCl, pH 8.6 and the electrophoresis is conducted using 50 mM Tris HCl, pH 8.6, buffer conditions which will ensure LDH is carrying a negative charge and will migrate towards the anode. The electrophoresis is conducted at 100 V until the bromophenol blue marker dye (see Chapter 8, section 8.2.1) migrates close to the anode (~45–60 min).

Activity staining

The LDH activity stain relies on a product of the LDH activity generating a colour change:

$$\text{pyruvate} + \text{NADH} + \text{H}^+ \rightleftharpoons \text{lactate} + \text{NAD}^+$$

When assayed in the reverse direction, LDH will generate NADH which donates its electrons to an electron carrier meldola blue, which in turn reduces thiazolyl blue tetrazolium (MTT) to generate a visible insoluble blue product. The activity stain comprises 100 mM sodium lactate; 0.2% NAD^+ (w/v); 0.6% (w/v) thiazolyl blue tetrazolium (MTT), and 0.005% (w/v) meldola blue. Following electrophoresis, the surface of the activity gel is covered with the activity stain and incubated at 37°C for up to 30 min to allow the stain to develop.

Sample data for 7.3.1

The data in Table 7.6 were obtained during the purification of rabbit muscle LDH. Complete the table for this LDH purification.

Table 7.6 Purification table for rabbit muscle LDH

Step	Volume (mL)	Protein content (mg mL^{-1})	Total protein (mg)	Total LDH activity (µmol min^{-1})	Yield (expressed as %)	Specific activity (µmol min^{-1} mg^{-1})	Purification Factor
Cell extract	45	40		1380			
Dialysed ammonium sulphate precipitate	10	57		1100			
PEP eluant	20	1.15		5			
NAD^+ eluant	10	0.23		950			

SELF TEST ?

Check that you have mastered the key concepts at the start of this section by attempting the following question.

ST 7.6 Using the information in section 7.3.1, calculate the quantities of the components you will need for a typical 1 mL LDH assay.

Stock solution	Final concentration in LDH assay (mM)	Calculated volume of stock solution required for 1-mL LDH assay
0.5 M potassium phosphate, pH 7.0	50	
10 mM NADH	0.15	
10 mM pyruvate	1	

ST 7.7 Using the information in section 7.3.1, calculate the mass of each of the activity stain components to add to 10 mL 100 mM sodium lactate to make 10 mL activity stain.

Answers

ST 7.6 50/500 × 1 = 0.1 mL or 100 µL 0.5 M potassium phosphate, pH 7.0; 0.15/10 × 1 = 0.015 mL or 15 µL 10 mM NADH; 1/10 × 1 = 0.1 mL or 100 µL 10 mM pyruvate. Water (785 µL) is added to make the volume up to 1 mL.

ST 7.7 100% (w/v) = 100 g/100 mL
NAD^+, 0.2 g/100 mL ⇒ 0.02 g in 10 mL
MTT, 0.6 g/100 mL ⇒ 0.06 g in 10 mL
Medola blue, 0.005 g/100 mL ⇒ 0.5 mg in 10 mL

7.3.2 Alcohol dehydrogenase (yeast)

Theory

ADH (EC 1.1.1.1) catalyses the oxidation of ethanol to ethanal (acetaldehyde):

$$\text{ethanol} + NAD^+ \rightleftharpoons \text{ethanal (acetaldehyde)} + NADH + H^+$$

Details are given in the Practical activity 7.2, which is available for lecturers in the Online Resource Centre for this text to support teaching of this topic.

ADH is a cytosolic enzyme which has been isolated from many sources including rat liver, horse liver, and the bakers yeast, *Saccharomyces cerevisiae*. This enzyme requires zinc as a cofactor and displays broad substrate specificity. ADH from *S. cerevisiae* exists as a tetramer with an overall molecular mass of 145 kDa, whereas mammalian ADH exists as a dimer with an overall molecular mass of 80 kDa, as shown in Fig. 7.13. Like LDH (described in section 7.3.1), ADH occurs as a number of isoenzymes; in liver, there are two subunit forms (E and S) which combine to form the isoenzymes EE, ES, and SS.

Purification rationale

Yeast is a source which is rich in ADH. As yeast cells have a tough cell wall, sonication is employed to rupture the cells. Initially, an ammonium sulphate step removes some contaminating proteins and enriches the ADH content of the extract. Prior to a series of affinity purification steps, dialysis is used to remove high concentrations of salt in an attempt to promote the binding of ADH to the

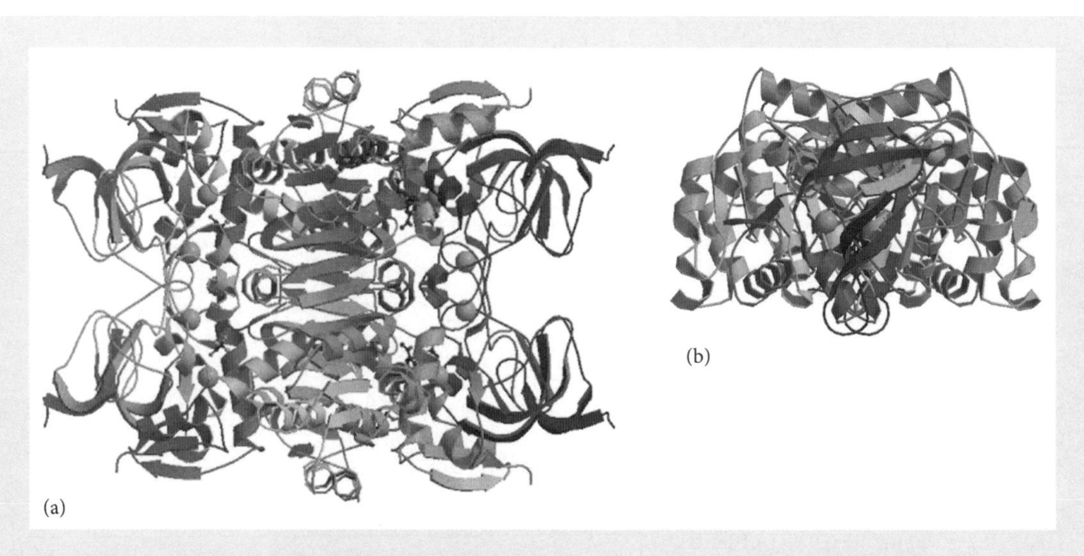

Fig. 7.13 Schematic representation of the structure of alcohol dehydrogenase from (a) *Saccharomyces cerevisiae*, which exists as a tetrameric form (PDB ID:2HCY) and (b) horse liver, which exists as a dimeric form (PDB ID:1YE3).

affinity column. As a member of the NAD^+-binding domain superfamily, the presence of a nucleotide binding site in ADH can be exploited for purification purposes. Cibacron Blue affinity columns have been shown to bind proteins containing a nucleotide binding fold specifically. ADH is initially eluted from a Cibacron blue affinity column using a KCl gradient which removes some, but not all, contaminating proteins. A second Cibacron blue affinity step is then used to elute ADH specifically using the substrate NAD^+.

Cibacron Blue F3GA exhibits specific binding for nicotinamide adenine dinucleotide (NAD^+)-dependent enzymes, whereas Procion Red HE3B binds nicotinamide adenine dinucleotide phosphate ($NADP^+$)-dependent enzymes. As a result both affinity dyes have been used in the purification of a number of dehydrogenases.

Experimental design

This experiment outlined in Fig. 7.14 describes the efficient purification of ADH from the bakers' yeast, *S. cerevisiae* (adapted from Lian *et al.*, 1996).

Steps 1 and 2

Harsh sonication conditions are used to rupture cell wall of yeast.

Step 3

A centrifugation step is used to remove cell debris.

Step 4

Ammonium sulphate fractionation removes a number of contaminating proteins and enriches the ADH activity of the extract.

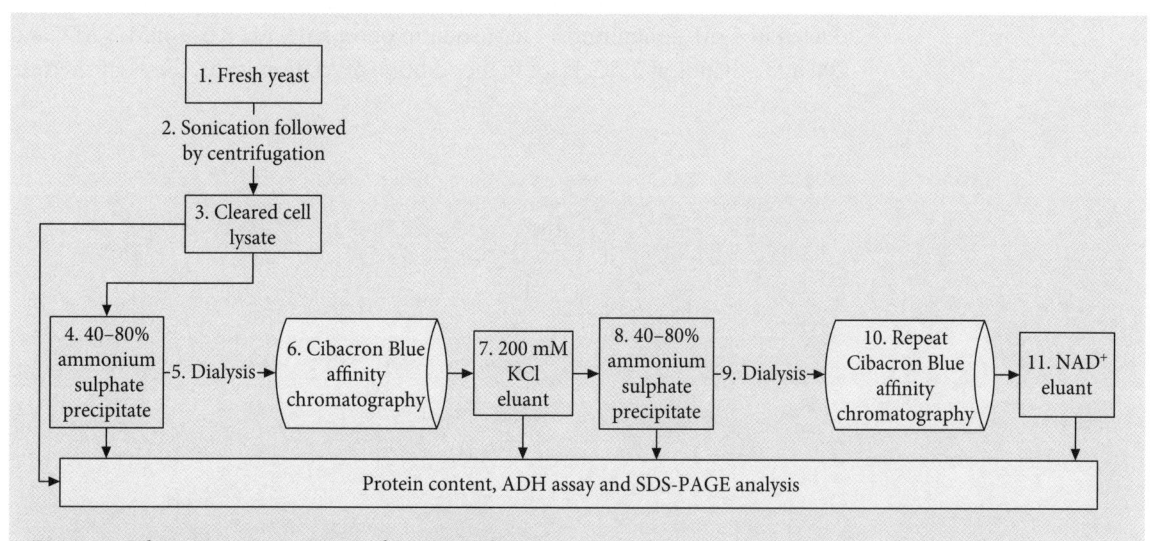

Fig. 7.14 Schematic representation of ADH purification.

Steps 5 and 6

Dialysis removes ammonium sulphate from the extract to allow ADH to bind to the Cibacron Blue affinity column.

Steps 7–9

First of two affinity chromatography procedures. Cibacron Blue affinity chromatography is used, exploiting the specific interaction between the NAD^+-binding domain of ADH and the affinity ligand, Cibacron Blue F3GA (Lian *et al.*, 1996) (see Fig. 7.15). ADH is eluted from the affinity column using KCl to generate an extract with enriched ADH content; however, KCl also elutes a number of contaminating proteins and a second chromatography step is required.

Steps 10 and 11

In a second affinity chromatography step, a pulse of NAD^+ is used to elute ADH from the Cibacron Blue affinity column. The specificity of the ADH–Cibacron Blue interaction, together with the selectivity of the NAD^+ elution gives rise to a 200-fold purification factor while maintaining a yield of around 50%.

Activity assay

The reaction catalysed by ADH is:

$$\text{ethanol} + NAD^+ \rightleftharpoons \text{ethanal (acetaldehyde)} + NADH + H^+$$

The assay for ADH relies on monitoring the formation of NADH, which absorbs light at 340 nm, by measuring the rate of increase of A_{340}. The assay is typically conducted in 1 mL containing 33 mM sodium phosphate, pH 8.0, 1 mM NAD^+ and 200 mM ethanol, at 25°C. Prior to the addition of ADH extracts, the assay mixture

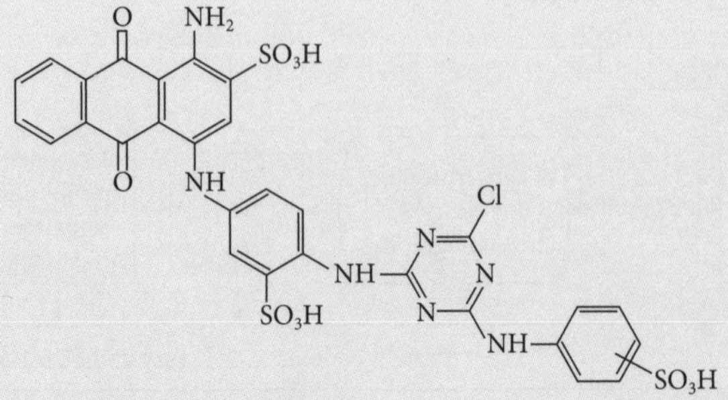

Fig. 7.15 Structure of Cibacron Blue F3GA.

should have an A_{340} value of zero. The assay is started by the addition of ADH extracts, diluted by 20 mM sodium phosphate, pH 8.0 containing 50 mM DTT such that the initial rate of increase of A_{340} is less than 0.1 min^{-1} but greater than 0.01 min^{-1}. The absorbance change per minute should be calculated for the undiluted ADH extract and subsequently converted into a rate expressed as µmol per min, using the molar absorption coefficient (ε) for NADH at 340 nm which is 6220 M^{-1} cm^{-1}.

Sample data for 7.3.2

The data in Table 7.7 were collected during the purification of yeast ADH.

Complete the table for this ADH preparation.

Table 7.7 Purification table for yeast ADH

Step	Volume (mL)	Protein content (mg mL^{-1})	Total protein (mg)	Total ADH activity (µmol min^{-1})	Yield (expressed as %)	Specific activity (µmol min^{-1} mg^{-1})	Purification factor
Cell extract	100	25		6250			
KCl eluant	50	20		4000			
NAD$^+$ eluant	40	0.15		3000			

SELF TEST ?

Check that you have mastered the key concepts at the start of this section by attempting the following question.

ST 7.8 If an ADH activity assay of the NAD$^+$ eluant produces an A_{340} change per minute of 0.08:

(a) calculate the rate, expressed as a concentration change per minute

(b) knowing that the assay was conducted in 1 mL, calculate the rate of NADH formation, expressed as µmol min^{-1}

(c) if the amount of protein required to produce this change was 0.035 µg, calculate the specific activity of the NAD$^+$ eluant, expressed in terms of µmol min^{-1} mg^{-1}.

Answer

ST 7.8

(a) An A_{340} of 6220 corresponds to 1 M NADH, therefore an absorbance change of 0.08 min^{-1} corresponds to 12.9 µM NADH min^{-1}.

(b) 12.9 µM NADH min^{-1} corresponds to 12.9 µmol L^{-1} min^{-1} NADH. In a 1-mL assay, this is 12.9 nmol min^{-1}.

(c) 12.9 nmoles min^{-1} was produced by 0.035 µg protein, thus 1 mg would produce 12.9/0.035 × 1000 = 369 µmol min^{-1} mg^{-1}.

7.3.3 Isocitrate dehydrogenase

Theory

ICDH (EC 1.1.1.41) catalyses the oxidative decarboxylation of isocitrate and 2-oxoglutarate:

$$\text{isocitrate} + NAD^+ \rightleftharpoons \text{2-oxoglutarate} + CO_2 + NADH + H^+$$

Details are given in the Practical Activity 7.3, which is available for lecturers in the Online Resource Centre for this text to support teaching of this topic.

There are two types of ICDH, utilizing NAD^+ and $NADP^+$, respectively. As a TCA cycle enzyme, the NAD^+-dependent ICDH occurs exclusively in the mitochondrial matrix, whereas the $NADP^+$-dependent ICDH occurs predominantly in the cytosol. ICDH has been isolated from beef heart, and muscle and heart from human and rat sources.

Purification rationale

Bakers' yeast is a source which is rich in ICDH and we describe the partial purification of ICDH with a view to exploring the unusual kinetic properties of this enzyme (Chapter 9, section 9.6). An initial bulk step generates calcium phosphate which binds many contaminating proteins and the calcium phosphate-contaminant precipitate is removed by centrifugation. The remaining supernatant, containing ICDH, is subjected to ammonium sulphate precipitation to remove further contaminating proteins and enrich the extract with ICDH. Following this ammonium sulphate step, dialysis is used to remove excess ammonium sulphate to generate the low salt conditions required to optimise protein–ion-exchange interactions on the subsequent chromatography step. Finally, ICDH is selectively eluted from the anion exchanger, DEAE-cellulose, by increasing the ionic strength of the elution buffer.

Experimental design

This protocol describes the partial purification of the NAD^+-dependent ICDH from the bakers' yeast, *S. cerevisiae* (Illingworth, 1972). This purification involves two precipitation steps followed by an ion-exchange chromatography procedure, see Figure 7.16.

Steps 1 and 2

The tough cell wall of yeast cells is disrupted mechanically by grinding with glass beads.

Steps 2 and 3

Centrifugation is used to remove cell debris.

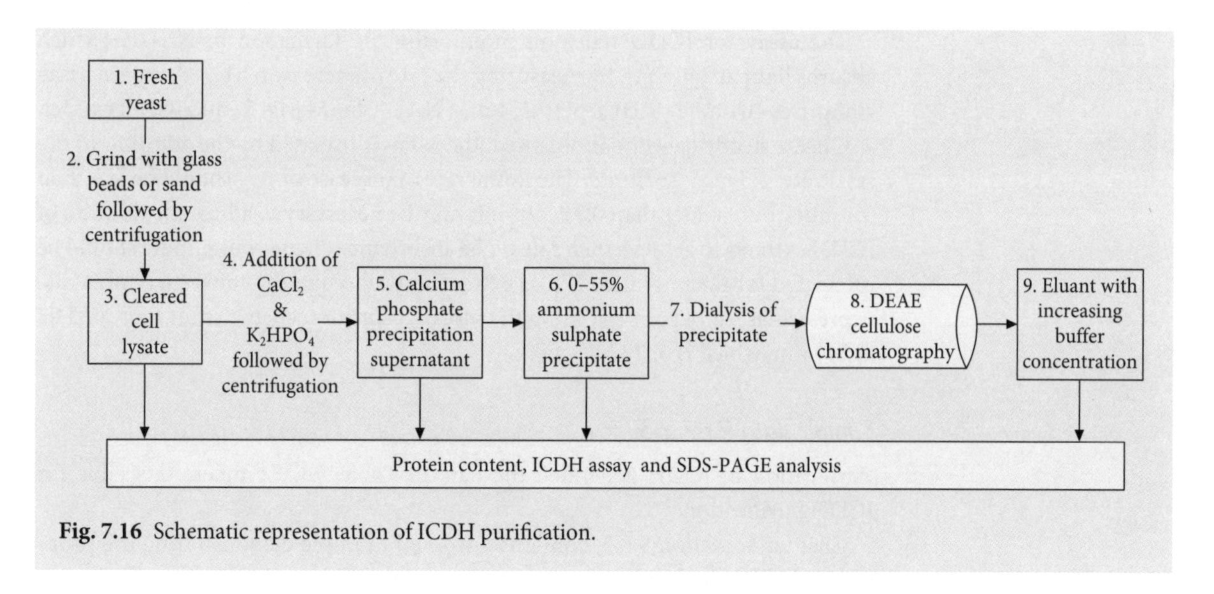

Fig. 7.16 Schematic representation of ICDH purification.

Steps 4 and 5

The first precipitation step involves the formation of a calcium phosphate precipitate, which binds contaminating proteins which are subsequently removed by centrifugation. The supernatant is retained for Step 6.

Step 6

The second precipitation step promotes the selective precipitation of NAD^+-dependent ICDH using ammonium sulphate.

Step 7

The ICDH-rich ammonium sulphate precipitate is resuspended and dialysed to remove excess ammonium sulphate.

Step 8

The dialysed extract is applied to the anion exchanger, DEAE–cellulose, under conditions which promote the adsorption of NAD^+-dependent ICDH.

Step 9

NAD^+-dependent ICDH is selectively eluted from DEAE-cellulose by applying an elution buffer of increasing ionic strength.

Activity assay

The reaction catalysed by ICDH is:

$$\text{isocitrate} + NAD^+ \rightleftharpoons \text{2-oxoglutarate} + CO_2 + NADH + H^+$$

The assay for ICDH relies on monitoring the formation of NADH (which absorbs light at 340 nm), by measuring the rate of increase of A_{340}. A typical assay comprises 40 mM Tris.HCl, pH 7.5, 1 mM NAD^+, 4 mM $MgCl_2$, and ICDH extract.

Once a steady baseline is obtained, the assay is initiated by the addition of the substrate, 2.5 mM isocitrate. The initial rate of increase of A_{340} should be less than 0.1 min^{-1} but greater than 0.01 min^{-1}. It may be necessary to adjust the dilution of ICDH extracts to achieve such rates. The absorbance change per minute should be calculated for the undiluted ICDH extract and subsequently converted into a rate expressed as μmol per min, using the molar absorption coefficient (ε) for NADH at 340 nm, which is 6220 M^{-1} cm^{-1}.

Sample data for 7.3.3

Purification of ICDH generated the data in Table 7.8. Complete this table for ICDH purification.

Chapter 9, section 9.6.3, contains a worked example demonstrating the properties of ICDH in the absence and presence of the effector AMP.

Table 7.8 Purification table for ICDH

Step	Volume (mL)	Units mL^{-1}	Total units	Protein concentration ($mg\ mL^{-1}$)	Total protein (mg)	Specific activity (units mg^{-1})	Yield (%)
Cell lysate	150	1.1		15			
Calcium phosphate precipitation supernatant	148	1.0		12			
Dialysed ammonium sulphate precipitate	10	14		11			
CM-cellulose eluant	8	15		0.3			

1 unit of activity corresponds to 1 μmol min^{-1}.

? SELF TEST

Check that you have mastered the key concepts at the start of this section by attempting the following question.

ST 7.9 If 10 μL of a 50-fold dilution of an ICDH preparation produces a change in the A_{340} of 0.14 min^{-1}, calculate the dilution you would have to make to ensure that the addition of 10 μL of diluted preparation resulted in a change in the A_{340} per minute of 0.035.

Answer

ST 7.9 1:50 dilution gives ΔA_{340} min^{-1} of 0.14

Undiluted preparation gives ΔA_{340} min^{-1} of 7, therefore to obtain ΔA_{340} min^{-1} of 0.035, a 7/0.035 = 200-fold dilution is required.

7.3.4 Glutathione-*S*-transferase

Theory

The ability to isolate genes, to overexpress encoded proteins in heterologous systems and to produce fusion proteins has simplified the process of protein isolation and extended the range of proteins that have been studied. Numerous fusion partners are now employed in gene fusion technology ranging from His-tags (short peptide consisting of (usually) six consecutive histidine residues) to enzymatically active fusion partners such as glutathione-*S*-transferase (GST, EC 2.5.1.18).

GST fusion proteins are commonly produced in *E. coli* expression systems and these are readily purified using immobilized glutathione. The purification of GST from *E. coli* cells transformed with commercially available GST expression vectors highlights the ease of protein overexpression and purification.

GSTs perform a protective role in the cell by detoxifying endogenous compounds during oxidative stress, e.g. quinones and epoxides. GSTs are also responsible for detoxifying chemical carcinogens, environmental pollutants, and a range of pharmaceutical compounds, leading to drug resistance.

Details are given in the Practical activity 7.4, which is available for lecturers in the Online Resource Centre for this text to support teaching of this topic.

Purification rationale

In this example, we are purifying the protein of interest from a non-native source. GST is isolated from *E. coli* cells, containing a GST expression plasmid. This approach ensures a high concentration of GST in the starting material. The presence of a high concentration of GST in the cell extract acts effectively as a purification step, i.e. GST may be 10% of total cell protein in the overexpression system rather than <0.1% of the total cell protein in its native source. Following cell lysis, a one-step affinity chromatography step is used to purify GST. This chromatographic approach relies on the specificity of the interaction between GST and its substrate, glutathione. The GST affinity column, which contains immobilized glutathione, binds GST whereas most contaminating proteins fail to bind to the column. Glutathione, which is subsequently introduced into the column eluant, interacts with GST and promotes the specific elution of the enzyme.

Experimental design

The purification of GST and GST fusion proteins by glutathione affinity chromatography relies on the specific interaction between GST and immobilized glutathione. Optimal binding of GST to immobilized glutathione requires a low flow rate (≤ 1 mL min^{-1}) due to the relatively weak affinity of the enzyme for the substrate. Specific elution of GST is achieved by applying a solution of reduced glutathione. The overall strategy is depicted in Fig. 7.17.

Good levels of overexpression combined with efficient chromatography can produce a few milligrams of pure GST, or GST fusion protein, from a 50 mL culture in a one-step procedure.

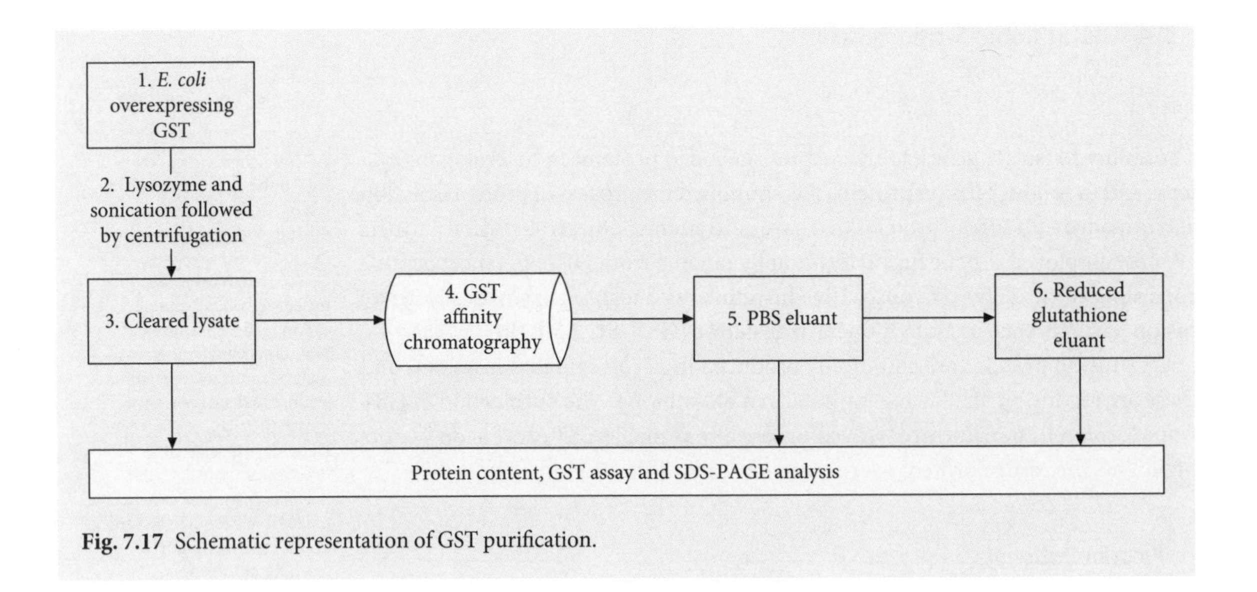

Fig. 7.17 Schematic representation of GST purification.

Step 1

E. coli cells containing the GST expression plasmid are grown, expression is induced and the cells are harvested by centrifugation.

Steps 2 and 3

The cell wall of *E. coli* is disrupted by treatment with lysozyme and sonication. Cell debris is removed by centrifugation.

Steps 4 and 5

E. coli cell lysate is applied to the GST-affinity column. Application of phosphate-buffered saline (PBS) promotes the removal of contaminating proteins but the GST remains bound to the affinity column.

Step 6

GST is specifically eluted from the column by adding reduced glutathione.

Assay of GST activity

The assay for GST is based on the reaction between glutathione (GSH) and 1-chloro-2,4-dinitrobenzene (CDNB). The thiol group of GSH acts as the nucleophile, displacing the chloride ion. Formation of the dinitrophenyl derivative of

GSH leads to an increase in absorbance at 350 nm, with an absorption coefficient of 9600 M^{-1} cm^{-1} at this wavelength.

The standard assay is conducted at 25°C in 0.1 M potassium phosphate buffer, pH 6.5, with final concentrations of GSH and CDNB of 5 and 1 mM, respectively, in a total volume of 1 mL. Stock solutions of GSH (50 mM) and CDNB (20 mM) are made up by weight in buffer and ethanol, respectively.

It is not necessary to remove the glutathione present in the protein fractions (from the chromatography step) prior to assay: only small volumes of the protein samples are added to the assay mixture, so that the concentrations of this substrate carried through are very small compared with the actual concentration added (5 mM)).

Sample data for 7.3.4

The data in Table 7.9 were collected for purification of GST. Complete the purification table for this enzyme.

Table 7.9 Purification table for glutatione-*S*-transferase

Purification step	Volume of sample (mL)	Protein concentration (mg mL^{-1})	Total protein (mg)	GST activity (ΔA_{350} min^{-1} per 10-µL sample)[¶]	Total activity (µmol min^{-1})	Specific activity (µmol min^{-1} mg^{-1})
Cell extract	10	1.53		0.12*		
Run through	10	0.96		0.12		
Wash	10	0.14		0.04		
GST eluant	4	1.1		0.18*		

[¶]Note that the volume used for assay of GST activity is 1 mL in each.

*10 µL of 1/5 dilution of cell extract and GST eluant were used in the activity measurements.

SELF TEST ?

Check that you have mastered the key concepts at the start of this section by attempting the following question.

ST 7.10 Calculate the weights of GSH (molecular mass 307.3 Da) and CDNB (molecular mass 202.6 Da) required to make up 5 mL of the required stock solutions of each compound (50 mM and 20 mM respectively), and the weight of KH_2PO_4 (molecular mass 136.1 Da) required to make up 1 L of 0.1 M buffer. What volumes of these stock solutions and buffer are required in the standard assay solution?

Answer

ST 7.10 In all, 76.8 mg GSH is dissolved in 5 mL buffer to give a 50 mM solution, 20.3 mg CDNB is dissolved in 5 mL ethanol to give a 20 mM solution. 13.6 g KH_2PO_4 would be weighed out, dissolved in about 800 mL water, the pH adjusted to 6.5 with KOH, and the volume made up to 1 L with water. The mixture for each assay contains 0.1 mL GSH solution, 0.05 mL CDNB solution, and 0.85 mL buffer.

7.3.5 Myoglobin

Theory

The oxygen-binding protein myoglobin (Mb) is present in abundance in muscle tissue where it stores oxygen, only releasing it under conditions of oxygen deficiency (partial pressure of O_2 less than 20 mm Hg). Myoglobin is a monomeric protein, Fig. 7.18, of molecular mass 17 kDa and contains a haem group with an iron at its centre. The iron is coordinated to four nitrogen atoms from the haem group and a fifth coordination site is provided by the imidazole-nitrogen atom of a histidine residue. The sixth coordination site is accessible to ligands, i.e. oxygen. In this protocol, we shall conduct a partial purification of oxy-myglobin ($Mb.O_2$) and met-myglobin ($Mb^+.H_2O$) from steak, a source rich in this protein. Spectral analysis will be used to monitor changes in the haem group of myoglobin, in the presence of an oxidizing agent and a reducing agent, to understand the changes which occur in myoglobin on the binding of O_2. In addition, we can use these spectral data to quantitate the relative amounts of oxy-myoglobin ($Mb.O_2$) and met-myoglobin ($Mb^+.H_2O$) present in the bovine muscle starting material.

Met-myoglobin is denoted $Mb^+.H_2O$ to show that Fe is in the Fe^{3+} state as opposed to the Fe^{2+} state in the deoxy- or oxy- forms.

Details are given in the Practical Activity 7.5, which is available for lecturers in the Online Resource Centre for this text to support teaching of this topic.

Purification rationale

A partial purification of myoglobin from bovine muscle is outlined with a view to monitoring the changes in the haem group of myoglobin on binding O_2. Myoglobin is isolated from bovine muscle as this is a natural, readily available, source in which myoglobin is abundant. Initially, bovine muscle is minced to disrupt tissue structure. Muscle cells are subsequently ruptured by mechanical

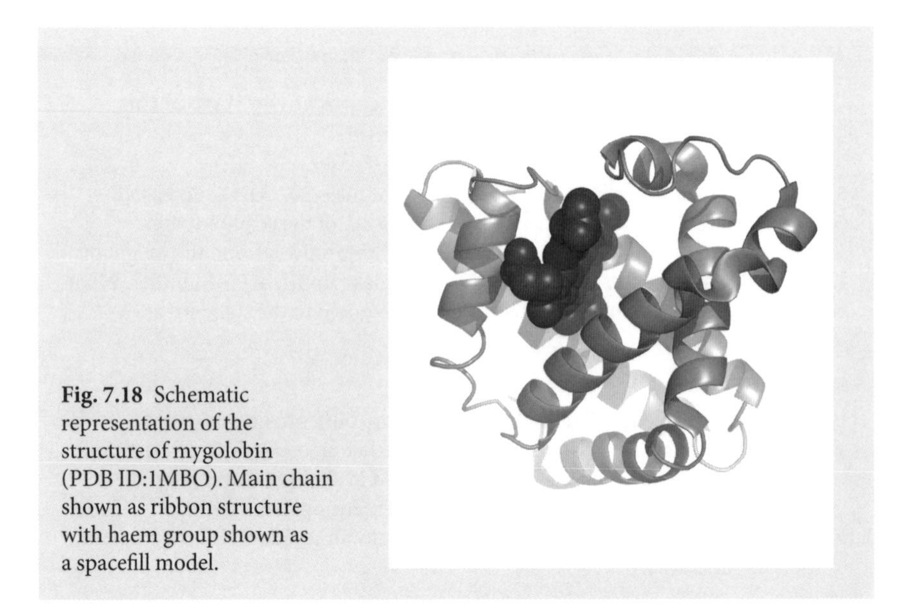

Fig. 7.18 Schematic representation of the structure of mygolobin (PDB ID:1MBO). Main chain shown as ribbon structure with haem group shown as a spacefill model.

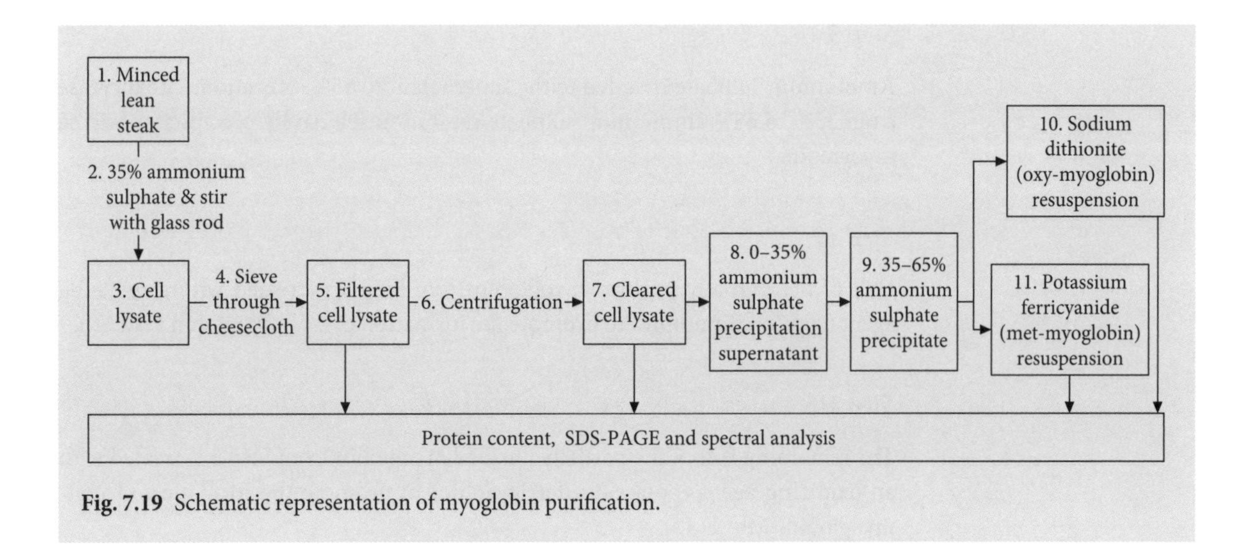

Fig. 7.19 Schematic representation of myoglobin purification.

agitation with a glass rod in the presence of 35% ammonium sulphate. Many contaminating proteins precipitate in presence of 35% ammonium sulphate and these contaminants are removed, along with cell debris, by a sieving and centrifugation step. A 35–65% ammonium sulphate precipitation step selectively precipitates myoglobin. This step generates a suitable quantity of partially purified myoglobin to monitor the redox changes of the haem group of myoglobin using absorption spectroscopy.

Experimental design

The protocol outlined in Fig. 7.19 relies on the solubility of myoglobin at high concentrations of ammonium sulphate (Bylkas and Andersson, 1997).

Step 1

The tissue structure is disrupted by mincing beef steak.

Steps 2 and 3

Cells are ruptured by mixing with a glass rod in the presence of 35% ammonium sulphate. A concentration of 35% ammonium sulphate saturation promotes the precipitation of most proteins present in bovine muscle but not myoglobin.

Steps 4–8

Contaminating proteins and cell debris are removed by sieving and centrifugation.

Step 9

Ammonium sulphate is added to the supernatant to 65% saturation. The increase from 35% to 65% ammonium sulphate saturation selectively precipitates bovine myoglobin.

Step 10

Half of the partially purified myoglobin from Step 9 is treated with a reducing agent (sodium dithionite) to promote the formation of oxy-myoglobin ($Mb.O_2$).

Step 11

The remaining half of the partially purified myoglobin from Step 9 is treated with an oxidizing agent (potassium ferricyanide) to promote the formation of met-myoglobin ($Mb^+.H_2O$).

Spectral analysis

The structural environment of the haem group in myoglobin differs in the oxy- and met- forms of this protein. It is possible to detect these structural differences using absorption spectroscopy with oxy- and met- myoglobin giving rise to the characteristic absorption peaks detailed in Table 7.10.

To monitor the nature of the environment around the haem group in the partially purified oxy- and met-myoglobin preparations, the absorption spectrum from 700 to 300 nm is recorded. Comparison of the absorption peaks with the data in Table 7.10 will indicate the presence or absence of oxy- and met-myoglobin.

Spectral analysis also enables quantitation of the relative amounts of oxy- and met-myoglobin in the bovine muscle cell extract. Following dilution of the cleared cell lysate, representing the cell extract, the absorption spectrum from 700 to 300 nm is recorded compared with the absorption spectra of oxy- and met-myoglobin. Using the absorption coefficients in Table 7.10, it should be possible to estimate the amount of myoglobin in the cell extracts.

Sample data for 7.3.5

Given that 1 in 200 dilution of steak lysate has $A_{417} = 0.64$ and $A_{409} = 0.45$, calculate the concentration of oxy- and met-myoglobin in the lysate.

Table 7.10 Absorption peaks for myoglobin complexes

Myoglobin complex	Absorption peaks (nm)
Oxy-myoglobin [Fe^{2+}–O_2–Mb]	580, 542, 417(128)* and 248
Met-myoglobin [Fe^{3+}–H_2O–Mb]	635, 504 and 409 (179)*

*Values in brackets are the absorption coefficients in units of mM^{-1} cm^{-1}.

SELF TEST ?

Check that you have mastered the key concepts at the start of this section by attempting the following question.

ST 7.11 If partially purified myoglobin exhibits spectral properties that are common to both oxy- and met-myoglobin, comment on the forms present.

Answer

ST 7.11 Both forms are present in the original tissue and the relative quantities of oxy- and met-myoglobin can reflect the 'age' of the meat, with the amounts of met-myoglobin increasing with the 'age' of the meat.

7.3.6 Immunoglobulin M

Theory

Immunoglobulins are key components of the immune system in vertebrates. There are five classes of immunoglobulins (IgM, IgA, IgD, IgE, and IgG). Each immunoglobulin is composed of four chains, two heavy (55–75 kDa) and two light (25 kDa). IgM, shown in Fig. 7.20, is a pentameric molecule rich in disulphide bonds, both within monomers and between monomers. The disulphide bonds of IgM interact specifically with mercaptopyridine, an interaction which is exploited

Details are given in the Practical Activity 7.6, which is available for lecturers in the Online Resource Centre for this text to support teaching of this topic.

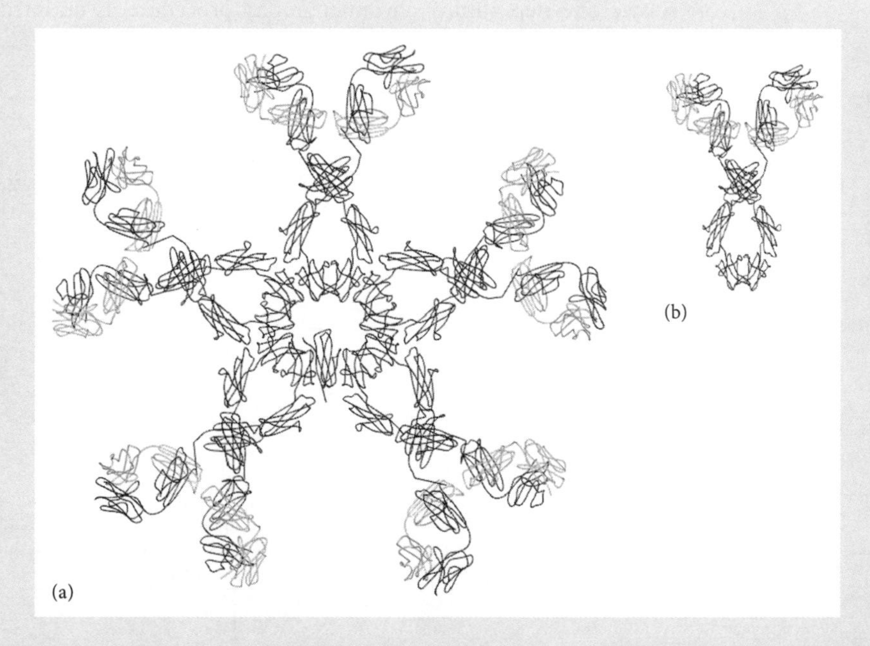

Fig. 7.20 Representation of the structure of IgM. IgM exists as a pentamer (a) with an overall mass of 950 kDa. Each IgM monomer (b) is composed of two heavy chains (black) and two light chains (grey), with an overall mass of 190 kDa. A series of disulphide bonds within each monomer link heavy chain to heavy chain and heavy chain to light chain. In addition, inter-monomer disulphide bond chains are required to retain the pentameric structure.

in the affinity purification of IgM. In this protocol, immunoglobulin is purified from pig serum by affinity chromatography using mercaptopyridine as the immoblized ligand.

Purification rationale

We present a protocol for a single affinity chromatography step purification of IgM from pigs' blood. Pigs' blood is a source that is readily available and rich in IgM. IgM is present in the serum component of blood and an overnight incubation of blood results in the formation of a dense aggregate of cells which can be collected by centrifugation and the resultant serum supernatant. Prior to the addition of this IgM-rich serum to the affinity chromatography column, the serum is diluted using buffer containing 0.8 M ammonium sulphate. This results in a reduction in the protein concentration and an increase in the ionic strength of the serum, conditions which promote the interaction of IgM with the mercaptopyridine ligand on the IgM affinity column. IgM is selectively eluted by lowering the ionic strength of the elution buffer, conditions that promote IgM–solvent interactions.

Experimental design

A simple one-step affinity chromatography procedure as outlined in Fig. 7.21 generates pure IgM.

Steps 1 and 2

Following the collection of a blood sample, a coagulation step occurs, in which a dense aggregate of cells forms.

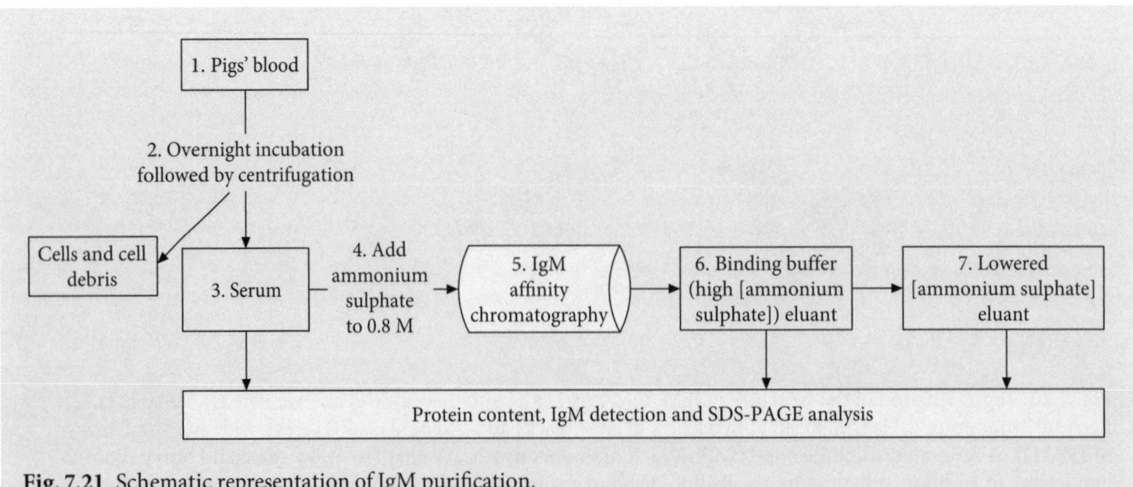

Fig. 7.21 Schematic representation of IgM purification.

Steps 2 and 3

Centrifugation is used to isolate the immunoglobulin-rich serum from the dense aggregate of cells.

Step 4

The serum is diluted by the addition of buffer containing 0.8 M ammonium sulphate. This promotes the hydrophobic interactions between IgM and the ligand mercaptopyridine on the IgM affinity chromatography column. In addition, serum has a high protein content (~30 mg mL^{-1}) which will require dilution to optimize IgM–mercaptopyridine interactions.

Step 5

Serum is applied to the IgM affinity chromatography column in the presence of 0.8 M ammonium sulphate.

Step 6

Non-specifically bound contaminants are removed by applying elution buffer containing 0.8 M ammonium sulphate.

Step 7

A gradient of decreasing ammonium sulphate concentration will promote IgM–solvent interactions resulting in the selective elution of porcine IgM from the affinity column.

Purification analysis

IgM has four chains, two heavy (55–75 kDa) and two light (25 kDa), which are covalently linked by disulphide bonds; therefore on a denaturing gel under reducing conditions, two bands indicative of the heavy and light chain should be observed, as demonstrated in Fig. 7.22(a). A suitable set of molecular mass markers should permit the estimation of the molecular mass of the components of the purified protein. The presence of a 55- to 75-kDa band and a 25-kDa band will support the presence of the heavy and light chains which are characteristic of IgM. The absence of contaminating proteins on reducing SDS-PAGE should also provide an indication of the success of the purification.

SDS-PAGE analysis of IgM under non-reducing conditions should produce a single band, similar to the representation in Fig. 7.22(b) Under non-reducing conditions, the disulphide bonds of human IgM, both within monomers and between monomers, remain intact resulting in the presence of a single pentameric protein, which appears as a single band on non-reducing SDS-PAGE. Further evidence to

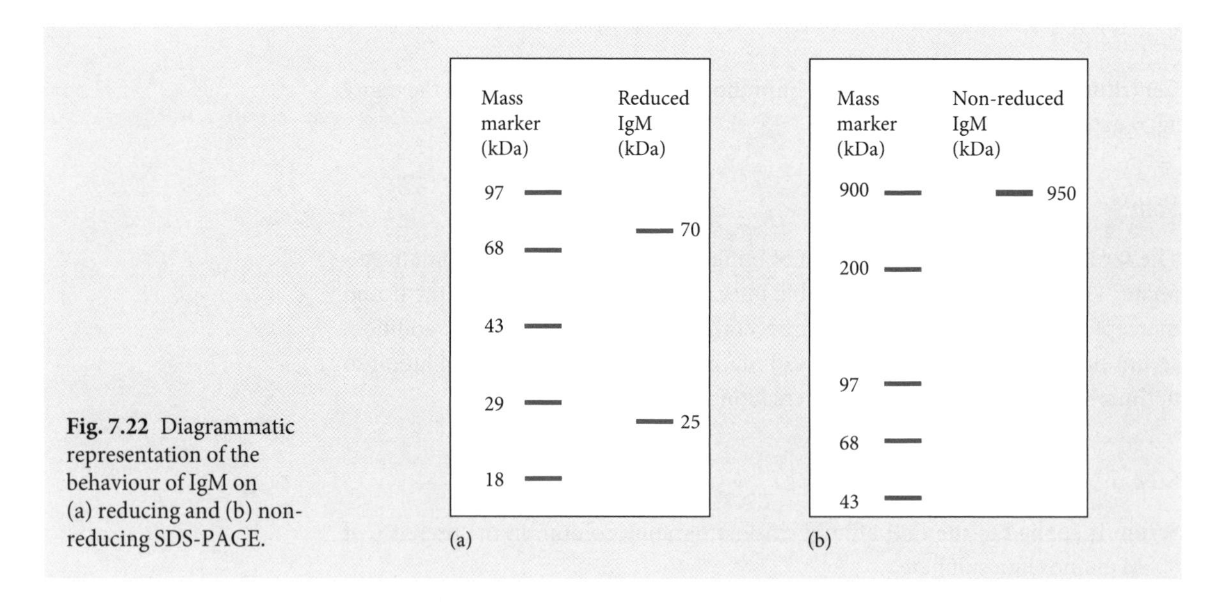

Fig. 7.22 Diagrammatic representation of the behaviour of IgM on (a) reducing and (b) non-reducing SDS-PAGE.

support the presence of IgM can be provided by Western blot analysis of both reducing and non-reducing SDS-PAGE gels, using an anti-porcine IgM antibody.

Sample data for 7.3.6

SDS-PAGE analysis, under reducing conditions, of purified IgM reveals the presence of two protein species with apparent molecular mass values of 25.5 and 60 kDa plus a less abundant species of 58 kDa, as shown in Fig. 7.23. All three bands cross-react with an anti-porcine IgM antibody. Comment on the nature of all three species and describe the experiments you would conduct to confirm this.

7.3.7 Green fluorescent protein

Theory

Details are given in the Practical Activity 7.7, which is available for lecturers in the Online Resource Centre for this text to support teaching of this topic.

GFP is a naturally occurring 27-kDa protein that occurs in the jellyfish *Aequorea victoria*. Biologists have exploited the fluorescent properties of GFP, using it as a reporter molecule in molecular biology and developmental biology. Illumination of GFP with UV light results in fluorescence from the fluorophore, *p*-hydroxybenzylidene-imidazolidone, which is formed by three amino acids (Gly, Tyr, and Thr or Ser) buried in the centre of its hydrophobic barrel structure. The Gly–Tyr–Thr/Ser sequence reacts to form the unusual five-membered ring of the fluorophore as shown in Fig. 7.24.

GFP exhibits unique spectral properties that can be exploited to detect and quantitate GFP throughout this purification. GFP has two absorption maxima at 395 and 475 nm with absorption coefficients of 30 000 and 7000 M^{-1} cm^{-1}, respectively. In the case of non-UV illuminated GFP, the major absorption peak is

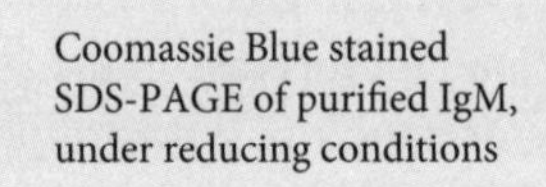

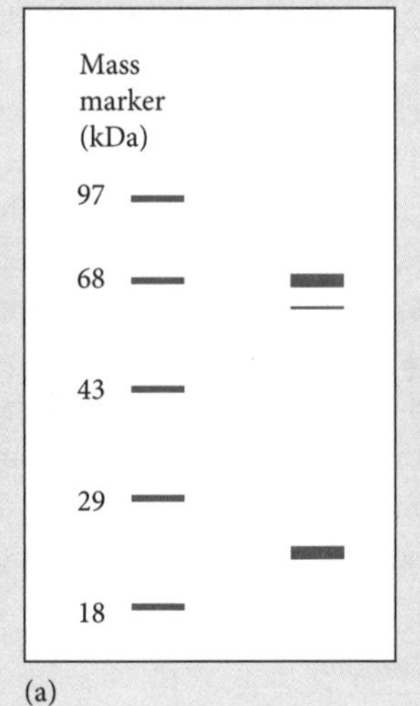

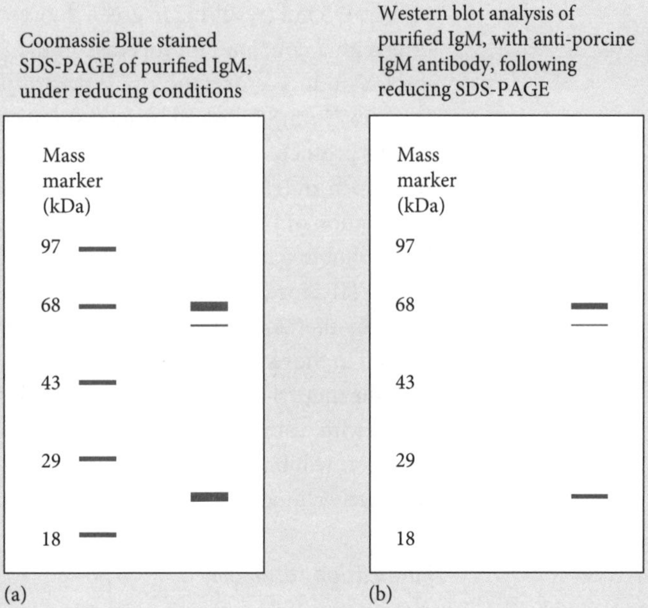

Fig. 7.23 Representation of SDS-PAGE analysis, under reducing conditions, of purified porcine IgM (a) stained with Coomassie Blue and (b) probed with anti-porcine IgM antibody on a Western blot.

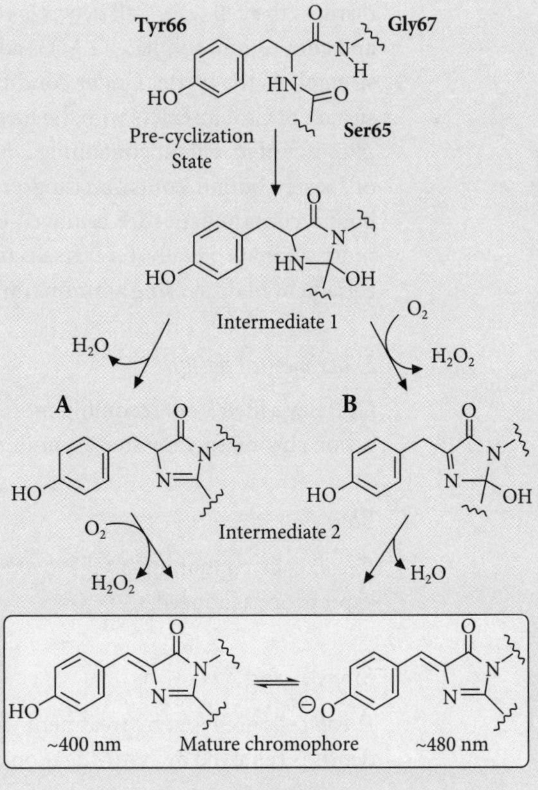

Fig. 7.24 Mechanism of GFP fluorophore formation. From Wachter (2006).

at 395 nm. However, GFP illuminated with UV light exhibits a major absorption peak at 475 nm (resulting in green fluorescence at 510 nm), which diminishes over time with a concomitant increase in the 395-nm peak. This pattern arises due to the UV-induced formation of a de-protonated anionic form of the fluorophore (λ_{max} = 475 nm), followed by its restoration to a neutral state (λ_{max} = 395 nm).

This protocol describes the purification of recombinant GFP from a strain of *E. coli* which overexpresses GFP (a construct that is available commercially). The purification of this protein relies on the interaction between GFP, which has many hydrophobic patches on its surface, and the hydrophobic interaction chromatography HIC medium, methyl-HIC. The success of this one-step purification is facilitated by the fact that GFP is much more hydrophobic than most of the proteins present in the expression host, *E. coli*. As a result, host proteins will interact poorly with the methyl-HIC column under high salt buffer conditions and will be readily eluted with a small reduction in the salt concentration of the elution buffer. GFP, however, will bind methyl-HIC efficiently under high salt conditions and will be selectively eluted under very low salt conditions.

Purification rationale

E. coli cells containing a GFP overexpression plasmid are grown, induced to generate large quantities of GFP, and then harvested by centrifugation. An *E. coli* cell lysate is prepared by treatment with lysozyme, followed by freeze thawing which disrupts the cell wall. Cell debris is removed from the cell lysate by centrifugation and ammonium sulphate (2 M) is added to the cleared lysate to increase the ionic strength of the lysate. Under conditions of high ionic strength, the hydrophobic surface of GFP interacts with the methyl ligand of the HIC column. Continued application of an eluant containing 2 M ammonium sulphate results in the removal of loosely bound contaminating proteins from the HIC column. More tightly bound contaminants are removed by applying eluant containing 1.3 M ammonium sulphate. Finally, GFP is selectively eluted from the HIC column by applying a gradient of decreasing ammonium sulphate (1.3–0 M).

Experimental design

GFP is purified from recombinant *E. coli* cells using a one-step hydrophobic interaction chromatograpy step as outlined in Fig. 7.25.

Step 1

E. coli cells containing a GFP overexpression plasmid are grown and then GFP expression is induced.

Steps 2 and 3

E. coli cells are lysed by treatment with lysozyme and freeze thawing, and the cell debris is removed by centrifugation.

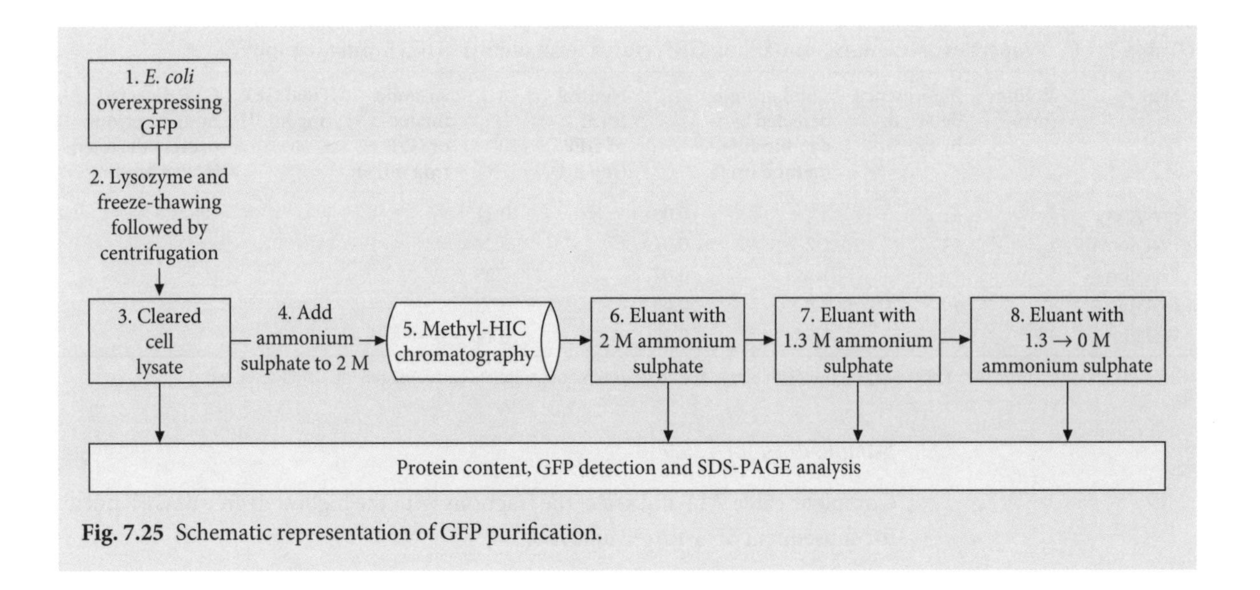

Fig. 7.25 Schematic representation of GFP purification.

Step 4

Ammonium sulphate (2 M) is added to the cell lysate to promote the hydrophobic interaction between GFP and the methyl group of the HIC column in Steps 5–7.

Steps 5–7

GFP binds to the HIC column and non-specifically bound contaminating proteins are removed by applying elution buffers containing 2 M and then 1.3 M ammonium sulphate.

Step 8

GFP is specifically eluted by lowering the ammonium sulphate concentration (1.3 – 0 M) of the eluant thus promoting GFP–solvent interactions.

Purification analysis

The fluorescent properties of GFP simplify the detection of this protein during its expression and purification; the protein is readily detected by eye using a UV light source. The unique absorbance spectrum of GFP can also be used to detect and quantitate the protein during its purification.

Fractions collected from the methyl-HIC column exhibiting the greatest specific GFP content, should be analysed by SDS-PAGE to determine the purity of the sample and to confirm the presence of an intact 27-kDa protein.

Table 7.11 Properties of fractions, containing GFP, eluted form methyl-HIC chromatography

Step	Volume (mL)	Fluoresence detected by eye	Total protein detected by dye-binding method (mg)	A_{395}	Neutral form of GFP (mg mL^{-1})	A_{475}	Anionic form of GFP (mg mL^{-1})	Total GFP (mg mL^{-1})	GFP detected by absorption/ protein detected by dye binding
Fraction 2	1	+	4	0.08		0.11			
Fraction 4	1	++	1.2	0.07		0.28			
Fraction 6	1	+++	1.3	0.07		0.29			
Fraction 8	1	+	0.5	0.03		0.12			
Fraction 10	1	+	1.5	0.03		0.14			

*Fractions are eluted from the methyl-HIC column with buffer devoid of ammonium sulphate and subsequently UV illuminated.

Sample data for 7.3.7

Complete Table 7.11 and select the fractions with the highest degree of GFP purity for subsequent SDS-PAGE analysis.

7.3.8 His-tagged protein

Theory

Recombinant DNA technology enabled the expression of proteins in heterologous systems. Subsequently, the purification of recombinant proteins was simplified by fusing purification tags to recombinant proteins (Chapter 5, section 5.4.2). There are a number of purification tags, as outlined in the earlier description of affinity chromatography in section 7.2.2. The most commonly used purification tag is the His-tag which is composed typically of six consecutive histidine residues. Histidine, cysteine, and tyrosine residues are known to interact specifically with divalent metal ions such as Ni^{2+}, Cu^{2+}, Co^{2+}, and Zn^{2+} and it is this specific interaction which forms the basis of the purification of recombinant proteins with a His-tag. Divalent metal ions, most commonly Ni^{2+}, are immobilized on a chromatographic support presenting a ligand, which will bind the His-tagged recombinant protein. Selective elution is achieved by the addition of imidazole, which acts as a histidine mimic and competes for the Ni^{2+}-binding sites.

Details are given in the Practical Activity 7.8, which is available for lecturers in the Online Resource Centre for this text to support teaching of this topic.

Purification rationale

In this protocol, we present a general method for the isolation of a His-tagged protein from an overexpressing recombinant *E. coli* strain. The abundance of the His-tagged protein in the starting material, together with the ease of a single Ni^{2+} affinity chromatograpy step, makes this a favourable approach. *E. coli* cells are grown, induced to generate large quantities of the His-tagged protein, and then harvested by centrifugation. The action of lysozyme and sonication are needed to disrupt the tough cell wall present in *E. coli*. A second centrifugation step is used to remove cell debris and generate a clear lysate which can be applied to a Ni^{2+} affinity chromatograpy column. The lysate is applied under high salt, low imidazole, conditions to promote protein solubility and His-tagged protein-immoblized Ni^{2+} interactions. Non-specifically bound proteins are removed by retaining the high salt, low imidazole, elution conditions. The His-tagged protein is selectively eluted by applying an increasing concentration of imidazole. Imidazole, a histidine mimic, binds to the Ni^{2+} coordination sites of the affinity column and displaces the His-tagged protein.

Experimental design

This protocol presents a general method for the isolation of His-tagged proteins from *E. coli*. However, it should be noted that the conditions may require optimization by varying the pH, ionic strength and/or imidazole concentration of the binding and elution buffers (see Fig. 7.26).

Step 1

E. coli cells are grown and protein expression induced.

Step 2

E. coli cells are ruptured by treatment with lysozyme and sonication.

Step 3

A centrifugation step is used to remove cell debris.

Step 4

Purification of His-tagged proteins by Ni^{2+} affinity chromatography requires optimal binding of the His-tagged protein to the immobilized Ni^{2+}. This is achieved by dilution of cell lysate into binding buffer (low concentration of imidazole, 0.5 M NaCl, 20 mM sodium phosphate buffer, pH 7.4–7.6).

Step 5

Non-specifically bound contaminating proteins are eluted from the affinity column by continuing to apply binding buffer.

Step 6

An increasing imidazole concentration should selectively release a good yield of pure His-tagged protein.

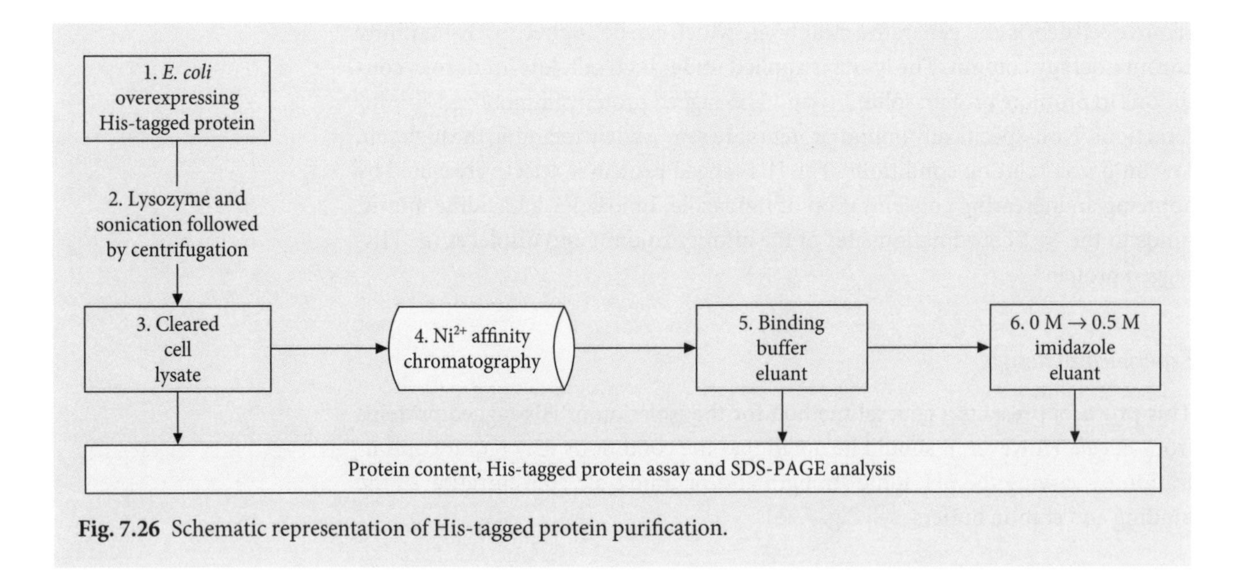

Fig. 7.26 Schematic representation of His-tagged protein purification.

Purification analysis

The A_{280} values of the fractions collected following His-tagged protein elution should indicate the presence of a protein peak. It is worth noting that imidazole absorbs to a small extent at 280 nm and so imidazole-containing elution buffer should be used as the reference solution for absorption measurements. Fractions that represent the start, mid-point, and end of the putative protein peak should be analysed by SDS-PAGE. If the protein fused to the His-tag is an enzymatic protein, the activity of each of the fractions should be analysed to identify the fractions with the highest specific activity. Alternatively, a band on an SDS-PAGE gel with a relative molecular mass similar to the predicted theoretical molecular mass of the His-tagged protein may suggest the presence of the His-tagged protein. However, such deductions should be treated with some caution. One method routinely used to detect the presence of His-tagged proteins is Western blotting with an anti-His-tag antibody. This approach can be used to detect and quantify the purified His-tagged protein.

Sample data for 7.3.8

The process of overexpressing a protein enhances the proportion of protein present within the cell and effectively acts as a purification process. For example, an overexpressed His-tagged protein may constitute 10% of the total cell protein, whereas in its native state and native source it may only constitute <0.02% of the total cell protein, which equates to a 500-fold enhancement in the specific activity of the cell lysate. As a result it is not entirely appropriate to construct a purification table for overexpressed proteins. However, it is a useful tool to analyse the efficiency of subsequent purification steps.

Table 7.12 Purification table for His-tagged chorismate mutase

Step	Total protein (mg)	Total activity (units)	Specific activity (units mg^{-1})	Yield (%)	Purification factor
Cell lysate	525	210			
Imidazole eluant	5.3	160			

In this example, the enzyme chorismate mutase (EC 4.2.3.5) from *Neurosporsa crassa* was overexpressed with a poly-histidine tag at the C-terminus. The cell lysate was applied to a metal affinity column and the His-tagged chrorismate mutase was selectively eluted with 0.5 M imidazole. Complete Table 7.12 and calculate what percentage of the total cell protein is tagged chorismate mutase in the cell lysate given that the specific activity of the purified native form of the enzyme is 32 units mg^{-1}.

SELF TEST ?

Check that you have mastered the key concepts at the start of this section by attempting the following question.

ST 7.12 Calculate the quantities of NaCl (molecular mass 58.4 Da) and imidazole (molecular mass 68.1 Da) to add to 20 mM sodium phosphate buffer, pH 7.4 to make:

(a) 50 mL binding buffer: 20 mM sodium phosphate buffer, pH 7.4, 0.5 M NaCl, 40 mM imidazole

(b) 5 mL elution buffer, 20 mM sodium phosphate buffer, pH 7.4, 0.5 M NaCl, 0.5 M imidazole.

Answer

ST 7.12

(a) NaCl: $0.5 \times 58.4 \times 0.05 = 1.46$ g

Imidazole: $0.04 \times 68.1 \times 0.05 = 0.136$ g

(b) NaCl $0.5 \times 58.4 \times 0.005 = 0.146$ g

Imidazole $0.5 \times 68.1 \times 0.005 = 0.17$ g

7.4 Conclusions

The purification procedures outlined in this chapter provide a small sample of the diverse range of techniques which exploit the unique biochemical characteristics of individual proteins to facilitate their purification from complex biochemical sources. A good purification protocol will generate a high yield of pure protein which retains maximum activity. The success of a novel purification protocol can be improved by:

1) Understanding the biochemical properties of proteins and how these can be exploited for purification purposes (section 7.2).
2) Learning lessons from previous studies of closely related proteins, e.g. optimal stability conditions and suitable chromatography media.
3) Exploiting amino acid sequence information, either of the target protein or a closely related protein. This will provide the predicted properties such as molecular mass, pI value, domain types, and possible post-translational modifications, see Chapter 1, section 1.3.2. These predicted properties are extremely useful in selecting suitable purification procedures.
4) An ability to calculate specific activities and to complete purification tables (Chapter 3, sections 3.8 and 3.9).

References for Chapter 7

Bylkas, S.A. and Andersson, L.A. (1997) *J. Chem. Educ.* **74**, 426–30.

Hearn, M.T. and Acosta, D. (2001) *J. Mol. Recog.* **14**, 323–69.

Illingworth, J.A. (1972) *Biochem. J.* **129**, 1119–24.

Lian, D.-J., Li, L., and Xu, G.-J. (1996) *Acta Biochim. Biophys. Sinica* **28**, 396–403.

North M.J. and Beynon R.J. (2001) Prevention of unwanted proteolysis. In: *Proteolytic enzymes*, 2nd edn, Benyon, R. and Bond, J.S., eds. Oxford University Press, Oxford, pp. 211–32.

Scopes, R.K. (1977) *Biochem. J.* **161**, 252–63.

Tomokiyo, K., Yano, H., Imamura, M., Nakano, Y., Nakagaki, T., Ogata, Y., Terano, T., Miyamoto, S., and Funatsu, A. (2003) *Vox Sanguinis* **84**, 54–64.

Wachter R.M. (2006) *Phytochem. Phytobiol.* **82**, 339–44.

Solutions for sample data in Chapter 7

Sample data 7.3.1

The following data were obtained during the purification of rabbit muscle LDH. Table 7.6 is completed as follows.

Step	Volume (mL)	Protein content (mg mL^{-1})	Total protein (mg)	Total LDH activity (μmol min^{-1})	Yield (expressed as %)	Specific activity (μmol min^{-1} mg^{-1})	Purification factor
Cell extract	45	40	1800	1380	100	0.8	1
Dialysed ammonium sulphate precipitate	10	57	570	1100	80	1.9	2.4
PEP eluant	20	1.15	23	5	0.4	0.2	—
NAD^+ eluant	10	0.23	2.3	950	69	413	217

Sample data 7.3.2

The following data were collected during the purification of yeast ADH. Table 7.7 is completed as follows.

Step	Volume (mL)	Protein content (mg mL^{-1})	Total protein (mg)	Total ADH activity (μmol min^{-1})	Yield (expressed as %)	Specific activity (μmol min^{-1} mg^{-1})	Purification factor
Cell extract	100	25	2500	6250	100	2.5	1
KCl eluant	50	20	1000	4000	64	4	1.6
NAD^+ eluant	40	0.15	6	3000	48	500	125

Sample data 7.3.3

Purification of ICDH generated the data in the following table. Table 7.8 is completed as follows.

Step	Volume (mL)	U mL^{-1}	Total Units	mg mL^{-1}	Total protein (mg)	Specific activity (U mg^{-1})	Yield (%)
Cell lysate	150	1.1	165	15	2250	0.073	100
Calcium phosphate precipitation supernatant	148	1.0	148	12	1776	0.083	89
Dialysed ammonium sulphate precipitate	10	14	140	11	110	1.27	85
CM-cellulose eluant	8	15	120	0.3	2.4	50	72

Sample data 7.3.4

Table 7.9 is completed as follows.

Purification step	Volume of sample (mL)	Protein concentration (mg mL^{-1})	Total protein (mg)	GST activity (ΔA_{350} min^{-1} per 10 μL sample)	Total activity (μmol min^{-1})	Specific activity (μmol min^{-1} mg^{-1})
Cell extract	10	1.53	15.3	0.12*	62.5	4.1
Run through	10	0.96	9.6	0.12	12.5	1.3
Wash	10	0.14	1.4	0.04	4.1	2.9
GST eluant	4	1.1	4.4	0.18*	37.5	8.5

*10 μL of 1/5 dilution of cell extract and GST eluant were used in the activity measurements.

As an example of the calculation of activity, consider the data for the cell extract. A ΔA_{350} min^{-1} of 0.12 corresponds to a rate of product formation of 0.12/9600 M/min, i.e. 12.5 μM min^{-1}. In 1 mL reaction volume, this is equivalent to 0.0125 μmol min^{-1}. This is for 10 μL of a 1/5 dilution, so in 10 mL of undiluted cell extract, the total activity is $0.0125 \times 5 \times 10/0.01$ μmol min^{-1} = 62.5 μmol min^{-1}. The total protein in the cell extract is 10×1.53 mg = 15.3 mg. Hence the specific activity is 62.5/15.3 = 4.1 μmol min^{-1} mg^{-1}. It should be noted that the purification procedure only leads to a 2.1-fold increase in specific activity compared with the cell extract; however, it must be remembered that the extract is highly enriched in GST, which is overexpressed in this system. For further details about purification tables see Chapter 3, section 3.9.

Sample data 7.3.5

Given that 1 in 200 dilution of steak lysate has $A_{417} = 0.64$ and $A_{409} = 0.45$, calculate the concentration of oxy- and met-myoglobin in the lysate.

Answer

If $A_{417} = 0.64$ following a 1:200 dilution, undiluted absorbance value is 128. 1 mM oxy-myoglobin has an absorption coefficient of 128, suggesting that the lysate contains 1 mM oxy-myoglobin.

If $A_{409} = 0.45$ following a 1:200 dilution, undiluted absorbance value is 90.

1 mM met-myoglobin has an absorption coefficient of 179, suggesting that the lysate contains $90/179 = 0.5$ mM met-myoglobin.

Sample data 7.3.6

SDS-PAGE analysis, under reducing conditions, of purified IgM reveals the presence of two protein species with apparent molecular mass values of 25.5 and 60 kDa plus a less abundant species of 58 kDa. All three bands cross-react with an anti-porcine IgM antibody. Comment on the nature of all three species and describe the experiments you would conduct to confirm this.

Solution

The heavy chain of IgM has a molecular mass of 55–75 kDa, implying that the band at 60 kDa could be the heavy chain.

The light chain of IgM has a molecular mass of 25 kDa, implying that the band at 25.5 kDa could be the light chain.

The fact that both the 60 and 25.5 kDa bands cross-react with the anti-IgM antibody also supports this.

The less abundant species at 58 kDa also cross-reacts with anti-IgM suggesting that this species may represent a degraded form of the heavy chain. This could be confirmed by mass spectroscopy.

Sample data 7.3.7

Table 7.11 is completed as follows.

Step	Volume (mL)	Fluorescence detected by eye	Total protein detected by dye-binding method (mg)	A_{395}	Neutral form of GFP (mg mL^{-1})	A_{475}	Anionic form of GFP (mg mL^{-1})	Total GFP (mg mL^{-1})	GFP detected by absorption/ protein detected by dye binding
Fraction 2	1	+	4	0.08	0.07	0.11	0.42	0.49	0.12
Fraction 4	1	++	1.2	0.07	0.06	0.28	1.08	1.14	0.95
Fraction 6	1	+++	1.3	0.07	0.06	0.29	1.12	1.18	0.91
Fraction 8	1	+	0.5	0.03	0.03	0.12	0.46	0.49	0.98
Fraction 10	1	+	1.5	0.03	0.03	0.14	0.54	0.57	0.38

*Fractions are eluted from the methyl-HIC column with buffer devoid of ammonium sulphate and subsequently UV illuminated.
The fluorescence scoring system ranges from +++ to +, with +++ being the strongest fluorescence.
On the basis of these data, Fractions 4, 6 and 8 should be selected for further analysis.

Sample data 7.3.8

Table 7.12 is completed as follows.

Step	Total protein (mg)	Total activity (Units)	Specific activity (Units mg^{-1})	Yield (%)	Purification factor
Cell lysate	525	210	0.4	100	1
Imidazole eluant	5.3	160	30.2	76	75

If there are 210 Units in the cell extract, then we can determine that 210/32 mg of tagged chorismate mutase in the extract, i.e. 6.5 mg.

% total cell protein = (6.5/525) × 100% = 1.2% total cell protein is tagged chorismate mutase.

Appendix

Appendix 7.1 Table for ammonium sulphate precipitation

The table indicates the correct amount of ammonium sulphate (grams) to be added to 1 L of solution to produce a desired change in the percentage saturation of ammonium sulphate. For example, to find out how much ammonium sulphate is to be added to bring a solution which is 25% saturation up to 50% saturation, we look down the left hand column to 25, then along the row to the heading 50; the result is 158 g L^{-1}.

Initial	Final percentage saturation of ammonium sulphate																			
	10	15	20	25	30	33	35	40	45	50	55	60	65	70	75	80	85	90	95	100
0	56	84	114	144	176	196	209	243	277	313	351	390	430	472	516	561	610	662	713	767
10		28	57	86	118	137	150	183	216	251	288	326	365	406	449	494	540	592	640	694
15			28	57	88	107	120	153	185	220	256	294	333	373	415	459	506	556	605	657
20				29	59	78	91	123	155	189	225	262	300	340	382	424	471	520	569	619
25					30	49	61	93	125	158	193	230	267	307	348	390	436	485	533	583
30						19	30	62	94	127	162	198	235	273	314	356	401	449	496	546
33							12	43	74	107	142	177	214	252	292	333	378	426	472	522
35								31	63	94	129	164	200	238	278	319	364	411	457	506
40									31	63	97	132	168	205	245	285	328	375	420	469
45										32	65	99	134	171	210	250	293	339	383	431
50											33	66	101	137	176	214	256	302	345	392
55												33	67	103	141	179	220	264	307	353
60													34	69	105	143	183	227	269	314
65														34	70	107	147	190	232	275
70															35	72	111	153	194	237
75																36	74	115	155	198
80																	38	77	117	157
85																		39	77	118
90																			38	77
95																				39

8 Determination of protein structure

8.1 Introduction: scope of chapter

The structure-based drug design approach involves the use of structural information on the target protein or enzyme to guide the development of drugs that are likely to interact with the target. This is in contrast with the random screening of large numbers of compounds to find one which may act as a useful drug.

Good introductory accounts of X-ray crystallography and NMR are given in the book by Sheehan, 2000.

In Chapter 7 we explored the various methods used for purification of proteins and for establishing the homogeneity of the preparations obtained. This is a prerequisite for studying the properties of the protein and hence understanding the basis of its biological activity in molecular terms. This type of understanding leads to the possibility of manipulating the activity either by altering the nature of key amino acids in the enzyme (see Chapter 9, section 9.10) or by designing molecules that can alter the activity by binding at key sites. The latter approach is the basis of structure-based drug design, which has led to the development of many therapeutically useful compounds such as captopril for treatment of hypertension, simvastatin for control of blood cholesterol levels, and inhibitors of HIV protease.

There has been spectacular progress in the field of structural biology over the past two decades so that many tens of thousands of three-dimensional structures of proteins at atomic resolution have now been deposited in the Protein Data Bank (PDB; http://www.rcsb.org) (see Chapter 5, section 5.8.1); the large majority of these structures have been determined by X-ray crystallography, the remainder by nuclear magnetic resonance (NMR). These methods are becoming more accessible to the biochemical community by developments both in experimental protocols and in computer-based data analysis. However, these methods are beyond the scope of this book as they are too complex to form the basis of undergraduate laboratory practicals or of relatively easy data-handling exercises. In this chapter, we will explore various aspects of protein structure, which allow us to gain insights into the size and shapes of proteins, as well as the stability of the folded structure itself. Chapters 9 and 10 will deal with functional properties of proteins—the catalytic power of enzymes in Chapter 9 and the binding properties of proteins in Chapter 10.

The questions that we wish to answer about protein structure are as follows:

- What is the molecular mass of the protein?
- How many subunits (i.e. individual polypeptide chains) does it contain, and are they identical?

- What can we learn about the secondary, tertiary, and quaternary structures of the protein?
- How stable is the native folded structure of the protein?

The different levels of protein structure are defined in Chapter 1, sections 1.3–1.6.

8.2 Molecular mass determination

KEY CONCEPTS

- Knowing the different methods available for determining the molecular masses of proteins
- Understanding the strengths and weaknesses of each of these methods
- Explaining how post-translational modifications of proteins can be detected

The molecular mass of a protein can be determined using a number of methods, including polyacrylamide gel electrophoresis, gel filtration, analytical ultracentrifugation, and mass spectrometry. Each method differs according to:

- The amount of protein required for analysis
- The accuracy (resolution)
- The extent of retention of non-covalent interactions, which may allow the mass of oligomeric structures to be determined
- The molecular mass range.

A comparison of properties of the methods that can be used to determine the molecular mass of a protein is given in Table 8.1.

In the following sections, we shall describe each of the methods that can be used to determine the molecular mass of proteins and explore the structural information that can be gleaned from data generated by these methods.

8.2.1 SDS-PAGE

Sodium dodecylsulphate (SDS)-PAGE (Chapter 6, section 6.2.1) is one of the most routinely used methods for determining the molecular mass of protein subunits, in particular, the method developed by Laemmli (1970). SDS-PAGE separates proteins in a polyacrylamide gel according to their molecular mass. Proteins destined for SDS-PAGE analysis are treated with the anionic detergent SDS, which has two effects:

1) The proteins adopt a uniform rod shape.
2) The proteins have a constant negative charge:size ratio.

Table 8.1 Properties of methods employed to determine the molecular mass of proteins (the quantities in bracket are those for a 50-kDa protein)

Method	Scale	Resolution	Molecular mass range	Retention of non-covalent interactions
Polyacrylamide gel electrophoresis	Coomassie Brilliant Blue stain 2–40 pmol/lane (0.1–2 μg) Silver stain 20 fmol/lane (1 ng) Western blot 2 fmol/lane (0.2 ng)	1–2% (0.5–1 kDa)	SDS-PAGE 1–200 kDa Non-denaturing 10–1500 kDa	Not retained in SDS-PAGE Retained in non-denaturing PAGE
Gel filtration	20–100 nmol (1–5 mg)	Globular 2–10% (1–5 kDa) Non-globular >10% (>5 kDa)	1–10 000 kDa	Retained in the absence of denaturants Not retained in the presence of denaturants
Analytical ultracentrifugation	0.2–2 nmol (10–100 μg)	1–5% (0.5–2.5 kDa)	10–10 000 kDa	Retained in the absence of denaturants Not retained in the presence of denaturants
Mass spectrometry	Electrospray 10–100 pmol (0.5–5 μg) MALDI 1–10 pmol (50–500 ng)	<0.01% (<5 Da)	Electrospray less than 100 kDa MALDI less than 500 kDa	Typically, not retained due to ionization techniques

Bromophenol blue is a dye which is applied to SDS-PAGE samples to enable visualization of the sample during gel loading and to track the progress of the protein samples during electrophoresis as it migrates ahead of the protein samples.

As a result, when an electric field is applied to an acrylamide gel loaded with SDS-treated proteins, the proteins migrate towards the anode according to their size only, with the log (molecular mass) related in a linear fashion to the relative distance migrated in the polyacrylamide gel (see Fig. 6.8). The mobilities (distances migrated) are typically expressed relative to the distance migrated by bromophenol blue; the ratio of the mobility of a protein to that of the bromophenol blue is denoted by R_f.

The steps followed to determine the molecular mass of an unknown protein are:

1) SDS-PAGE analysis of the unknown protein alongside a series of proteins of known molecular mass, known as marker proteins. It is important to select a polyacrylamide gel with a suitable pore size to optimize the resolution of the protein of interest and the marker proteins (i.e. the gel can separate two proteins, which differ in mass by more than 0.5–1 kDa). This can be achieved by

varying the relative proportions of acrylamide and bis-acrylamide in the gel mixture, with effective gel separation ranges and acrylamide concentration detailed in Table 6.2, Chapter 6.

It is also important to select marker proteins which span mobilities greater than and less than the unknown protein to allow an accurate estimate of the molecular mass in steps 2 and 3.

2) construction of a calibration plot showing the mobilities of the marker proteins against their log (molecular mass).

3) measurement of the mobility of the unknown protein in the acrylamide gel and use of the calibration plot generated in step 2 to determine the molecular mass of the unknown protein.

WORKED EXAMPLE 

Three peripheral membrane proteins, namely α-SNAP, β-SNAP, and γ-SNAP, were extracted by treating bovine brain cell membranes with high concentrations of KCl. The mobility of each of these proteins in SDS-PAGE gels was determined, together with marker proteins, see Table 8.2 below. Calculate the apparent molecular masses of α-SNAP, β-SNAP, and γ-SNAP.

Table 8.2 SDS-PAGE mobility properties of marker proteins and α-SNAP, β-SNAP, and γ-SNAP

Protein	Mobility (cm)	Molecular mass (kDa)
Phosphorylase *b*	1.25	97
Bovine serum albumin	2.10	66
Ovalbumin	3.70	43
Carbonic anhydrase	6.10	31
Trypsin inhibitor	8.20	21.5
α-SNAP	5.30	
β-SNAP	5.20	
γ-SNAP	4.85	

Table 8.3 SDS-PAGE mobility properties and log (molecular mass) of marker proteins and α-SNAP, β-SNAP, and γ-SNAP

Protein	Mobility (cm)	Molecular mass (kDa)	Log (molecular mass)
Phosphorylase b	1.25	97	1.99
Bovine serum albumin	2.10	66	1.82
Ovalbumin	3.70	43	1.63
Carbonic anhydrase	6.10	29	1.46
Trypsin inhibitor	8.20	21.5	1.33
α-SNAP	5.30	35	1.54
β-SNAP	5.20	36	1.56
γ-SNAP	4.85	39	1.59

STRATEGY

Using the data for the marker proteins, plot mobility against log (molecular mass).

SOLUTION

The values of mobility and log (molecular mass) for the marker proteins are included in Table 8.3; these are plotted in Fig. 8.1. It would appear that α-SNAP, β-SNAP, and γ-SNAP each has a unique molecular mass of 35, 36, and 39 kDa, respectively, showing that the three proteins have distinctly different structures.

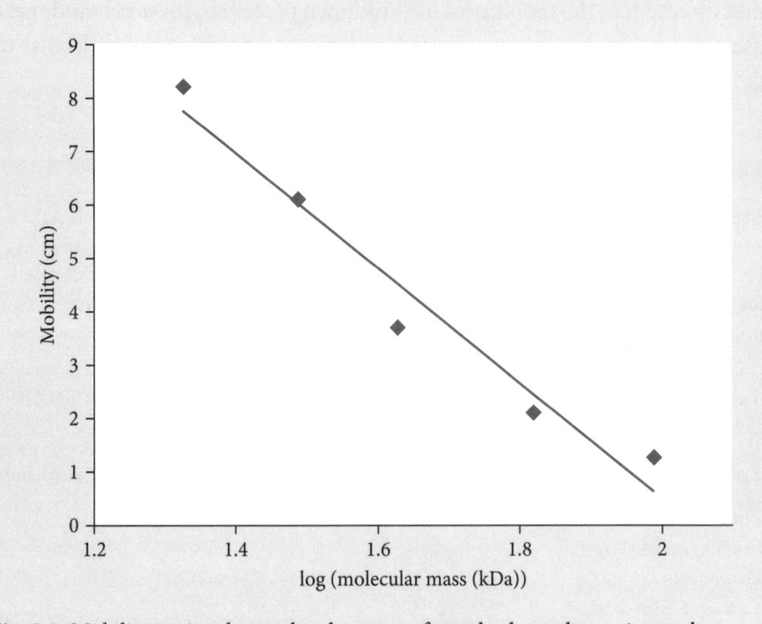

Fig. 8.1 Mobility against log molecular mass of standards used to estimate the molecular mass of α-SNAP, β-SNAP, and γ-SNAP.

Following gel electrophoresis, protein bands are detected routinely by Coomassie Brilliant Blue staining. Coomassie Brilliant Blue binds to proteins in the gel allowing the rapid detection and visualization of as little as 0.1–2 μg protein, see Chapter 6, section 6.2.1.

SDS-PAGE combined with Coomassie Brilliant Blue detection of proteins is a very convenient method for establishing the molecular mass of protein subunits with a margin of error of 1–2%.

In primary structure terms, 1–2% of a 50-kDa protein equates to 4.5–9 amino acids (see Chapter 2, section 2.1). Thus, SDS-PAGE combined with Coomassie Brilliant Blue staining could also be used to detect the reduction in mass associated with the loss of four or more amino acids due to post-translational processing or due to unwanted proteolysis. Phosphoglycerate mutase from *Saccharomyces cerevisiae* has a 14 amino acid C-terminal tail, which is highly flexible and susceptible to proteolysis. Fig. 8.2 indicates that it is possible to monitor the loss of this

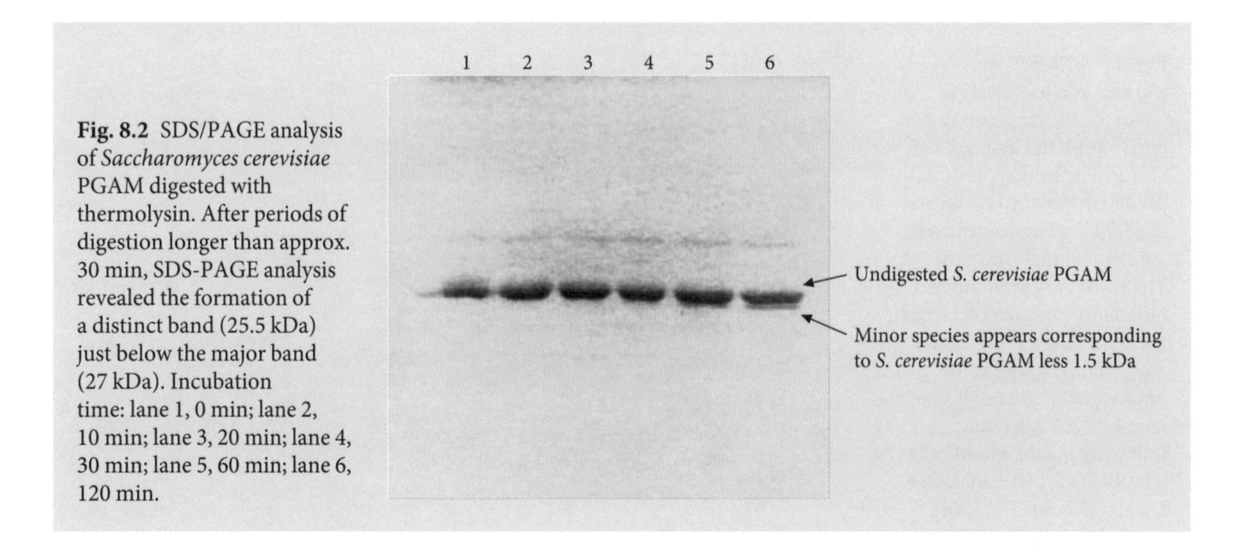

Fig. 8.2 SDS/PAGE analysis of *Saccharomyces cerevisiae* PGAM digested with thermolysin. After periods of digestion longer than approx. 30 min, SDS-PAGE analysis revealed the formation of a distinct band (25.5 kDa) just below the major band (27 kDa). Incubation time: lane 1, 0 min; lane 2, 10 min; lane 3, 20 min; lane 4, 30 min; lane 5, 60 min; lane 6, 120 min.

C-terminal tail, following treatment with thermolysin, using SDS-PAGE combined with Coomassie Brilliant Blue staining (Walter *et al.*, 1999).

SDS-PAGE combined with more specific detection methods can be used to monitor a range of post-translational modifications. While these changes result in a mass change, the mass difference between the modified and unmodified protein cannot usually be resolved by a Coomassie Brilliant Blue-stained SDS-PAGE gel but requires SDS-PAGE combined with a post-translational-specific detection method.

Coomassie Brilliant Blue is also known as Coomassie Blue. The Coomassie dye was developed as a wool dye and named to commemorate the British occupation in 1896 of the city now known as Kumasi in Ghana during the Anglo-Ashanti wars.

In gel (in situ) detection methods

There are a number of in-gel detection methods that can be used to detect specific post-translational modifications, such as phosphorylation, glycosylation, and lipid addition. The detection reagents react with the specific post-translational modification groups of proteins separated by SDS-PAGE and generate a colour change that can be detected visually, for example detection of glycoproteins using the Gel Code Glycoprotein Detection Kit (Pierce). Initially, glycoproteins are treated with an oxidizing agent (periodic acid) resulting in aldehyde formation. Subsequent addition of staining reagent followed by a reducing agent results in the development of magenta bands where glycoproteins are present.

Immunoblot detection methods

The availability of antibodies that can detect post-translationally modified amino acids makes it possible to combine SDS-PAGE with immunoblotting (Chapter 6, section 6.2.1) to characterize proteins, which may be ubiquitinated, acetylated, O- or N-glycosylated, or have a phosphorylated tyrosine residue. Antibodies which detect poly-histidine residues are used to detect the presence of His-tagged

Immunoblotting is also known as Western blotting.

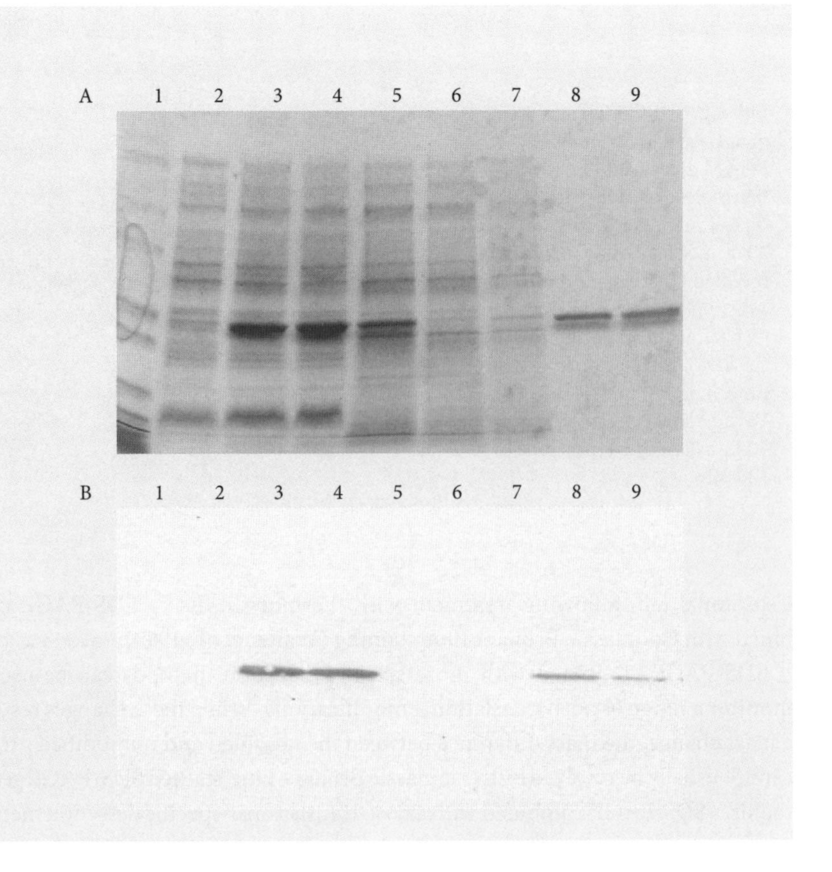

Fig. 8.3 SDS-PAGE (A) and Western analysis (B) of the overexpression and purification of His-tagged α-SNAP. The primary antibody, used to detect the His-tagged α-SNAP, is a mouse anti-His tag monoclonal antibody. The secondary antibody is a rabbit anti-mouse polyclonal antibody. Lane 1, molecular mass markers; lane 2, uninduced *E. coli* cell lysate; lane 3, *E. coli* cell lysate following incubation of cells with IPTG for 60 min; lane 4, *E. coli* cell lysate following incubation of cells with IPTG for 120 min; 5, Ni-NTA eluent with 5 mM imidazole; lane 6, Ni-NTA eluent with 20 mM imidazole; lane 7, Ni-NTA eluent with 50 mM imidazole; lane 8, Ni-NTA eluent with 100 mM imidazole; lane 9, Ni-NTA eluent with 150 mM imidazole.

proteins overexpressed in heterologous systems (Chapter 5, section 5.4.2). This is particularly useful if the activity of the overexpressed protein is difficult to assay, for example the α-soluble NSF attachment protein (α-SNAP) which is involved in membrane fusion. The expression of α-SNAP with a His-tag at the C-terminus, combined with the use of an anti-His-tag antibody, allows the rapid detection of this protein (see Fig. 8.3).

Typically, standard molecular mass estimations by SDS-PAGE involve the presence of a reducing agent, such as 2-mercaptoethanol. The reducing agent promotes the reduced state of cysteine side chains and minimizes the formation of disulphide bonds within, or between, subunits. However, analysis of native protein–protein interactions, involving disulphide bonds, requires that SDS-PAGE is performed both in the presence and in the absence of reducing agents. Reducing SDS-PAGE will generate linearized protein momomers, with fully reduced cysteine side chains, which will provide an estimate of the molecular mass of the monomer. Non-reducing SDS-PAGE will generate proteins with the retention of disulphide bonds within, and between subunits (see Fig. 8.4). This can be a useful tool in determining the nature of the disulphide bonds and characterizing the quaternary structure of proteins. An example of this is the use of reducing and non-reducing SDS-PAGE to explore the properties of immunoglobuins (see Fig. 7.22, Chapter 7).

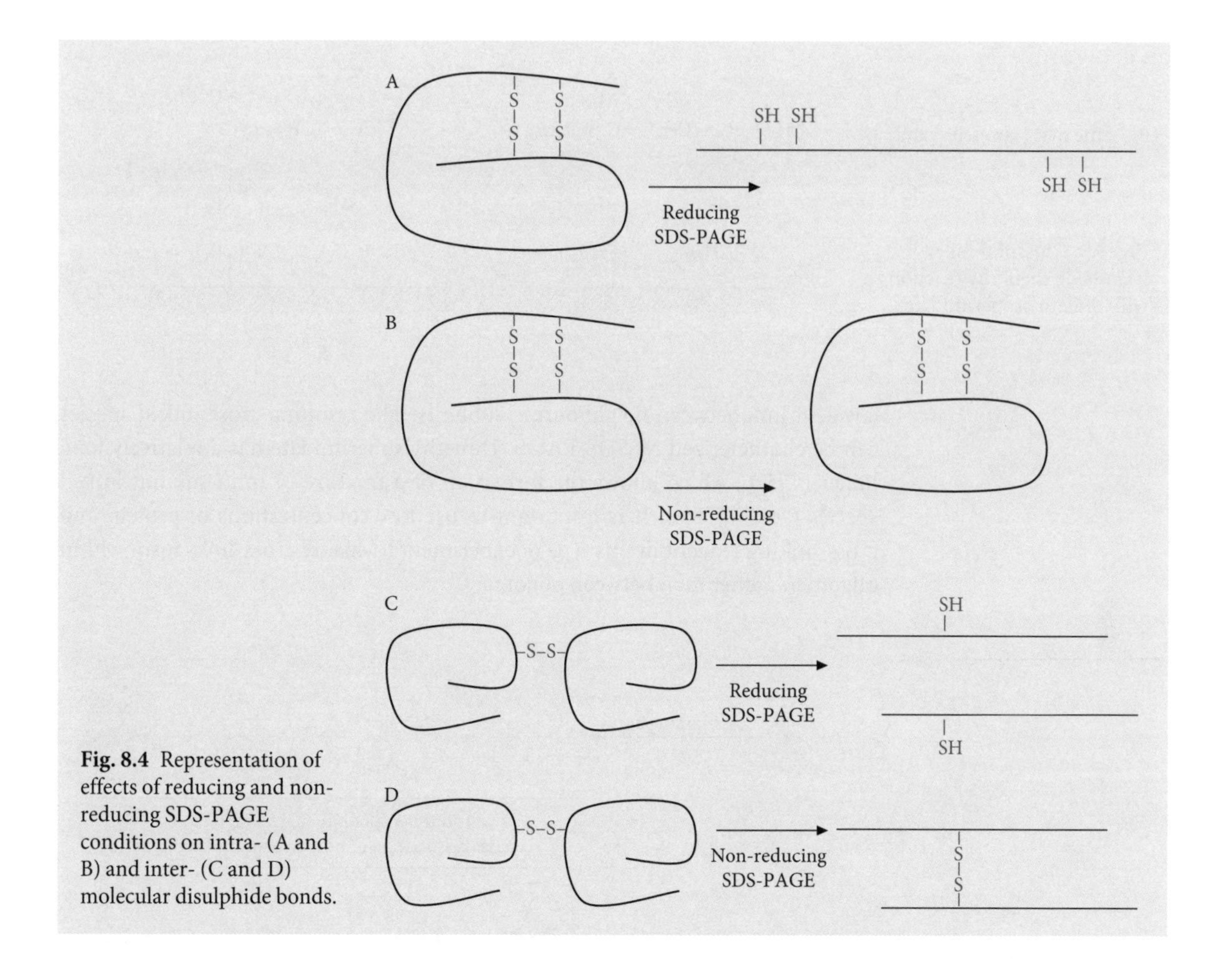

Fig. 8.4 Representation of effects of reducing and non-reducing SDS-PAGE conditions on intra- (A and B) and inter- (C and D) molecular disulphide bonds.

SDS-PAGE combined with protein cross-linking can be used to estimate the native molecular mass of oligomeric proteins (Chapter 1, section 1.6). Under appropriate conditions, a cross-linking reagent will react with the side chain on one subunit and a side chain on a neighbouring subunit to create a covalent link between the subunits. The most reactive side chains in proteins are the thiol group of cysteine residues and the amino group of lysine residues. A number of cross-linking reagents that will react with thiol or amino groups are available (Lundblad, 1995; Coggins, 1996) and these can have:

- Identical cross-linking groups connected via a simple carbon chain (linker)
- Non-identical cross-linking groups connected via a simple carbon chain
- Cross-linking groups connected via a cleavable linker (this enables the cross-linking to be reversed and facilitates the identification of the cross-linked residues).

Tris buffer contains reactive amino groups, which will react with dimethylsuberimidate. It is therefore important to use an alternative buffer system, such as triethanolamine.

One of the most commonly used reagents in the estimation of the native molecular mass of oligomeric proteins is dimethylsuberimidate. This cross-linker reacts with the amino groups of lysine residues (see Fig. 8.5) to generate an irreversible

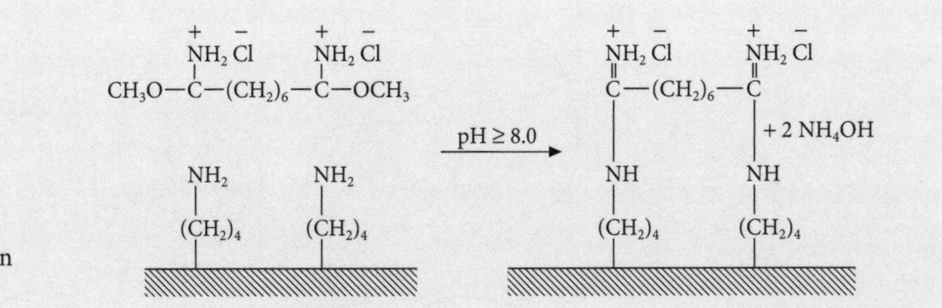

Fig. 8.5 Cross-linking of lysine side chains by reaction with dimethylsuberimidate.

covalent link between neighbouring subunits. The resulting cross-linked species can be characterized by SDS-PAGE. Dimethylsuberimidate has a relatively long linker $(CH_2)_6$, which allows the formation of a mixture of intra-subunit-linked species (see Fig. 8.6). It is important to use low concentrations of protein and cross-linking reagent in this type of experiment to ensure cross-links form within oligomers rather than between oligomers.

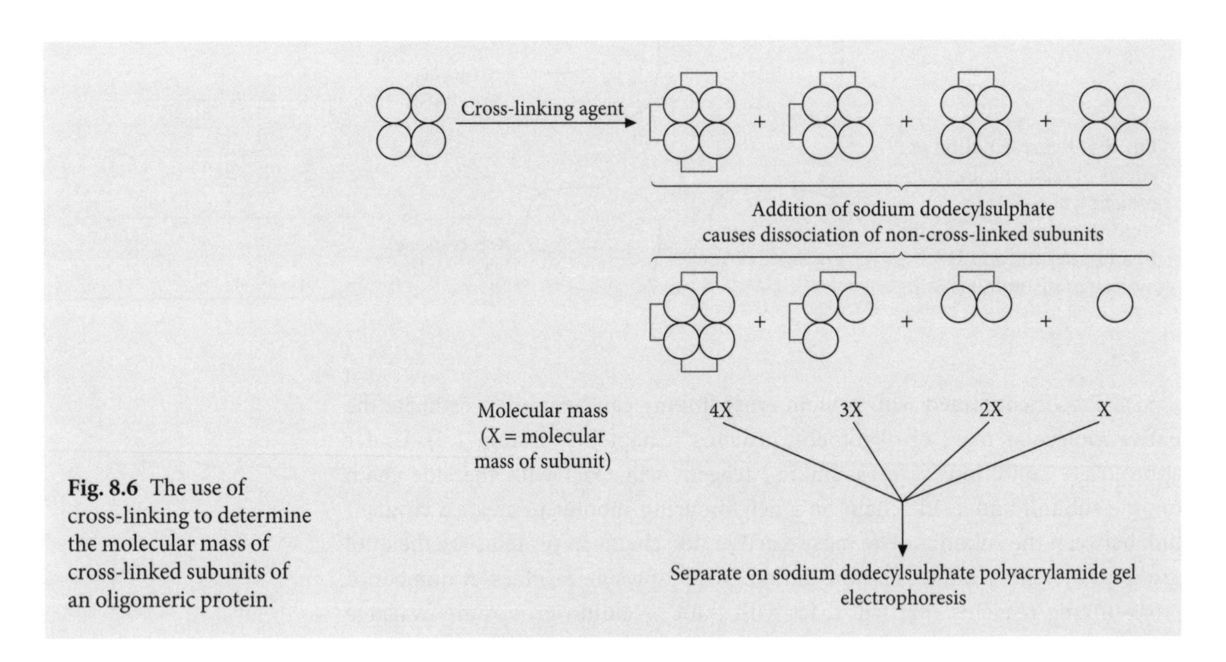

Fig. 8.6 The use of cross-linking to determine the molecular mass of cross-linked subunits of an oligomeric protein.

✓ WORKED EXAMPLE

The enzyme aldolase was treated with the cross-linking reagent dimethylsuberimidate and subjected to SDS-PAGE. The resultant gel exhibited four bands with R_f values (expressed as fractions) indicated in Table 8.4. Estimate the molecular mass of the four bands, and suggest the native molecular mass of this oligomeric enzyme and its possible quaternary structure.

Table 8.4 R_f values of protein standards and cross-linked aldolase

Sample	R_f value	Molecular mass (kDa)
Myosin	0.10	205
β-Galactosidase	0.35	116
Phosphorylase *b*	0.42	97
Fructose-6-phosphate kinase	0.48	84
BSA	0.56	66
Glutamate dehydrogenase	0.62	55
Lactate dehydrogenase	0.76	36
Carbonic anhydrase	0.90	29
Band 1	0.20	
Band 2	0.34	
Band 3	0.47	
Band 4	0.72	

Table 8.5 R_f and log (molecular mass) values of protein standards and cross-linked aldolase

Sample	R_f value	Molecular mass (kDa)	Log molecular mass
Myosin	0.10	205	2.31
β-Galactosidase	0.35	116	2.06
Phosphorylase b	0.42	97	1.99
Fructose-6-phosphate kinase	0.48	84	1.92
BSA	0.56	66	1.82
Glutamate dehydrogenase	0.62	55	1.74
Lactate dehydrogenase	0.76	36	1.56
Carbonic anhydrase	0.90	29	1.46
Band 1	0.22	160	2.21
Band 2	0.34	120	2.08
Band 3	0.47	80	1.92
Band 4	0.72	40	1.62

STRATEGY

Use the R_f values of the protein standards to construct a plot of R_f against log (molecular mass). This should generate a straight line that can be used to calculate the molecular masses of bands 1–4. To do this, use the mobility values of each band to determine log (molecular mass) and then using the anti-log function, calculate the molecular mass in Daltons (or kDa according to how the plot was done).

SOLUTION

The values of R_f and log (molecular mass) for the marker proteins are shown in Table 8.5; these are plotted in Fig. 8.7.

Cross-linking SDS-PAGE analysis suggests that bands 1–4 have mass values of 160, 120, 80, and 40 kDa, respectively. As the smallest species is 40 kDa, this is

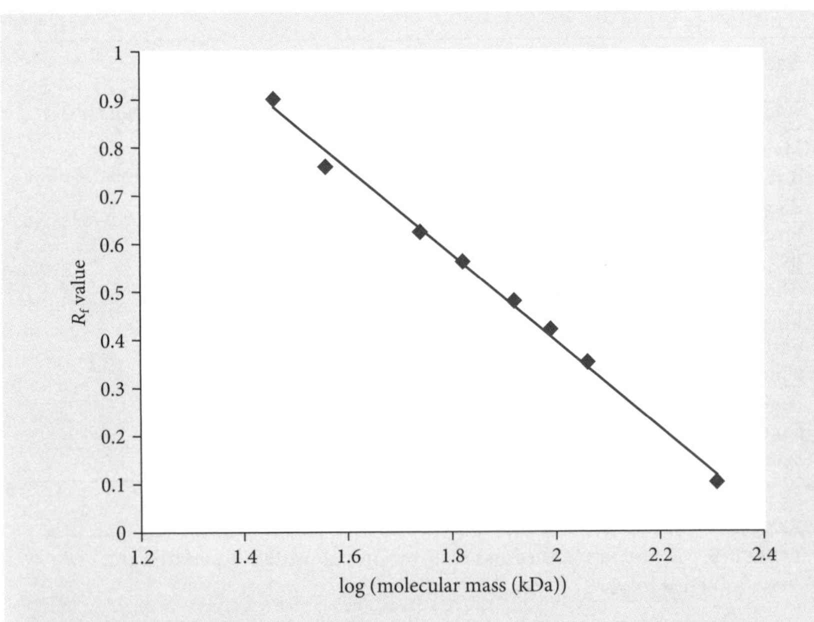

Fig. 8.7 R_f against log (molecular mass) to estimate the molecular mass of the aldolase cross-linked species.

indicative of a monomer with a molecular mass of 40 kDa. Assuming the subunits are identical in aldolase, the data suggest the presence of species with molecular masses, which equate to multiples of 40 kDa, i.e. a dimer of 80 kDa, a trimer of 120 kDa, and a tetramer of 160 kDa. As the largest species is 160 kDa it is likely that aldolase exists as tetramer of identical subunits. This could be confirmed by the use of a Ferguson plot (section 8.2.2) or gel permeation studies (section 8.2.3) of native aldolase.

8.2.2 Ferguson plots

Based on the analysis of proteins of known three-dimensional structure and disulphide bonding arrangement, 97% of oligomeric proteins do not contain disulphide bonds (Kartik *et al.*, 2006).

SDS-PAGE is a useful tool in determining the mass of protein subunits and a subset of oligomeric proteins, i.e. those with subunits that are linked covalently via disulphide bonds or cross-linking reagents. However, the majority of oligomeric proteins and protein complexes associate via non-covalent interactions. The denaturing conditions of SDS-PAGE destroy these interactions, making it impossible to gain native molecular mass estimations of these oligomeric proteins and protein complexes.

Non-denaturing gel electrophoresis analysis, combined with Ferguson plots (Chapter 6, section 6.2.2), can be used to determine the molecular mass of oligomeric proteins and protein complexes in which the subunits associate non-covalently. This technique involves the following steps:

1) Determine the relative mobility of calibration proteins, and the protein of interest, in a non-denaturing gel.

2) Repeat step 1 using a series of polyacrylamide gels, with different acrylamide concentrations.
3) For each calibration protein, plot log (relative mobility) against gel concentration to generate a straight line.
4) Plot the value of the slope of the straight line for each calibration protein, obtained in step 3, against the molecular mass of the calibration protein to create a Ferguson plot.
5) From the Ferguson plot prepared in Step 4, use the value of the slope generated by the protein of interest to provide an estimate of its molecular mass. An example of a Ferguson plot is given in Chapter 6, Fig. 6.9.

WORKED EXAMPLE 

Construct a Ferguson plot to calculate the molecular mass of the native form of aldolase using the R_f data (expressed as %) given in Table 8.6.

Table 8.6 R_f data of aldolase and marker proteins in non-denaturing SDS-PAGE, using gels of varying acrylamide gel concentrations

Protein	Molecular mass (kDa)	4% gel	6% gel	8% gel	10% gel
BSA monomer	66	—	83.0	57.5	38.9
BSA dimer	132	63.1	41.7	27.5	17.8
Catalase	232	45.7	21.9	11.0	5.4
Ferritin	443	33.9	15.1	6.3	—
Thyroglobulin	669	28.2	8.9	3.6	—
Aldolase		46.8	28.2	17.4	10

STRATEGY

Construct a Ferguson plot of log R_f values against the % acrylamide gel for each of the standard proteins and the aldolase. Then construct a plot of the slopes from the Ferguson plots against the molecular mass of each of the standard proteins.

SOLUTION

Table 8.7 shows the log R_f values at each % gel; these are plotted in Fig. 8.8.

Table 8.7 Log R_f values of aldolase and marker proteins in non-denaturing SDS-PAGE

Protein	Molecular mass (kDa)	4% gel	6% gel	8% gel	10% gel	Slope
BSA monomer	66	—	1.92	1.76	1.58	0.085
BSA dimer	132	1.80	1.62	1.44	1.25	0.092
Catalase	232	1.66	1.34	1.04	0.73	0.150
Ferritin	443	1.53	1.18	0.80	—	0.182
Thyroglobulin	669	1.45	0.95	0.55	—	0.220
Aldolase		1.67	1.45	1.24	1.05	0.103

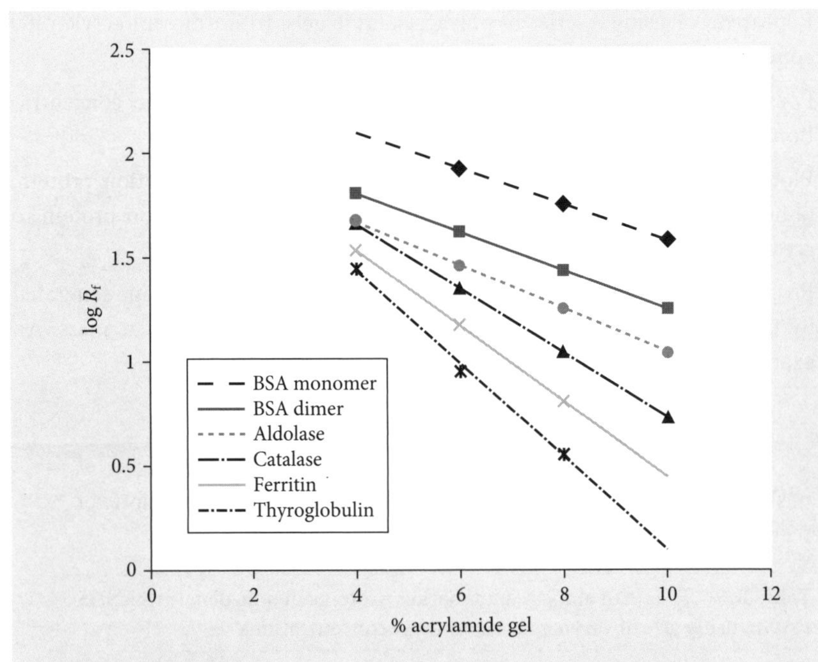

Fig. 8.8 Ferguson plot.

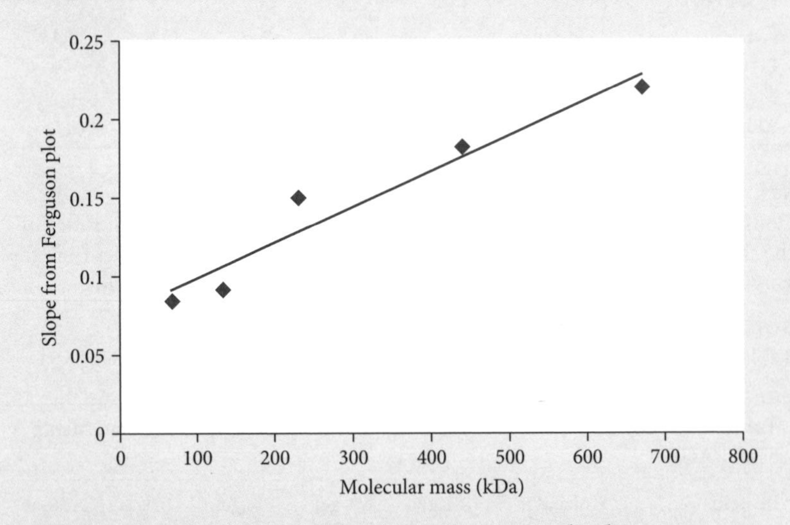

Fig. 8.9 Plot of slopes of Ferguson plots (Fig. 8.8) against molecular mass.

The slopes of the lines in the Ferguson plot are plotted against molecular mass in Fig. 8.9. For aldolase, the slopes of the plot of log R_f against % gel is 0.11, which corresponds to a molecular mass of 160 kDa. This value is in good agreement with the value obtained in the cross-linking worked example in section 8.2.1.

8.2.3 Gel filtration

Gel filtration chromatography separates proteins according to their Stokes radius and can be used to estimate the molecular mass of proteins. The basis of gel filtration as an analytical tool is the linear relationship between the logarithm of the molecular mass of a protein and its distribution coefficient, K_d. (The mathematical definition of the distribution coefficient, K_d, is given in Chapter 7, section 7.2.2.) Large globular proteins, with large Stokes radius values, fail to penetrate the gel filtration matrix beads and have low distribution coefficients. Small globular proteins, with small Stokes radius values, enter the matrix of gel filtration beads and as a result have high distribution coefficients.

The term 'Stokes radius', R_s, is used to describe the size of a protein and is a function of the shape and mass of a protein. The R_s value of a protein molecule is its average radius in solution; a large globular protein will have a larger R_s value than a small globular protein. Thus, a large globular protein will have a smaller K_d value than a small globular protein. In the case of two proteins of identical molecular mass, but different structures, we find that non-globular proteins (e.g. elongated structures) have larger R_s values, and hence smaller K_d values, than more compact globular proteins.

The stability of gel filtration media together with the wide range of possible elution conditions, make it possible to analyse proteins under either:

- *Native conditions*, with the use of buffers and solutes which promote the native state of the protein of interest. This allows the accurate determination of the molecular mass of both monomeric and oligomeric proteins

or

- *Denaturing conditions*, in which the presence of denaturants such as guanidinium chloride, urea, or detergents, promote the disruption of inter-subunit interactions, as well as intra-subunit interactions, leading to a random coil state (lack of ordered structure). Under these conditions, it is possible to obtain accurate molecular mass estimates for individual subunits

Determination of the molecular mass of a protein, under native or denaturing conditions, requires measuring the distribution coefficient (K_d value) of a range of proteins of known molecular mass, under native and denaturing conditions, respectively. Construction of a plot of the K_d values of the standard proteins against log (molecular mass) generates a standard curve, which can be used to extrapolate the molecular mass of the unknown protein from its K_d value.

Accurate molecular mass determination by gel filtration requires:

1) A gel filtration column with a suitable fractionation range.
   It is important to select a gel filtration medium, which will exhibit a linear relationship between the logarithm of the molecular mass of a range of appropriate protein standards and their distribution coefficient, K_d, values (see Fig. 8.10). Table 7.5 of Chapter 7, gives an indication of the fractionation ranges of some of the most commonly used gel filtration media.
2) A range of calibration proteins, which have mass values greater than and less than the protein of interest.
3) In the case of native gel filtration, elution buffers which promote protein stability, while minimizing protein–column interactions, are required (for example 50 mM Tris HCl, 0.15 M NaCl, pH 8.0). Identical elution conditions should be used for the standards and the unknown protein.

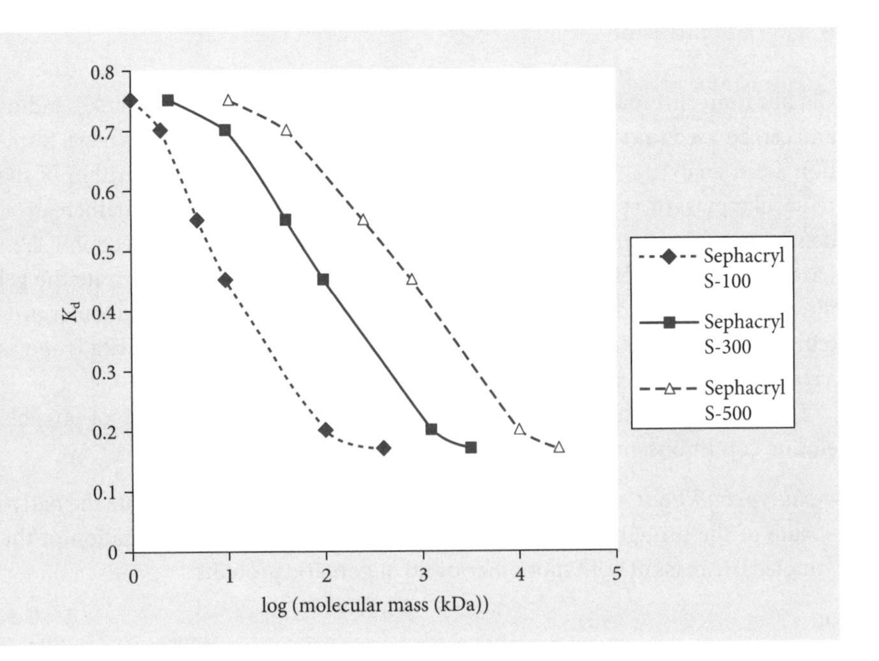

Fig. 8.10 The fractionation range of Sephacryl S-100, S-300, and S-500. The linear portion of each of the plots indicates the suitable fractionation range for each filtration medium.

4) In the case of denaturing conditions, elution buffers containing denaturant (for example 6 M guanidinium chloride) should be used for both the standard proteins and the unknown protein.

✓ WORKED EXAMPLE

The oligomeric structure of the molecular chaperone, prefoldin, was studied using gel filtration. Prefoldin was applied to a Sephacryl S-200 column, using the eluant 20 mM Tris HCl, 100 mM NaCl, pH 8.0.

A. Using the calibration data in Table 8.8, calculate the native molecular mass of prefoldin

B. SDS-PAGE analysis of prefoldin indicates the presence of two species, 15.5 and 13.5 kDa. The 13.5-kDa band is twice as intense as the 15.5-kDa band. Taking the SDS-PAGE analysis data together with the gel filtration data, suggest a possible oligomeric structure for prefoldin.

Table 8.8 K_d **values of prefoldin and protein standards obtained following Sephacryl S-200 gel filtration**

Protein	Molecular mass (kDa)	K_d
Catalase	232	0.15
Bovine serum albumin	67	0.38
Ovalbumin	43	0.47
Chymotrypsinogen	29	0.56
RNase A	14	0.70
Prefoldin		0.36

STRATEGY

Calculate log (molecular mass) of the standard proteins and then proceed to plot the K_d values against log (molecular mass) values to generate a standard curve. The molecular mass of prefoldin can be extrapolated from the standard curve.

SOLUTION

A standard curve of K_d values against log (molecular mass) is given in Fig. 8.11.

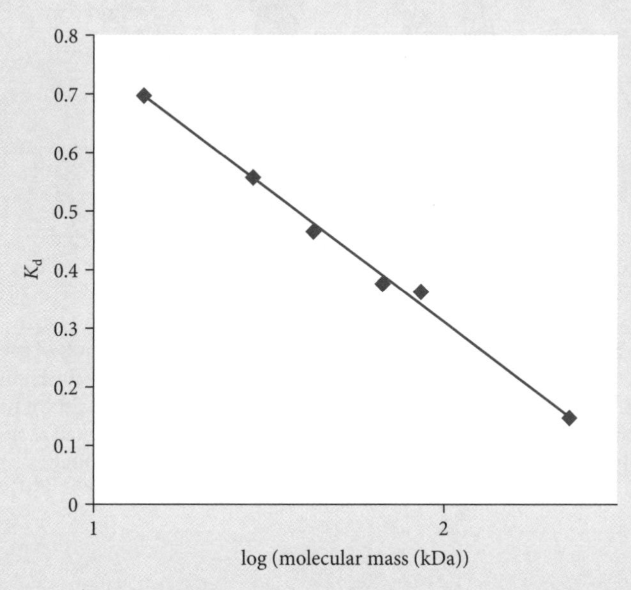

Fig. 8.11 Standard curve of K_d values against log (molecular mass) to estimate the molecular mass of prefoldin.

A. Prefoldin has a K_d of 0.36 which corresponds to a log (molecular mass) value of 1.93. Taking the anti-log of 1.93, we obtain a native molecular mass of 85 kDa.

B. The SDS-PAGE data suggest a 2:1 ratio of 13.5 kDa:15.5 kDa subunits.

If prefoldin exists as a trimer with two 13.5-kDa subunits and one 15.5-kDa subunit, this would suggest an overall molecular mass of (2 × 13.5) + 15.5 = 42.5 kDa.

If prefoldin exists as a hexamer with four 13.5-kDa subunits and two 15.5-kDa subunits this would suggest an overall molecular mass of (4 × 13.5) + (2 × 15.5) = 85 kDa. Given the estimated native molecular mass of prefoldin is 85 kDa (part (a)), this suggests that we have a hexamer of two subunits of 15.5 kDa and four subunits of 13.5 kDa.

Indeed, the structure of prefoldin from the archaeum *Methanobacterium thermoautotrophicum* has been solved by X-ray crystallography and refined to 2.3 Å resolution (Siegert *et al.*, 2000), revealing a hexameric structure ($\alpha_2\beta_4$) (see Fig. 8.12).

The designation of quaternary structures in terms of the types of polypeptide chains present is discussed in section 8.5.1.

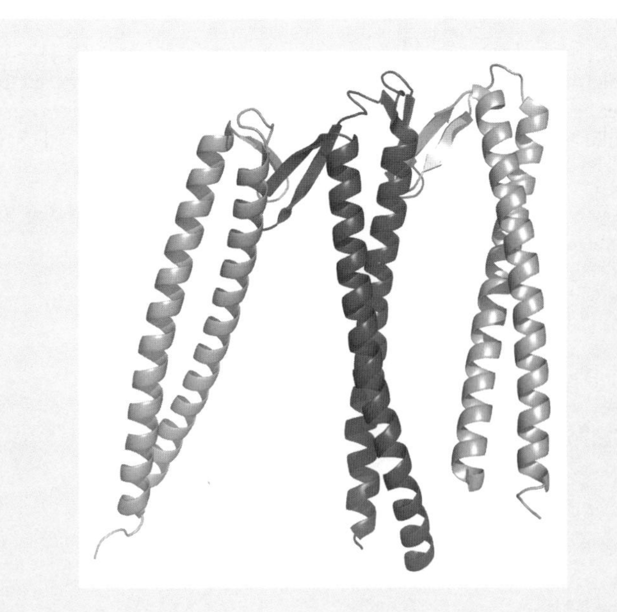

Fig. 8.12 Schematic representation of the asymmetric unit (i.e. trimer) of prefoldin (PDB ID: 1FXK). The functionally active prefoldin molecule is a hexamer, containing two trimers. The α subunit is the darker subunit (15.5 kDa) and consists of two α-helices connected by β hairpins. The β subunits (13.5 kDa), indicated as the lighter subunits, also consist of two α-helices connected by a β hairpin.

8.2.4 Ultracentrifugation

The ultracentrifuge consists of a rotor that can be spun in an evacuated chamber at a very high constant speed (up to about 80 000 revolutions per minute (rpm)). Samples to be analysed are placed in specially constructed cells that are fitted into the rotor. The ultracentrifuge can generate forces on a molecule in solution, which can be up to several hundred thousand times that due to gravity. Forces of this magnitude will cause macromolecules such as proteins to sediment (i.e. move under a centrifugal field towards the bottom of the cell). We can use the behaviour of a macromolecule under such conditions to determine various properties, including its molecular mass, overall shape, and association–dissociation properties. There are two major advantages of using the ultracentrifuge; firstly it is used to study (often complex) macromolecular species and assemblies in solution, and, secondly, it is an absolute method, i.e. does not rely on comparisons with the behaviour of other molecules. The two principal applications of the ultracentrifuge are sedimentation velocity and sedimentation equilibrium.

The sharpness of the boundary between the protein and the solvent depends on the rate of diffusion of the protein. A very large protein would have a low diffusion rate and the boundary would be much sharper than that for a small protein.

In *sedimentation velocity*, the rotor is run at very high speeds so that the process of sedimentation outweighs that of diffusion. As a result, the macromolecule moves away from the axis of rotation as a (reasonably) sharp boundary (Fig. 8.13).

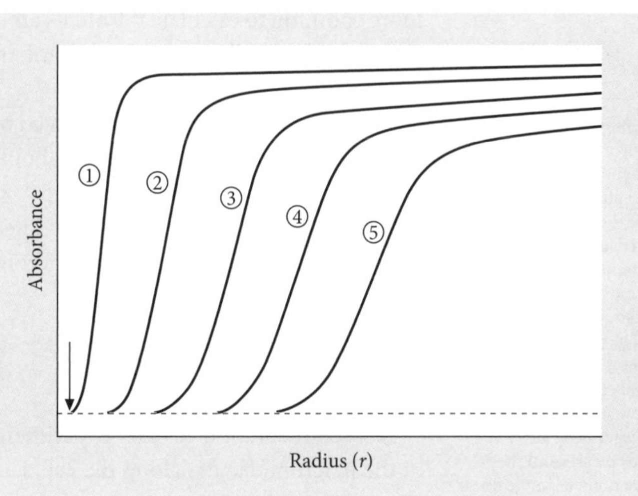

Fig. 8.13 Sedimentation of a macromolecule in a sedimentation velocity experiment. The concentration is shown as a function of radius from the axis of rotation at successive time intervals (1–5). The dashed line represents zero concentration, and the vertical arrow denotes the position of the air–solution meniscus.

In combination with other data such as the diffusion coefficient, we can use sedimentation velocity data to determine the molecular mass of a protein and to learn something about the overall shape of the protein.

In *sedimentation equilibrium*, the rotor speed is lower, so that there is a balance between processes of sedimentation and diffusion. This leads to a stable (equilibrium) distribution of the macromolecule within the cell. From the sedimentation equilibrium experiment, we can determine the molecular mass of the protein.

8.2.4.1 Sedimentation velocity

In sedimentation velocity ultracentrifugation, there are three types of force that are exerted on the protein, namely the *centrifugal force* due to the gravitational field, the *buoyancy force* due to the solution displaced by the protein, and the *frictional force* due to the viscous drag on the protein as it moves through the solution.

Considering the overall effect of these three forces on the protein leads to eqn. 8.1; the derivation of this and subsequent equations in this section is described in more detail in Appendix 8.1.

$$M(1-\bar{v}\rho)/Lf = v/\omega^2 r \quad \textbf{8.1}$$

where M is the molecular mass of the protein, $\bar{v}$ is the partial specific volume of the protein (see note below), ρ is the density of the solution, L is Avogadro's number, f is the frictional coefficient of the protein, v is the velocity at which the protein moves through the solution, ω is the angular velocity of rotation (radians per second), and r is the distance of the protein from the axis of rotation.

A radian (abbreviated rad) is defined as being $(1/2\pi)$ times a complete revolution, i.e. 1 rad = $(360/2\pi)° = 57.3°$. Hence $\omega = 2\pi \times$ (revolutions per second (rps)), or $0.1047 \times$ (rpm).

The partial specific volume ($\bar{v}$) of a solute is the increase in volume when 1 g of the protein is added to a large volume of solution. While it is possible to measure $\bar{v}$ for a protein directly from rather elaborate measurements of density, it is much more common to calculate it from its amino acid composition, assuming that each amino acid contributes to the volume in an additive manner (van Holde *et al.*, 1998).

The sizes of large biological molecular assemblies are often denoted in terms of their *S* values. Examples include the 20S proteasome involved in protein degradation and the ribosome involved in protein synthesis. In prokaryotes and eukaryotes, ribosomes are denoted as 70S and 80S, respectively. It must be borne in mind that sedimentation velocity depends on size and shape so there is not a simple linear relationship between the *S* value and the molecular mass.

The term $v/\omega^2 r$ in eqn. 8.1 is replaced by s, the *sedimentation coefficient* which is in units of seconds (s). Values of about 1×10^{-13} s are common for biological macromolecules such as proteins, so 1×10^{-13} s is known as 1 Svedberg unit (S) after the Swedish physical chemist The Svedberg who developed the ultracentrifuge for the study of biological macromolecules (see Chapter 6, section 6.4.1.2). This gives eqn. 8.2:

$$M(1 - \bar{v}\rho)/Lf = s \quad (8.2)$$

In a sedimentation velocity experiment, the rate of movement of the boundary of the macromolecules along the cell, i.e. away from axis of rotation is measured (Fig. 8.13). At a given time t, the distance of the boundary (in practice, the midpoint of the boundary) from the axis of rotation is r. The value of the sedimentation coefficient s can be determined from a graph of $\ln r$ vs t, the slope of which is $\omega^2 s$. Since we know the rotor speed and therefore the value of ω, we can calculate s.

Eqn. 8.2 shows that we can calculate the molecular mass, M, from the sedimentation coefficient provided we know the partial specific volume (which we can calculate from the amino acid composition), the solution density ρ and the frictional coefficient f.

For an ideal solution, f can be obtained from the diffusion coefficient D using eqn. 8.3:

$$f = RT/LD \quad (8.3)$$

where R is the gas constant and T is the temperature in K. D can be measured directly from observing the rate of diffusion of the macromolecule in a column or indirectly by the techniques of dynamic light scattering or from diffusion measurements using NMR.

By substitution and rearrangement (see Appendix 8.1) we obtain eqn. 8.4, which is usually known as the Svedberg equation:

$$M = RTs/(D(1 - \bar{v}\rho)) \quad (8.4)$$

The Svedberg equation allows us to calculate the molecular mass of the protein, if we know its sedimentation coefficient, diffusion coefficient, partial specific volume, and the solution density.

In practice, the values of s for monomeric proteins (i.e. those proteins that do not associate to give well-defined dimers, tetramers, etc.) will depend to a small extent on the concentration of the macromolecule. This is a reflection of the non-ideal behaviour of the solute due to weak molecular interactions; there is usually a small decrease in s with increasing concentration. Hence s values are obtained over a range of concentrations and then extrapolated to zero concentration to give values of s^0 (the corresponding diffusion coefficient would be denoted D^0). For those proteins which associate at higher concentrations to give dimers, tetramers, etc., the value of *s* will increase with increasing concentration up to a certain point and then decline as for non-associating proteins. The data can be analysed to give information on the association constants for the interactions between the subunits (Byron, 1996; van Holde *et al.*, 1998).

In addition, the values of *s* and *D* are usually corrected for the effects of viscosity, which is highly temperature dependent, and to the behaviour in water, rather than buffer. The values under these standard conditions are then designated as $s_{20,w}$ and $D_{20,w}$ (where 20 denotes the temperature in °C, and *w* denotes pure water as the solvent).

If we know the value of the molecular mass, *M*, from other measurements, we can use the values of *M* and *s* to give the value of *f*, the frictional coefficient. The value of *f* tells us about the resistance of the protein to movement through the solution, and depends on two factors, namely the shape and the degree of hydration of the macromolecule. The shape factor is related to the degree of asymmetry, i.e. the deviation from a perfect sphere and the degree of hydration will influence the actual mass of the moving particle. A value of 0.3 g bound water per g protein is a typical value for most proteins.

Although a detailed account of this topic is beyond the scope of this book, these types of measurements have been used to explore the shapes of proteins in solution. For example, one of the subunits (Cdc27) of the DNA polymerase δ from the fission yeast *Schizosaccharomyces pombe* had been proposed to be a dimer in solution on the basis of its behaviour on gel filtration (see section 8.2.3). Using sedimentation velocity data, it was shown that in fact the subunit was monomeric and had a markedly elongated shape, accounting for its unusual hydrodynamic behaviour (Bermudez *et al.*, 2002).

The shapes of asymmetric molecules are often described as ellipsoids, which are termed prolate (rugby or American football shaped) or oblate (squashed balloon shaped). Values of the frictional coefficients (relative to that for a sphere) have been calculated for various assumed ratios of the major and minor axial ratios of these ellipsoids. These can be compared with experimental data to give insights into molecular shape.

WORKED EXAMPLE 

The data in Table 8.9 were obtained for the proteins ribonuclease (RNase), bovine serum albumin (BSA), fibrinogen, and urease. In all cases, it can be assumed that the solution density (ρ) = 0.998 g mL^{-1}. What are the molecular masses of these proteins?

Table 8.9 Sedimentation velocity and diffusion coefficient data for different proteins

Protein	T (°C)	$s^0 \times 10^{13}$ (s)	$D^0 \times 10^7$ (cm^2 s^{-1})	$\bar{v}$ (mL g^{-1})
RNase	20	1.82	11.9	0.728
BSA	20	5.10	6.97	0.734
Fibrinogen	20	7.90	2.02	0.706
Urease	20	18.5	3.46	0.730

STRATEGY

The calculations use the Svedberg equation (eqn. 8.4) to calculate *M*. It is very important to make sure that the correct units are employed in the calculations. The centimetre–gram–second system (rather than the metre–kilogram–second system) is normally used in these equations and measurements, and hence the appropriate value of *R* is 8.31×10^7 erg K^{-1} mol^{-1}. The value of *M* will then be in units of g mol^{-1}.

SOLUTION

As an example the calculation is shown for RNase. In this case,

$$T = 293 \text{ K and } (1 - \bar{v}\rho) = 0.2735$$

Hence

$$M = (8.31 \times 10^7 \times 293 \times 1.82 \times 10^{-13})/(11.9 \times 10^{-7} \times 0.2735) \text{ g mol}^{-1} = 13\,610 \text{ g mol}^{-1}$$

Using a similar approach, the values of *M* for BSA, fibrinogen, and urease can be calculated to be 66 600, 322 400, and 479 500 g mol^{-1} (i.e. Da), respectively. It is clear that although the values of *s* generally increase with molecular mass, there are clearly other factors at work. In fact, fibrinogen is a highly elongated molecule, which impedes its movement through the solution as shown by the lower than expected values of *s* and *D*.

The observed value of the frictional coefficient for fibrinogen would be consistent with a prolate ellipsoid with an axial ratio (major axis/minor axis) of about 30:1.

8.2.4.2 *Sedimentation equilibrium*

We shall not deal with the derivations of the various equations involved in sedimentation equilibrium here. More detailed accounts are given by Byron (1996) and van Holde *et al.* (1998).

In the sedimentation equilibrium experiment, a stable distribution of macromolecules along the rotor cell is formed since there is an exact balance of the forces leading to sedimentation and diffusion. This distribution depends on the molecular mass of the macromolecule but is independent of its shape and the viscosity of the solution. It is therefore possible to obtain accurate values of molecular mass and also to characterize the behaviour of associating systems. For a detailed account of this topic, see van Holde *et al.* (1998).

The equation describing the equilibrium gradient of concentration (*c*) is:

$$M = (2RT/((1 - \bar{v}\rho)\omega^2)) \times d(\ln c)/dr^2 \qquad \mathbf{8.5}$$

where the terms *R*, *T*, $\bar{v}$, ρ and *r* have the same meaning as for sedimentation velocity.

Since for a given protein being studied under specified conditions, *M*, *R*, *T*, $\bar{v}$, ρ and ω are all constants, a plot of ln *c* vs r^2 will give a straight line, whose slope equals $(M(1 - \bar{v}\rho)\omega^2)/2RT$. This allows the value of *M* to be calculated (eqn. 8.6):

$$M = (\text{slope} \times 2RT)/((1 - \bar{v}\rho)\omega^2) \qquad \mathbf{8.6}$$

The concentration distribution of the macromolecule along the rotor cell is usually determined by absorbance measurements, usually at 280 nm for proteins. Since only the slope of the graph is required, it is sufficient to plot the (natural) logarithm of the absorbance (A) rather than of the actual concentration vs r^2.

This point can be shown by noting that absorbance (A) is proportional to c, i.e. $A = kc$. Taking logarithms of both sides, $\ln A = \ln k + \ln c$ (note the properties of logarithms described in Chapter 2, section 2.3). If we plot $\ln A$ vs r^2, instead of $\ln c$ vs r^2, we will not alter the slope of the graph, merely its intercept on the y-axis.

Eqn. 8.5 applies to an ideal solution of a macromolecule of defined mass M. In practice the effects of non-ideality (i.e weak interactions between solute molecules) will lead to a small effect of concentration on the observed molecular mass (usually there will be a small decrease in the observed mass with increasing concentration). It is usual therefore to determine M at several concentrations and extrapolate the results to give M at zero concentration.

If the macromolecule undergoes self-association to give dimers, tetramers, etc. the graph of $\ln c$ vs r^2 will show upward curvature. It is possible to use curve fitting procedures to fit the data to a proposed model of association and to determine the appropriate equilibrium constants for these processes, and hence the strengths of the interactions between the protein subunits (van Holde *et al.*, 1998).

Sedimentation equilibrium is thus a very powerful method for determining the molecular mass of a protein in solution. One possible disadvantage is that it can take several hours for the distribution of macromolecule in the rotor cell to reach equilibrium, which may make the study of unstable proteins difficult. However, a number of solutions to these problems have been proposed, including the use of very small cells (with the height of the liquid column in the range 1–3 mm).

WORKED EXAMPLE 

A sedimentation equilibrium experiment was used to characterize the type II dehydroquinase from *Mycobacterium tuberculosis*, the polypeptide chain of which has a mass of 16 kDa. The values in Table 8.10 of absorbance (A_{280}) as a function of the distance, r, from the axis of rotation were obtained.

Table 8.10 Absorbance as a function of r for dehydroquinase

Radius (cm)	A_{280}
6.928	0.067
6.954	0.093
6.979	0.128
7.005	0.177
7.030	0.245
7.055	0.338
7.080	0.467
7.106	0.645
7.131	0.891
7.155	1.231

In the experiment, the rotor speed was 9000 rpm, the temperature was 25°C, the partial specific volume ($\bar{v}$) of the protein was 0.737 mL g^{-1}, and the buffer density was 0.998 g mL^{-1}. What can you conclude about the molecular mass and quaternary structure of the enzyme? ($R = 8.31 \times 10^7$ erg K^{-1} mol^{-1}).

STRATEGY

A graph is plotted of ln A vs r^2. From the slope of the graph, M can be calculated using eqn. 8.7.

SOLUTION

The data are used to construct Table 8.11.

Table 8.11 Values of r^2 and ln A_{280} for dehydroquinase

Radius (cm)	r^2 (cm^2)	A_{280}	ln A_{280}
6.928	47.997	0.067	−2.703
6.954	48.358	0.093	−2.375
6.979	48.706	0.128	−2.056
7.005	49.070	0.177	−1.732
7.030	49.421	0.245	−1.406
7.055	49.773	0.338	−1.085
7.080	50.126	0.467	−0.761
7.106	50.495	0.645	−0.439
7.131	50.851	0.891	−0.115
7.155	51.194	1.231	0.208

The slope of the plot (ln A vs r^2) (Fig. 8.14) is 0.91 cm^{-2}. $(1 - \bar{v}\rho) = 0.264$, $\omega = 942$ rad s^{-1} and $T = 298$ K. Hence

$$M = (0.91 \times 2 \times 8.31 \times 10^7 \times 298)/(0.264 \times 942^2)\ \text{g mol}^{-1} = 192\,400\ \text{g mol}^{-1}$$

Thus, the molecular mass of the enzyme is 192.4 kDa; hence it is likely to consist of 12 subunits (12 × 16 kDa = 192 kDa).

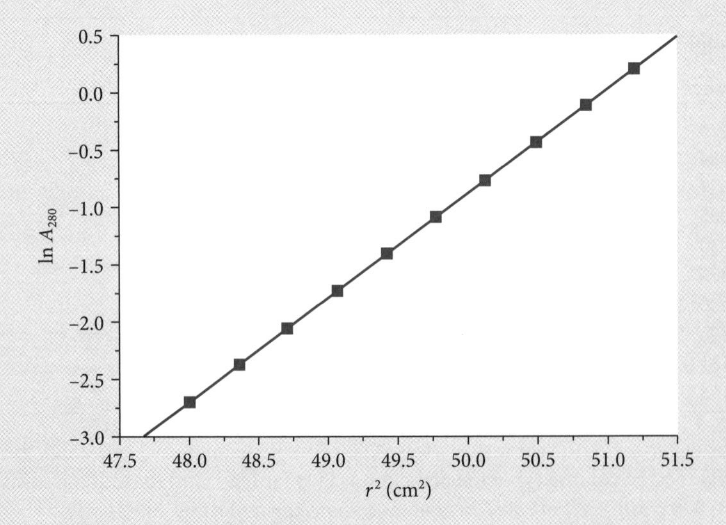

Fig. 8.14 Plot of ln A_{280} vs r^2 for dehydroquinase.

8.2.5 Mass spectrometry

Mass spectrometry is a technique that separates molecules according to their mass and charge. Since its invention around 1920, mass spectrometry has been developed to determine the mass of elements, identify isotopes, and characterize small organic compounds. However, it was not until the development of soft (i.e. low energy) ionization techniques in the late 1980s that this tool could be used to characterize biological macromolecules including proteins and peptides. Molecular mass estimations by mass spectrometry are several orders of magnitude more accurate than those generated by SDS-PAGE, Ferguson plots, gel permeation, or ultracentrifugation, with an error of less than 0.01%, i.e. less than 5 Da for a 50-kDa protein. This level of accuracy allows us not only to determine the mass of a given protein, but also to detect mass differences associated with proteolysis, post-translational modifications and chemical modifications.

While the soft ionization techniques allow the study of proteins with mass values as large as 1000 kDa (using electrospray ionization), the most commonly used mass spectrometry techniques are limited to the study of individual subunits as oligomeric proteins will usually dissociate into their constituent monomers during ionization.

Mass spectrometers are composed of three parts: an ionization source, an analyser, and a detector. The mass analyser separates charged protein ions according to their mass (m) to charge (z) ratios (abbreviated as m/z). Before separation, however, protein ions are generated within the ionization source; the most commonly used ionization techniques, in the determination of protein mass, are electrospray ionization (ESI) and matrix-assisted laser desorption ionization (MALDI).

8.2.5.1 Electrospray ionization

ESI requires proteins to be dissolved in volatile, polar solvents such as acetonitrile plus formic acid to generate positive ions (MH^+), or acetonitrile plus NH_4OH to generate negative ions (M^-). Typically, proteins are analysed as positive ions, with protonation sites provided by the side chains of basic amino acids (lysine, arginine, and histidine) and the N-terminal amino acid. Once dissolved in a suitable volatile, polar solvent, the protein is pumped through a narrow capillary. A high voltage, between 3 and 4 kV, is applied across the tip of the capillary such that as the protein sample emerges, an aerosol of charged droplets forms (Fig. 8.15 (a)). A flow of nitrogen is applied along the tip of the capillary to direct the emerging charged droplets into the mass spectrometer, and in the process evaporate the droplets causing ion formation (Fig. 8.15 (b)).

Peptides with a molecular mass less than about 1.2 kDa generate charged ions carrying a single charge, giving rise to a single predominant species on a mass spectrum (% abundance against m/z). Electrospray mass analysis of angiotensin II provides a good example of a singly charged ion in Fig. 8.16(a). Proteins with molecular mass values greater than 1.2 kDa tend to generate ions carrying multiple

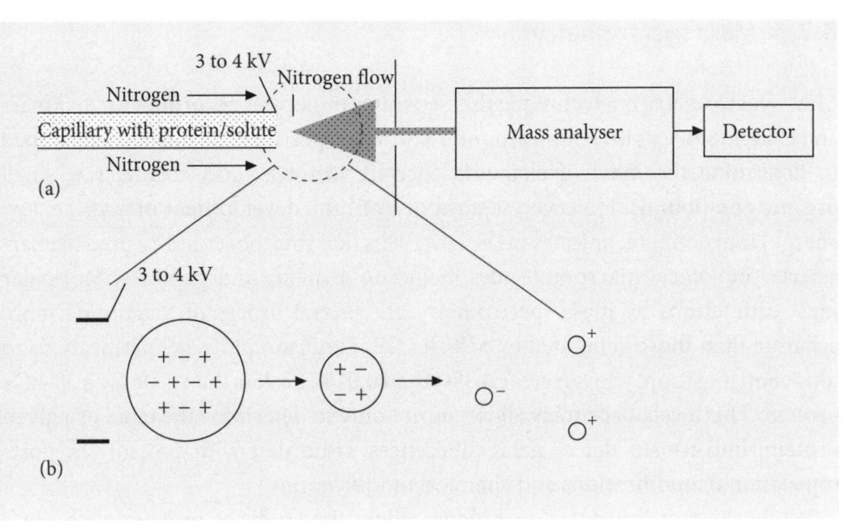

Fig. 8.15 Representation of ionization process of electrospray mass spectrometry (a) protein sample is pumped through a capillary which has a highly charged tip. As protein ion droplets emerge from the capillary they are evaporated, generating protein ions, which are then directed towards the mass analyser. (b) The evaporation of the protein droplets to generate charged protein ions carrying single or multiple charges.

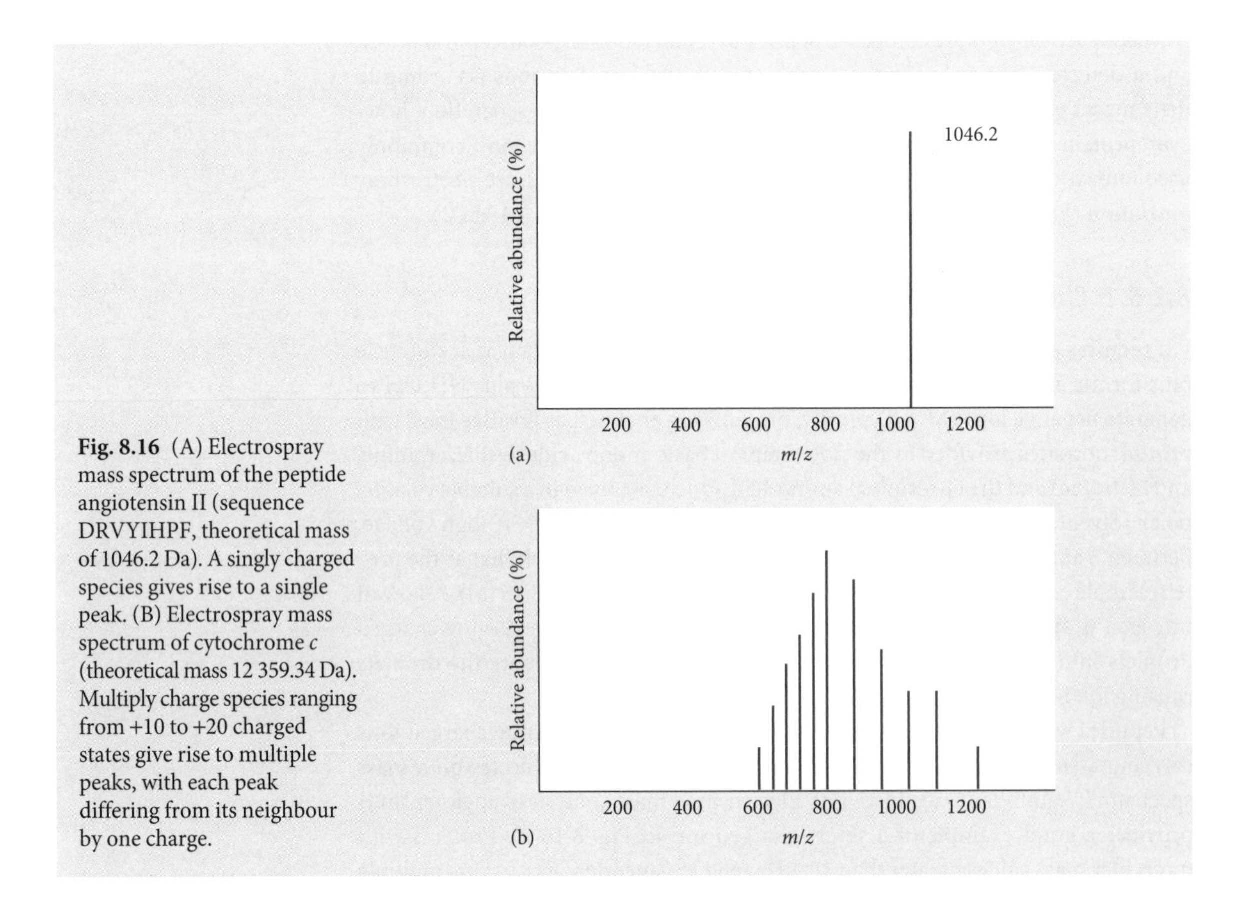

Fig. 8.16 (A) Electrospray mass spectrum of the peptide angiotensin II (sequence DRVYIHPF, theoretical mass of 1046.2 Da). A singly charged species gives rise to a single peak. (B) Electrospray mass spectrum of cytochrome *c* (theoretical mass 12 359.34 Da). Multiply charge species ranging from +10 to +20 charged states give rise to multiple peaks, with each peak differing from its neighbour by one charge.

charges. The resulting spectrum exhibits a Gaussian-type (bell shaped) distribution of charged ions. Each peak on the mass spectrum differs from the neighbouring peak by one charge. Electrospray mass analysis of cytochrome *c* (whose calculated molecular mass is 12 359.34 Da) would generate a mass spectrum similar to that shown in Fig. 8.16(b).

In a situation where the number of charges on an ion is known, the m/z values can be used to calculate the molecular mass of the protein using the following equation:

$$m/z = (\text{molecular mass} + n \times 1.0073)/n \qquad \textbf{8.7}$$

where n = the number of charges on the ion and 1.0073 is the mass of a proton (in Da).

Rearranging this equation gives:

$$\text{molecular mass} = (m/z - 1.0073) \times n \qquad \textbf{8.8}$$

Thus, knowing the charges of all the cytochrome *c* ion species in Fig. 8.16(b), we can use eqn. 8.8 to calculate the molecular mass of cytochrome *c*, as indicated in Table 8.12.

Usually, the number or charges on an ion is unknown. However, it is possible to calculate the number of charges assuming that ion species giving rise to adjacent peaks on a spectrum differ by one charge. In the case of cytochrome *c*, if the peak with an m/z value of 825.02 has a charge of n, the neighbouring peak with an m/z value of 773.5 will have a charge of $n + 1$. We can use simultaneous equations (eqn. 8.7) to calculate the value of n for the cytochrome *c* peak with a m/z value of 825.02, and then progress to calculate the molecular mass value using eqn. 8.8.

Table 8.12 Molecular mass values of cytochrome *c*, calculated from the m/z values of the ions generating the tallest spectral peaks

Charged state	m/z	Molecular mass (Da) (calculated from (m/z – 1.0073) × charged state)
20	619.02	12 360.25
19	651.54	12 360.12
18	687.69	12 360.29
17	728.11	12 360.75
16	773.50	12 359.88
15	825.02	12 360.19
14	883.88	12 360.22
13	951.80	12 360.30
12	1031.03	12 360.27
11	1124.67	12 360.29
10	1237.03	12 360.23

$$m/z = (\text{molecular mass} + n \times 1.0073)/n$$

For the peak at 825.02:

$$825.02 = (\text{molecular mass} + n \times 1.0073)/n \quad (8.9)$$

For the peak at 773.5:

$$773.5 = (\text{molecular mass} + (n + 1) \times 1.0073)/(n + 1) \quad (8.10)$$

Rearranging eqns 8.9 and 8.10 gives:

$$\text{molecular mass} = n \times (825.02) - (n \times 1.0073)$$

$$\text{molecular mass} = (n + 1) \times 773.5 - ((n + 1) \times 1.0073)$$

Therefore:

$$n \times (825.02) - (n \times 1.0073) = (n + 1) \times 773.5 - ((n + 1) \times 1.0073) \quad (8.11)$$

$$n \times (825.02) = n \times (773.5) + 773.5 - 1.0073$$

$$n \times (825.02 - 773.5) = 773.5 - 1.0073$$

$$n = 15$$

i.e. the ion species giving rise to the m/z value of 825.02 carries a charge of 15.
Using these values to solve eqn. 8.8:

$$\text{molecular mass} = (825.02 - 1.0073) \times 15 = 12\,360.19 \text{ Da}$$

Fortunately, the process of calculating the molecular mass from spectra of multiply charged ions is performed by software, which extrapolates the mass of the species carrying no net charge. This software is essential for the interpretation of mass spectra generated by samples containing a number of proteins or peptides.

8.2.5.2 Matrix-assisted laser desorption ionization

MALDI requires proteins to be dissolved in a volatile solvent. Positive ion formation (M^+) also requires the presence of formic acid, whereas negative ion formation (M^-) requires the presence of NH_4OH. Once dissolved the protein sample is mixed with an organic matrix such as cinnamic acid (for peptide mass analysis) or sinapinic acid (for protein mass analysis) and dried onto a support. The presence of the matrix is required to moderate the ionization process and prevent degradation of the target protein. A laser pulse directed at the dried protein results in ion formation (Fig. 8.17). Unlike ESI, MALDI tends to generate singly charged

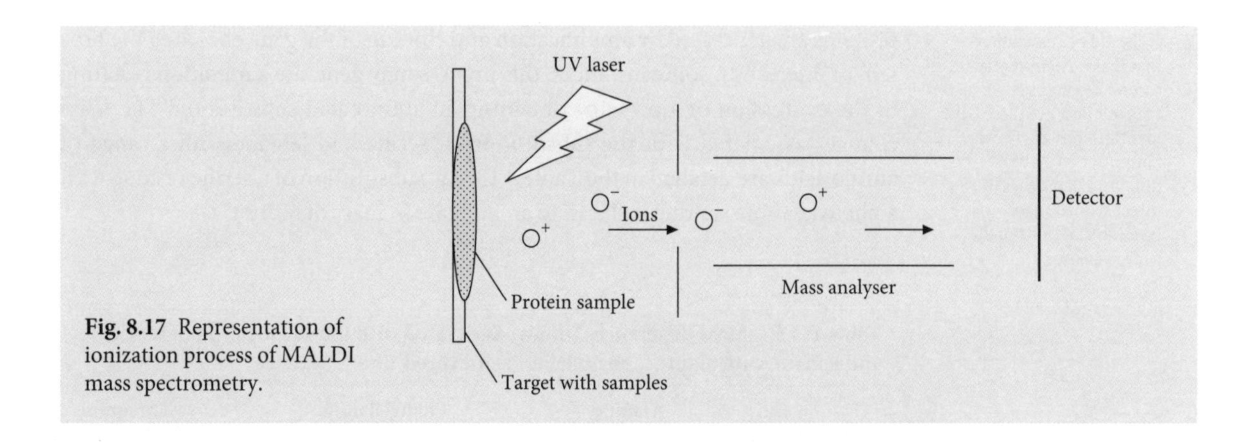

Fig. 8.17 Representation of ionization process of MALDI mass spectrometry.

ions, thus the spectra are much simpler and the process of spectral interpretation is easier.

The accuracy of mass spectrometry (0.01% i.e. within ±5 Da for a 50-kDa protein) makes it possible not only to establish the molecular mass of a protein, but also allows the measurement of mass differences associated with proteolysis, amino acid substitutions, post-translational, and chemical modifications. This approach requires knowledge of the theoretical mass of the protein being analysed. Web-based tools such as ProtParam on the ExPASy server (Chapter 5, section 5.8.2) can be used to calculate the theoretical molecular mass of complete proteins or protein fragments.

Differences between the theoretical mass and experimentally determined mass of a protein may be indicative of proteolysis, amino acid substitutions, post-translational, or chemical modifications.

8.2.5.3 Proteolysis

Characterization of proteolysis resulting from cellular processing, or unwanted proteolysis during isolation, requires knowledge of the mass of individual amino acids. Table 1.1 in Chapter 1, details the mass (in Da) of the amino acids commonly found in proteins. The molecular mass of a proteolytic fragment can be calculated by adding the mass values of the constituent amino acids plus that of 1 molecule of H_2O (18.02) as the mass values do not include the mass of water. Alternatively, the theoretical mass of parent proteins and their peptide fragments can be calculated using web-based tools such as ProtParam on the ExPASy server (Chapter 5, section 5.8.2).

The one extra water molecule arises from the additional H atom and the OH group at the N- and C-termini, respectively.

8.2.5.4 Amino acid substitutions

Amino acid substitutions can be readily detected by mass spectrometry. This is a particularly useful tool in the analysis of recombinant proteins. Recombinant

It should be remembered that mass spectrometry cannot distinguish between Leu and Ile since these have identical masses. These amino acids can be differentiated either by direct sequence analysis (section 8.3) or from the gene sequence.

proteins are generated by amplification and cloning of the gene encoding the protein of interest. In some instances, this process may generate a mutation resulting in the expression of a protein containing an amino acid substitution. The mass changes associated with the substitution of serine and leucine with a range of amino acids are detailed in the Table 8.13, e.g. substitution of a serine residue with a phenylalanine residue will lead to an increase in mass of 60.10 Da.

Table 8.13 Mass differences (in Da) associated with the substitution of serine and leucine with alanine, phenylalanine, or threonine

	Alanine	Phenylalanine	Threonine
Serine	−16.00	+60.10	+14.03
Leucine	−42.08	+34.02	−12.05

 **WORKED EXAMPLE**

Non-specific lipid carrier protein has a calculated mass of 13 282.2 Da. Mass spectrometry analysis of recombinant non-specific lipid carrier protein indicates the presence of a single species with an apparent molecular mass of 13 342.3 Da. Using the mass changes associated with amino acid substitutions, suggest a possible reason for the discrepancy between the theoretical mass and the experimentally determined mass.

STRATEGY

To characterize the nature of the recombinant non-specific lipid carrier protein, we must first determine the difference between the theoretical mass and the experimentally determined mass. This mass difference can then be used to characterize the nature of the recombinant protein by considering mass differences associated with amino acid substitutions.

SOLUTION

The difference between the theoretical mass and the experimentally determined mass is 13 342.3 − 13 282.2 = +60.1 Da. An increase in mass of 60.1 Da could be indicative of a serine residue being substituted with a phenylalanine residue.
It should be noted that cysteine to tyrosine (+60.04) and alanine to methionine (+60.11) would give a similar increase in mass, so additional information would be needed to distinguish between these possibilities.

8.2.5.5 Post-translational and chemical modifications

Characterization of the nature of post-translational modifications (Chapter 5, section 5.6.3) and chemical modifications (Chapter 9, section 9.10.2) requires knowledge of the mass of modifying groups. Some of the more common post-translational modifications are listed in Table 8.14 with their associated molecular masses.

Table 8.14 Mass changes associated with some post-translational modifications

Post-translational modification	Mass change (Da)
Acetylation	+42.0106
ADP-ribosylation	+541.0610
Disulphide bond formation	−2.000
Lipoylation	+188.033
Methylation	+14.0157
Myristolyation	+210.1984
Addition of *O*-GlcNAc	+203.0794
Phosphorylation	+79.9663
Proteolysis	Sequence-specific values*
Addition of pyridoxal phosphate	+229.014
Sulphation	+79.9568

*See section 8.2.5.3 on use of mass spectrometry to characterize proteolysis.

WORKED EXAMPLE 

Western blot analysis, using anti-phosphoserine antibodies, indicated the presence of phosphoserine residues in human β-adrenergic receptor. Human β-adrenergic receptor was analysed by mass spectrometry to determine the stoichiometry and location of the phosphoserine sites. Both β-adrenergic receptor protein and tryptic digests of β-adrenergic receptor were analysed by mass spectrometry. Using the theoretical and experimentally determined masses in the Table 8.15, suggest the possible stoichiometry and location of the phosphorylation sites.

STRATEGY

Calculate the mass differences between the theoretical mass and the measured mass. As addition of a phosphate group would lead to an increase in mass of 80 Da, mass increases of 80 Da, or multiples of 80 Da, could indicate the addition of phosphate groups. The location of the phosphate groups can be determined from characterization of the tryptic peptides.

Table 8.15 Calculated and experimentally determined masses of β-adrenergic receptor and tryptic peptides of β-adrenergic receptor

Protein/peptide	Sequence	Calculated mass (Da)	Measured mass(es) (Da)
β-Adrenergic receptor	Full length	45 557	45 637 45 717 45 797
29–60	DEVWVVGMGIVMSLIVLAIV FGNVLVITAIAK	3 370	3 370
98–131	MWTFGNFWCEFWTSIDVLCV TASIETLCVIAVDR	3 956	3 956
240–253	FHVQNLSQVEQDGR	1 657	1 737
306–328	EVYILLNWIGYVNSGFNPLI YCR	2 746	2 746
345–348	SSLK	434	514 594

SOLUTION

β-Adrenergic receptor exhibits mass values 80, 160, and 240 Da greater than the theoretical mass, indicative of up to three phosphorylation sites.

Peptide 240–253 appears to have one phosphate group, and with only one serine residue in this peptide, we can assume that the phosphate group is attached to this serine residue.

Peptide 345–348 appears to have up to two phosphate groups. With two serine residues in this sequence, we can assume that the phosphate groups are attached to the serine residues.

While the incorporation of many side-chain-specific chemical modification reagents can be monitored spectrophotometrically, it is also possible to use mass spectrometry to characterize chemically modified proteins in more detail. This is particularly useful in confirming the stoichiometry of the modification and in identifying which part of the sequence has been modified. Identification of the location of the modification requires the proteolysis of the modified and non-modified versions of the protein of interest followed by identification of the modified peptide by an appropriate increase in mass. Some mass changes associated with the chemical modification reagents, described in more detail in Chapter 9, section 9.10.2, are given in Table 8.16.

Table 8.16 Mass changes associated with some chemical modifications of amino acid side chains

Chemical modification reagent	Modified residue	Mass change (Da)
5′,5-Dithiobis(2-nitrobenzoic acid) (DTNB, Nbs_2)	Cys	+198
2,4,6-Trinitrobenzenesulphonic acid (TNBS)	Lys	+218
N-Bromosuccinimide (NBS)	Trp	+16

Shikimate kinase (SK) from *Erwinia chrysanthemi* has a calculated mass of 18824 Da. This enzyme contains only one lysine residue and chemical modification with TNBS results in inactivation of SK. The mass of the TNBS-modified enzyme, measured using mass spectrometry, was 19 042 Da.

Only one chymotryptic peptide (position 8–37 with the sequence VGARGCGKTTVGRELARALG YEFVDTDIFM) of TNBS-modified SK had an experimentally determined mass value (3402.6 Da) which differed from the theoretical mass (3184.6 Da).

Determine the stoichiometry, and location of the modification of SK from *E. chrysanthemi* by TNBS.

STRATEGY

If one residue is modified by TNBS we would expect an increase in mass of 218 Da in the intact SK. However, if more than one residue of SK is modified we would observe an increase in mass of multiples of 218 Da, for example if there were three modifications there would be an increase of 3 × 218 Da.

The location of the peptide can be determined from the chymotryptic peptide information. A peptide containing a residue that is modified by TNBS will have an associated increase in mass (218 Da) that can be detected using mass spectrometry.

SOLUTION

The mass increase associated with TNBS modification of SK is 19 042 – 18 824 = 218 Da. This is indicative of only one modification site per enzyme subunit. The location of this modification site appears to be within the peptide at position 8–37 (VGARGCGKTTVGRELARALG YEFVDTDIFM).

In a manner similar to the intact SK, only one modification site is present within this peptide as the mass increase equates to 3402.6 – 3184.6 = 218 Da. Analysis of the sequence of this peptide reveals the presence of a lysine residue (Lys 15) suggesting that this residue is the modified by TNBS.

SELF TEST 

Check that you have mastered the key concepts at the start of this section by attempting the following questions.

ST 8.1 Which method would be most appropriate to determine the molecular mass of:

(a) an intact oligomeric protein

(b) an individual subunit

(c) a post-translational modification of a protein

ST 8.2 Calculate the mass changes associated with:

(a) substituting a histidine residue with a glutamate residue in a polypeptide

(b) acetylation of an N-terminal serine residue

(c) loss of an N-terminal peptide with the sequence Arg–Ala–Arg–Gly–Ala–Lys.

Which methods could be employed to characterize the mass changes in each case?

ST 8.3 An ultracentrifuge rotor is being spun at 25 000 rpm. What is the angular velocity in terms of radians s^{-1}?

ST 8.4 The molecular mass of alcohol dehydrogenase from horse liver was studied by equilibrium ultracentrifugation. The rotor speed was 25 000 rpm and the temperature was 20°C. The partial specific volume ($\bar{v}$) is 0.732 mL g^{-1} and the density of the buffer is 0.998 g mL^{-1}. A plot of ln A vs r^2 is linear with a slope of 3.05 cm^{-2}. What is the molecular mass of the enzyme? (The value of the gas constant (R) is 8.31×10^7 erg K^{-1} mol^{-1}).

Answers

ST 8.1

(a) A non-denaturing method such as Ferguson plots, gel filtration, or analytical ultracentrifugation

(b) In order of improving resolution: gel filtration, analytical centrifugation, SDS-PAGE, and mass spectrometry. Gel filtration and analytical centrifugation conditions may require the presence of a denaturant to dissociate the subunit from a larger oligomeric structure.

(c) Mass spectrometry could be used to detect the change in mass associated with the modification of the protein.

ST 8.2

(a) His to Glu substitution would result in a mass reduction of 9 Da.

(b) Acetylation will result in an increase in mass of 42 Da.

(c) The mass of the parent protein will be decreased by 657.77 Da on losing the N-terminal sequence Arg–Ala–Arg–Gly–Ala–Lys.

Mass spectrometry could be used to detect these mass changes.

ST 8.3 The angular velocity is $2\pi \times 25\,000/60$ rad s^{-1} = 2618 rad s^{-1}.

ST 8.4 Substituting the values into eqn. 8.6, the molecular mass of liver alcohol dehydrogenase is calculated as 80.4 kDa.

Attempt Problems 8.1–8.6 at the end of the chapter.

8.3 Primary structure

KEY CONCEPTS

- Knowing the different methods available for determining the sequences of proteins
- Understanding the strengths and weaknesses of each of these methods

The amino acid sequence of a protein can be determined via one of four possible routes:

- Direct Edman sequencing of the protein or peptide fragments of the protein
- Direct sequencing by mass spectrometry of the protein or peptide fragments of the protein (MS/MS sequencing)
- Indirectly from the nucleotide sequence of the gene encoding the protein, using the genetic code to translate nucleotide sequence into amino acid sequence
- Predicted from peptide fragment mass values. The peptide masses are matched against a database of peptide masses derived from known protein sequences (peptide fingerprinting) to give the probable identity of the protein, and hence its sequence

The choice of method is determined by a number of factors, including the availability of equipment, whether the gene encoding the protein has been isolated, the presence of post-translational modifications and the availability of similar sequences within protein databases. In the following sections, we shall consider the quality of the amino acid sequence information that can be derived from Edman and MS/MS sequencing, translating nucleotide sequences, and peptide fingerprinting.

8.3.1 Edman sequencing

Edman sequencing (Edman and Begg, 1967) involves the sequential degradation of a protein, one amino acid at a time, from the N-terminus (Fig. 8.18). Currently, only 5–50 pmol of protein (0.25–2.5 μg of a 50-kDa protein) are required for N-terminal Edman sequencing.

Coomassie Brilliant Blue staining of a polyacrylamide gel can readily detect 2 pmol of protein (0.1 μg of a 50-kDa protein) therefore a single band on a PAGE gel, detected by Coomassie Brilliant Blue staining, contains enough material for Edman sequencing.

Initially, the N-terminal amino acid is modified by phenyl isothiocyanate under basic conditions to generate a phenylthiocarbamyl (PTC) derivative. The

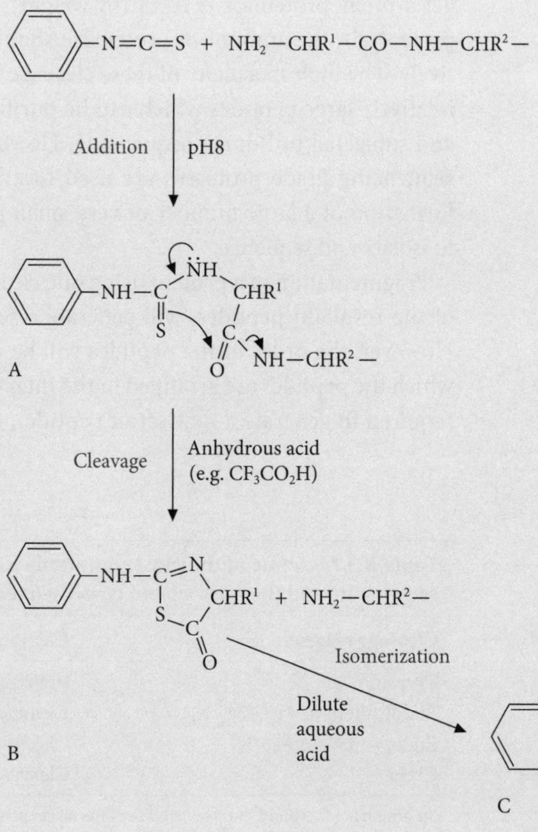

Fig. 8.18 Edman sequencing using sequential degradation of a peptide chain by reaction with phenyl isothiocyanate. A, B, and C are the phenylthiocarbamyl, anilinothiazolinone, and phenylthiohydantoin derivatives of the amino acid, respectively.

PTC–protein is then selectively cleaved, in the presence of trifluoroacetic acid, to generate a PTC-N-terminal amino acid plus parent protein. The PTC-N-terminal amino acid forms an unstable anilinothiazolinone which then forms a phenylthiohydantoin (PTH) derivative of the N-terminal amino acid. The identity of the PTH derivative of the N-terminal amino acid is determined by comparing its retention time on reverse phase chromatography with a set of standard PTH-modified amino acids. The automation and optimization of this process means that each cycle is around 98% efficient, allowing accurate characterization of 30–40 PTH-modified N-terminal amino acids in a given sequence. Beyond 30–40 cycles, the accumulation of parent protein molecules of varying length, resulting from incomplete degradation, makes it difficult to identify the next true N-terminal amino acid. Further details of the reactions involved are described in Price and Stevens (1999).

Most proteins contain between 100 and 500 amino acids, therefore to obtain the amino acid sequence of the entire protein it must be fragmented into smaller peptides which are then individually sequenced by Edman degradation (internal Edman sequencing). The most common methods of fragmentation and their characteristics are listed in Table 8.17.

5–100 pmol of protein (0.25–5 μg of a 50-kDa protein) is required for internal Edman sequencing and in the case of proteolytic digestion, the typical ratio of target protein: proteinase is 100:1 (by weight). In the case of limited or non-specific proteolysis, the amount of proteinase should be increased or decreased, respectively. The high specificity of these cleavage methods generates a small number of relatively large peptides which can be purified by reverse phase chromatography and subjected to Edman sequencing. However, it is important that highly pure, sequencing grade proteases are used to avoid non-specific proteolysis and the formation of a large number of very small peptides, which would prove difficult to isolate and sequence.

Fragmentation of a protein using one cleavage method, followed by sequencing of the resultant peptides, will generate a series of peptides of known sequence. However, the order of the peptides will be unknown. To determine the order in which the peptides are arranged in the intact protein, a second cleavage method is required to generate a new set of peptides, e.g. if a series of peptides are initially

Table 8.17 Some of the most commonly used cleavage reagents for Edman sequencing and their associated target sequences

Cleavage reagent	Cleavage target*
Trypsin	Preferentially cleaves at Arg and Lys in position P_1
Endoproteinase Lys-C	Cleaves at Lys in position P_1
Endoproteinase Asp-N	Cleaves at Asp or Glu in position P_1'
CNBr	Cleaves at Met in position P_1

*The amino acids around the cleavage site of the target protein are labelled $-P_3-P_2-P_1 \equiv P_{1'}-P_{2'}-P_{3'}$, with the peptide bond which is cleaved indicated by the symbol ≡.

generated using trypsin, a second series of peptides could be generated using endoproteinase Asp-N. Knowing the sequence of both sets of peptides, it should be possible to determine regions of overlap and align the peptides into a complete sequence. An example of this approach is given in Fig. 8.19. Alternatively, peptide alignment can be achieved if the sequence of a related protein is available.

One problem routinely encountered with sequencing the N-terminus of a protein is the occurrence of N-terminal post-translational modifications, e.g. acetylation or formylation. Such modifications prevent Edman degradation and the sequence is said to be N-terminally blocked. It is estimated that 40–70% of all naturally occurring proteins are N-terminally blocked. Some success has been

α-subunit human haemoglobin

↓ Treatment with trypsin

11 peptides ranging from 4 to 29 amino acids in length

Two of the trypsin generated peptides are:

Peptide T1: TYFPHFDLSHGSAQVK

Peptide T2: GHGK

The order of the peptides in the overall sequence cannot be determined from the tryptic digest alone. Overlapping peptides, generated by an alternative cleavage reagent, are required. For example, endoproteinase Asp-N treatment:

α-subunit human haemoglobin

↓ Treatment with Asp-N

8 peptides ranging from 5 to 41 amino acids in length

One of the Asp-N generated peptides is:

Peptide C1: DLSHGSAQVKGHGKKVA

Thus we can use the overlap information to determine the order of the peptides in the sequence:

TYFPHFDLSHGSAQVKGHGK
← T1 →← T2 →

DLSHGSAQVKGHGKKVA
← C1 →

This information allows us to deduce this portion of the sequence of α-subunit human haemoglobin:

TYFPHFDLSHGSAQVKGHGKKVA

Fig. 8.19 Alignment of peptide sequences generated by trypsin and endoproteinase Asp-N treatment of the α-subunit human haemoglobin.

achieved in deblocking N-termini by chemical modification or using enzymatic hydrolysis. Examples of deblocking enzymes include:

- Acylaminoacyl peptidases (EC 3.4.19.1) which cleave N-acetyl or N-formyl amino acids from the N-terminus of a polypeptide
- Pyroglutamyl amino peptidases (EC 3.4.19.3), which cleave pyroglutamyl groups from the N-terminus of polypeptide.

However, deblocking attempts are not always successful. In these cases an alternative approach must be devised to characterize the N-terminus, including amino acid composition analysis and peptide mass analysis (section 8.3.3).

8.3.2 Sequencing by mass spectrometry (MS/MS sequencing)

MS/MS sequencing involves the fragmentation of peptides into a series of peptide ions, which differ from each other by one amino acid (see Fig. 8.20). The mass spectrum of the resultant series of peptide ions (Fig. 8.21) can be interpreted, knowing the mass of individual amino acids, to generate a sequence of 5–10 amino

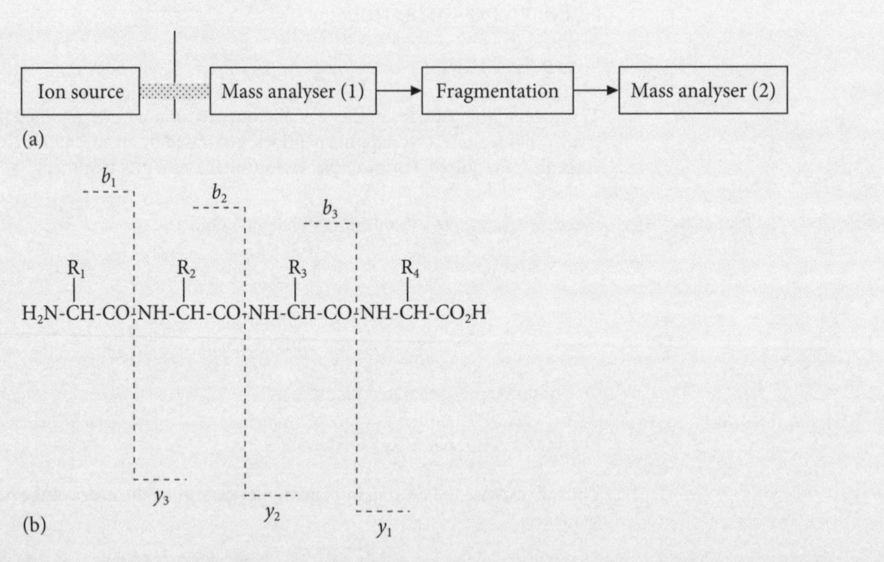

Fig. 8.20 Fragment ions generated during MS/MS sequencing of a short peptide. Peptide ions are analysed and selected using mass analyser (1) for collision-induced dissociation to generate fragment ions that are subsequently analysed in mass analyser (2). Fragmentation at the peptide bonds generates a series of *b* ions (each containing the N-terminal amino group) and a series of *y* ions (each containing the C-terminal carboxyl group). Each member of the *b*-series and *y*-series differs from the adjacent series member by one amino acid, e.g. b_3 is one amino acid longer than b_2 and y_3 is one amino acid longer than y_2. Only *b* and *y* fragment ions are shown as these are the most common, however, it is possible to generate *a* and *x* ions (fragmentation between C_α and carboxyl group) and *c* and *z* ions (fragmentation between C_α and amino group).

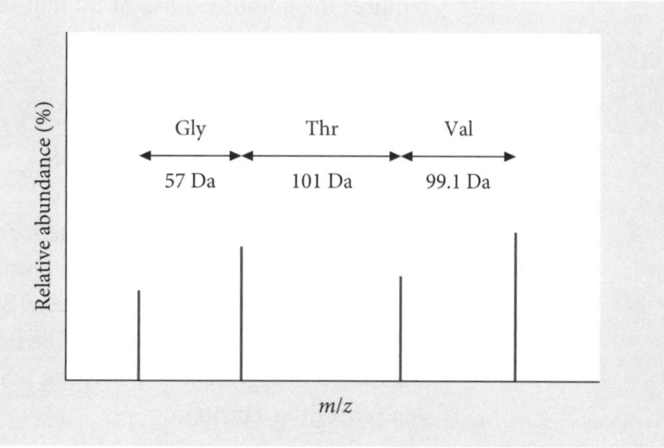

Fig. 8.21 Mass spectrum of ion fragments generated during MS/MS sequencing. Each ion species differs from its neighbour by one amino acid and thus the mass difference between each ion peak can be used to determine the amino acid sequence, e.g. a mass difference of 101 Da corresponds to Thr.

acids. In a manner similar to the approach in Edman sequencing, peptides generated by alternative cleavage methods are required to determine regions of overlap and align the peptides into a complete sequence.

The amount of information which can be generated by MS/MS sequencing is limited by difficulties associated with:

- The complexity of mass spectra resulting from peptide fragmentation
- Some amino acids having the same mass (Lys and Gln both have a mass of 128.1 Da, and Ile and Leu both have a mass of 113.1 Da)
- Post-translationally modified residues having non-standard amino acid masses.

Nevertheless, MS/MS sequencing is the method of choice in generating short amino acid sequences (termed sequence 'tags') which can be used to confirm the identity of a protein. A sequence tag can be identified by submitting the sequence to a suitable database, such as NCBI dbEST (http://www.ncbi.nlm.nih.gov/dbEST/index.html). Indeed, with ever-expanding databases, identification is becoming more efficient.

The sensitivity of MS/MS sequencing allows the analysis of sub-picomole amounts of protein. While the ability to analyse this amount of protein is useful, there are a number of technical problems associated with analysing such small amounts of protein including difficulty in distinguishing peptides derived from the protein of interest from protease autolysis products, and the presence of peptides from contaminant proteins such as human keratins (skin proteins which may be introduced, unwittingly, into the sample during protein purification or preparation for MS/MS sequencing).

Unlike Edman sequencing, MS/MS sequencing does not require the purification of peptide fragments prior to sequencing and it is possible to use high throughput analysis to characterize multiple peptides from one protein, or to characterize many proteins. However, complete primary structure characterization requires the combined use of Edman sequencing, MS/MS sequencing, and mass analysis of peptide fragments.

8.3.3 DNA-derived sequences

It is possible to determine the sequence of a protein indirectly from the nucleotide sequence of the gene encoding the protein. Details of gene isolation and sequencing are described in a number of texts (Brown, 2001; Sambrook and Russell, 2006). However, the general strategy for the isolation, cloning, and sequencing of a gene is outlined in Fig. 8.22. Two starting points are possible at the outset of isolating a gene encoding a protein:

1) The protein of interest has been purified. In this situation, the protein is sequenced by Edman degradation and/or MS/MS sequencing (steps 1 and 2). The resultant amino sequence is used to design degenerate oligonucleotides. For example, a tryptic digest of catalase from the actinomycete *Streptomyces coelicolor* generates around 30 peptides and two peptide sequences are detailed below. Oligonucleotides are designed from the peptide sequences (step 5), using the genetic code, taking care to include degeneracy where appropriate, e.g. glycine (G) is encoded by GGA, GGC, GGG, or GGT, therefore a mixture of oligonucleotides is prepared to include the sequences GGA, GGC, GGG, or GGT.

Degeneracy arises due to the nature of the genetic code in which one amino acid may be encoded by a number of codons, e.g. glycine (G) may be encoded by GGA, GGC, GGG, or GGT. When designing oligonucleotides, for use as PCR primers or DNA probes, it is preferable to select amino acid regions, which will be specified by oligonucleotides with lower degeneracy, e.g. contain tryptophan (W), which is encoded by only one codon (TGG).

```
Peptide 1    G     E     S     F     F     V     K
Oligo 5'   G G A G A A T C A T T T T T T G T A A A A 3'
               C     G A G C     C     C     C     G
               G         G                 G
               T         T                 T

Peptide 2    T     E     V     F     L     R
Oligo 5'   A C A G A A G T A T T T C T A C G A 3'
               C     G     C     C A   C A   C
               G           G           G     G
               T           T           T     T
```

2) The protein of interest has not been purified but a number of closely related proteins have been isolated, and sequenced. An alignment of related proteins (steps 3 and 4) will pinpoint amino acids which are highly conserved due to their structural and/or functional importance (Chapter 5, section 5.8.3), and it is highly likely that these amino acids will occur in the protein of interest. Thus, conserved amino acid sequences are used to design degenerate oligonucleotides that can be used to isolate the gene encoding the protein of interest (step 5). An example of an alignment is provided in Chapter 5, Fig. 5.20. In this example, an alignment of haemocyanin sequences indicates that residues located around the functionally important histidine residues, which coordinate to copper, are highly conserved. Therefore, if we wanted to isolate the gene encoding

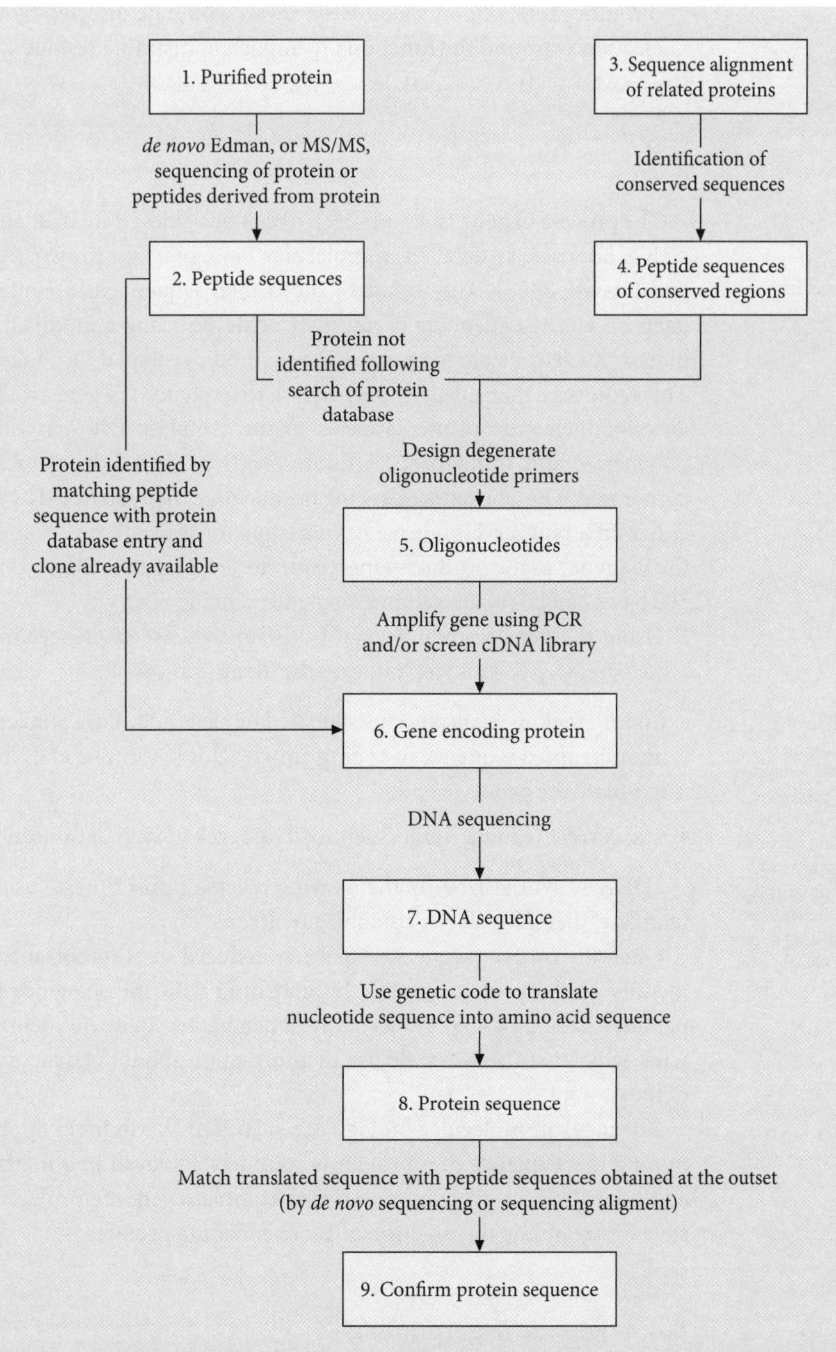

Fig. 8.22 General strategies used to determine the sequence of a protein via the indirect sequencing of the gene encoding the protein. Once the nucleotide sequence of the gene has been determined, the genetic code is used to translate the nucleotide sequence into amino acid sequence. Matching the deduced protein sequence with the original peptide sequence obtained by de novo sequencing, or sequence alignments, is required to verify the protein sequence.

haemocyanin, oligonucleotide synthesis would be directed by the amino acid sequence around the functionally important histidine residues. For example:

Conserved sequence	H	H	W	H	W	H	
Oligo 5′	C A T	C A T	T G G	C A T	T G G	C A T	3′
	C	C		C		C	

The process of gene isolation (step 6) can be achieved by PCR and/or screening cDNA libraries, as detailed in molecular biology texts (Brown, 2001; Sambrook and Russell, 2006). The isolated gene is then sequenced to confirm its identity (step 7). DNA sequencing is routinely achieved using a modified version of the dideoxynucleotide chain termination method developed by Sanger *et al.* (1977). This requires a thermostable polymerase to sequence the gene by thermal cycling: an oligonucleotide primer anneals to the template DNA, is extended by the polymerase and, in the process, fluorescently labelled dideoxynucleotides are incorporated. The labelled sequencing products are then separated by electrophoresis either on a polyacrylamide gel or in a capillary, and a fluorescent detector 'reads' the sequence as the products emerge for the gel, or the capillary. Typically, around 500 bases can be sequenced in a single sequencing reaction.

Using the genetic code we can translate the base sequences into amino acid sequences, step 8. This step requires the identification of:

The coding region of a gene contains a series of successive, non-overlapping codons which specify the sequence of amino acids within the protein encoded by the gene. The start of a coding region contains the start codon ATG, whereas the end of the coding region contains a stop codon, which may be TAA, TAG, or TGA.

- Coding regions, typically characterized by a start codon, a stop codon, and an uninterrupted sequence of coding amino acids (assuming cDNA has been used to isolate the gene)
- The correct reading frame, identified by a lack of stop or nonsense codons

cDNA is synthesized by the enzyme reverse transcriptase, using mRNA as a template, therefore cDNA contains no introns.

Once the protein sequence has been deduced, it is important to confirm the identity of the protein initially by matching with the sequence of the original peptides (step 2 or step 4). Secondly, a database search will identify similar proteins, which in turn, will yield useful information about the structure and function of the protein.

Advances in molecular biology are such that the indirect approach of determining the sequence of a protein is routinely achieved in a matter of weeks. In addition, the ever-growing number of complete genome projects makes it possible to streamline the isolation of genes encoding proteins.

8.3.4 Peptide fingerprinting

It is possible to determine the probable sequence of a protein from the mass of the peptides generated by a specific cleavage method, most commonly proteolysis. This approach relies on the fact that the sequence of a protein determines the

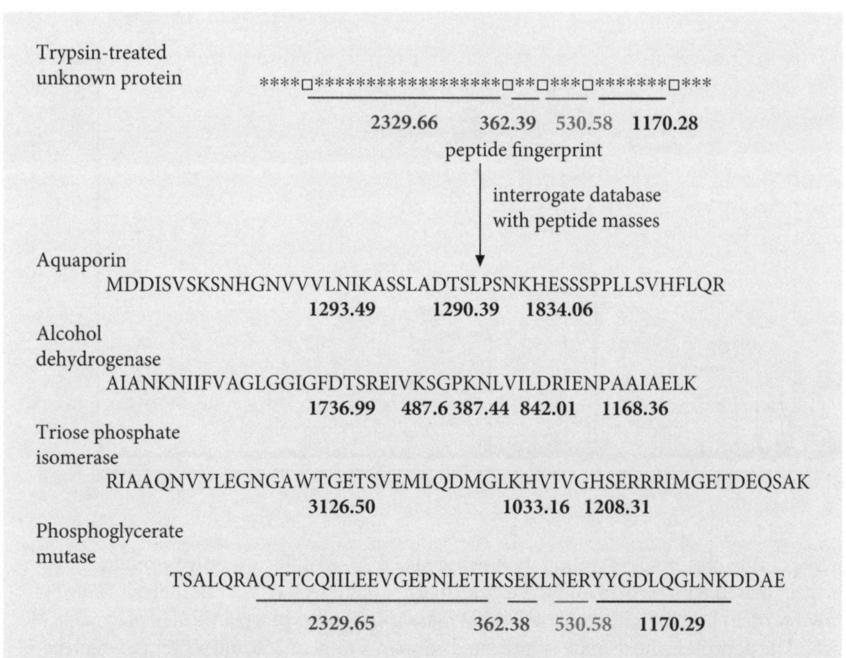

Fig. 8.23 Identification of a protein from the mass of its peptides (peptide mass fingerprinting). The unknown protein is treated with trypsin, which preferentially cleaves at Arg (R) and Lys (K), indicated by the symbol □. A series of peptides are generated (in this case, 2329.66, 362.39, 530.58, and 1170.28 Da), which is termed a peptide fingerprint. A peptide fingerprint is unique to a protein sequence as the sequence dictates the location of the trypsin cleavage sites and the mass of the resultant fragments. The peptide fingerprint of the unknown protein is then matched against a protein database containing sequences that have been treated, theoretically, with trypsin. The database search produces a list of proteins which generate peptides most similar to the peptide masses to the unknown protein. In this case, our protein shares no peptide mass similarities with a number of proteins including aquaporin, alcohol dehydrogenase, and triosephosphate isomerase. However, our protein generates peptides with the same mass as phosphoglycerate mutase, implying that the unknown protein may be phosphoglycerate mutase.

location of the cleavage sites and the mass of the resultant peptides. Thus if a similar sequence, stored within a database, is cleaved theoretically using an identical cleavage method to that used for the protein of interest, their peptide masses should be similar (see Fig. 8.23).

Mass analysis of digested proteins for peptide fingerprinting tends to use MALDI MS (section 8.3.2) due to the speed of the data acquisition and the fact that the resulting spectra are relatively easy to interpret. A standard peptide fingerprinting protocol involves SDS-PAGE of the protein followed by in-gel digestion of the protein and then mass analysis of the resultant peptides. An example of

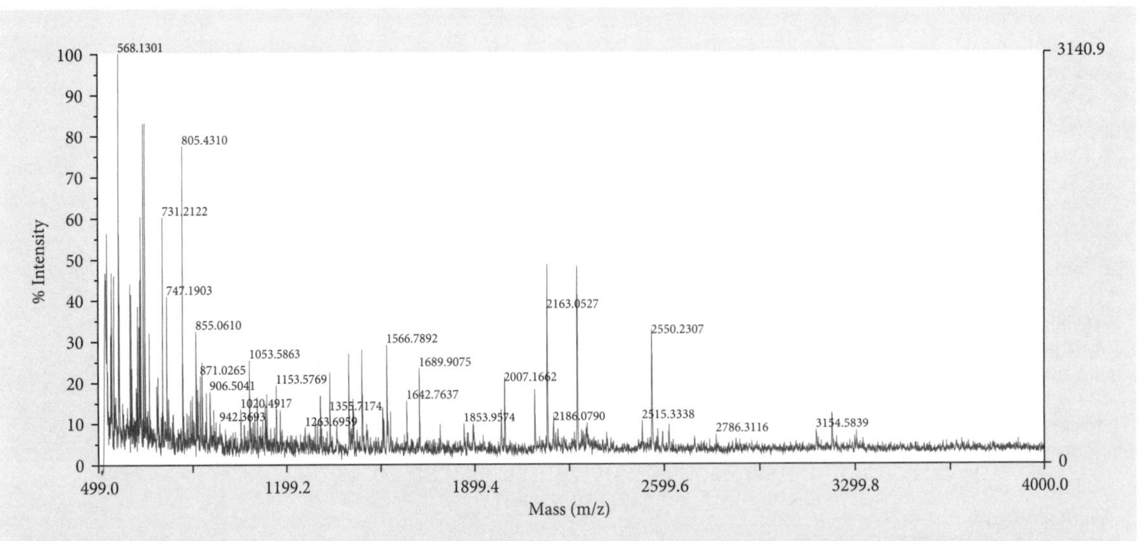

Fig. 8.24 Protein identification by peptide mass fingerprinting. Following SDS-PAGE of the unknown protein, a single band was excised from the gel, treated with trypsin and the trypic digest was subjected to mass analysis. The peptide fingerprint of the unknown protein best matches the theoretical mass values of tryptic peptides of glycogen phosphorylase from rabbit muscle. The peptides whose mass is indicated contain a total of 250 amino acids, which is equivalent to 30% of full-length glycogen phosphorylase.

peptide mass fingerprinting is detailed in Fig. 8.24. In this case, the database search was restricted to proteins with an apparent mass of between 80 and 100 kDa as SDS-PAGE analysis of the unknown protein indicates that it has a mass of around 90 kDa. It is also possible to refine searches using apparent pI values. However, caution should be exercised when restricting database searches due to the possibility that predicted masses and pI values may vary from the actual values (leading to matches being missed due to inappropriate exclusion of proteins from databases).

Peptide fingerprinting is a very sensitive and powerful technique, requiring only 100 ng–1 μg of protein (for a 50-kDa protein this equates to 2–20 pmol) and generally the mass of six peptides is sufficient to identify a protein. Chemical modification of side chains can be used to indicate the presence of particular amino acids within a peptide, for example alkylation of cysteine residues. Therefore, it is possible to refine searches for peptides containing particular amino acids. This makes it possible to identify a protein knowing the mass of less than six peptides. In the situation where a database search with a peptide fingerprint fails to produce a significant match, de novo sequencing by the Edman (section 8.3.1) or MS/MS (section 8.3.2) approaches will be required to confirm the identity of an unknown protein.

SELF TEST ?

Check that you have mastered the key concepts at the start of this section by attempting the following questions.

ST 8.5 A peptide generated by the tryptic digest of a protein is sequenced by MS/MS. Explain why an internal amino acid with a mass of 128.1 is more likely to be Gln rather than Lys.

ST 8.6 Following SDS-PAGE, a Coomassie Brilliant Blue-stained protein band was digested with trypsin and the resultant peptides were sequenced by MS/MS sequencing. Describe the control experiment you would conduct to eliminate the possibility that the MS/MS sequence was due to contaminating peptides derived from trypsin autolysis or from keratin.

Answers

ST 8.5 Trypsin preferentially cleaves at Arg and Lys in position P_1, therefore the Lys is less likely to exist within a peptide. Under optimal tryptic digest conditions, we would expect to find it at the C-terminus of a peptide.

ST 8.6 Prepare a control SDS-PAGE gel, without protein. Proceed to treat this gel in exactly the same way as the experimental gel, i.e. stain with Coomassie Blue, treat with trypsin, and analyse resulting MS/MS spectrum. This will permit the identification of contaminant ion fragments that can be subsequently eliminated from the experimental ion fragments.

Attempt Problem 8.7 at the end of the chapter.

8.4 Secondary and tertiary structure

KEY CONCEPTS

- Explaining how chemical modification can be used to assess the accessibility of amino acid side chains in proteins
- Explaining how fluorescence can be used to give information about the tertiary structures of proteins
- Explaining how circular dichroism can give information about the secondary structures of proteins

The atomic resolution structure of a protein can only be revealed by the methods of X-ray crystallography and NMR; both of these techniques are time consuming and require considerable amounts of purified protein. In the case of X-ray crystallography, it is necessary to be able to crystallize the protein and for the crystals obtained to diffract X-rays well enough to allow the data to be analysed. In the case of NMR the protein sample will also require specific isotope labelling (usually with ^{13}C and ^{15}N) to assign the observed peaks to specific nuclei in the protein. Such methods are not suitable for undergraduate teaching laboratory experiments.

However, it is possible to obtain some important information on the three-dimensional structural features of a protein using much simpler approaches. In this section, we will outline the use of some of these methods.

8.4.1 Exposure of amino acid side chains

Two types of approaches can be used to show which amino acids are on the surface of a folded polypeptide chain, namely chemical modification, which is described in this section and fluorescence, which is described in section 8.4.2.1.

A detailed account of the chemical modification of amino acid side chains is given in Chapter 9, section 9.10.2. For the present purposes, we should note that a number of chemical reagents have been developed that show moderate-to-high degrees of selectivity for certain types of side chains. These reagents can be used under mild conditions (temperature, pH, and low or zero concentrations of organic solvents) which do not compromise the stability of the protein under study. In a native protein it is likely that not all the side chains of a certain type of amino acid will be available for reaction, since some may be buried either partially or completely within the interior of the protein or involved in specific interactions in the tertiary structure.

Experimental details of the modification of tryptophan side chains in lysozyme by reaction with *N*-bromosuccinimide are given in Practical activity 8.1, which is available for lecturers in the Online Resource Centre for this text to support teaching of this topic.

Experimental details of the modification of cysteine side chains in aldolase by reaction with 5,5′-dithiobis-(2-nitrobenzoic acid) (Nbs_2) are given in Practical activity 8.2, which is available for lecturers in the Online Resource Centre for this text to support teaching of this topic.

Determination of the extent of reaction with a reagent will indicate how many amino acids of that type are exposed. This type of information is of considerable value if there is only one or a very small number of this type of amino acid, since we can make definite statements about particular amino acid residues. If there are many of this type of amino acid present in the protein, it is necessary to perform detailed analysis of the modified protein (usually involving digestion by proteases, separation of peptides, and sequence analysis by Edman or mass spectrometric procedures; see section 8.3).

✓ WORKED EXAMPLE

Ellman's reagent (5,5′-dithiobis-(2-nitrobenzoic acid); usually abbreviated as Nbs_2) reacts in a stoichiometric fashion with the thiol group of cysteines to give the thionitrophenolate ion. This product has an absorption maximum at 412 nm, with an absorption coefficient at this wavelength of 13 600 M^{-1} cm^{-1}.

The reaction of aspartate aminotransferase (AAT) from pig heart at a concentration of 0.43 mg mL^{-1} with a large molar excess of Nbs_2 was monitored by the increase in A_{412}. Reactions were performed in cuvettes of pathlength 1 cm. Data given in Table 8.18 were obtained using the native enzyme and enzyme which had been incubated for 30 min in 6 M urea.

What can you deduce about the extent of reaction of AAT with Nbs_2 in the two cases? AAT is a dimeric enzyme with a subunit molecular mass of 46.3 kDa.

STRATEGY

This problem involves calculating the concentrations of enzyme and thionitrophenolate formed in molar terms and then using the ratio to deduce the stoichiometry of reaction. In this problem, it is the extent of reaction (the limiting value of A_{412}) which is important rather than the rate of the reaction.

Table 8.18 Time course of increase in A_{412} for reaction of Ellman's reagent with aspartate aminotransferase

Time (s)	A_{412}	
	Native AAT	AAT + 6 M urea
0	0	0
30	0.141	0.302
60	0.204	0.454
120	0.238	0.576
180	0.245	0.624
240	0.247	0.633
300	0.247	0.635

SOLUTION

The time courses of the increase in A_{412} are shown in Fig. 8.25. The subunit concentration of AAT can be obtained by dividing the concentration in mg mL^{-1} by the molecular mass in Da (i.e. 0.43/46 300 M = 9.3 μM). Under native conditions the limiting value of A_{412} = 0.247; the concentration of thionitrophenolate is obtained using the absorption coefficient (i.e. 0.247/13 600 M = 18.2 μM). Thus, the extent of reaction is 18.2/9.3, i.e. 1.96 or very nearly two Cys side chains per subunit.

In the presence of 6 M urea, the limiting value of A_{412} = 0.635, corresponding to a concentration of thionitrophenolate of 46.7 μM. Thus, the extent of reaction is five Cys side chains per subunit.

From this we can deduce that the enzyme has a total of five Cys side chains per subunit. Of these two are accessible for reaction with Nbs_2 in the native enzyme. When the enzyme is unfolded by incubation in urea, all the Cys side chains can react.

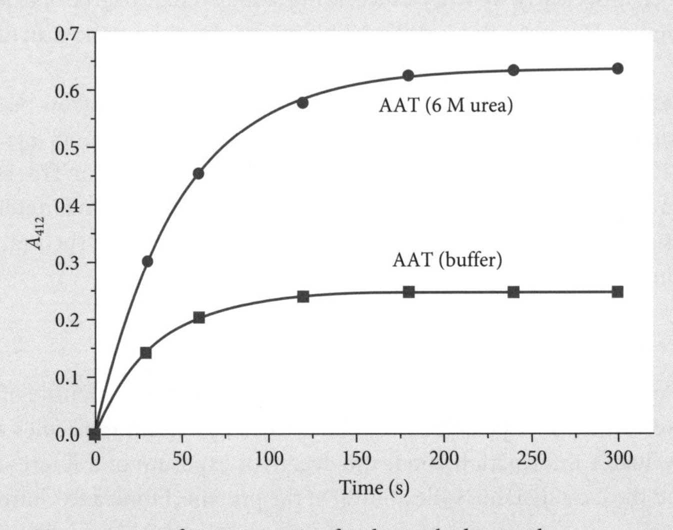

Fig. 8.25 Time courses of increase in A_{412} for the worked example.

8.4.2 Spectroscopic methods to study secondary and tertiary structure

In this section, we will describe applications of fluorescence and circular dichroism (CD) to study some of the overall aspects of the structure of proteins.

8.4.2.1 Fluorescence

The physical basis of fluorescence has been discussed in Chapter 6, section 6.4.2.2. In proteins, fluorescence is due to the chemical properties of the side chains of Trp and Tyr. Usually the fluorescence due to Trp is much greater than that due to Tyr and the latter is only readily observed in proteins such as bovine ribonuclease which do not contain Trp.

The importance of fluorescence is that it is sensitive to the environment of the fluorescent group (or fluorophore). In the case of Trp, the wavelength of maximum emission is an excellent measure of polarity and this will reflect the extent to which a Trp side chain is either on the surface of the protein (and hence accessible to the aqueous solvent) or is buried in the non-polar interior of the protein. Thus, when excited at 290 nm, the emission maximum of azurin is 308 nm, indicating that the single Trp in this protein is deeply buried in a non-polar environment. By contrast the emission maximum in glucagon is 352 nm, which is not very different from the value of 358 nm observed for a model Trp compound in water. Hence the single Trp in glucagon can be regarded as very nearly completely exposed to the solvent (Eftink and Ghiron, 1976). The basis of this sensitivity to the environment of the Trp side chain is explained in Chapter 6, section 6.4.2.2.

We can use the wavelength of emission maximum to give information on the degree of exposure of Trp side chains in proteins. If a particular protein has multiple Trp side chains, clearly we will be measuring some average degree of exposure of these residues. However, this is still valuable information about the structure of the protein.

It should be noted that fluorescence intensity is measured in arbitrary units. For example, if the intensity of the excitation light were increased, the intensity of emission would increase. Under defined conditions (wavelengths, slit widths, etc.) with a given fluorimeter, the intensities should be very similar from day to day.

On excitation at 280 nm, Tyr fluorescence is observed as a narrow peak with a maximum at 306 nm. The wavelength of the emission maximum of Tyr is much less sensitive to the polarity of the environment than is the case for Trp. However, there can be substantial changes in the intensity of fluorescence in different environments so that Tyr fluorescence can still be used to report on structural features of proteins.

Fluorescence quenching

Experimental details of the quenching of fluorescence of lysozyme and aldolase are given in Practical activity 8.3, which is available for lecturers in the Online Resource Centre for this text to support teaching of this topic.

In addition to measurements of the emission wavelength and intensity of fluorescence, we can use the quenching of fluorescence by certain molecules and ions to give valuable information about the degree of exposure of a fluorescent side chain and the local dynamics (flexibility) of the protein (Eftink and Ghiron, 1976, 1984).

Quenching of fluorescence occurs when the excited state of a fluorophore (denoted E^*) loses energy either by colliding with an external added molecule (quencher, Q) or forming a long-lasting complex with the quencher; these are termed dynamic and static quenching, respectively.

Internal quenching usually occurs when vibrational levels of the ground state and excited state are of similar energies. This will often be the case for flexible molecules, and is the major reason why relatively few molecules do act as fluorophores. Most fluorescent molecules contain rigid aromatic systems with delocalized electrons.

The dependence of the degree of quenching on the concentration of quencher is described by the eqns 8.12 and 8.13 below; the derivation of these equations is given in Appendix 8.2 to this chapter.

The excited state (E^*) can undergo three possible fates:

1) Fluorescence (to give the ground state and emit light).
2) Internal quenching (structural rearrangement in the excited state during which energy is lost).
3) External quenching by the quencher.

In the presence of the quencher (Q) the measured fluorescence, F is given by the Stern–Volmer equation (eqn. 8.12):

$$F_0/F = 1 + K_{SV}\,[Q] \qquad 8.12$$

where F_0 is the measured fluorescence in the absence of quencher, $[Q]$ is the concentration of quencher, and K_{SV} is a constant known as the Stern–Volmer constant. A plot of F_0/F vs $[Q]$ (Fig. 8.26) will give a straight line of slope K_{SV} with a y-axis intercept of 1.

It should be noted that both the dynamic and static quenching processes give linear Stern–Volmer plots but can be distinguished by measuring the quenching over a range of temperatures and solvent viscosities.

An extension of the Stern–Volmer equation can be made to deal with the situation where a certain fraction (f_a) of the observed fluorescence is accessible to externally added quenchers and the remaining part is not accessible at all (Lehrer, 1971). These two categories of fluorophores are assigned Stern–Volmer constants of K_{SV} and 0, respectively. The equation derived (see Appendix 8.2) is usually known as the Lehrer equation:

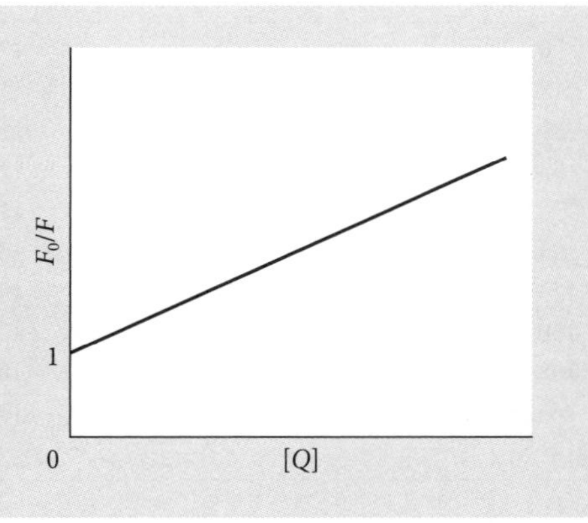

Fig. 8.26 Stern–Volmer plot of fluorescence quenching data. The slope of the line gives the Stern–Volmer constant (K_{SV}), see eqn. 8.12.

It is possible to derive equations for a model in which the various fluorophores in a protein each have different degrees of accessibility to external quenchers. However, these equations are considerably more complex than those derived here for the two category model, and require much more experimental data to allow a complete analysis of the proposed model.

$$F_0/\Delta F = (1 + K_{SV}[Q])/(f_a K_{SV}[Q]) = 1/f_a K_{SV}[Q] + 1/f_a \qquad 8.13$$

where F_0 is the measured fluorescence in the absence of quencher, ΔF is the decrease in fluorescence in the presence of quencher, f_a is the fraction of fluorescence accessible to quencher, K_{SV} is the Stern–Volmer constant for the accessible fluorescence, and $[Q]$ is the concentration of quencher.

Hence a plot of $F_0/\Delta F$ vs $1/[Q]$ (Fig. 8.27) will give a straight line of slope $1/f_a K_{SV}$ and a y-axis intercept of $1/f_a$. From these the values of f_a (equal to $1/y$-axis intercept) and K_{SV} (equal to the ratio intercept/slope) can be determined.

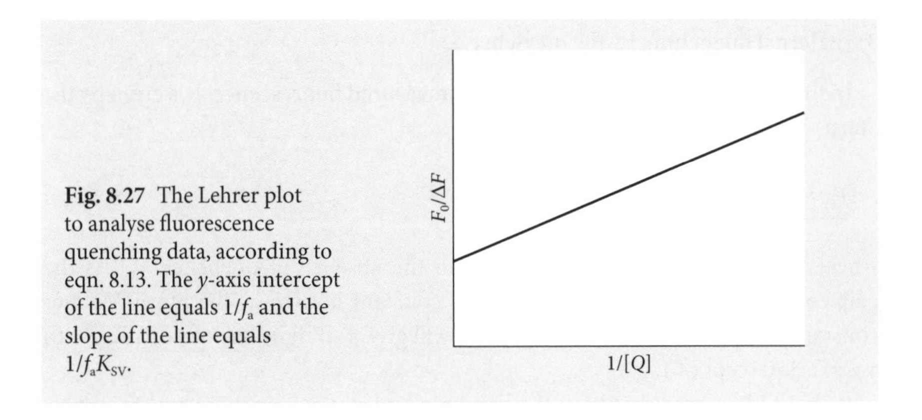

Fig. 8.27 The Lehrer plot to analyse fluorescence quenching data, according to eqn. 8.13. The y-axis intercept of the line equals $1/f_a$ and the slope of the line equals $1/f_a K_{SV}$.

WORKED EXAMPLE

When excited at 290 nm, the fluorescence emission maximum of AAT is at 330 nm. The quenching of the fluorescence by succinimide was studied and the results in Table 8.19 were obtained.

Table 8.19 Quenching of fluorescence of aspartate aminotransferase by succinimide

[Succinimide] (mM)	Fluorescence
0	124.0
19.8	119.7
39.2	116.0
58.3	112.8
76.9	110.0
95.2	107.5
113.2	105.3
130.8	103.3
148.1	101.6
165.1	100.0
181.8	98.5

Analyse the data to calculate the fraction of the Trp side chains accessible to the quencher.

STRATEGY

This is an application of eqn. 8.13; $F_0/\Delta F$ is plotted against $1/[Q]$ and from the slope and intercept the values of f_a and K_{SV} can be calculated.

SOLUTION

The data are used to construct Table 8.20.

Table 8.20 Fluorescence quenching data rearranged for the Lehrer plot

[Q] (mM)	1/[Q] (M^{-1})	F	ΔF	$F_0/\Delta F$
0	—	124.0	0	—
19.8	50.5	119.7	4.3	28.9
39.2	25.5	116.0	8.0	15.6
58.3	17.2	112.8	11.2	11.1
76.9	13.0	110.0	14.0	8.9
95.2	10.5	107.5	16.5	7.5
113.2	8.8	105.3	18.7	6.6
130.8	7.6	103.3	20.7	6.0
148.1	6.8	101.6	22.4	5.5
165.1	6.1	100.0	24.0	5.2
181.8	5.5	98.5	25.5	4.9

The plot of $F_0/\Delta F$ vs $1/[Q]$ (Fig. 8.28) is a straight line with a y-axis intercept ($=1/f_a$) of 1.92, from which $f_a = 0.52$. The slope of the line ($=1/f_a K_{SV}$) is 0.53, from which $K_{SV} = 3.6\ M^{-1}$. Thus, approximately 50% of the Trp residues are accessible to the quencher; this quenching is characterized by a Stern–Volmer constant of $3.6\ M^{-1}$.

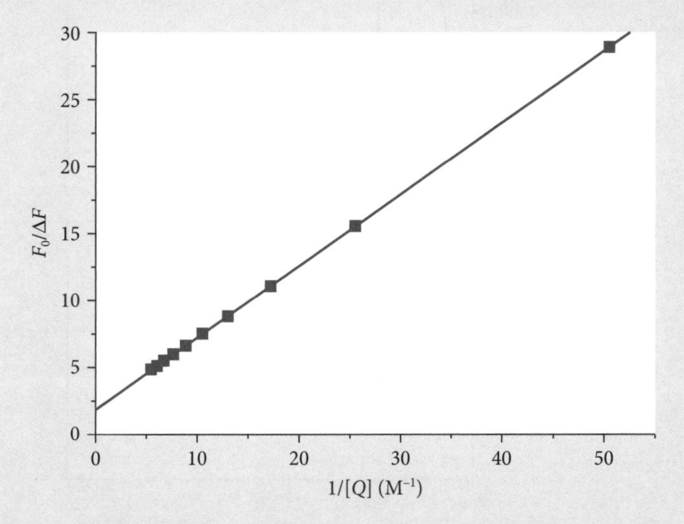

Fig. 8.28 Lehrer plot of data from worked example according to eqn. 8.13.

8.4.2.2 Circular dichroism

Chirality refers to the situation when an object is not superimposable on its mirror image.

As mentioned in Chapter 5, section 5.2.2, the technique of CD depends on the difference in absorbance between left and right circularly polarized light beams. CD will be observed in the region of wavelengths where chiral molecules (or parts of molecules) absorb. For further details on the application of CD to the study of proteins see Kelly *et al.* (2005).

Of particular interest in the present context is the use of CD to explore the secondary structures of proteins. The relevant spectral region is in the far UV, i.e. from 240 nm down to 180 nm where the peptide bond absorbs. In this region, the different types of regular secondary structure such as α-helix and β-sheet show characteristic spectral patterns. Fig. 8.29 shows the characteristic far UV CD spectra of α-helix, β-sheet, and random coil. The α-helix spectrum shows two negative minima at 208 and 222 nm, and a positive maximum at 193 nm. The β-sheet spectrum shows a single negative minimum at about 215 nm and a positive maximum at 196 nm (both of these are much smaller than the signals for α-helix). The random coil shows only a very small signal above 210 nm and a small negative minimum at about 198 nm.

The units in which CD signals are reported require some explanation. The signals are reported as either the difference in absorbance between the two polarized light

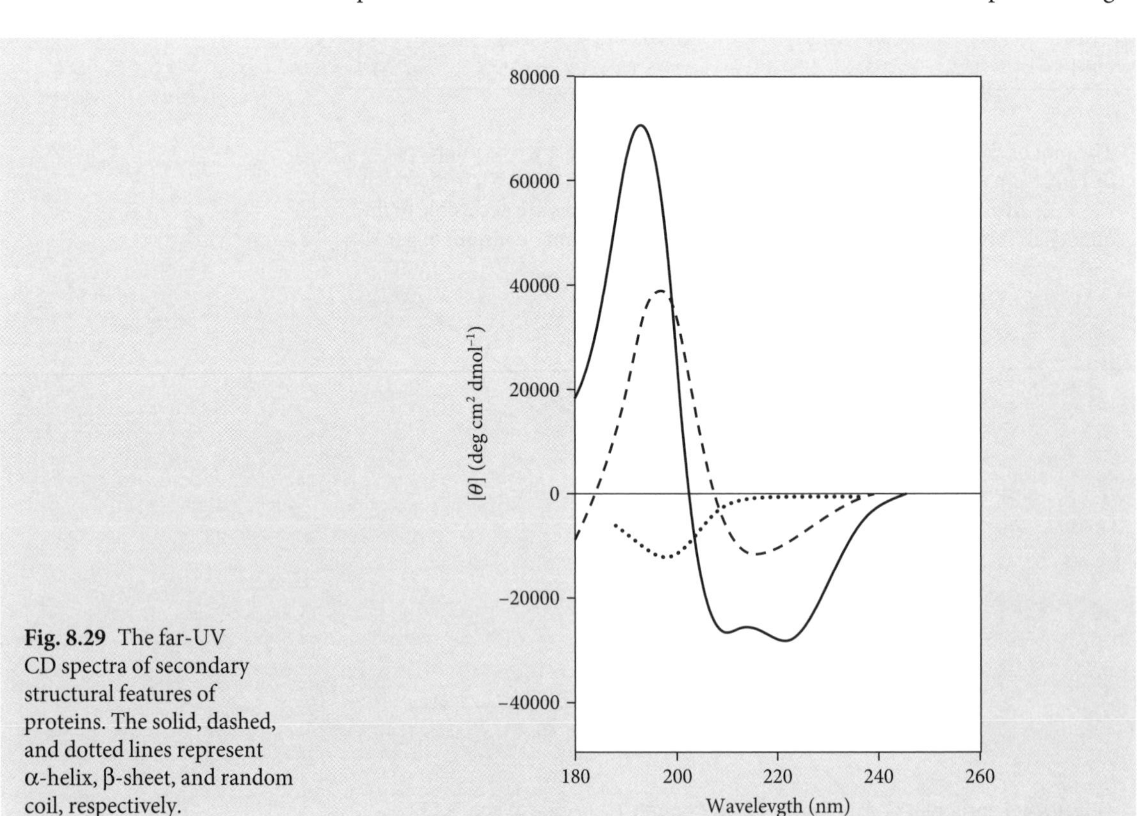

Fig. 8.29 The far-UV CD spectra of secondary structural features of proteins. The solid, dashed, and dotted lines represent α-helix, β-sheet, and random coil, respectively.

beams ($\Delta A = A_L - A_R$, where A_L and A_R refer to the absorbances of the left and right polarized beams, respectively), or in terms of ellipticity (θ) measured in degrees. In practice the actual differences in absorbance are very small, of the order of 1 part in 10 000. There is a simple numerical relationship between ΔA and θ (in degrees), namely:

$$\theta = 32.98\Delta A \quad (8.14)$$

For quantitative investigations it is necessary to normalize CD signals to a molar basis; this can be done for either the whole protein molecule or the repeating unit. In the far UV, where the absorbance of the peptide bond is being measured, the repeating unit is the peptide bond. The so-called mean residue weight (MRW), which can be thought of as the average molecular mass of the amino acids in the polypeptide chain, is calculated from the equation (eqn. 8.15)

$$\text{MRW} = M/(N-1) \quad (8.15)$$

where M is the molecular mass (in Da) of the polypeptide chain and N is the number of amino acid residues. (We subtract 1 from N because there is always one less peptide bond than the number of amino acids to be joined together.) For most proteins the value of MRW is 110 ± 5 Da.

The mean residue ellipticity at a given wavelength (λ) is given by the equation:

$$[\theta]_{\text{mrw},\lambda} = (\text{MRW} \times \theta_\lambda/(10 \times d \times c) \quad (8.16)$$

where θ_λ is the observed ellipticity (degrees) at wavelength λ, d the pathlength of sample cell (cm), and c the concentration of protein (g mL^{-1}).

The units of mean residue ellipticity are deg cm^2 $dmol^{-1}$.

For data in absorbance units, it is usual to calculate the molar differential absorption coefficient, $\Delta\varepsilon$. If a solution of molar concentration m (expressed in terms of the MRW for far UV CD data) in a cell of pathlength d (cm) has an observed ΔA at a certain wavelength (λ), then $\Delta\varepsilon$ can be calculated from the relationship:

$$\Delta\varepsilon_\lambda = \Delta A/(m \times d) \quad (8.17)$$

Although a reliable quantitative analysis of far UV CD spectra involves the use of sophisticated computer-based algorithms, it is nevertheless possible to deduce some features of overall secondary structure relatively simply. Since the CD signals from α-helix are relatively large (Fig. 8.29), we can use the data at wavelengths of 208 and 222 nm to estimate the helical content of proteins (Barrow *et al.*, 1992):

$$\%\ \alpha\text{-helix} = (([\theta]_{208} + 4000)/(-29\,000)) \times 100\% \quad (8.18)$$

$$\%\ \alpha\text{-helix} = (([\theta]_{222} - 3000)/(-39\,000)) \times 100\% \quad (8.19)$$

The origin of the term 'ellipticity' is from the fact that if the two light beams, which would be of different intensities after passage through the sample, were recombined they would constitute elliptically, rather than plane-polarized light. The value of the ellipticity (θ) is obtained from the relationship, $\tan\theta = b/a$ where b and a are the minor and major axes of the ellipse, respectively (see Kelly *et al.*, 2005).

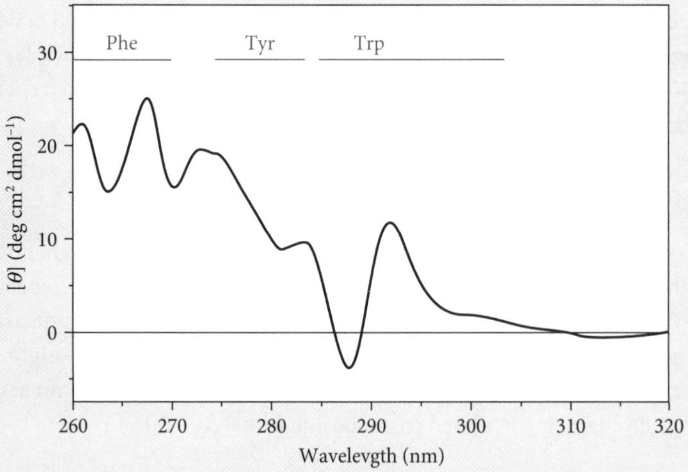

Fig. 8.30 The near UV CD spectrum of type II dehydroquinase showing the spectral regions where signals from the aromatic amino acids can be detected.

When inspecting published near UV CD data, it is important to establish which molar unit has been used for normalizing the observed data. Some workers prefer to use the repeating unit of the protein (i.e. the MRW); others prefer to use the entire protein (i.e. the molecular mass of the polypeptide chain).

CD signals from proteins can also be observed in the near UV region (260 to 320 nm); these arise from the aromatic side chains (Phe, Tyr, and Trp). Although there is a degree of overlap between the absorption bands of these amino acids, it is reasonable to say the signals in the region 255–270 nm arise from Phe, those in the region 275–282 nm from Tyr and those in the region 285–310 nm from Trp. As an example, Fig. 8.30 shows the near UV CD spectrum of type II dehydroquinase. Chirality (and hence the CD signal) arises predominantly from the immobilization of these side chains within the overall structure of the protein, and hence the more mobile the side chain, the smaller the CD signal. The near UV CD spectrum can be used as a 'fingerprint' of the tertiary structure of a protein and thus indicate whether amino acid mutations lead to structural changes; however, it is not possible at present to derive any precise three-dimensional structural information from the spectrum.

✓ WORKED EXAMPLE

Calcitonin-gene-related peptide (CGRP) is a 37 amino acid neuropeptide which has potent vasodilatory effects. The structure of a solution of CGRP (0.1 mg mL^{-1}) was studied by CD in the far UV under two conditions: (i) in sodium phosphate buffer at pH 7.0 and (ii) in buffer containing 50% (vol/vol) trifluoroethanol (TFE). The spectra obtained are shown in Fig. 8.31; the ellipticity values at 208 and 222 nm are shown in Table 8.21. What can you conclude from these data? The cell pathlength was 0.05 cm and the MRW of CGRP is 105 Da.

STRATEGY

The CD data are to be converted into mean residue ellipticity values using eqn. 8.16; eqns. 8.18 and 8.19 can then be used to estimate the % α-helix.

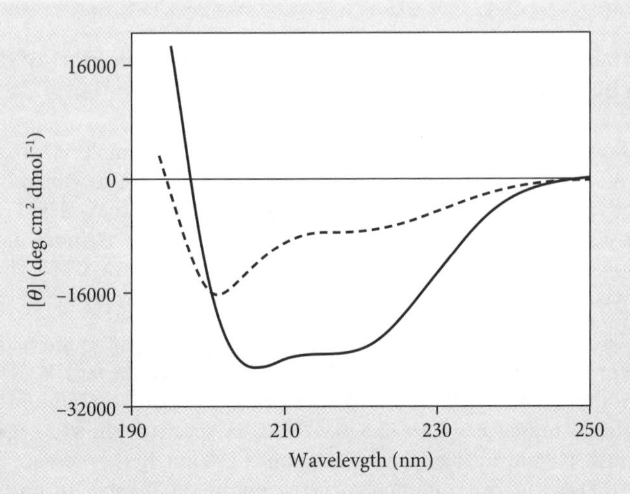

Fig. 8.31 The far-UV CD spectra of CGRP in buffer (dashed line) and buffer + trifluoroethanol (solid line).

Table 8.21 Observed far UV CD data for CGRP

Sample	θ_{208} (mdeg)	θ_{222} (mdeg)
CGRP in buffer	−5.5	−3.1
CGRP in buffer + TFE	−12.6	−11.9

SOLUTION

Noting that in eqn. 8.16, MRW = 105, $d = 0.05$ cm, and $c = 0.1 \times 10^{-3}$ g mL^{-1}, we obtain the values for the mean residue ellipticities shown in Table 8.22.

Table 8.22 Mean residue ellipticities for CGRP

Sample	$[\theta]_{208}$ (deg cm^2 dmol^{-1})	$[\theta]_{222}$ (deg cm^2 dmol^{-1})
CGRP in buffer	−11 550	−6 510
CGRP in buffer + TFE	−26 460	−24 990

The appearance of the spectra in Fig. 8.31 indicates that in buffer there is some α-helical content, but β-sheet and random coil structure are also clearly present in significant quantities. In the presence of TFE, the spectrum changes to one characteristic of a large amount of α-helical structure.

Using eqns. 8.18 and 8.19, in buffer alone, the % α-helix in CGRP is 26% (from 208 nm data) and 24% (from 222 nm data). The average value is 25%.

In the presence of TFE, the % α-helix in CGRP is 78% (from 208 nm data) and 72% (from 222 nm data). The average value is 75%.

TFE is a solvent which is widely used to assess the structure-forming potential of peptides. It promotes secondary structure formation by weakening the hydrogen bond interactions with water and encouraging hydrogen bonds within the peptide.

? SELF TEST

Check that you have mastered the key concepts at the start of this section by attempting the following questions.

ST 8.7 Bovine liver glutamate dehydrogenase has a subunit molecular mass of 56 kDa. A solution of enzyme (0.83 mg mL^{-1}) was reacted in a cuvette of 1-cm pathlength with a large molar excess of tetranitromethane (TNM). TNM reacts with Tyr side chains to give 3-nitrotyrosine, which absorbs at 428 nm with an absorption coefficient of 4100 M^{-1} cm^{-1}. The increase in A_{428} was 0.058. How many Tyr side chains are modified per subunit of glutamate dehydrogenase?

ST 8.8 The structure of a 21 amino acid peptide (representing amino acids 106–126 of the prion protein) was studied by CD. In water, the far UV CD spectrum showed no signal above 210 nm and a small negative minimum at 197 nm. In the presence of 0.1% (3.5 mM) SDS, the spectrum shows a negative minimum at 218 nm and a positive maximum at 198 nm. In the presence of 0.8% (27.8 mM) SDS, the spectrum shows negative minima at 222 and 207 nm and a positive maximum at 192 nm. What can you conclude about the structures of the peptide under these conditions?

Answers

ST 8.7 The molar concentration of enzyme is 0.83/56 000 M = 14.8 μM. The concentration of 2-nitrotyrosine formed is 0.058/4100 M = 14.1 μM. Thus, 0.95 (i.e. essentially one) Tyr are modified per subunit. There are 16 Tyr in the sequence of the polypeptide chain. Hence, only one of these is accessible for reaction; the others are buried in the enzyme.

ST 8.8 In water the peptide appears to have little stable secondary structure; the CD spectrum is typical of a random coil. In 0.1% SDS, the peptide adopts a β-sheet structure and at the higher concentration of SDS, the structure is α-helical. The data indicate that the peptide can adopt a variety of structures under different conditions. This may be related to the ability of the protein to form insoluble fibrils which occur in various neurodegenerative diseases.

Attempt Problems 8.8–8.10 at the end of the chapter.

8.5 Quaternary structure

KEY CONCEPTS

- Understanding how molecular mass information can be used to determine the number of subunits of oligomeric proteins
- Explaining how information on the spatial arrangement of subunits can be obtained by cross-linking

Many proteins in their native, functional state consist of more than one polypeptide chain, i.e. they are said to be oligomeric (or multimeric in the case of large numbers of polypeptide chains). The term 'quaternary structure' is used to describe such proteins. This encompasses two main aspects, namely the numbers and types of such polypeptide chains (or subunits) in a protein and secondly the spatial arrangement of subunits in a protein.

8.5.1 Numbers and types of polypeptide chains in a protein

In order to derive the numbers of polypeptide chains, we need to measure the molecular masses of the native protein and of the individual polypeptide chains. The methods available for doing this have been described in detail in section 8.2. The most widely used methods to study native proteins are analytical ultracentrifugation (principally sedimentation equilibrium) and gel filtration.

As also described in section 8.2, the molecular masses of the individual polypeptide chains in a protein are usually derived by SDS-PAGE or mass spectrometry. In the case of expressed recombinant proteins we can also calculate the molecular mass from the amino acid sequence, although it may be necessary to take into account the effects of post-translational modifications, such as limited proteolysis, glycosylation, phosphorylation, etc.

If the subunits are identical (as shown by the presence of a single band on SDS-PAGE or a single peak in mass spectrometry), the protein is termed homo-oligomeric and the number of subunits is merely the ratio of intact mass divided by the mass of the subunits; thus liver alcohol dehydrogenase, muscle lactate dehydrogenase, and liver glutamate dehydrogenase have 2, 4, and 6 subunits, respectively. The chaperone protein GroEL, which is involved in assisting the folding of proteins in *Escherichia coli*, has 14 subunits, arranged as two stacked rings of seven subunits. (The X-ray structure of GroEL can be obtained from the Protein Data Bank under the entry code 1kp8.)

It can be more difficult to deduce the subunit composition in the case of hetero-oligomeric proteins, which contain different types of polypeptide chains. It may be relatively straightforward if the number of subunits is small to check which combination of subunit molecular masses gives the native mass of the protein. Thus, G-proteins, which are involved in many signalling processes, can be seen to consist of one copy of each of three different types of polypeptide chain, i.e. they are hetero-trimeric, and would be designated $\alpha\beta\gamma$.

The Greek letters used to designate quaternary structure of a protein denote whether subunits are identical or not. Thus, lactate dehydrogenase and glutamate dehydrogenase would be designated α_4 and α_6, respectively. Haemoglobin would be designated $\alpha_2\beta_2$ because it contains two copies of two different types of polypeptide chain.

There can be major problems establishing the quaternary structure in systems that contain several types of subunits, especially if some of them are small. A good example is the mitochondrial F1 ATPase (the enzyme which is driven in the direction of ATP synthesis by the proton gradient generated by the electron transport chain). The enzyme consists of five different types of polypeptide chain with molecular masses (in the case of the bovine mitochondrial enzyme) of 53, 50, 33, 17, and 7.5 kDa and is designated $\alpha_3\beta_3\gamma\delta\varepsilon$; the overall molecular mass is about 370 kDa.

In such cases, chemical analysis can help to establish the numbers of subunits. We could, for example, determine the numbers of the various N- or C-terminal amino acids present in each molecule of the native enzyme by Edman degradation or carboxypeptidase digestion, respectively (section 8.3).

8.5.2 Spatial arrangement of polypeptide chains in a protein

The most detailed information comes from high-resolution structural studies such as X-ray crystallography. However, a good deal of useful information can be obtained by using cross-linking reagents.

In section 8.2.1, it was noted that cross-linking of pairs of Lys side chains using the bis-(imido ester) dimethylsuberimidate followed by SDS-PAGE could be used to establish the number of subunits in the enzyme aldolase. Cross-linking can also be used to investigate which parts of the interacting protein partners are in close proximity to each other; this depends on identifying the amino acid side chains which react with the cross-linking agent.

A large range of cross-linking agents have been devised (Lundblad, 1995). These vary in terms of:

- The chemical nature of the reactive groups; this allows the types of amino acid side chains which can react to be varied
- The distance between the functional groups (by varying, for example the number of – (CH_2)– groups in a spacer arm. It is also possible to perform the so-called 'zero-length' cross-linking by using a water-soluble carbodiimide to promote the reaction of the amino group of a Lys side chain with the carboxyl group of an Asp or Glu side chain
- Whether or not the cross-linking can be reversed, e.g. if a cross-linking agent contained a disulphide bond, this could be subsequently cleaved by treatment with a thiol compound such as dithiothreitol.

Examples of the use of cross-linking to explore the interactions between proteins include the following:

- The structures of complexes involved in muscle contraction were investigated. Using a zero-length cross-linking approach, it was possible to isolate a species in which parts of the polypeptide chains of the two head groups of heavy meromyosin in the complex with F-actin were cross-linked. This demonstrated that there was direct contact between the two head groups in this complex (Onishi *et al.*, 1989)
- In the Ca^{2+}- and Mg^{2+}-activated ATPase in the inner membrane of *E. coli*, it was shown that the three α and three β subunits were arranged alternately in a ring around a central γ subunit (Fig. 8.32). This conclusion was reached because it was possible to form the cross-linked complexes $\alpha\beta$, $\alpha\gamma$, and $\beta\gamma$; however, no $\alpha\alpha$ or $\beta\beta$ cross-links could be formed (Bragg and Hou, 1975). This conclusion was subsequently confirmed by X-ray crystallography.

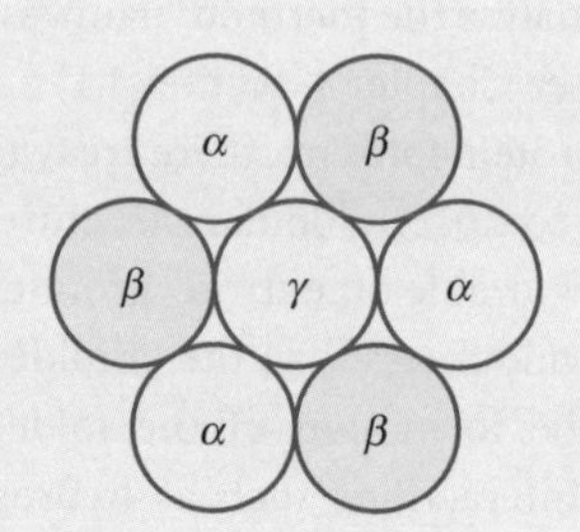

Fig. 8.32 The arrangement of subunits in ATPase as determined by cross-linking experiments.

SELF TEST ?

Check that you have mastered the key concepts at the start of this section by attempting the following question.

ST 8.9 The *E. coli* ribosome consists of a large and a small subunit, each of which contains ribosomal RNA and several proteins. In total, there are three different RNA chains and over 50 different proteins. How would you use cross-linking to investigate the three-dimensional structure of the ribosome?

Answer

ST 8.9 Cross-linking can be used to examine which proteins are in close proximity. It is useful to employ reversible cross-linking agents, e.g. those which incorporate a disulphide bond in their structure. After cross-linking, the product can be separated into the constituent components by incubation with dithiothreitol. The identities of these components can be established using immunodetection methods after SDS-PAGE (see section 8.2.1), provided that a range of specific antibodies is available. It should also be noted that it is possible to devise cross-linking agents which cross-link proteins with the ribosomal RNA chains, giving additional information.

8.6 How stable is the folded structure of a protein?

KEY CONCEPT

- Understanding how studies of the unfolding of proteins can be used to assess their stability

The three-dimensional structure of a protein is the key to its biological function. Any changes in the experimental parameters such as pH, temperature, or the addition of organic solvents can lead to loss of the native three-dimensional structure and hence loss of biological activity.

In thermodynamic terms the stability of the native folded structure (N) of a protein, relative to that of a fully unfolded state (U) in which there are no stable interactions between either main chain or side-chain atoms could be described in terms of the free energy.

In solution, a protein shows significant flexibility in its structure, as shown by studies of hydrogen/deuterium exchange, NMR, polarization of fluorescence, etc. It is therefore more appropriate to describe the structure of a protein in solution as being a time-averaged structure of all the different conformational states accessible to the protein.

It must be remembered that although the thermodynamic stability of a protein may be small, there is usually a large kinetic barrier to unfolding (i.e. unfolding has a large energy of activation, see Chapter 4, section 4.2), so proteins do no generally unfold spontaneously under normal physiological conditions. However, under unusual conditions, e.g. at the elevated temperatures associated with high fevers, the kinetic stability of proteins may be compromised.

We can analyse the thermodynamics of the process U $\rightleftharpoons$ N, using the eqn. $\Delta G = \Delta H - T\Delta S$ (see Chapter 4, section 4.1).

When a protein folds up, there are two competing forces at work. The formation of a highly compact globular structure from an extended polypeptide chain is highly unfavourable on entropy grounds, since the folded state has a much lower degree of randomness than the unfolded state (i.e. ΔS will be large and negative). However, the formation of the folded state will involve a large number of favourable interactions such as hydrogen bonds, ionic bonds, van der Waals' forces, etc., which will be associated with large negative values of ΔH. It is thought that the burial of non-polar side chains in the interior of the protein away from solvent (i.e. hydrophobic interactions, see Chapter 1, section 1.7.4) plays a major part in the overall energy balance favouring the folding of proteins.

The net result is that the overall ΔG for the process of folding represents a small difference between two large quantities, i.e. in thermodynamic terms, the stability of the folded state of a protein is usually small.

8.6.1 Determination of the stability of the folded state of a protein

Urea and guanidinium chloride are examples of chaotropic agents, which disrupt protein structures by altering the structure of water. This has the effect that all the side chains are nearly equally soluble in the solution of the denaturing agent; consequently proteins will unfold because this is entropically favourable.

Estimates of the thermodynamic stability of proteins can be made from measuring the thermal unfolding of proteins or by monitoring the unfolding of proteins by denaturing agents such as urea or guanidinium chloride. Both the experimental requirements and the equations involved are usually rather simpler for the second approach, so this will be described here.

In these experiments, the protein of interest is incubated with a range of concentrations of the denaturing agent, and after equilibrium has been achieved the extent of unfolding is measured, usually by changes in a spectroscopic parameter, which distinguishes the folded and unfolded states. Changes in the CD or fluorescence spectra of the protein are often used to monitor the unfolding. At each concentration of the denaturing agent we can therefore evaluate the fraction, f, of protein which is unfolded. It is particularly important to obtain data over a substantial proportion of the unfolding curve, generally between 10 and 90% unfolding (i.e. $0.1 < f < 0.9$).

Some proteins unfold by more complex mechanisms, which may involve one or more intermediate steps. It is possible to analyse data for such systems, although the equations involved are considerably more complex than those described here.

In the simplest case the protein behaves simply according to the equation N $\rightleftharpoons$ U, i.e. the two states interconvert reversibly. This can be demonstrated by showing that the unfolding is indeed reversible so that as the denaturant concentration is lowered there is a refolding curve, superimposable on the unfolding curve.

At each denaturant concentration we can work out a value for the equilibrium constant K:

$$K = [U]/[N] = f/(1-f) \qquad 8.20$$

This can be expressed as a free energy for the unfolding process (ΔG^0) under these conditions, using the eqn. $\Delta G^0 = -RT \ln K$ (see Chapter 4, section 4.1):

$$\Delta G^0 = -RT \ln (f/(1-f)) \qquad 8.21$$

Experimentally, it is usually found that there is a linear relationship (eqn. 8.22) between the values of ΔG^0 and the [denaturant]:

$$\Delta G^0 = \Delta G^0(H_2O) - m\,([\text{denaturant}]) \qquad 8.22$$

so that a plot of ΔG^0 vs [denaturant] can be extrapolated to the y-axis (i.e. where [denaturant] = 0) to give the ΔG^0 of the folded state under native conditions, i.e. in the absence of denaturant (Fig. 8.33).

While there is a certain amount of theoretical justification for the linear extrapolation procedure, in some cases it has been found necessary to include terms in $[\text{denaturant}]^2$ in the equation for ΔG^0 to make the extrapolation (and hence the estimate of $\Delta G^0(H_2O)$) more reliable.

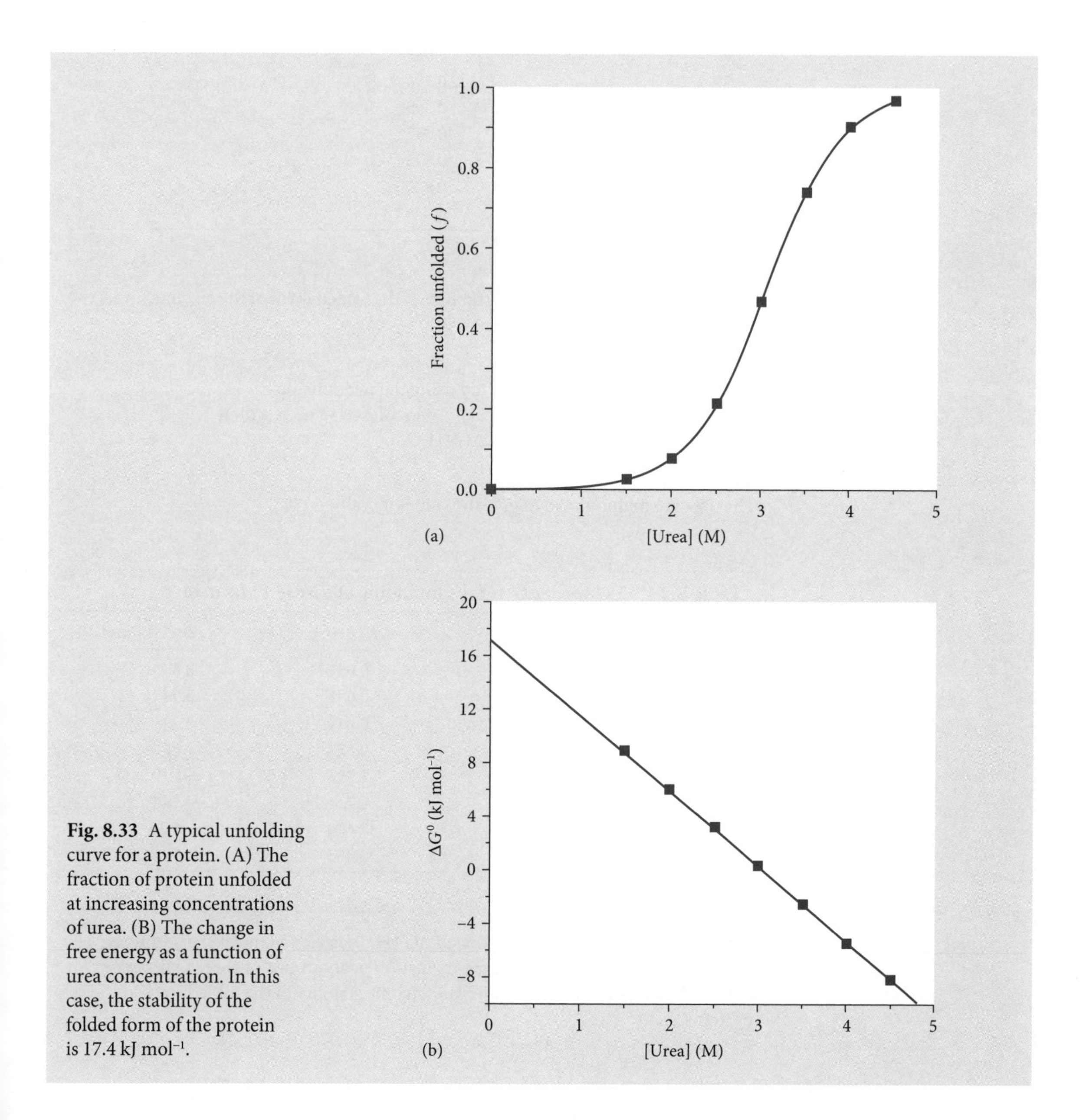

Fig. 8.33 A typical unfolding curve for a protein. (A) The fraction of protein unfolded at increasing concentrations of urea. (B) The change in free energy as a function of urea concentration. In this case, the stability of the folded form of the protein is 17.4 kJ mol^{-1}.

The unfolding of the enzyme RNase T1 from the fungus *Aspergillus oryzae* by urea can be conveniently monitored by changes in the fluorescence due to tyrosine residues (there are no tryptophan residues in the protein). The data in Table 8.23 were obtained at 25°C.

Table 8.23 Unfolding of RNase T1 by increasing concentrations of urea

[Urea] (M)	Fraction unfolded (f)
3.5	0.033
4.0	0.083
4.5	0.196
5.0	0.395
5.5	0.637
6.0	0.825
6.5	0.927
7.0	0.971

Analyse the data to estimate the stability of the folded form of the enzyme under native conditions (i.e. when [urea] = 0).

STRATEGY

At each [urea], the ΔG^0 is calculated using the relationship $\Delta G^0 = -RT \ln (f/(1-f))$ (eqn. 8.21). A plot of ΔG^0 vs urea is then extrapolated to the y-axis to give $\Delta G^0(H_2O)$.

SOLUTION

The data are manipulated to give the values in Table 8.24.

Table 8.24 Values of ΔG^0 for the unfolding of RNase T1 by urea

[Urea] (M)	f	$f/(1-f)$	ΔG^0 (kJ mol^{-1})
3.5	0.033	0.034	8.37
4.0	0.083	0.091	5.94
4.5	0.196	0.244	3.49
5.0	0.395	0.653	1.06
5.5	0.637	1.755	−1.39
6.0	0.825	4.714	−3.84
6.5	0.927	12.698	−6.29
7.0	0.971	33.483	−8.69

From the graph (Fig. 8.34) the value of $\Delta G^0(H_2O)$ is 25.5 kJ mol^{-1}. This value is equivalent in energetic terms to the formation of about two or three hydrogen bonds, showing the very small thermodynamic stability of the folded states of proteins.

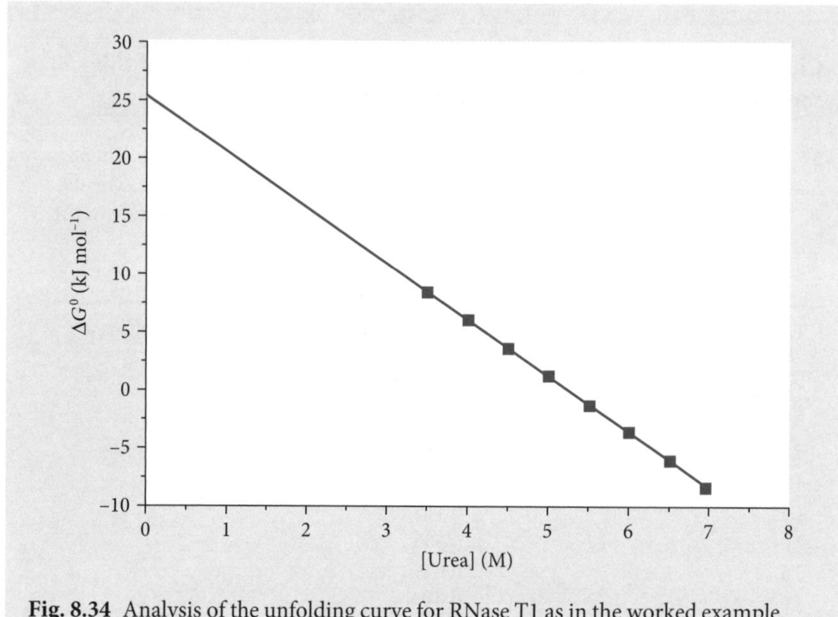

Fig. 8.34 Analysis of the unfolding curve for RNase T1 as in the worked example.

Studies of the type explained in the worked example have shown that the thermodynamic stabilities of the folded states of proteins (i.e. the values of $\Delta G^0(H_2O)$) are generally in the range 20–60 kJ mol^{-1}, so represents the value of relatively few weak interactions (see Chapter 1, section 1.7) and is much less than the strength of a typical covalent bond (about 400 kJ mol^{-1}). As stated earlier, this illustrates the fact that the stability of the folded state of a protein represents the small difference between two large energies with the ΔH and $T\Delta S$ contributions working in opposition. The importance of interactions formed by individual side chains in a protein can be assessed by the production of suitably designed site-directed mutants. Examples of the application of these are given in the self-test question ST 8.10 and the Problem 8.11 at the end of the chapter.

? SELF TEST

Check that you have mastered the key concepts at the start of this section by attempting the following question.

ST 8.10 The urea-induced unfolding of a mutant of RNase T1 in which Glu 58 had been replaced by Ala was monitored by changes in Tyr fluorescence. The data in Table 8.25 were obtained at 25°C. What is the stability of the folded state of the enzyme?

Table 8.25 Unfolding of RNase T1 by increasing concentrations of urea

[Urea] (M)	Fraction unfolded (f)
3.0	0.033
3.5	0.088
4.0	0.206
4.5	0.415
5.0	0.665
5.5	0.848
6.0	0.937
6.5	0.976

Answer

ST 8.10 From the extrapolated graph of ΔG^0 vs [urea], the $\Delta G^0(H_2O)$ is 23.3 kJ mol^{-1}. Thus the mutant is about 2.2 kJ mol^{-1} less stable than the wild-type enzyme described in the worked example in Section 8.6. Loss of the negatively charged side chain of Glu 58, with consequent disruption of hydrogen bonds and ionic interactions, does not greatly affect the stability of the folded state of the enzyme.

Attempt Problem 8.11 at the end of the chapter.

8.7 Problems

Full solutions to odd-numbered problems are available to all in the student section of the Online Resource Centre at www.oxfordtextbooks.co.uk/orc/price/. Full solutions to even-numbered problems are available to lecturers only in the lecturer section of the Online Resource Centre.

8.1 The eukaryotic 26S proteasome is responsible for the degradation of proteins via the ubiquitin–proteasome pathway. The 26S proteasome has an estimated molecular mass of 2600 kDa and is composed of 48 subunits. The 26S proteasome is composed of a 20S core proteolytic particle and a 19S regulatory particle. The 20S proteolytic particle was analysed using non-denaturing gel electrophoresis. The molecular mass markers used were: alcohol dehydrogenase (150 kDa), apoferritin (443 kDa), thyroglobuin (669 kDa), and 26S proteasome (2600 kDa). Using the log R_f values below and Ferguson plot analysis, estimate the molecular mass of the 20S particle.

Protein	**Molecular mass (kDa)**	**2.5% gel**	**3% gel**	**3.5% gel**	**4% gel**
Alcohol dehydrogenase	150	1.98	1.94	1.92	1.89
Apoferritin	443	1.82	1.74	1.67	1.58
Thyroglobulin	669	1.73	1.61	1.49	1.38
26S proteasome	2600	1.57	1.16	0.78	0.41
20S proteasome		1.69	1.56	1.44	1.33

8.2 Bacterioferritin, a haem-containing iron carrier, was isolated from the nitrogen fixing bacterium *Azotobacter vinelandii*. The molecular mass of the protein was analysed by gel filtration under native and denaturing conditions. Gel filtration with Sephacryl S-300, equilibrated with buffer A (50 mM sodium phosphate buffer, 100 mM NaCl, pH 7.2), suggested a native molecular mass of 435 kDa. Gel filtration with Sephacryl S-200, equilibrated with buffer A plus 6 M guanidinium chloride, resulted in the elution of a single species with a K_d value of 0.65. Using the calibration data obtained in the presence of 6 M guanidinium chloride in the table below:

(a) Calculate the apparent molecular mass of the bacterioferritin subunits

(b) Suggest the oligomeric structure of *A. vinelandii* bacterioferritin

Protein	**Molecular mass (kDa)**	K_d
Bovine serum albumin	66	0.20
Ovalbumin	45	0.33
Chymotrypsinogen A	25	0.53
Myoglobin	17.8	0.67
Cytochrome *c*	12.4	0.80
Bacterioferritin		0.65

8.3 The molecular mass of the enzyme glyceraldehyde-3-phosphate dehydrogenase from rabbit muscle was studied by sedimentation velocity. The experiment was performed at 20°C using a protein concentration of 0.5 mg mL^{-1}. Once the rotor had reached a constant speed of 60 000 rpm, the distance (r) of the boundary from the axis of rotation was recorded over a period of 30 min with the following results.

Time (s)	r (cm)
0	6.2215
360	6.2909
720	6.3611
1080	6.4321
1440	6.5045
1800	6.5771

Given that the partial specific volume ($\bar{v}$) of the enzyme is 0.729 mL g^{-1}, the density of the buffer is 0.9982 g mL^{-1}, and the diffusion coefficient of the enzyme is 4.75×10^{-7} cm^2 s^{-1}, what is the molecular mass of the enzyme? The appropriate value of the gas constant, R, to be used is 8.31×10^7 erg K^{-1} mol^{-1}.

8.4 Positive mode electrospray mass spectrometry analysis of apomyoglobin (myoglobin without the haem group) gave the following m/z values. These values represent the most abundant adjacent peaks. Calculate the molecular mass of apomyoglobin.

m/z
848.55
893.16
942.73
998.12

8.5 A subunit of ATPase from *E. coli* was purified and analysed by mass spectrometry. Two species were identified: the full-length protein with a mass of 15 068.2 Da and a second species with an apparent mass of 14 865.9 Da. Using the sequence below, together with amino acid mass values in Table 1.1, Chapter 1 identify the nature of the species with the lower molecular mass.

MAMTYHLDVVSAEQQMFSGLVEKIQVTGSEGELGIYPGHAPLLTAIKPG
MIRIVKQHGHEEFIYLSGGILEVQPGNVTVLADTAIRGQDLDEARAMEAK
RKAEEHISSSHGDVDYAQASAELAKAIAQLRVIELTKKAM

8.6 From the molecular mass values of a lysozyme (14 700.6 Da) and its precursor prelysozyme (16 537.0 Da), calculate the mass of the N-terminal signal peptide of lysozyme. Using the sequence details for prelysozyme below and the calculated signal peptide mass, determine the sequence of the signal peptide. Note that the mass values given are for the reduced forms of lysozyme and prelysozyme.

MKALIVLGLVLLSVTVQGKVFERCELARTLKRLGMDGYRGISLANWMCL
AKWESGYNTRATNYNAGDRSTDYGIFQINSRYWCNDGKTPGAVNACH
LSCSALLQDNIADAVACAKRVVRDPQGIRAWVAWRNRCQNRDVRQYV
QGCGV

8.7 Glycogen phosphorylase (residues 2–843) has a theoretical molecular mass of 97 158 Da. The experimentally determined mass of glycogen phosphorylase is 97 280 Da. Two tryptic peptides of glycogen phosphorylase have molecular mass values, which differ from the expected values as indicated below. Determine the nature and position of the post-translational modification of glycogen phosphorylase.

Tryptic peptide	Position	Calculated mass (Da)	Experimentally determined mass (Da)
SRPLSDQEK	2–10	1059.5	1101.5
QISVR	13–17	602.4	682.4

8.8 Ellman's reagent (5,5′-dithiobis-(2-nitrobenzoic acid); Nbs_2) reacts with Cys side chains in proteins to give stoichiometric quantities of thionitrophenolate, which absorbs at 412 nm. The absorption coefficient of thionitrophenolate at this wavelength is 13 600 M^{-1} cm^{-1}. When a 0.35 mg mL^{-1} solution of aldolase from rabbit muscle was reacted with a large molar excess of Nbs_2 in a cuvette of 1-cm pathlength, there was a rapid increase in A_{412} to a value of 0.115. This was followed by a slower reaction which gave rise to a total increase in A_{412} to 0.342. When the experiment was carried out in the presence of 4 M guanidinium chloride the limiting increase in A_{412} was 0.817. The molecular mass of each subunit of aldolase is 40 kDa. What can you conclude about the Cys side chains in aldolase?

8.9 The enzyme SK from the plant pathogen *E. chrysanthemi* has a single Trp residue (Trp 54). On excitation at 290 nm, the fluorescence emission maximum from this Trp is at 346 nm. The quenching of the fluorescence by succinimide was studied for the enzyme in buffer and in the presence of a substrate (shikimate) or product (ADP). Parallel studies of the effect of succinimide of the fluorescence of the model compound *N*-acetyltryptophanamide (NATA) were made; in this case inclusion of the substrate molecules has no effect on the pattern of quenching. The results obtained are shown below; because the substrates cause quenching of the protein fluorescence, in each case the data have been normalized so that in the absence of succinimide the fluorescence is shown as 100 Units. What can you conclude about the degree of exposure of the Trp residue in SK and the effect of the ligands on this residue?

[Succinimide] (mM)	Fluorescence			
	NATA	SK alone	SK + 2 mM shikimate	SK + 2 mM ADP
0	100	100	100	100
19.8	84.0	91.3	96.6	92.0
39.2	72.7	84.2	93.4	85.3
58.3	64.1	78.1	90.5	79.6
76.9	57.5	73.0	87.8	74.7
95.2	52.2	68.6	85.4	70.5
113.2	47.9	64.8	83.1	66.8
130.8	44.3	61.4	80.9	63.5
148.1	41.3	58.4	79.0	60.5
165.1	38.7	55.8	77.1	57.9
181.8	36.4	53.4	75.3	55.6

8.10 The enzyme structure of the enzyme citrate synthase was studied by CD. A 0.43 mg mL^{-1} solution of enzyme in sodium phosphate buffer at pH 7.0 was studied in a cell of pathlength 0.02 cm. The far UV CD spectrum showed two negative minima at 222 and 208 nm and a large positive maximum at 193 nm. The values of the ellipticities at 222 and 208 nm were −13.7 and −15.1 mdeg, respectively. What can you conclude about the structure of the enzyme? (The MRW of citrate synthase is 112 Da).

8.11 The urea-induced unfolding of a mutant of RNase T1 in which Asp 76 was replaced by Asn was studied at 25°C. The following results were obtained:

[Urea] (M)	Fraction unfolded (f)
0.5	0.041
1.0	0.127
1.5	0.328
2.0	0.620
2.5	0.846
3.0	0.948

What is the stability of the folded state of the enzyme?

References for Chapter 8

Barrow, C.J., Yasuda, A., Kenny, P.T., and Zagorski, M.G. (1992) *J. Mol. Biol.* **225**, 1075–93.

Bermudez, V.P., MacNeill, S.A., Tappin, I., and Hurwitz, J. (2002) *J. Biol. Chem.* **277**, 36853–62.

Bragg, P.D. and Hou, C. (1975) *Arch. Biochem. Biophys.* **167**, 311–21.

Brown, T.A. (2001) *Essential Molecular Biology*, 2nd edn. Oxford University Press, Oxford.

Byron, O. (1996) *Analytical Ultracentrifugation*. In: *Proteins Labfax*, ed. Price, N.C. Bios Scientific Publishers, Oxford, 318 pp, Chapter 16A.

Coggins, J.R. (1996) *Cross-linking Reagents for Proteins*. In: *Proteins Labfax*, ed. Price, N.C. Bios Scientific Publishers, Oxford, Chapter 26.

Edman, P. and Begg, G. (1967) *Eur. J. Biochem.* **1**, 80–91.

Eftink, M.R. and Ghiron, C.A. (1976) *Biochemistry* **15**, 672–80.

Eftink, M.R. and Ghiron, C.A. (1984) *Biochemistry* **23**, 3891–99.

Giletto, A. and Pace, C.N. (1999) *Biochemistry* **38**, 13379–84.

Kartik, V.J., Lavanya, T., and Guruprasad, K. (2006) *Int. J. Biol. Macromolecules* **38**, 174–79.

Kelly, S.M., Jess, T.J., and Price, N.C. (2005) *Biochim. Biophys. Acta* **1751**, 119.

Krell, T., Coggins, J.R., and Lapthorn, A.J. (1998) *J. Mol. Biol.* **278**, 985–97.

Laemmli, U.K. (1970) *Nature* **227**, 680–85.

Lehrer, S.S. (1971) *Biochemistry* **10**, 3254–63.

Liu, H.L., Zhou, H.N., Xing, W.M., Zhao, J.F., Li, S.X., Huang, J.F., and Bi, R.C. (2004) *FEBS Lett.* **573**, 93–98.

Lundblad, R.L. (1995) *Techniques in Protein Modification*. CRC Press, Boca Raton, FL, USA, 288 pp, Chapter 15.

Onishi, H., Maita, T., Matsuda, G., and Fujiwara, K. (1989) *Biochemistry* **28**, 1905–12.

Price, N.C. and Stevens, L. (1999) *Fundamentals of Enzymology*, 3rd edn. Oxford University Press, Oxford.

Sambrook, J. and Russell, D.W. (2006) *Condensed Protocols from Molecular Cloning: A Laboratory Manual*. Cold Spring Harbor Laboratory Press, Cold Spring Harbor, NY.

Sanger, F., Nicklen, S., and Coulson, A.R. (1977) *Proc. Natl. Acad. Sci. USA* **74**, 5463–67.

Sheehan, D. (2000) *Physical Biochemistry: Principles and Applications*. John Wiley and Sons, Chichester, Chapter 6.

Siegert, R., Leroux, M.R., Scheufler, C., Hartl, F.U., and Moarefi, I. (2000) *Cell* **103**, 621–32.

van Holde, K.E., Johnson, W.C., and Ho, P.S. (1998) *Principles of Physical Biochemistry*. Prentice-Hall, Upper Saddle River, NJ, 657 pp, Chapters 5 and 13.

Walter, R.A., Nairn, J., Duncan, D., Price, N.C., Kelly, S.M., Rigden, D.J., and Fothergill-Gilmore, L.A. (1999) *Biochem. J.* **337**, 89–95.

Appendix

Appendix 8.1 The Svedberg equation for sedimentation velocity ultracentrifugation

A radian (abbreviated rad) is defined as being $(1/2\pi)$ times a complete revolution, i.e. 1 rad = $(360/2\pi)° = 57.3°$. Hence $\omega = 2\pi \times$ (revolutions per second (rps)), or $0.1047 \times$ (rpm).

The partial specific volume ($\bar{v}$) of a solute is the increase in volume when 1 g of the protein is added to a large volume of solution. While it is possible to measure $\bar{v}$ for a protein directly from rather elaborate measurements of density, it is much more common to calculate it from its amino acid composition assuming that each amino acid contributes to the volume in an additive manner (van Holde *et al.*, 1998).

In sedimentation velocity ultracentrifugation, there are three types of force that are exerted on the protein.

The first of these is the centrifugal force (F_c) which is equal to $\omega^2 rm$ (where ω is the angular velocity in radians per second), r is the distance from the axis of rotation and m is the mass of the macromolecule.

The second force is that of buoyancy (F_b) resulting from the mass of the solution (m_o) displaced by the protein; this is equal to $-\omega^2 rm_o$. The mass of solution displaced (m_o) is given by the product of the mass of the macromolecule (m), its partial specific volume ($\bar{v}$), and the density of the solution (ρ).

The third force is a frictional force (F_f) which arises from the viscous drag on the protein moving at a velocity v; this is equal to $-fv$, where f is known as the frictional coefficient.

We can derive the equations governing sedimentation velocity by noting that in the steady state the sum of the three forces ($F_c + F_b + F_f$) will be zero, i.e. the protein will be moving at a constant speed in the gravitational field. When the various terms are multiplied by Avogadro's number (L) to bring them to a molar basis we obtain eqn. A.8.1, where M is the molecular mass:

$$M(1 - \bar{v}\rho)/Lf = v/\omega^2 r \quad \text{A.8.1}$$

The term $v/\omega^2 r$ in eqn. A.8.1 is replaced by s, the sedimentation coefficient which is in units of seconds (s). Values of about 1×10^{-13} s are common for biological macromolecules such as proteins, so 1×10^{-13} s is known as 1 Svedberg unit (S) after the Swedish physical chemist Svedberg who developed the ultracentrifuge for the study of biological macromolecules (see Chapter 6, section 6.4.1.2). This gives eqn. A.8.2:

$$M(1 - \bar{v}\rho)/Lf = s \quad \text{A.8.2}$$

In a sedimentation velocity experiment, the rate of movement of the boundary of the macromolecules along the cell (i.e. away from axis of rotation) is measured.

At a given time t, the distance of the boundary (in practice, the mid-point of the boundary) from the axis of rotation is r. The value of the sedimentation coefficient s can be determined from a graph of $\ln r$ vs t, the slope of which is $\omega^2 s$. Since we know the rotor speed and therefore the value of ω, we can calculate s.

Since $v = dr/dt$, we can substitute in the definition of s, to give $dr/dt = \omega^2 rs$. Rearranging, this gives $dr/r = \omega^2 s.dt$. On integration between time zero and time t, we obtain $\ln r_t - \ln r_o = \omega^2 s(t - t_o)$. Hence a plot of $\ln r$ vs t is a straight line of slope $\omega^2 s$.

Eqn. A.8.2 shows that we can calculate the molecular mass, M, from the sedimentation coefficient provided we know the partial specific volume (which we can calculate from the amino acid composition), the solution density ρ and the frictional coefficient f. For an ideal solution, f can be obtained from the diffusion coefficient D using eqn. A.8.3:

$$f = RT/LD \quad \text{A.8.3}$$

where R is the gas constant and T the temperature in K. D can be measured directly from observing the rate of diffusion of the macromolecule in a column or indirectly by the techniques of dynamic light scattering or from diffusion measurements using NMR.

Further details about the relationship between the diffusion coefficient and frictional coefficient can be found in van Holde *et al.* (1998).

Substituting the value of f from eqn. A.8.3 into eqn. A.8.2 and simplifying, we obtain eqn. A.8.4:

$$MD(1 - \bar{v}\rho)/RT = s \quad \text{A.8.4}$$

which on rearrangement gives eqn. A.8.5:

$$M = RTs/(D(1 - \bar{v}\rho)) \quad \text{A.8.5}$$

Eqn. A.8.5 is usually known as the Svedberg equation.

Appendix 8.2 The Stern–Volmer and Lehrer equations for fluorescence quenching

Quenching of fluorescence occurs when the excited state of a fluorophore (denoted E^*) loses energy either by colliding with an external added molecule (quencher, Q) or forming a long-lasting complex with the quencher; these termed dynamic and static quenching, respectively.

The excited state (E^*) can undergo three possible fates; these processes are assigned rate constants k_1, k_2, and k_3:

1) Fluorescence (to give the ground state and emit light); the rate of this reaction is $k_1[E^*]$.
2) Internal quenching (structural rearrangement in the excited state during which energy is lost); the rate of this reaction is $k_2[E^*]$.
3) External quenching by the quencher; the rate of this reaction is $k_3[E^*][Q]$.

Internal quenching usually occurs when vibrational levels of the ground state and excited state are of similar energies. This will often be the case for flexible molecules, and is the major reason why relatively few molecules do act as fluorophores. Most fluorescent molecules contain rigid aromatic systems with delocalized electrons.

Thus in the presence of the quencher the relative fluorescence, F (i.e. relative to the fluorescence if the excited state emitted fluorescence with 100% efficiency) is given by dividing the rate of process 1 by the sum of the rates of processes 1, 2, and 3, i.e.

$$F = k_1[E^*]/(k_1[E^*] + k_2[E^*] + k_3[E^*][Q]) \quad \text{A.8.6}$$

Division of each term in the numerator and denominator of eqn. A.8.6 by $[E^*]$ gives:

$$F = k_1/(k_1 + k_2 + k_3[Q]) \quad \text{A.8.7}$$

In the absence of quencher, the fluorescence (designated as F_0) is given by setting $[Q] = 0$ in eqn. A.8.7:

$$F_0 = k_1/(k_1 + k_2) \quad \text{A.8.8}$$

The ratio of F_0 to F is thus given by dividing eqn. A.8.8 by eqn. A.8.7:

$$F_0/F = (k_1 + k_2 + k_3[Q])/(k_1 + k_2)$$

which can be rearranged to give eqn. A.8.9:

$$F_0/F = 1 + (k_3[Q]/(k_1 + k_2)) \quad \text{A.8.9}$$

Since k_1, k_2, and k_3 are constants for a given fluorophore, this can be expressed as the Stern–Volmer equation (A.8.10):

$$F_0/F = 1 + K_{SV}[Q] \quad \text{A.8.10}$$

The quantity $1/(k_1 + k_2)$ is known as the life time (τ) of the excited state. The Stern–Volmer equation (A.8.10) can thus be also expressed as $F_0/F = 1 + k_3\tau[Q]$, with K_{SV} equal to $k_3\tau$.

where K_{SV} is a constant known as the Stern–Volmer constant.

A plot of F_0/F vs $[Q]$ (Fig. 8.26) will give a straight line of slope K_{SV} with a y-axis intercept of 1.

It should be noted that both the dynamic and static quenching processes give linear Stern–Volmer plots but can be distinguished by measuring the quenching over a range of temperatures and solvent viscosities.

An extension of the Stern–Volmer equation can be made to deal with the situation where a certain fraction (f_a) of the observed fluorescence is accessible to externally added quenchers and the remaining part is not accessible at all (Lehrer, 1971). These two categories of fluorophores are assigned Stern–Volmer constants of K_{SV} and 0, respectively.

In the absence of quencher, the fluorescence (F_0) can be regarded as the sum of the accessible fluorescence and the non-accessible fluorescence:

$$F_0 = f_aF_0 + (1 - f_a)F_0 \quad \text{A.8.11}$$

In the presence of quencher at a concentration $[Q]$, according to eqn. A.8.10 the accessible fluorescence will be reduced by a factor $(1 + K_{SV}[Q])$; the non-accessible fluorescence will be unaffected. The overall fluorescence, F, will thus be given by:

$$F = f_a F_0/(1 + K_{SV}[Q]) + (1 - f_a)F_0 \qquad \text{A.8.12}$$

The difference in fluorescence (ΔF) caused by addition of quencher is obtained by subtracting eqn. A.8.12 from eqn. A.8.11:

$$\begin{aligned}\Delta F &= F_0 - F \\ &= f_a F_0 - (f_a F_0/(1 + K_{SV}[Q]))\end{aligned}$$

which can be rearranged to give:

$$\Delta F = f_a F_0 K_{SV}[Q]/(1 + K_{SV}[Q]) \qquad \text{A.8.13}$$

Hence, by dividing eqn. A.8.11 by eqn. A.8.13, $F_0/\Delta F$ can be expressed as:

$$\begin{aligned}F_0/\Delta F &= (1 + K_{SV}[Q])/(f_a K_{SV}[Q]) \\ &= 1/f_a K_{SV}[Q] + 1/f_a \qquad \text{A.8.14}\end{aligned}$$

which is usually known as the Lehrer equation.

A plot of $F_0/\Delta F$ vs $1/[Q]$ (Fig. 8.27) will give a straight line of slope $1/f_a K_{SV}$ and a y-axis intercept of $1/f_a$. From these the values of f_a (equal to $1/y$-axis intercept) and K_{SV} (equal to the ratio intercept/slope) can be determined.

9 Enzyme activity and mechanism

9.1 Introduction

KEY CONCEPTS

- Knowing what is meant by the mechanism of an enzyme-catalysed reaction
- Explaining the importance of studying enzyme kinetics

Describing the mechanism of an enzyme-catalysed reaction involves providing answers to the following questions:

- What is the sequence of enzyme-containing complexes as the substrate is converted to product?
- How quickly are these complexes interconverted?
- How is the breaking and making of bonds facilitated by the enzyme?

From this we can see that both kinetic and structural information is necessary to provide an understanding of the mechanism. In this chapter we shall focus mainly on the kinetic experiments as these are more amenable to routine laboratory investigations. Some of the structural approaches which can be used, such as the mapping of active sites and identification of amino acids involved in catalysis or binding, will also be mentioned. We shall review the basic theory underpinning each of the topics and illustrate the approaches using worked examples and specimen data. Details of experiments that can be used in teaching laboratories are given in the Practical activities 9.1–9.5, which are available for lecturers in the On-line Resource Centre for this text to support teaching of this topic. The high-resolution structural methods (X-ray crystallography and NMR) which can elucidate structures of enzymes at the atomic level are much more specialized and beyond the scope of this book.

Good introductory accounts of X-ray crystallography and NMR are given in the book by Sheehan, 2000.

There are at least three good reasons why a study of enzyme kinetics is important: firstly, it is observing the enzyme doing its job, i.e. actually catalysing a reaction, secondly, it can provide very important information about the mechanism, and thirdly, it helps us to understand the physiological role of the enzyme in an organism.

9.2 How to assay the activity of an enzyme

KEY CONCEPTS

- Designing a suitable assay for a given enzyme
- Knowing how to ensure that reliable kinetic data are obtained

Many of the key points about enzyme assays have been covered in Chapters 5 and 6 (sections 5.3.1 and 6.3, respectively). The following recaps some of the main points. In all cases, to design an enzyme assay, it is necessary to identify some property that can distinguish between the substrate and the product of the reaction, and can be easily measured. There are three key approaches that we can use.

1) The stop and sample approach
   In principle, any enzyme can be assayed by simply mixing enzyme with substrate at a defined time point and then, after specified time intervals, stopping the reaction quickly and analysing the mixture to ascertain how much substrate has been converted to product (see Chapter 6, section 6.3.1).
2) The continuous, direct approach
   It is much easier to assay an enzyme if there is some readily measurable change so that the reaction can be monitored continuously, e.g. by spectrophotometry (see Chapter 6, section 6.3.1). Obtaining the data continuously will also make the analysis more reliable, since the effect of sampling errors is eliminated.
3) The continuous, indirect approach
   In cases where there is no convenient difference between the properties of the substrates or products of the reaction itself, it may be possible to couple the reaction of interest to one or more further enzyme-catalysed reaction so that a change can be measured in a continuous fashion (see Chapter 6, section 6.3.3). For such an assay system to work efficiently, any coupling substrate and enzyme must be added in sufficient quantities so that the coupling step is not limiting. This can be checked by establishing that the observed rate of the reaction of interest is proportional to the amount of enzyme added to the assay mixture. It is also essential that the coupling enzyme must be highly purified; in particular it must not contain any detectable amount of the enzyme we are trying to assay.

Some examples of how different types of enzymes can be assayed using these approaches are described in sections 9.2.1–9.2.4. In section 9.2.5, we indicate how to establish the reliability of the enzyme kinetic data obtained.

9.2.1 Assays of hydrolases

In vivo many hydrolases (enzymes catalysing hydrolysis reactions) act on macromolecular substrates such as proteins, nucleic acids, or polysaccharides. In general, it is not very convenient to use these substrates for detailed mechanistic studies,

Enzymes that catalyse the hydrolysis of macromolecular substrates cannot generally be monitored in a continuous fashion. One exception is lysozyme, which can be assayed by its action on a suspension of the Gram-positive bacterium *Micrococcus lysodeikticus*; lysis of the bacterial cells leads to a decrease in turbidity, which can be monitored in a spectrophotometer as an apparent decrease in absorbance at 450 nm.

mainly because the detectable change in properties (for example, release of low molecular mass fragments which are soluble under conditions which precipitate the macromolecule) cannot be ascribed to cleavage of a single bond. It is therefore more useful to study the action of the enzymes on small model substrates which contain the important structural features of the natural substrates; the action of enzymes on these substrates represents a defined chemical event. Some examples of this approach are described in section 9.9.3.

It is often possible to design small artificial substrates with desirable spectroscopic properties for hydrolytic enzymes. For example, 4-nitroanilides are convenient model substrates for proteases (see section 9.9.3), since they contain an amide bond and the product (4-nitroaniline) absorbs strongly at 405 nm (ε = 9500 M^{-1} cm^{-1} at pH 7.6), while the substrate absorbs only weakly at this wavelength. Hydrolysis of 4-nitrophenyl esters can be conveniently monitored at 400 nm where the product, 4-nitrophenol, absorbs (ε = 13 500 M^{-1} cm^{-1} at pH 7.6). On hydrolysis, umbelliferyl esters yield the highly fluorescent product umbelliferone (7-hydroxycoumarin) and can thus be used as very sensitive substrates for the assay of a number of hydrolases, including esterases, proteases, and lipases.

Many hydrolysis reactions proceed with the liberation or uptake of a proton. For example, the hydrolysis of esters (catalysed by esterases and proteases) generates an acid, which will be ionized at around neutral pH and an alcohol, with the release of a H^+. If the reaction is carried out in an unbuffered (or very lightly buffered) solution, the release of the H^+ can be detected by using an autotitration device, which will measure the amount of OH^- required to maintain a constant pH. This type of approach could also be used, for example to measure the activity of ATPase (which catalyses the hydrolysis of ATP to give ADP + P_i (inorganic phosphate) with release of a H^+) or urease (which catalyses the hydrolysis of urea to give CO_2 + $2NH_3$ accompanied by uptake of a H^+). This approach can be very useful when there is no convenient spectroscopic change on reaction.

Assays that involve the formation of fluorescent or radioactive products can be made very sensitive indeed. For example, it is possible to detect the activity of single molecules of a number of enzymes such as LDH taking advantage of the fact that NADH is fluorescent, whereas NAD^+ is not (Xue and Yeung, 1995).

9.2.2 Assays of dehydrogenases

Dehydrogenases catalyse redox reactions in which $NAD(P)^+$ and NAD(P)H are interconverted (e.g. in the case of lactate dehydrogenase (LDH) the reaction is lactate + $NAD^+ \rightleftharpoons$ pyruvate + NADH + H^+). The redox reaction can readily be monitored spectrophotmetrically because NAD(P)H absorbs at 340 nm (ε = 6220 M^{-1} cm^{-1}), whereas $NAD(P)^+$ does not absorb at the wavelength (see Chapter 7, section 7.3.1).

WORKED EXAMPLE 

LDH was assayed in the direction of lactate formation (i.e. pyruvate + NADH + H^+ → lactate + NAD^+). 10 μL of a solution of LDH (10.4 μg mL^{-1}) was added to an assay mixture of total volume 2 mL, in a cell of 1-cm pathlength. The decrease in A_{340} min^{-1} was 0.153. What is the specific activity of the LDH in units of μmol^{-1} min^{-1} mg^{-1}?

STRATEGY

This problem involves using the absorption coefficient (6220 M^{-1} cm^{-1}) to calculate the rate of NADH consumption (in units of μmol min^{-1}) and the amount of enzyme added (in mg). Division then gives the specific activity.

SOLUTION

The change in NADH concentration is 0.153/6220 M min^{-1} = 24.6 μM min^{-1}. With a 2 mL reaction volume, this is (2/1000) × 24.6 = 0.0492 μmol min^{-1}. The amount of enzyme added to the assay mixture is (10/1000) × 10.4 μg = 0.104 μg = 0.104 × 10^{-3} mg. Thus, the specific activity of the LDH is 0.0492/(0.104 × 10^{-3}) μmol min^{-1} mg^{-1} = 473 μmol min^{-1} mg^{-1}. (The calculation of specific activity has been discussed in Chapter 3, section 3.8.)

9.2.3 Assays of kinases

Kinases catalyse reactions in which the γ-phosphoryl group of ATP is transferred to a second substrate, with the formation of ADP. Examples include hexokinase (acts on glucose to yield glucose-6-phosphate) and creatine kinase (acts on creatine to yield phosphocreatine).

The phosphorus atoms of nucleotides such as ATP are labelled sequentially from the point of attachment to the 5′-carbon atom of the ribose, as α, β, and γ. (For the sake of simplicity, only the bridging oxygen atoms are shown.)

$$\text{Adenine-ribose–O–}\overset{\alpha}{\text{P}}\text{–O–}\overset{\beta}{\text{P}}\text{–O–}\overset{\gamma}{\text{P}}$$

It is the γ-phosphoryl group of ATP that is transferred to an acceptor substrate by the action of a kinase. By contrast, it is the base-sugar-α-phosphoryl grouping that is added to a growing DNA or RNA chain by the action of the appropriate polymerase enzyme.

For the vast majority of kinases we can monitor the ADP produced by a double-coupled reaction involving the pyruvate kinase and LDH reactions, resulting in a conversion of NADH to NAD^+, which can be monitored by a decrease in absorbance at 340 nm (section 9.2.2). The coupling substrates (phosphoenolpyruvate (PEP) and NADH) and the coupling enzymes (pyruvate kinase and LDH) must be added in sufficient quantities so that the coupling step is not limiting.

creatine + ATP → phosphocreatine + ADP	creatine kinase
ADP + PEP → ATP + pyruvate	pyruvate kinase
pyruvate + NADH + H^+ → lactate + NAD^+	lactate dehydrogenase

In the case of hexokinase, we could also monitor the glucose-6-phosphate produced by coupling to the glucose-6-phosphate dehydrogenase (G6PDH) reaction, resulting in the conversion of $NADP^+$ to NADPH (increase in A_{340}).

In a few cases, e.g. pyruvate kinase, the equilibrium lies very much to the side of ATP synthesis (PEP + ADP → pyruvate + ATP) and we cannot monitor ADP production via a double coupled assay.

In the pyruvate kinase-catalysed reaction, the reaction can be monitored directly since there is a small decrease in absorbance in the far-UV as PEP is converted to pyruvate. (At 230 nm, $\Delta\varepsilon = -2304\ M^{-1}\ cm^{-1}$ (Pon and Bondar, 1967). However, this relatively small change is often not convenient for detailed studies since some components of the assay system may absorb strongly at this wavelength. A better approach is to couple the pyruvate kinase reaction to the reaction catalysed by LDH so that the pyruvate formed is immediately converted to lactate with concomitant oxidation of NADH to NAD^+ (monitored by the decrease in absorbance at 340 nm).

ADP + PEP → ATP + pyruvate	pyruvate kinase
pyruvate + NADH + H^+ → lactate + NAD^+	lactate dehydrogenase

9.2.4 Assays of enzymes acting on DNA

DNA polymerase catalyses the synthesis of DNA chains by addition of deoxynucleotides from deoxynucleoside triphosphate precursors (dATP, dCTP, etc.) to a pre-existing (or 'primer') strand of DNA. There is no convenient change in spectroscopic or other physical properties which would permit a continuous assay of the enzyme. DNA polymerase can be assayed by a 'stop and sample' approach in which trichloroacetic acid (TCA) is added to samples withdrawn at known times to inactivate the enzyme. We can also exploit the fact that the macromolecular product (DNA) is insoluble in TCA, whereas the substrates (of low molecular mass) are soluble. The standard assay of DNA polymerase is therefore based on the incorporation of radioactivity from [^{3}H] dTTP into TCA-insoluble DNA.

The key concepts concerning radioactivity have been discussed in Chapter 4, section 4.6.

DNA polymerase is assayed by measuring the incorporation of [^{3}H]dTTP into DNA, in a mixture containing all four dNTPs, Mg^{2+} ions, and both template and primer strands of DNA. One unit of enzyme activity is defined as the incorporation of 1 nmol dTTP into DNA in 60 min. In an assay the total volume was 100 μL and included 5 mM [^{3}H]dTTP of specific radioactivity 74 GBq mol^{-1} and 10 μL of a sample of an extract of *Escherichia coli* containing 5.4 mg mL^{-1} protein. After 1 h, the radioactivity incorporated into TCA-insoluble material (i.e. DNA) amounted to 3130 cpm. The efficiency of counting was 15%. What was the specific activity of the DNA polymerase in the extract?

STRATEGY

The amount of dTTP (in mol) is calculated from the observed radioactivity (cpm) incorporated by using the values of efficiency of counting and specific radioactivity. Division of the activity by the amount of protein gives the specific activity of the enzyme.

SOLUTION

The radioactivity incorporated is 3130 cpm = (100/15) × 3130 dpm = 20 867 dpm = 348 dps (Bq). Using the specific radioactivity, this amounts to 348/(74 × 10^9) mol = 4.7 nmol, i.e. 4.7 Units of activity. The amount of protein in the assay is (10/1000) × 5.4 mg = 0.054 mg. Thus, the specific activity = 4.7/0.054 Units mg^{-1} = 87 Units mg^{-1}.

A number of reactions involve alteration of the topology of DNA chains, e.g. uncoiling of supercoiled chains or recombination processes where segments are excised and rejoined to generate new sequences, etc. In such cases it is often possible to take advantage of the fact that the different forms of DNA will migrate at different rates on electrophoresis. Separation of the reaction products, followed by staining of the gel with a fluorescent dye such as ethidium bromide to detect the different forms of DNA, can be used to measure the rate of formation of a particular product. The same electrophoresis-based approach can be used to assay the activity of restriction endonucleases, which cleave both strands of double stranded DNA at specific sequences of bases.

The most widely studied group of restriction enzymes are the type II enzymes. These cleave within the recognition sequence, which is almost always palindromic (i.e. shows two-fold rotational symmetry). Type II enzymes are used extensively for cutting and joining DNA fragments to create recombinant DNA.

9.2.5 General guidelines for obtaining reliable enzyme kinetic data

The main points of these guidelines have been described in Chapter 6, section 6.3.4, and are summarized below:

- Substrates and buffer components should be of high purity; they should be checked for any potential inhibitors (e.g. heavy metals) which might be present in trace amounts
- Suitable control experiments should be performed to check that the reaction does not occur in the absence of enzyme
- The enzyme should be stable under the conditions of the assay
- Temperature and pH should be controlled by suitable thermostat and buffer systems, respectively
- The rate should be constant over the time period of interest and should be proportional to the amount of enzyme added
- The initial rate should be measured, to avoid complications caused by depletion of substrate, product inhibition, occurrence of the reverse reaction, etc.

- In order to obtain reliable values for kinetic parameters (K_m, V_{max}) it is important to cover an adequate proportion of the saturation curve (see section 9.3.1).

? SELF TEST

Check that you have mastered the key concepts at the start of the section by attempting the following questions.

ST 9.1 How would you assay the enzyme glucose-6-phosphate dehydrogenase (G6PDH), which catalyses the reaction glucose-6-phosphate + $NADP^+ \rightleftharpoons$ D-glucono-1,5-lactone-6-phosphate + NADPH + H^+?

ST 9.2 G6PDH was assayed in a total volume of 1 mL in a cuvette of pathlength 1 cm. When 0.08 μg enzyme was added, the increase in A_{340} was 0.172 min^{-1}. The absorbance coefficient of NADPH at 340 nm is 6220 M^{-1} cm^{-1}. What is the specific activity of the G6PDH expressed as μmol min^{-1} mg^{-1}?

ST 9.3 How might you assay hexokinase which catalyses the reaction glucose + ATP $\rightleftharpoons$ glucose-6-phosphate + ADP?

Answers

ST 9.1 G6PDH can be assayed by monitoring absorbance changes at 340 nm where NADPH absorbs, but $NADP^+$ does not.

ST 9.2 The specific activity of G6PDH is 346 μmol min^{-1} mg^{-1}.

ST 9.3 There is no convenient spectrophotometric change during the reaction; however, the production of glucose-6-phosphate can be linked via the G6PDH reaction (see ST 9.1) to the formation of NADPH, which is monitored at 340 nm. Alternatively, the ADP produced can be linked via the pyruvate kinase and lactate dehydrogenase reactions to the consumption of NADH, which is also monitored at 340 nm.

Attempt Problems 9.1–9.3 at the end of the chapter.

9.3 Kinetics of enzyme-catalysed reactions: basic concepts

KEY CONCEPTS

- Knowing the Michaelis–Menten equation describing the variation of reaction rate (v) with substrate concentration ([S])
- Understanding the significance of the parameters K_m, V_{max}, k_{cat}, and k_{cat}/K_m

In Chapter 4, section 4.3.3, the Michaelis–Menten equation was shown to provide a convenient description of the kinetics of reactions involving one substrate. Although a large majority of enzymes catalyse reactions involving two (or more) substrates (see section 9.5), many of the concepts developed for single substrate reactions can be applied to these more complex reactions, in particular the significance of the kinetic parameters (K_m, V_{max}) and the distinction between the various types of reversible inhibition. This section will deal with the kinetic parameters; section 9.4 will deal with inhibition of enzyme-catalysed reactions.

9.3.1 The Michaelis–Menten equation

The Michaelis–Menten equation (eqn. 9.1) for a single substrate reaction describes how the rate (velocity, v) of a reaction varies with substrate concentration ([S]). This behaviour is illustrated in Fig. 9.1:

$$v = \frac{V_{max} \times [S]}{K_m + [S]} \quad (9.1)$$

where V_{max} is the maximum (or limiting) velocity at saturating concentrations of substrate, and K_m is the Michaelis constant for the substrate. When analysing experimental data to determine K_m and V_{max} it is important to cover an adequate proportion of the saturation curve. In practice, this means covering a range of substrate concentrations from $[S] \approx K_m/3$ to $[S] \approx 3 \times K_m$ (i.e. a range of velocities from about 25 to 75% V_{max}), or a greater range if possible.

As explained in Chapter 4, section 4.4, K_m is equal to [S] when $v = 0.5\ V_{max}$ and V_{max} can be expressed as a turnover number (k_{cat}) when the molar concentration of enzyme is taken into account (see Chapter 4, section 4.5). The significance of the parameters K_m, V_{max}, k_{cat}, and the k_{cat}/K_m ratio will be discussed further in sections 9.3.3–9.3.5.

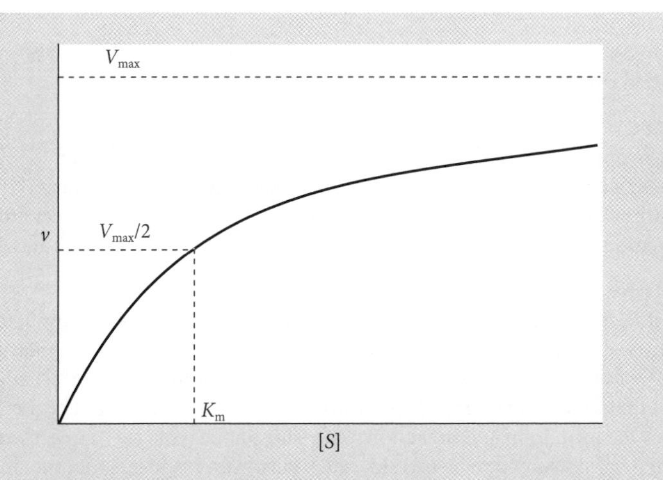

Fig. 9.1 Variation of v with [S] according to the Michaelis–Menten equation.

Elevated levels of alkaline phosphatase in serum can be indicative of certain disease states, including obstructive jaundice and bone disease. The tissue of origin can be determined by measurements of heat stability and effects of inhibitors.

Experimental details are given in Practical activity 9.1, which is available for lecturers in the Online Resource Centre for this text to support teaching of this topic.

9.3.2 Measurement of the kinetic parameters of alkaline phosphatase

Alkaline phosphatases are a group of enzymes which catalyse the hydrolysis of a very wide variety of phosphomonoesters and show maximal catalytic activity at high pH. Although their physiological function is not well understood (except in bone where it is clear that the enzyme is involved in calcification processes), there is considerable interest in the assays of alkaline phosphatase in serum samples as an indicator of certain disease states.

Alkaline phosphatase is an excellent enzyme to study in a laboratory practical because of the ease of measuring the 4-nitrophenol, formed by hydrolysis of 4-nitrophenylphosphate (4NPP), using a spectrophotometer (Price and Newman, 2000).

Data obtained for the action of alkaline phosphatase on 4NPP and glucose-1-phosphate (G1P) under certain conditions are shown in Table 9.1.

Table 9.1 Rate of alkaline phosphatase-catalysed hydrolysis of different substrates

[Substrate] (mM)	Rate (μmol min^{-1} mg^{-1} enzyme)	
	4NPP	G1P
0.2	1.06	0.20
0.5	2.04	0.48
1.0	2.94	0.88
2.0	3.79	1.50
4.0	4.42	2.32
8.0	4.82	3.19
12.0	4.97	3.64

What are the values of K_m and V_{max} for the two substrates and what can be concluded from these values?

STRATEGY

This is an application of the methods used to derive the parameters in the Michaelis–Menten eqn. 9.1. This can be done using one of the linear transformation methods, such as the Lineweaver–Burk plot, or by direct fitting to the equation using non-linear regression (see Chapter 2, sections 2.5.2 and 2.6.5).

SOLUTION

For 4NPP, the values of K_m and V_{max} are 0.8 mM and 5.3 μmol min^{-1} mg^{-1}; for G1P, the values of K_m and V_{max} are 4.8 mM and 5.1 μmol min^{-1} mg^{-1}. The similarity of the V_{max} values (which is also observed for a number of other substrates) might suggest that there is a common rate-limiting step for all substrates. The mechanism involves the formation and breakdown of a phosphoenzyme (E-P) intermediate, the latter was presumed to be this slow step. However, it appears that the slow step is in fact a structural change in the E-P intermediate preceding its hydrolysis.

9.3.3 Significance of K_m values

As mentioned in Chapter 4, section 4.4, in a broad sense the K_m value represents a measure of the affinity of an enzyme for a given substrate (although it is not a true K_d except under very special circumstances; see Chapter 4, Appendix 4.1). In an operational sense, however, K_m is important in indicating the range of substrate concentrations over which enzyme activity will change significantly in response to changes in substrate concentration.

WORKED EXAMPLE ✓

Calculate the effect of doubling the [substrate] on the velocity of a reaction, when (a) the initial [substrate] is 0.5 K_m, and (b) the initial [substrate] is 5 K_m.

STRATEGY

The Michaelis–Menten equation is used to calculate the velocity (v) as a fraction of the V_{max} for each reaction.

SOLUTION

In case (a), when $[S] = 0.5\ K_m$, $v = 0.333\ V_{max}$; when $[S] = K_m$, $v = 0.5\ V_{max}$, i.e. there is a 50% increase in rate ((0.167/0.333) × 100%).

In case (b), when $[S] = 5\ K_m$, $v = 0.833\ V_{max}$; when $[S] = 10\ K_m$, $v = 0.909\ V_{max}$, i.e. there is a 9.1% increase in rate.

Clearly the rate is more sensitive to changes in [S], when [S] is of the order of K_m, rather than being much higher than K_m.

In many cases, the value of K_m can be used to understand the physiological role of an enzyme. Examples are provided by the isoenzymes of hexokinase and of aldehyde dehydrogenase (AlDH).

Isoenzymes are different forms of an enzyme (encoded by distinct genes). The small changes in amino acid sequence will usually result in changes in the kinetic parameters of the different forms.

9.3.3.1 Isoenzymes of hexokinase

There are four principal isoenzymes (designated I–IV) of hexokinase that catalyse the reaction between glucose and ATP to form glucose-6-phosphate and ADP; this is the first step in the metabolism of glucose. Isoenzymes I–III are similar to each other; they have a wide tissue distribution and a low K_m for glucose (about 40 μM). By contrast, isoenzyme IV (also known as glucokinase) has a K_m for glucose (about 10 mM) and is principally found in the liver and the β-cells of the pancreas (the cells responsible for secretion of the hormone insulin). When a patient has had an overnight fast, the blood glucose level is typically about 4 mM; after a meal this will rise quickly to about 9 mM, before falling again.

Isoenyzme IV does not obey strict Michaelis–Menten kinetics; the K_m value quoted represents the [glucose] at which the velocity is 0.5 V_{max}. This does not affect the validity of the conclusions reached here.

The marked response of hexokinase isoenzyme IV to the increase in [glucose] ensures that (a) the extra glucose can be taken up by the liver for storage as glycogen, and (b) the rise in glucose can be sensed by the pancreatic β-cells, leading to secretion of insulin. Insulin not only stimulates the synthesis of glycogen in liver and muscle, but also the uptake of glucose into muscle and adipose tissue. By contrast, isoenzymes I–III show essentially no response to the increased [glucose] as their active sites are effectively already saturated at the fasting levels of blood glucose. This is important for the activity of the brain, which is normally completely dependent on glucose as a metabolic fuel.

9.3.3.2 Isoenzymes of aldehyde dehydrogenase

Ethanal is associated with symptoms of nausea, facial flushing, and tachycardia (rapid heart beat). The mutant form of theAlDH-2 isoenzyme is widespread in many Asian populations, accounting for the increased sensitivity of affected individuals to alcoholic drinks.

AlDH catalyses the oxidation, by NAD^+, of a number of aldehydes, including ethanal (acetaldehyde). Ethanal is produced by the oxidation of ethanol, by NAD^+, catalysed by alcohol dehydrogenase (ADH). Two isoenzymes of AlDH have been identified, a cytoplasmic form with a high K_m for ethanal and a mitochondrial form with a low K_m (designated AlDH-1 and AlDH-2, respectively). A single mutation in AlDH-2 (replacement of Glu 487 by Lys) leads to loss of essentially all activity.

The drug disulfiram is a powerful inhibitor of the AlDH-2 isoenzyme. It has been widely used for the treatment of alcohol abuse because of the unpleasant symptoms associated with the consequent accumulation of ethanal when ethanol is ingested.

9.3.4 Significance of k_{cat} values

The value of k_{cat} provides a measure of the timescale of catalytic events on a molecular timescale; it measures the rate of the slowest step of the reaction mechanism. In practice values of k_{cat} range from about 10 s^{-1} for complex biosynthetic reactions to as high as 10^6 s^{-1} for simple chemical reactions such as the hydration of carbon dioxide catalysed by carbonic anhydrase or the breakdown of hydrogen peroxide catalysed by catalase. In some cases, it has been possible to show that structural changes in the enzyme required, for example for release of products are the slowest step in the catalytic cycle.

9.3.5 Significance of k_{cat}/K_m ratios

The k_{cat}/K_m ratio can be used in two main ways. In the first instance, it provides a measure of catalytic efficiency which can be achieved by an enzyme. From rapid reaction studies (see section 9.9.4) it is known that the rate constant for association of enzyme with substrate can in many cases approach the value at which the molecules collide by diffusion, i.e. about 10^9 M^{-1} s^{-1}. The enzymes fumarase and triosephosphate isomerase have k_{cat}/K_m ratios of 1.6×10^8 and 2.4×10^8 M^{-1} s^{-1}, respectively, and so these enzymes are operating close to the limit of efficiency (this has been referred to as having almost achieved 'catalytic perfection'). The use of the k_{cat}/K_m ratio to compare the catalytic efficiencies of the same enzyme from different sources is illustrated in Problem 9.4 at the end of this chapter.

Secondly, the k_{cat}/K_m ratio can be used to assess the preference of a given enzyme for different substrates and so to assess the important structural features required for recognition by the active site. The examples of chymotrypsin and ADH are described below.

The data in Table 9.2 show that as the bulkiness of the side chain increases, the efficiency of the enzyme towards the substrate increases markedly. Hence, the nature of the active site of chymotrypsin must be such as to recognize these features.

Table 9.2 Preferences for different ester substrates by chymotrypsin

Amino acid in ester	Amino acid side chain	k_{cat}/K_m (M^{-1} s^{-1})
Glycine	–H	0.13
Valine	$-CH(CH_3)_2$	2.0
Norleucine	$-(CH_2)_3CH_3$	3×10^3
Phenylalanine	$-CH_2$-phenyl	1×10^5

Yeast ADH provides a convenient laboratory example to investigate the specificity of an enzyme towards different substrates such as ethanol, propanol, and 2-propanol (Green *et al.*, 1993). The purification of yeast ADH is described in Chapter 7, section 7.3.2.

Typical kinetic data for yeast ADH are given in Table 9.3.

Table 9.3 Kinetic parameters for the action of yeast alcohol dehydrogenase on different alcohols

Substrate	Ethanol	Propanol	2-Propanol	Butanol	Benzyl alcohol
K_m (mM)	17	27	190	55	n.a.
k_{cat} (s^{-1})	340	120	4.8	51	n.a.
k_{cat}/K_m (M^{-1} s^{-1})	20 000	4400	25	930	n.a.

n.a., no activity observed.

Details are given in Practical activity 9.2, which is available for lecturers in the Online Resource Centre for this text to support teaching of this topic.

From the data in Table 9.3 it is clear that the active site of yeast ADH can best accommodate primary alcohols with both the affinity and the catalytic efficiency decreasing with chain length. Branched chain alcohols are not dealt with efficiently. The bulky alcohol benzyl alcohol shows no detectable activity (less than 1×10^{-5} of the standard assay with ethanol) under these conditions. Interestingly, very little activity is shown towards methanol, presumably the absence of a $-CH_2-$ group compared with ethanol and the consequent lack of interactions with the enzyme means that this substrate cannot be precisely positioned in the active site to allow efficient catalysis. In section 9.8.3, the structural features of the active sites of ADH from yeast and mammalian liver are compared.

? SELF TEST

Check that you have mastered the key concepts at the start of the section by attempting the following questions.

ST 9.4 What is the effect on the activity of hexokinases I–III and of hexokinase IV when the blood [glucose] increases from 4 to 9 mM after a meal?

ST 9.5 What would be the effect on alcohol metabolism of the Glu487 → Lys mutation in AlDH-2 (isoenzyme 2 of AlDH)?

Answers

ST 9.4 For isoenzymes I–III there is a 0.56% increase in velocity (from 0.9901 V_{max} to 0.9956 V_{max}); for isoenzyme IV the increase is 65.8% (from 0.2857 V_{max} to 0.4737 V_{max}.

ST 9.5 If the AlDH-2 isoenzyme is inactivated, only the high K_m form (AlDH-1) will be active. Thus, ethanal will be less rapidly oxidized and will accumulate to a greater extent on ingestion of ethanol.

Attempt Problem 9.4 at the end of the chapter.

9.4 Inhibition of enzyme-catalysed reactions

KEY CONCEPTS

- Understanding the main types of reversible inhibition of enzymes
- Being able to analyse inhibition data and to derive appropriate kinetic parameters

Inhibition of enzymes can be divided into reversible and irreversible types. This section will deal with reversible inhibition, where the inhibitor does not form a stable covalent bond with the enzyme. Irreversible inhibition of enzymes by chemical modification is discussed in section 9.10.2.

In Chapter 4, section 4.4, some of the key concepts relating to reversible inhibition of enzymes were discussed. There are three limiting cases of such inhibition; the terms introduced apply to reactions involving one or more substrates.

Competitive inhibition (V_{max} unchanged, K_m increased)

The most likely explanation for competitive inhibition is that the inhibitor binds to the active site, preventing binding of substrate. Competitive inhibitors are widely used to define the active site in crystallographic studies, and as potential therapeutic drugs directed towards identified enzyme targets. Competitive inhibition is most commonly recognized by its effect on the Lineweaver–Burk plot, where the lines intersect on the $1/v$ axis, showing that V_{max} is unchanged in the presence of inhibitor (Fig. 9.2). K_m is increased by a factor $(1 + ([I]/K_{EI}))$.

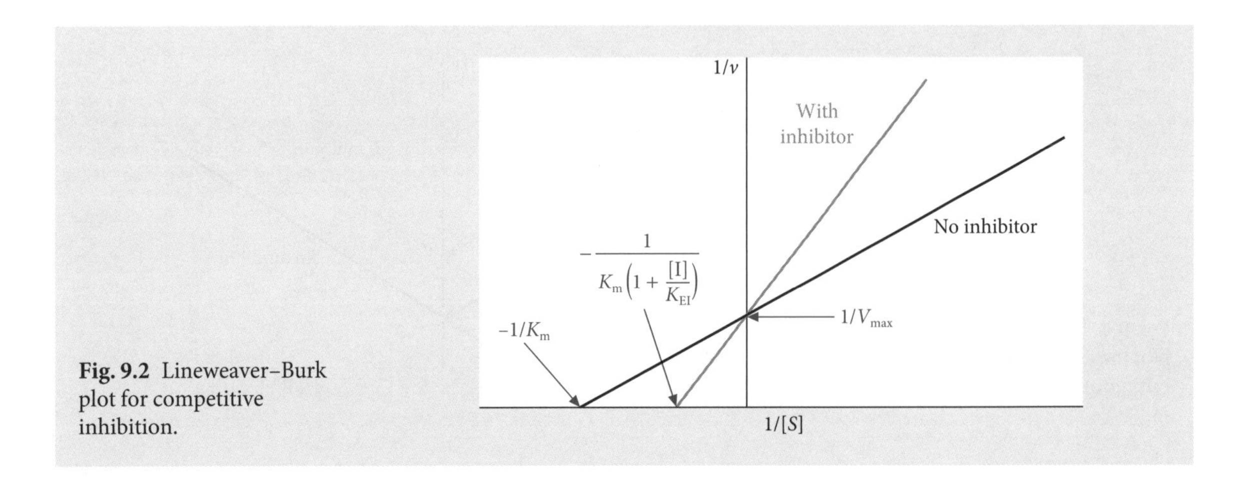

Fig. 9.2 Lineweaver–Burk plot for competitive inhibition.

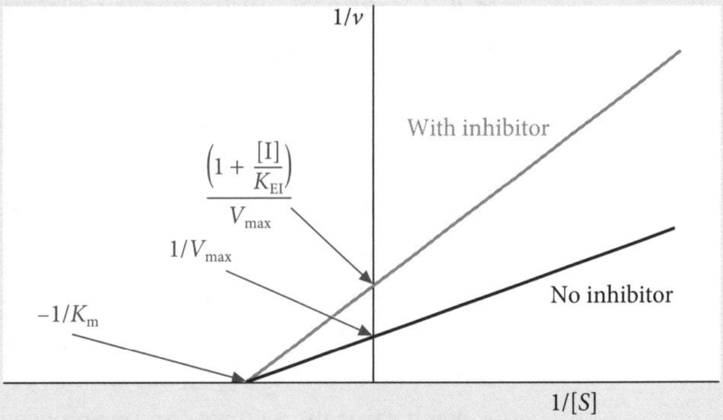

Fig. 9.3 Lineweaver–Burk plot for non-competitive inhibition.

Non-competitive inhibition (V_{max} decreased, K_m unchanged)

A non-competitive inhibitor binds equally well to the free enzyme as to the enzyme–substrate complex. Thus such an inhibitor must bind to a distinct site; presumably the binding of inhibitor leads to a conformational change at the active site and loss of activity. The effect on the Lineweaver–Burk plot is shown in Fig. 9.3, where the lines intersect on the 1/[S] axis, showing that K_m is unchanged in the presence of inhibitor. V_{max} is decreased by a factor $(1 + ([I]/K_{EI}))$.

Uncompetitive inhibition (V_{max} and K_m both decreased)

An uncompetitive inhibitor does not bind to the enzyme but only to the enzyme–substrate complex. Binding of substrate must therefore cause a conformational change, which leads to the formation of the inhibitor binding site. The effect on the Lineweaver–Burk plot is shown in Fig. 9.4. The lines are parallel to each other with both V_{max} and K_m decreased by the same factor $(1 + ([I]/K_{ESI}))$ in the presence of inhibitor.

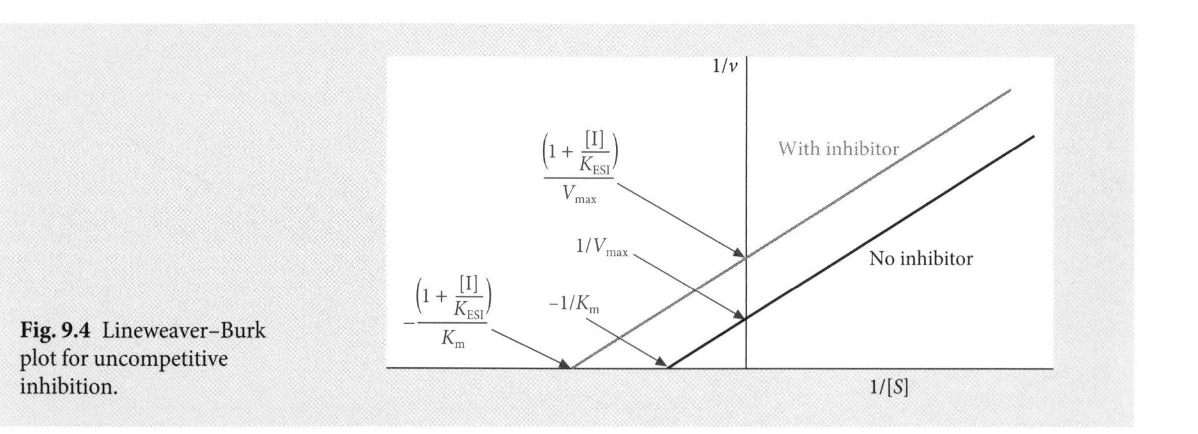

Fig. 9.4 Lineweaver–Burk plot for uncompetitive inhibition.

9.4.1 General guidelines for obtaining reliable data to analyse enzyme inhibition

The ideal is to obtain saturation (i.e. velocity vs substrate concentration) curves over a good range of [*S*] values in the absence of inhibitor and in the presence of at least two different concentrations of inhibitor. If the inhibitor concentration is too great, it may be difficult to measure remaining activity; if the inhibitor concentration is too small, the changes in activity due to the inhibitor may be too small to measure. Ideally the range of [*I*] should be such as to give between 20 and 80% inhibition at different [*S*], depending on the type of inhibition observed.

Each saturation curve can be analysed using either a linear transformation plot such as the Lineweaver–Burk plot or non-linear regression analysis (see Chapter 2, sections 2.5.2 and 2.6.5). One advantage of non-linear regression analysis is that it can give more reliable estimates of the magnitudes of the standard errors in the kinetic parameters K_m and V_{max}. This will give a measure of the degree of confidence in concluding which of the parameters have changed in the presence of the inhibitor.

The changes in kinetic parameters can be used to determine the type of inhibition and determine the inhibition constant (K_{EI} or K_{ESI} as appropriate). In any given case, for different concentrations of inhibitor, the values of K_{EI} or K_{ESI} should be constant. If they are not, it can be concluded that the mechanism of inhibition is more complex than outlined here, for example the inhibition is of a mixed type (i.e. not one of the three limiting cases described) or there are multiple modes of binding for the inhibitor.

If it is not practicable to perform inhibition experiments over a large range of substrate concentrations, the Dixon plot may be useful (Fig. 9.5). Plots of $1/v$ vs [*I*] are constructed at each of a number of substrate concentrations. If the resulting lines intersect above the *x*-axis, the inhibition is competitive, whereas if they intersect on the *x*-axis, the inhibition is non-competitive. In each case the value of K_{EI}

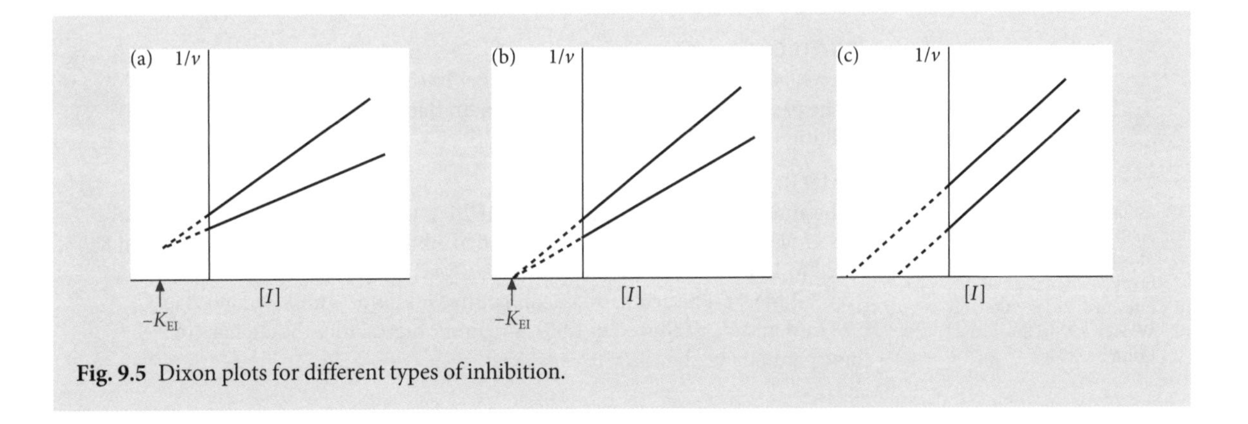

Fig. 9.5 Dixon plots for different types of inhibition.

can be read off from the *x* coordinate of the point of intersection. In the case of uncompetitive inhibition, the lines are parallel and no information on the magnitude of K_{ESI} can be obtained. For more details see Dixon (1953).

The different types of inhibition are illustrated in the examples described in sections 9.4.2–9.4.4.

9.4.2 Inhibition of alkaline phosphatase by phosphate and by L-Phe

The effects of phosphate and L-Phe on alkaline phosphatase are conveniently studied in laboratory exercises (see section 9.3.2 and Practical activity 9.1 which is available for lecturers in the Online Resource Centre). Specimen data sets are given in the worked example below.

WORKED EXAMPLE 

The activity of alkaline phosphatase towards 4NPP was studied in the absence of inhibitors and in the presence of 2.5 mM phosphate and 1 mM L-Phe (Table 9.4). What type of inhibition is being shown in each case?

Table 9.4 Inhibition of the alkaline phosphatase-catalysed hydrolysis of 4NPP

[4NPP] (mM)	Rate (μmol min^{-1} mg^{-1} enzyme)		
	No inhibitor	+2 mM phosphate	+1 mM L-Phe
0.2	1.06	0.54	0.78
0.5	2.04	1.17	1.19
1.0	2.94	1.92	1.45
2.0	3.79	2.82	1.63
4.0	4.42	3.68	1.74
8.0	4.82	4.34	1.80
12.0	4.97	4.62	1.82

STRATEGY

Lineweaver–Burk plots (or one of the other linear transformation methods) can be used to are used to analyse the kinetic data and hence deduce what type of inhibition is being observed.

SOLUTION

In the absence of inhibitors, the K_m for 4NPP is 0.8 mM and the V_{max} is 5.3 μmol min^{-1} mg^{-1}. Phosphate acts as a competitive inhibitor with V_{max} unchanged and K_m raised to 1.76 mM. The increase in K_m (2.2-fold) is equal to $(1 + ([I]/K_{EI}))$; hence $K_{EI} = 1.7$ mM. L-Phe acts as an uncompetitive inhibitor with K_m reduced to 0.28 mM and V_{max} reduced to 1.86 μmol min^{-1} mg^{-1}. From the changes in these parameters, the $K_{ESI} = 0.54$ mM.

L-Phe appears to act as an inhibitor of alkaline phosphatase by blocking breakdown of the E-P intermediate (Fernley and Walker, 1970; Ghosh and Fishman, 1966).

9.4.3 Inhibition of inositol monophosphatase by lithium ions

The enzyme inositol monophosphatase (IP_1ase) plays a key role in the metabolism of inositol-based signalling molecules. The enzyme from bovine brain is inhibited by lithium ions, as illustrated in the worked example below.

WORKED EXAMPLE

The data in Table 9.5 were obtained for the inhibition of IP_1ase by Li^+ ions. Deduce the type of inhibition shown by Li^+, and the value of the appropriate inhibitor constant. In brain cells the steady state $[IP_1]$ is estimated to be approximately 45 μM. What would be the rate of the IP_1ase reaction under these conditions? When taking a high therapeutic dose of Li^+, the intracellular $[Li^+]$ is 4.2 mM, and the $[IP_1]$ rises to 1.5 mM. What is the rate of the IP_1ase reaction under these conditions?

Table 9.5 Inhibition of IP_1ase by Li^+ ions

$[IP_1]$ (μM)	Rate (μmol min^{-1} mg^{-1})	
	No Li^+	+1.5 mM Li^+
20	0.118	0.100
50	0.250	0.182
100	0.400	0.250
200	0.571	0.308
500	0.769	0.357

Lithium continues to be used extensively for the treatment of manic-depressive disorders. It is likely that it is the selective uncompetitive nature of the inhibition of IP_1ase by Li^+ that is the key to its use as a drug. The very large increase in $[IP_1]$ in the presence of Li^+ perturbs the recycling of inositol and synthesis of phosphatidylinositol. This interferes with second messenger signalling in many cells, including neurons.

STRATEGY

The data in the presence and absence of Li^+ can be analysed by the Lineweaver–Burk plot to obtain values of K_m and V_{max}, and hence the inhibition constant for Li^+. Knowing these, we can calculate the rate of the reaction under any specified conditions using the Michaelis–Menten equation. In the presence of Li^+, the lines in the Lineweaver–Burk plot are parallel, showing that both K_m and V_{max} are reduced by the same factor.

SOLUTION

The inhibition is of the uncompetitive type with K_m reduced from 140 to 60 μM and V_{max} reduced from 0.96 μmol min^{-1} mg^{-1} to 0.42 μmol min^{-1} mg^{-1} in the presence of Li^+. Since the reduction in both K_m and V_{max} is by the factor $(1 + ([I]/K_{ESI}))$, the value of K_{ESI} is 1.15 mM. In the absence of Li^+, the $[IP_1] = 45$ μM, the $K_m = 140$ μM, and $V_{max} = 0.96$ μmol min^{-1} mg^{-1}; the rate of the reaction is 0.23 μmol min^{-1} mg^{-1}. In the presence of Li^+, the $[IP_1] = 1500$ μM and the K_m is reduced to 30 μM and V_{max} to 0.21 μmol min^{-1} mg^{-1}; the rate is now 0.206 μmol min^{-1} mg^{-1}.

9.4.4 Inhibition of lactate dehydrogenase by products

As will be discussed in section 9.5, the pattern of inhibition of an enzyme-catalysed reaction by products can be used to indicate the mechanism of the reaction. An example is provided by LDH, which catalyses the reaction:

$$\text{lactate} + NAD^+ \rightleftharpoons \text{pyruvate} + NADH + H^+$$

The effects of the product (pyruvate) on the rate of the reaction were measured in two separate experiments, in each of which the concentration of one substrate was varied while the other was kept constant. A specimen data set is shown in the worked example below. The preparation of LDH from rabbit muscle is described in Chapter 7, section 7.3.1.

Experimental details of the inhibition of LDH by products are given in Practical activity 9.3, which is available for lecturers in the Online Resource Centre for this text to support teaching of this topic.

WORKED EXAMPLE 

The data in Table 9.6 were obtained for the effect of pyruvate on the LDH reaction. Deduce the type of inhibition shown in each case.

Table 9.6 Inhibition of LDH by pyruvate

At a fixed [NAD^+] (2.5 mM)

[Pyruvate] (mM)	Rate (μmol min^{-1} mg^{-1}) at stated [lactate]			
	1.5 mM	2.0 mM	3.0 mM	10.0 mM
0	135	170	223	418
40	76	97	135	302

At a fixed [lactate] (20 mM)

[Pyruvate] (mM)	Rate (μmol min^{-1} mg^{-1}) at stated [NAD^+]			
	0.5 mM	0.7 mM	1.0 mM	2.0 mM
0	240	282	324	390
30	191	225	259	312

STRATEGY

The data are analysed by the Lineweaver–Burk plot to determine the type of inhibition observed.

This type of experiment can be used to show that LDH follows an ordered mechanism in which NAD^+ binds first followed by lactate to form the catalytically active complex (E-NAD^+-lactate) (see section 9.5).

SOLUTION

In the first experiment, in the absence of pyruvate the K_m (lactate) and V_{max} are 5.7 mM and 625 μmol min^{-1} mg^{-1}, respectively. Pyruvate acts as a competitive inhibitor with respect to lactate, leaving V_{max} unchanged but increasing K_m (lactate) to 11 mM. This would give a value for K_{EI} of pyruvate of 43 mM. In the second experiment, in the absence of pyruvate the K_m (NAD^+) and V_{max} are 0.54 mM and 495 μmol min^{-1} mg^{-1}, respectively. Pyruvate acts as a non-competitive inhibitor with respect to NAD^+, leaving K_m unchanged but decreasing V_{max} to 397 μmol min^{-1} mg^{-1}. The K_{EI} (=K_{ESI}) can be calculated as 122 mM.

? SELF TEST

Check that you have mastered the key concepts at the start of the section by attempting the following question.

The shikimate pathway occurs in bacteria, fungi, and plants but not in mammals and thus the enzymes of this pathway represent attractive targets for inhibitor design. Glyphosate is a powerful (but biodegradable) herbicide and is among the world's best selling biotechnology products.

ST 9.6 The enzyme 5-enoylpyruvyl-shikimate-3-phosphate (EPSP) synthase catalyses the reaction between shikimate-3-phosphate (S3P) and PEP to give EPSP and inorganic phosphate. The reaction is part of the shikimate pathway, leading to the synthesis of aromatic amino acids (Phe, Tyr, and Trp). The enzyme is reversibly inhibited by glyphosate (*N*-phosphonomethylglycine), as shown below. Deduce the type of inhibition shown by glyphosate towards each substrate.

At a fixed [shikimate-3-phosphate] (180 μM)

[Glyphosate] (μM)	Rate (μmol min^{-1} mg^{-1}) at given [PEP]			
	2 μM	4 μM	5 μM	10 μM
0	29	44	50	67
2	14	25	29	45

At a fixed [PEP] (15 μM)

[Glyphosate] (μM)	Rate (μmol min^{-1} mg^{-1}) at given [shikimate-3-phosphate]			
	0.1 μM	0.2 μM	0.5 μM	1.0 μM
0	14	25	45	63
2	11.5	18	27	32

Answer

ST 9.6 In the absence of glyphosate, the values of K_m (PEP) and V_{max} are 4.7 μM and 95 μmol min^{-1} mg^{-1}, respectively. Glyphosate acts as a competitive inhibitor with respect to PEP, with V_{max} remaining unchanged and K_m increased to 11 μM, giving a value for the K_{EI} of glyphosate of 1.5 μM. In the absence of glyphosate the K_m (S3P) and V_{max} are 0.6 μM and 100 μmol min^{-1} mg^{-1}, respectively. Glyphosate acts as an uncompetitive inhibitor with respect to S3P, decreasing K_m to 0.26 μM and decreasing V_{max} to 42 μmol min^{-1} mg^{-1}; this gives a value for K_{ESI} of 1.5 μM. These results indicate that EPSP synthase follows an ordered mechanism with S3P binding first followed by PEP to give the catalytically active complex. Glyphosate can compete with PEP for the enzyme–S3P complex, forming an inactive enzyme–S3P–glyphosate complex. There are some limited structural similarities between glyphosate and PEP.

Attempt Problems 9.5 and 9.6 at the end of the chapter.

9.5 Kinetics of two-substrate reactions

KEY CONCEPTS

- Knowing the main categories of reactions involving two substrates
- Understanding how the various types of mechanism can be distinguished experimentally

The majority of enzyme-catalysed reactions involve more than one substrate. Although many of the concepts used to analyse reactions involving one substrate can be applied to multi-substrate reactions, the equations involved are inevitably more complex. In this section, we will describe in outline how to analyse two substrate reactions. For further details the more detailed discussions such as those by Cornish-Bowden (2004), Dickinson (1996) and Price and Stevens (1999a) should be consulted.

9.5.1 Theory

Reactions involving two substrates fall into two main categories. We shall designate the reaction as:

$$\mathrm{E + A + B \rightleftharpoons E + P + Q}$$

where A and B are the substrates and P and Q are the products.

One type of mechanism is the ternary complex, which involves the necessary formation of a ternary complex (EAB); electron movement (i.e. bond breaking and making) occurs in this complex to generate a complex with the two products bound (EPQ), which then dissociate to regenerate the enzyme:

$$\mathrm{E + A + B \rightleftharpoons EAB \rightleftharpoons EPQ \rightleftharpoons E + P + Q}$$

This type of mechanism can be subdivided into ordered and random ternary complex mechanisms according to whether or not there is a required order of substrate binding. In an ordered mechanism substrate A, for example, could bind to the free enzyme, but substrate B will only bind to the EA complex:

$$\mathrm{E + A \rightarrow EA} \quad \text{but } \mathrm{E + B \not\longrightarrow EB}$$

$$\mathrm{EA + B \rightarrow EAB}$$

$$\mathrm{EAB \rightarrow EPQ \rightarrow \rightarrow E + P + Q}$$

The second type of mechanism for two substrate reactions is the enzyme substitution (or ping-pong) mechanism in which the first substrate combines with the enzyme to generate the first product and a modified form of the enzyme (E*); this can then react with the second substrate to form the second product and regenerate the enzyme:

$$E + A \rightarrow E^* + P$$

$$E^* + B \rightarrow E + Q$$

The kinetic equations describing the initial rate of the reaction (v) in terms of the concentrations of substrates A and B for the two types of mechanism are given as eqns. 9.2 and 9.3:

Ternary complex mechanism

$$v = \frac{V_{max}[A][B]}{K'_A K_B + K_B[A] + K_A[B] + [A][B]} \quad 9.2$$

In terms of the mechanisms of the reactions, K'_A, K_A, and K_B represent combinations of the rate constants of individual steps in the reaction; their precise meaning will depend on the type of mechanism being considered.

where V_{max}, K_A, K_B, and K'_A are constants for a given reaction. V_{max} represents the maximum (limiting) velocity when substrates A and B are present at saturating concentrations. K_A and K_B represent the Michaelis constants for substrates A and B, respectively, at saturating concentrations of the other substrate. K'_A does not have any such simple meaning.

Enzyme substitution mechanism

$$v = \frac{V_{max}[A][B]}{K_B[A] + K_A[B] + [A][B]} \quad 9.3$$

Note for the ternary complex mechanism, $K'_A = 0$, so that the term $K'_A K_B$ in the denominator of eqn. 9.2 is missing from eqn. 9.3.

9.5.2 Analysis of kinetic data: determination of kinetic parameters

Kinetic experiments are performed by keeping the concentration of one substrate (say B) constant and determining the rate as $[A]$ is varied over a chosen range. This procedure is then repeated with other fixed values of $[B]$. A primary plot showing $1/v$ vs $1/[A]$ for the different values of $[B]$ is plotted (Fig. 9.6).

For the tertiary complex mechanism

$$\text{The slope of each line in the primary plot} = \frac{1}{V_{max}}\left[K_A + \frac{K'_A K_B}{[B]}\right]$$

$$\text{The intercept of each line on the y-axis} = \frac{1}{V_{max}}\left[1 + \frac{K_B}{[B]}\right]$$

Thus as $[B]$ increases, both the slope and the intercept will decrease.

The slopes and intercepts of the lines in the primary plots are then plotted against $1/[B]$ in secondary plots to give straight lines (Figs 9.7 and 9.8). From the

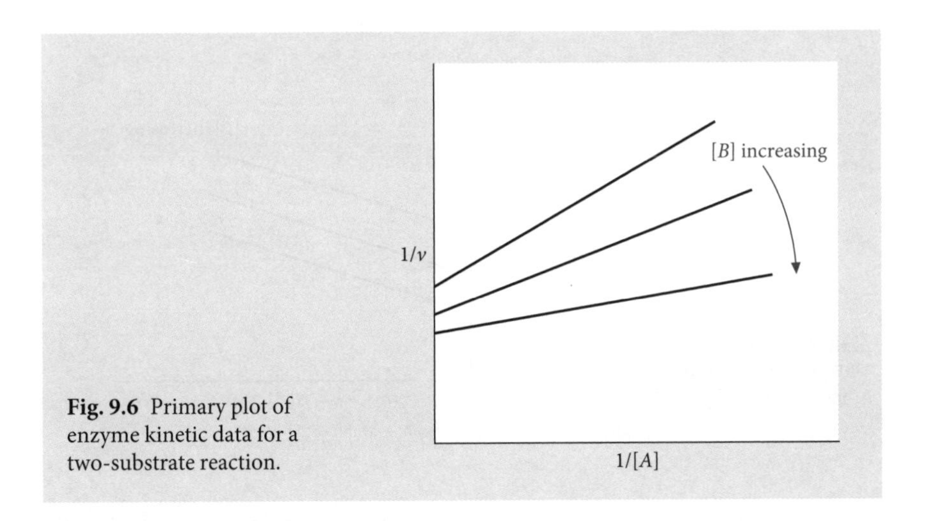

Fig. 9.6 Primary plot of enzyme kinetic data for a two-substrate reaction.

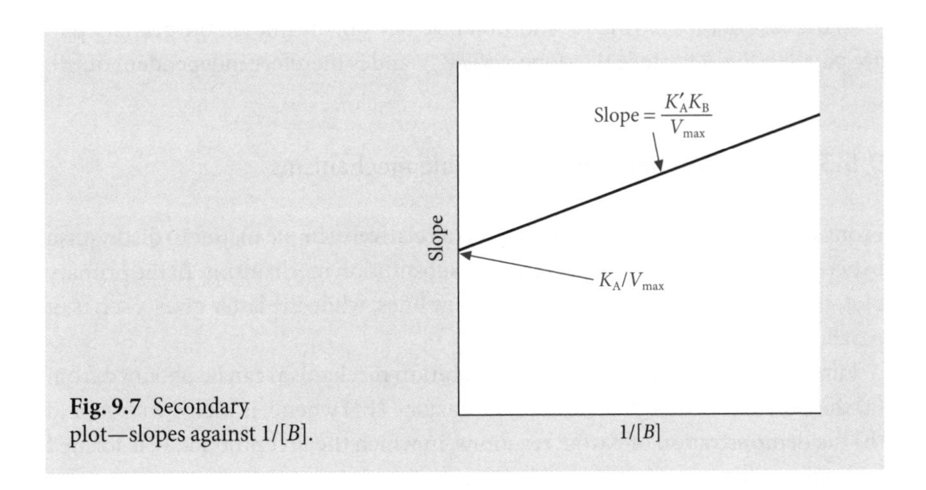

Fig. 9.7 Secondary plot—slopes against 1/[B].

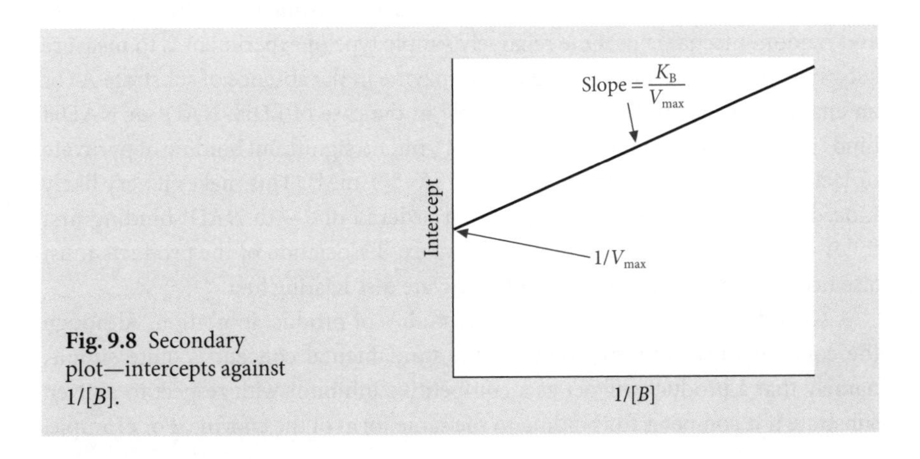

Fig. 9.8 Secondary plot—intercepts against 1/[B].

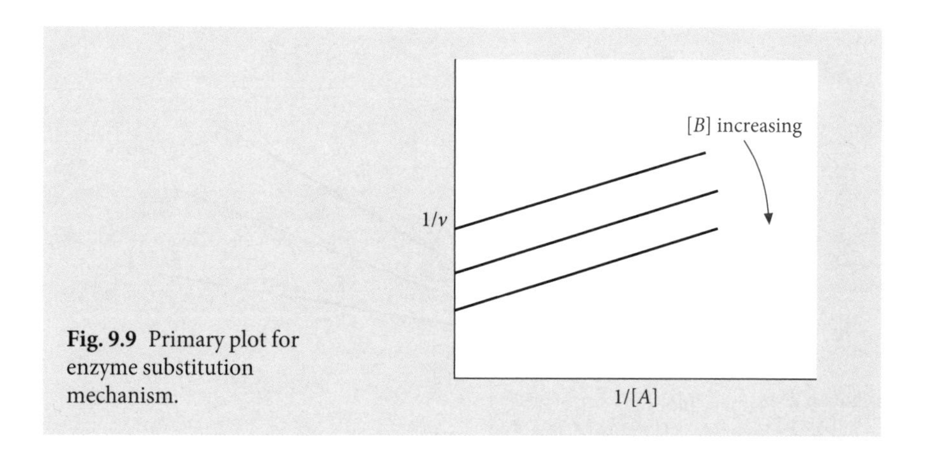

Fig. 9.9 Primary plot for enzyme substitution mechanism.

slopes and intercepts of these secondary plots it is possible to obtain the kinetic parameters.

In the case of the enzyme substitution mechanism, the lines in the primary plot are parallel (Fig. 9.9) since the slope $= K_A/V_{max}$ and is therefore independent of $[B]$.

9.5.3 Distinction between the possible mechanisms

From section 9.5.2, it can be seen that it is a relatively simple matter to distinguish between the ternary complex and enzyme substitution mechanisms. In the primary plot, the former gives a series of convergent lines, while the latter gives a series of parallel lines.

Further evidence for the enzyme substitution mechanism can be obtained from (a) the isolation of modified form of the enzyme (E*) when E is mixed with A, and (b) the demonstration of partial reactions, in which the first product, P, is formed without needing to add the second substrate.

Since all ternary complex mechanisms give rise to convergent lines in the primary plot, further measurements are needed to distinguish between ordered and random mechanisms. One relatively simple type of experiment is to measure substrate binding, i.e. does B bind to the enzyme in the absence of substrate A (or an analogue of substrate A)? For example in the case of LDH, NAD^+ or NADH binds tightly to the enzyme (K_d about 1 μM), but no significant binding of pyruvate or lactate to the enzyme can be measured ($K_d > 1$ mM). This makes it very likely indeed that the enzyme mechanism is an ordered one with NAD^+ binding first followed by lactate to form a ternary complex; dissociation of the products must also occur in an ordered fashion with pyruvate dissociating first.

A second type of experiment involves studies of product inhibition. Although the equations involved are complex, the fundamental concept is quite simple, namely that a product will act as a competitive inhibitor with respect to a given substrate if it competes for binding to the same form of the enzyme. For example, as shown in section 9.4.4, pyruvate acts as a competitive inhibitor with respect to lactate but is non-competitive with respect to NAD^+. This can be understood if the mechanism is ordered since pyruvate and lactate will compete for binding to the same form of the enzyme (with NAD^+ bound). By contrast, pyruvate will not compete with NAD^+ since the latter will bind to free enzyme.

9.5.4 Analysis of kinetic data for two-substrate reactions

The analysis of a two substrate reaction is illustrated in the worked example of creatine kinase below. A further example is given in Problem 9.7 at the end of the chapter.

WORKED EXAMPLE 

Creatine kinase catalyses the reaction creatine + ATP → phosphocreatine + ADP. The kinetic data in Table 9.7 were obtained.

Table 9.7 Kinetic data for the creatine kinase-catalysed reaction

[Creatine] (mM)	Rate (μmol min^{-1} mg^{-1})			
	0.46 mM ATP	0.62 mM ATP	1.23 mM ATP	3.68 mM ATP
6	22.6	27.8	39.6	58.1
10	33.3	40.7	57.0	78.5
20	50.7	60.3	80.3	108.2
40	70.8	82.7	103.1	137.7

Analyse the data to determine the kinetic parameters. What can you conclude about the mechanism of the enzyme?

STRATEGY

The data are analysed by constructing primary and secondary plots as indicated in section 9.5.2.

SOLUTION

The lines on the primary plot (e.g. $1/v$ vs 1/[creatine] for each fixed value of [ATP] are convergent. Thus the enzyme follows a ternary complex mechanism. From the secondary plots (slopes and intercepts of primary plots vs 1/[ATP]), the following parameters can be derived: V_{max} = 189 μmol min^{-1} mg^{-1}; K_{ATP} = 0.34 mM; $K_{creatine}$ = 9.1 mM; $K'_{creatine}$ = 41.8 mM. (Note if we had plotted the primary plots of $1/v$ vs 1/[ATP] at each fixed value of [creatine], then we would have obtained K'_{ATP} = 1.56 mM. There is a simple relationship $\frac{K_{ATP}}{K'_{ATP}} = \frac{K_{creatine}}{K'_{creatine}}$. Further experiments of the type described in section 9.5.3 show that the reaction is of the random ternary complex type.

In the case of creatine kinase, for each substrate the value of K' is greater than that of K. This means that the binding of one substrate strengthens the binding of the second substrate, a phenomenon known as *substrate synergism*.

SELF TEST

Check that you have mastered the key concepts at the start of the section by attempting the following question.

ST 9.7 Describe the main types of reactions involving two substrates and how they could be distinguished.

Answer

ST 9.7 The main types of two-substrate reactions are ternary complex and enzyme-substitution mechanisms; the former can be subdivided into ordered and random complex mechanisms. Distinction between the various mechanisms can be made as outlined in section 9.5.3.

Attempt Problem 9.7 at the end of the chapter.

9.6 Kinetics of allosteric enzymes

KEY CONCEPTS

- Using the Hill equation to analyse enzyme kinetic data
- Understanding the significance of the Hill coefficient

9.6.1 Introduction

The word 'allosteric' is made up from two Greek words meaning 'different solid', to convey the idea that the effector molecules are generally structurally unrelated to the substrate. The term 'sigmoidal' refers to the elongated S shape of the saturation curve (the Greek letter sigma is equivalent to S).

The kinetic behaviour (variation of rate with substrate concentration) of a number of enzymes does not conform to the typical Michaelis–Menten equation. The vast majority of these enzymes consist of multiple polypeptide chains and therefore contain multiple active sites. The unusual behaviour of these enzymes is presumed to arise from interactions between the active sites in a molecule, in an analogous fashion to that described in Chapter 4, section 4.4 to explain cooperative binding behaviour. Many of these enzymes occupy key positions in metabolic pathways, and their regulatory properties arise not only from the more marked sensitivity of rate to changes in substrate concentration, but also from regulation by ligands (effectors) which bind at sites distinct from the active sites. The term 'allosteric' is often used to describe these enzymes, and they are said to display cooperative or sigmoidal kinetics. The first well-documented examples of allosteric enzymes were found in biosynthetic pathways, where the first enzyme in the pathway was inhibited by the end product of the pathway (a form of negative feedback in which the activity of the overall pathway can be tailored to the supply of, and demand for, the end product).

Details of the models developed by Adair, Monod *et al.*, and Koshland *et al.* are given in a number of textbooks, e.g. Price and Stevens (1999b, Ch. 6); Fersht (1999b, Ch. 10).

Although a number of complex mathematical descriptions of the kinetics of allosteric enzymes have been developed, we shall confine our analysis to that using the Hill equation (see Chapter 4, section 4.3.2), since this simple approach can give valuable insights into the extent of interactions between sites in an allosteric enzyme.

9.6.2 Hill equation for enzyme kinetics

As described in Chapter 4, section 4.3.2, the Hill equation (eqn. 9.4) represents a modified form of the Michaelis–Menten equation in which the substrate concentration ([S]) is raised to the power h, where h is the Hill coefficient:

$$v = \frac{V_{max} \times [S]^h}{K + [S]^h} \quad 9.4$$

V_{max} is the maximum (limiting) rate when the active sites are saturated with substrate. As described below, K provides a measure (albeit indirect) of the affinity of the enzyme for its substrate and the range of substrate concentrations over which the rate is most sensitive to changes in substrate concentration.

As explained in Chapter 4, section 4.3.2, the value of h is a measure of the strength of interactions between the active sites. When $h = 1$, there are no interactions; a value >1 indicates positive cooperativity. It is important to note that although the Hill equation can be derived by assuming that there is an equilibrium between free protein and fully saturated protein (with no significant population of any partially saturated intermediates), it is more generally used as essentially an empirical description of saturation behaviour over a particular range of ligand concentrations; h can take non-integral values.

Positive cooperativity between binding sites arises when the binding of the first molecule of ligand facilitates binding of subsequent molecules. A value of $h < 1$ indicates negative cooperativity when the binding of subsequent molecules is hindered; negative cooperativity has been observed in a few cases, for example the binding of NAD^+ to glutamate dehydrogenase (see Price and Stevens (1999b).

The value of V_{max} can either be established using non-linear regression techniques to fit experimental data to the Hill equation or more directly estimated by making measurements of the rate at very high substrate concentrations.

Rearranging the Hill equation, we obtain:

$$\log\left(\frac{v}{V_{max} - v}\right) = h \log[S] - \log K \quad 9.5$$

so a plot of $\log\left(\frac{v}{V_{max} - v}\right)$ vs $\log[S]$ is a straight line of slope h, and with an intercept on the $y = 0$ axis equal to $(\log K)/h$ (Fig. 9.10). It should be noted that the value of this intercept corresponds to the log of the substrate concentration when the rate is half the maximum rate, denoted $\log[S]_{0.5}$.

9.6.3 Isocitrate dehydrogenase as an example of an allosteric enzyme

Isocitrate dehydrogenase (ICDH) catalyses the oxidative decarboxylation of isocitrate to give 2-oxoglutarate. Two different types of ICDHs are known, one using NAD^+, the other using $NADP^+$. These enzymes occur predominantly in the

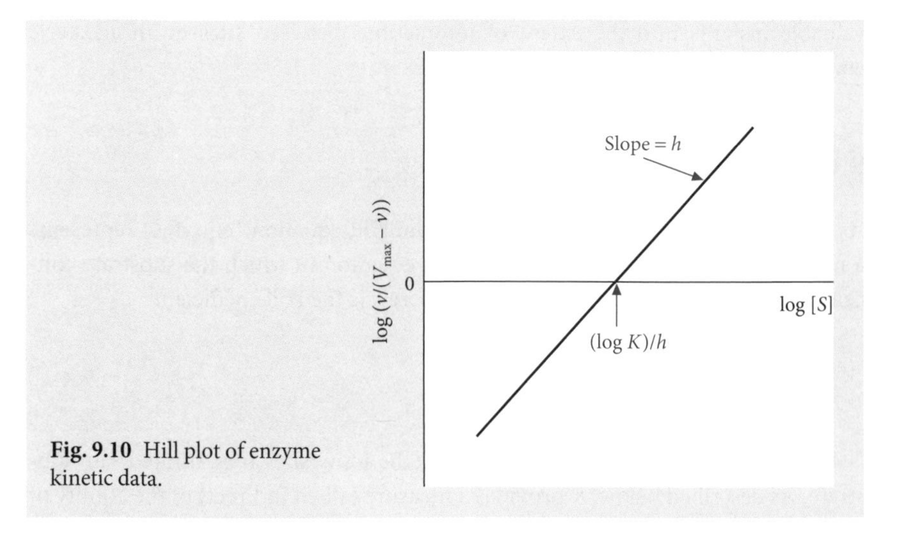

Fig. 9.10 Hill plot of enzyme kinetic data.

mitochondria and cytosol, respectively. The purification of the NAD^+-type ICDH from yeast is described in Chapter 7, section 7.3.3. A partially purified preparation of this enzyme can be used for kinetic studies (see Practical Activity 7.3 for experimental details; this is available for lecturers in the Online Resource Centre). A specimen data set for ICDH is given in the worked example below.

WORKED EXAMPLE

The variations of isocitrate dehydroginase (ICDH) activity with isocitrate concentration in the absence and presence of AMP are shown in Table 9.8. In separate experiments the V_{max} value was determined to be 0.230 μmol min^{-1} mg^{-1} in the absence and presence of AMP. Analyse the data in terms of the Hill plot and comment on the results.

STRATEGY

The data are plotted according to the rearranged Hill equation (eqn. 9.5).

SOLUTION

The Hill plot in the absence of AMP has a slope (Hill coefficient, h) of 2.9 and an intercept on the $y = 0$ axis ($\log[S]_{0.5}$) of −0.22, i.e. $[S]_{0.5} = 0.60$ mM. The value of K is thus 0.23, when the values of $[S]$ are expressed in mM units.

In the presence of AMP, the Hill coefficient is 1.8, and the intercept on the $y = 0$ axis ($\log[S]_{0.5}$) of −0.83, i.e. $[S]_{0.5} = 0.148$ mM. The value of K is thus 0.032.

From these values we can conclude that ICDH shows a significant degree of cooperativity between the active sites of the enzyme. In the presence of the activator AMP, the degree of cooperativity is reduced, and the range of substrate concentrations over which saturation is achieved is lowered, as indicated by the decrease in the value of $[S]_{0.5}$.

Table 9.8 Kinetic data for isocitrate dehydrogenase in the absence and presence of AMP

[Isocitrate] (mM)	Rate (μmol min^{-1} mg^{-1})	
	No AMP	+0.25 mM AMP
0.05	n.d.	0.030
0.075	n.d.	0.055
0.10	n.d.	0.080
0.15	n.d.	0.120
0.20	0.009	0.149
0.30	0.028	0.182
0.40	0.055	0.199
0.50	0.087	0.208
0.60	0.117	0.214
0.70	0.142	n.d.
0.80	0.162	n.d.
1.00	0.189	n.d.
1.50	0.215	n.d.

n.d., not determined.

SELF TEST 

Check that you have mastered the key concepts at the start of the section by attempting the following question.

ST 9.8 For a given enzyme assume that K in the Hill equation = 1 (in mM units). Use the equation to calculate the change in rate when [S] is increased from 0.5 to 1.5 mM for the cases when h = 1, 2, 3, and 4.

Answer

ST 9.8 The values of the changes in rate are shown in the table.

h	v/V_{max} (0.5 mM)	v/V_{max} (1.5 mM)	-Fold increase
1	0.33	0.60	1.8
2	0.20	0.69	3.5
3	0.11	0.77	7.0
4	0.06	0.84	14.0

The values indicate the considerable increase in sensitivity of the rate to changes in the substrate concentration at higher values of h. This is especially marked at values of [S] around the half-saturation point.

Attempt Problem 9.8 at the end of the chapter.

9.7 Effect of changes in pH and temperature on enzyme activity

KEY CONCEPTS

- Understanding why enzyme activity varies with changes in pH and temperature
- Being able to derive the parameters pK_a and activation energy from experimental data

9.7.1 Changes in pH

Although changes in pH can have a number of effects on enzyme-catalysed reactions, such as the unfolding and consequent loss of activity outside a defined range of pH values, or changes in the ionization state of the substrate(s), the effect of most interest in terms of the mechanism of the reaction is the ionization of amino acid side chains at the active site of the enzyme. It is hoped that any identified ionization processes could be ascribed to particular amino acid side chains and thus confirm their importance in either the catalytic process itself or in maintaining the structural integrity of the active site.

Analysis of a scheme in which the free enzyme (E) and the enzyme-substrate complex (ES) each possess two ionizing side chains shows that changes in V_{max} depend on ionizations of ES; changes in V_{max}/K_m depend on ionizations of E and changes in K_m depend on ionizations of both E and ES.

The linear nature of the plot of log V_{max} vs pH arises because pH is a logarithmic function of the H$^+$ concentration.

The effects of pH on the kinetics of enzyme-catalysed reactions can be complex since both the K_m and V_{max} can change. It is generally easier to analyse changes in V_{max} since this generally reflects changes in a single rate constant, whereas K_m is a more complex function of several rate constants.

If ionization of a single amino acid side chain is involved (e.g. the unprotonated (–NH_2) form of a single lysine side chain is required for activity), then the activity will increase with pH and reach a limiting (maximum) value above the pK_a of this lysine side chain. A plot of log V_{max} vs pH will consist of two linear portions of slope +1 and 0 (see Fig. 9.11); the point of intersection of these linear portions will correspond to the pK_a value (see Chapter 5, section 5.6.1). If the protonated form

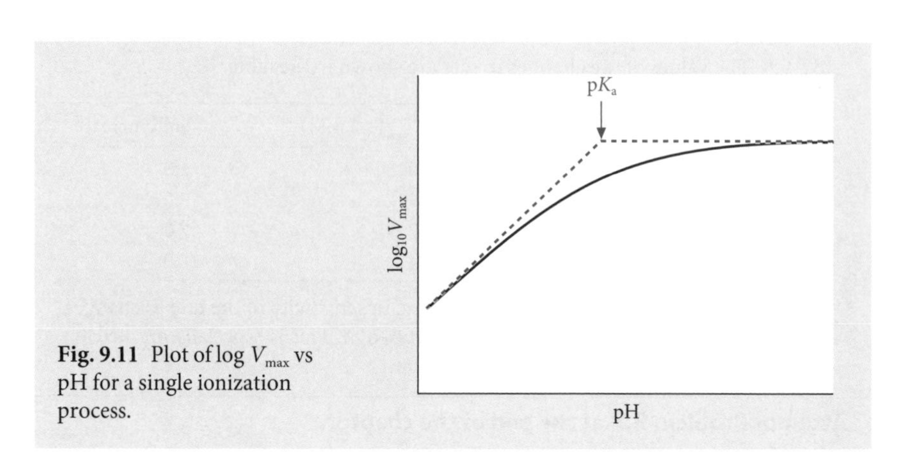

Fig. 9.11 Plot of log V_{max} vs pH for a single ionization process.

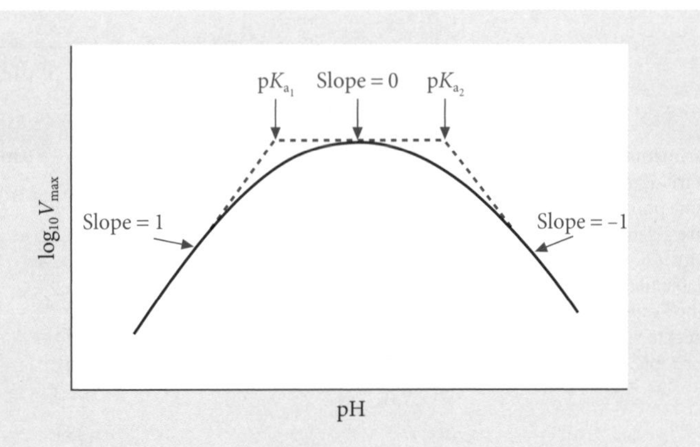

Fig. 9.12 Plot of log V_{max} vs pH for two ionization processes.

of a side chain is required for activity, the activity will decline with pH and the slopes of the two linear portions of the graph will be 0 and −1.

Many enzyme-catalysed reactions depend on the correct ionization state of two amino acid side chains. For example, only those enzyme molecules in which side chain A is protonated but side chain B is unprotonated will show activity. In such a case the plot of log V_{max} vs pH will show three linear portions of slopes +1, 0, and −1, and the pK_a values can be derived from the points of intersection of these portions (see Fig. 9.12).

If the two pK_a values are close to each other (in practice within 1.5 units) the processes cannot be considered as truly independent and a correction is required to obtain the true values.

Once the relevant pK_a value has been measured it is then necessary to identify the amino acid side chain involved. A good starting point is to look at the pK_a values of the free amino acids themselves (see Chapter 1, section 1.2.1). However, these values can only be regarded as indicative of the behaviour of side chains in the folded enzymes, since the equilibrium state of the ionization process can be markedly influenced by the environment. For example, the pK_a of the catalytically essential Glu 35 in lysozyme is estimated to be about 6, some 2 pH units above the value for free Glu (see Self test problem ST 9.9 at the end of this section).

WORKED EXAMPLE 

The variation of the rate of the trypsin-catalysed hydrolysis of the substrate *N*-benzoyl-L-arginine ethyl ester (BAEE) with pH was studied. What can you conclude from the data in Table 9.9?

STRATEGY

The data are plotted in the form log V_{max} vs pH.

SOLUTION

A plot of log V_{max} vs pH shows two linear portions of slope +1 and 0. Hence there is only one ionization process involving amino acid side chains at the active site; the point of intersection of the linear portions gives the pK_a as 6.25. Since the activity increases with increasing pH, i.e. as the amino acid side chain loses a proton, it is clear that the unprotonated form of this side chain is required for the enzyme to display activity.

Experimental details for the variation of activity of alkaline phosphatase with pH are given in Practical activity 9.1, which is available for lecturers in the Online Resource Centre for this text to support teaching of this topic.

Table 9.9 Variation of the rate of the trypsin-catalysed hydrolysis of BAEE with pH

pH	V_{max} (μmol min^{-1} mg^{-1})
4.0	1.68
4.5	5.22
5.0	16.2
5.5	45.6
6.0	108
6.5	192
7.0	255
7.5	284
8.0	295
8.5	298

9.7.2 Changes in temperature

In general, relatively limited information on the mechanism of an enzyme can be obtained from studies of the effect of temperature on the rates of reactions. However, such studies can yield information on the magnitude of the lowering of the energy barrier to reaction brought about by the enzyme and show the importance of structural transitions which can affect enzyme activity.

As mentioned in Chapter 4, section 4.2.3, the variation of the rate constant, k, of a reaction with temperature is usually described by the Arrhenius equation

$$k = Ae^{-\frac{E_a}{RT}} \quad 9.6$$

where A is known as the pre-exponential factor, R is the gas constant, T is the absolute temperature, and is E_a the activation energy for the reaction.

Taking (natural) logarithms of both sides of this equation, we obtain:

$$\ln k = \ln A - \frac{E_a}{RT} \quad 9.7$$

The transition state theory of reaction rates gives rise to an equation in which the variation of rate constant with temperature depends on the free energy of activation, $\Delta G^{\ddagger}$. This can be divided into the enthalpy ($\Delta H^{\ddagger}$) and entropy ($\Delta S^{\ddagger}$) of activation components. The activation energy $E_a = \Delta H^{\ddagger} + RT$. The value of the $\Delta S^{\ddagger}$ term can be derived from a plot of $T \ln(k/T)$ vs T, which has a slope of $\Delta S^{\ddagger}/R$.

Since $\ln A$, E_a, and R are all constants, a plot of $\ln k$ vs $1/T$ gives a straight line of slope $-E_a/R$ (Fig. 9.13). E_a can therefore be determined by multiplying the slope by $-R$. By extrapolation to $1/T = 0$ we can obtain the y-axis intercept (i.e. the value of $\ln A$).

When comparisons can be made between enzyme-catalysed reactions and the corresponding non-enzyme-catalysed reaction, the value of E_a is much less for the former. For example, the acid-catalysed hydrolysis of urea has an E_a of about 100 kJ mol^{-1}; whereas the same reaction catalysed by urease has an E_a of 42 kJ mol^{-1}.

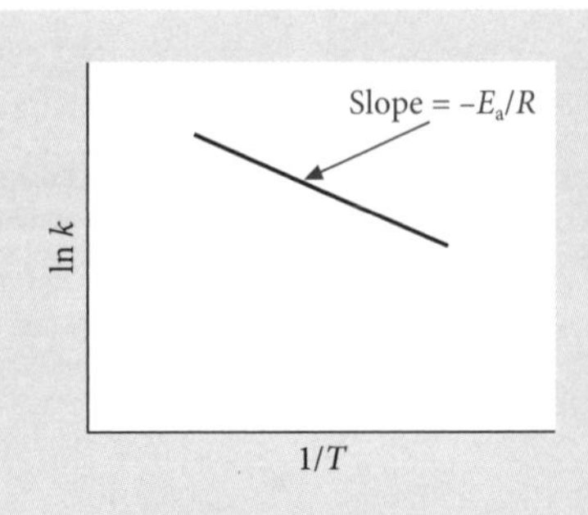

Fig. 9.13 Arrhenius plot.

WORKED EXAMPLE

The data in Table 9.10 were obtained for the hydrolysis of BAEE catalysed by trypsin at pH 7.5. What is the activation energy for this reaction?

Table 9.10 Variation of the rate of the trypsin-catalysed hydrolysis of BAEE with temperature

Temperature (°C)	Rate of BAEE hydrolysis (nmol s^{-1})
13	2.09
17	2.75
25	4.68
30	6.03
35	8.17

STRATEGY

The data are plotted in the form of the Arrhenius plot (ln (rate) vs $1/T$).

SOLUTION

The Arrhenius plot of these data (ln (rate) vs $1/T$, where T is the absolute temperature) is a straight line of slope −5450 K. Thus the value of E_a is 45 kJ mol^{-1}.

Above a certain temperature, the various weak forces stabilizing the compact three-dimensional structure of an enzyme will be broken and there will be a loss of catalytic activity. The Arrhenius plot will thus be typically of the form shown in Fig. 9.14. Although the term 'optimum temperature' is often used to denote the temperature at which activity is a maximum, it should be noted that the actual value will depend on the precise conditions under which experiments are carried out, including the length of incubation times used. Enzymes from mammalian sources often begin to lose activity at temperatures above about 45°C; however, those from thermophilic organisms such as *Pyrococcus furiosus* can retain activity even above 100°C.

The success of the polymerase chain reaction (PCR) used to amplify defined sequences of DNA depends on the stability of the enzyme DNA polymerase

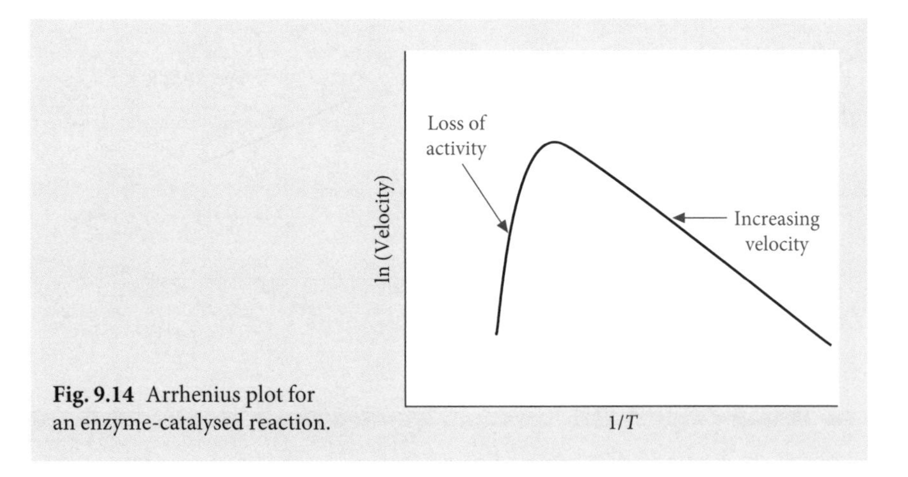

Fig. 9.14 Arrhenius plot for an enzyme-catalysed reaction.

Experimental details for the variation of the rate of hydrolysis of 4NPP catalysed by alkaline phosphatase and by OH^- with temperature are given in Practical activity 9.1, which is available for lecturers in the Online Resource Centre for this text to support teaching of this topic.

from a thermophilic bacterium (e.g. *Thermus aquaticus*) at the high temperatures (typically 95°C) used to separate the two strands of DNA.

In some cases the Arrhenius plots of enzyme-catalysed reactions can show two distinct linear portions (Fig. 9.15). Such biphasic plots usually point to a structural transition in the enzyme system. An example is provided in the worked example below.

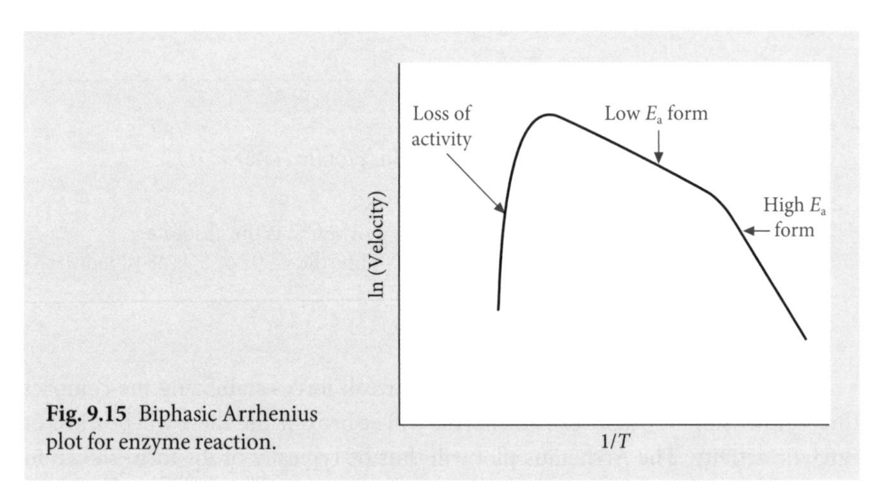

Fig. 9.15 Biphasic Arrhenius plot for enzyme reaction.

✓ WORKED EXAMPLE

The data in Table 9.11 were obtained for the variation of the rate of ATP hydrolysis by the muscle protein complex actomyosin. What can you conclude from the Arrhenius plot of these data?

STRATEGY

The data are plotted in the form of an Arrhenius plot (ln (rate) vs $1/T$).

Table 9.11 Variation of the rate of the actomyosin-catalysed hydrolysis of ATP with temperature

Temperature (°C)	Rate (μmol min^{-1} mg^{-1})
1	0.11
5	0.22
10	0.55
15	0.85
20	1.43
25	2.20
30	3.50

SOLUTION

The Arrhenius plot shows two distinct linear regions. Over the range from 1 to 10°C, the slope is 14 400 K, giving a value of 120 kJ mol^{-1} for E_a. Over the range from 15 to 30°C, the slope is −7000 K, giving a value of 58 kJ mol^{-1} for E_a. The discontinuity in the plot occurs at about 12°C and is ascribed to a structural change in the protein complex at this temperature.

SELF TEST ?

Check that you have mastered the key concepts at the start of the section by attempting the following questions.

ST 9.9 Explain why the side chain of Glu 35 in lysozyme (which is in a relatively non-polar environment) has a pK_a of about 6, some 2 pH units higher than that of the free amino acid.

ST 9.10 The following data were obtained for the variation of the rate of the fumarase-catalysed reaction with pH (fumarate + $H_2O \rightleftharpoons$ malate). What can you conclude from these data?

pH	V_{max} (μmol min^{-1} mg^{-1})
5.2	216
5.5	420
6.0	736
6.5	896
6.75	944
7.0	920
7.5	648
8.0	344
8.5	132

ST 9.11 By what factors are the rates of reactions with E_a values of 100 and 42 kJ mol^{-1} increased on going from 20°C to 30°C?

Answers

ST 9.9 Because of the relatively non-polar environment of Glu 35, the deprotonated, charged, form ($-CO_2^-$) is effectively destabilized, shifting the equilibrium towards the protonated, uncharged form ($-CO_2H$). This will have the effect of lowering K_a, and hence raising pK_a.

ST 9.10 A graph of log V_{max} vs pH for fumarase gives three linear portions of slopes +1, 0, and −1. There are two ionization processes; from the points of intersection of the linear portions, the two pK_as are 5.9 and 7.5, both of which could tentatively be ascribed to His side chains.

ST 9.11 Rearranging the Arrhenius equation we obtain:

$$\ln\left(\frac{k_{T_1}}{k_{T_2}}\right) = \frac{E_a}{R}\left(\frac{1}{T_2} - \frac{1}{T_1}\right)$$

In this example, set $T_1 = 303$ K and $T_2 = 293$ K. When $E_a = 100$ kJ mol^{-1}, this gives the value of ln (k_{T_1}/k_{T_2}), i.e. $k_{T_1}/k_{T_2} = 3.88$. When $E_a = 40$ kJ mol^{-1}, $k_{T_1}/k_{T_2} = 1.72$. (Remember to express the E_a in J mol^{-1} since R is given in units of J K^{-1} mol^{-1}).

Attempt Problems 9.9–9.11 at the end of the chapter.

9.8 Mapping of the active site by variation of substrate structure

KEY CONCEPTS

- Understanding how active sites can be mapped by varying substrate structure
- Understanding the basis of enzyme specificity

A number of features of the active sites of enzymes can be discerned by a substrate mapping approach, i.e. varying the structure of the substrate and studying the effect on the kinetic parameters of the enzyme-catalysed reaction. This can help to identify key interactions in the enzyme–substrate complex which are required for efficient catalysis.

In a classic study of this kind the hydrolysis of a large number of peptide substrates by papain (a thiol protease) was studied. A total of seven subsites around the site of cleavage were identified, as shown in Fig. 9.16.

Subsite S_2 interacts specifically with an L-Phe side chain in the substrate and S_1' shows specificity for L-amino acid side chains with a preference for non-polar side chains of Leu and Trp. The structural basis for the interactions at most of the subsites has been defined. The nomenclature shown for the subsites in both substrate and enzyme, relative to the site of hydrolysis, have been adopted for other enzymes with extended active sites.

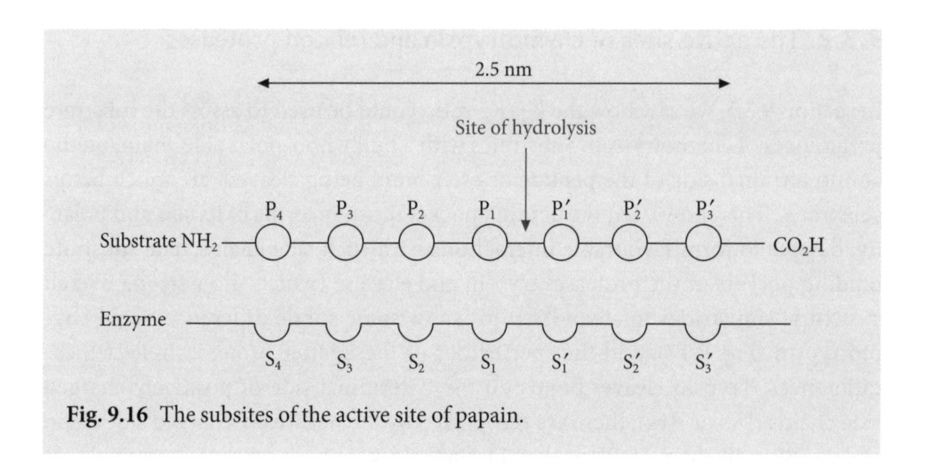

Fig. 9.16 The subsites of the active site of papain.

9.8.1 The active site of lysozyme

Lysozyme catalyses the hydrolysis of the copolymer which forms the polysaccharide component of the cell walls of Gram-positive bacteria, leading to lysis of the bacterial cells. The copolymer is made up of repeating *N*-acetylglucosamine (NAG)–*N*-acetylmuramate (NAM) units. Because of the difficulty of synthesizing small NAG–NAM copolymers, detailed kinetic studies have instead been undertaken with polymers of NAG ($(NAG)_n$), since these also act as substrates for the enzyme. By varying the size of the polymer, it has been possible to draw conclusions about the dimensions of the active site of lysozyme. The relative rates of hydrolysis of $(NAG)_n$ are shown in Table 9.12.

From these data it can be concluded that the active site of lysozyme (essentially a cleft running through the molecule) is capable of accommodating six monosaccharide units at subsites which are designated A–F. The rates do not increase when $n > 6$, showing that full activity depends on the occupation of all six subsites by substrate. An analysis of the products formed using the various $(NAG)_n$ substrates shows that cleavage occurs between subsites D and E.

Table 9.12 Lysozyme-catalysed hydrolysis of polymers of *N*-acetylglucosamine

Substrate	Relative rate of hydrolysis
NAG_2	0
NAG_3	1
NAG_4	8
NAG_5	4 000
NAG_6	30 000
NAG_8	30 000

9.8.2 The active sites of chymotrypsin and related proteases

In section 9.3.5, we saw how the k_{cat}/K_m ratio could be used to assess the substrate preferences of chymotrypsin. Substrates with a bulky non-polar side chain, on the amino terminal side of the peptide or ester bond being cleaved, are much better substrates. This shows that the binding pocket must, in terms of its size and polarity, be able to form favourable interactions with such side chains. The substrate binding pockets of the proteases trypsin and elastase (which show strong overall structural similarities to chymotrypsin) show some subtle differences from chymotrypsin (Fig. 9.17), and the specificities of these latter proteases reflect these differences. Trypsin cleaves bonds on the C-terminal side of positively charged side chains (Lys or Arg); these are recognized by the negatively charged side chain of Asp 189 at the base of the pocket. Elastase cleaves bonds on the C-terminal side of small, non-polar side chains such as Ala; larger side chains such as those of Val or Leu are unable to penetrate deeply into the pocket because of the bulky side chains (Val 216 and Thr 226) near the entrance to the pocket.

9.8.3 The active site of alcohol dehydrogenase

Alcohol dehydrogenase (ADH) is a Zn-dependent enzyme which catalyses the oxidation of a range of alcohols to the corresponding aldehydes, for example in the case of ethanol:

$$\text{ethanol} + \text{NAD}^+ \rightleftharpoons \text{ethanal} + \text{NADH} + \text{H}^+$$

The enzymes from yeast (*Saccharomyces cerevisiae*) and mammalian liver, which are tetrameric and dimeric respectively, show very different patterns of substrate specificity.

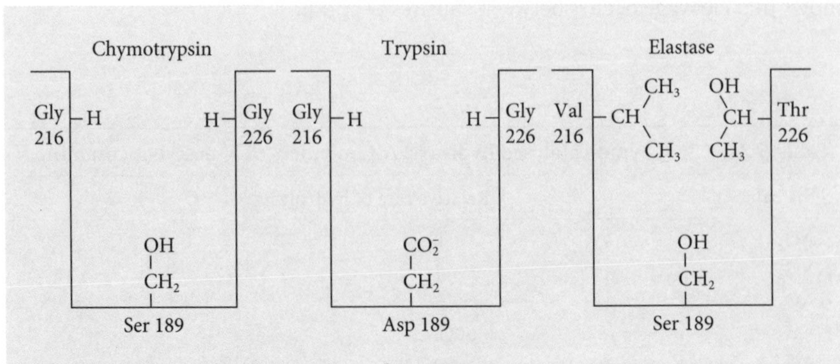

Fig. 9.17 Substrate binding pockets of serine proteases.

WORKED EXAMPLE ✓

Comment on the kinetic data shown in the Table 9.13.

Table 9.13 Kinetic data for alcohol dehydrogenases from yeast and horse liver

Substrate	Yeast ADH	Horse liver ADH
Ethanol		
k_{cat} (s^{-1})	340	7.8
K_m (mM)	17	0.8
k_{cat}/K_m (M^{-1} s^{-1})	20 000	9800
Propanol		
k_{cat} (s^{-1})	120	8.2
K_m (mM)	27	0.39
k_{cat}/K_m (M^{-1} s^{-1})	4400	21 000
Pentanol		
k_{cat} (s^{-1})	29	4.3
K_m (mM)	37	0.15
k_{cat}/K_m (M^{-1} s^{-1})	780	29 000
2-Methyl-1-propanol		
k_{cat} (s^{-1})	0.19	5.3
K_m (mM)	25	0.4
k_{cat}/K_m (M^{-1} s^{-1})	7.6	13 000
Benzyl alcohol		
k_{cat} (s^{-1})	NA	3.6
K_m (mM)	NA	0.046
k_{cat}/K_m (M^{-1} s^{-1})	NA	78 000

NA, no detectable activity ($<10^{-5}$ times the activity shown towards ethanol (Green *et al.*, 1993).

STRATEGY

The ratio k_{cat}/K_m is used as a measure of catalytic efficiency with respect to the different substrates for each enzyme.

SOLUTION

Using the k_{cat}/K_m values as a measure of catalytic efficiency towards a particular substrate, it is clear that in the case of the yeast enzyme, the activity diminishes with increasing chain length of primary alcohols and is very low indeed for branched chain or bulky alcohols. Conversely, the liver enzyme tends to show less variation in catalytic efficiency but there is a preference for longer, branched chain, or bulky alcohols.

Ingestion of alcohols such as methanol or ethylene glycol can be very dangerous, since the action of ADH can lead to poisonous oxidation products (e.g. formaldehyde (methanal) and formate (methanoate)). Formate, for example, can lead to metabolic acidosis and respiratory failure. The treatment is to administer large amounts of ethanol, which will compete effectively for the active sites of ADH and allow the other alcohols to be excreted.

The specificity patterns can be understood by comparing the sequences of the yeast and horse liver enzymes. A number of key amino acids in the active site of the enzyme are changed, including residues 48, 57, and 93 where the amino acids in the yeast enzyme (Thr, Trp, and Trp, respectively) are replaced by smaller amino acids in the horse liver enzyme (Ser, Leu, and Phe, respectively). The effect of these mutations is generally to increase the size of the active site of the horse liver enzyme, thereby enabling it to accommodate larger, branched substrates.

Details of experiments to explore the specificity of ADHs are given in Practical activity 9.2, which is available for lecturers in the Online Resource Centre for this text to support teaching of this topic.

? SELF TEST

Check that you have mastered the key concepts at the start of the section by attempting the following question.

ST 9.12 The active site of lysozyme contains six subsites (A–F) for binding of polysaccharide substrates. In this reaction, cleavage of substrates occurs between the D and E subsites. What products would you expect from hydrolysis of $(NAG)_5$ and $(NAG)_7$ substrates (NAG is *N*-acetylglucosamine)?

Answer

ST 9.12 The $(NAG)_5$ substrate of lysozyme could occupy subsites A to E or subsites B to F. In the former case, the products would be $(NAG)_4$ plus NAG; in the latter case, the products would be $(NAG)_3$ plus $(NAG)_2$. The $(NAG)_7$ substrate would occupy all six subsites with the extra NAG either 'before' subsite A or 'after' subsite F. In the former case, the products would be $(NAG)_5$ plus $(NAG)_2$; in the latter case, the products would be $(NAG)_4$ plus $(NAG)_3$.

Attempt Problem 9.12 at the end of the chapter.

9.9 Investigation of mechanism: evidence for intermediate formation

KEY CONCEPT

- Understanding the different approaches for gaining evidence for intermediates in reaction mechanisms

In terms of the energy profile for a reaction, an intermediate represents a local minimum (see Chapter 4, section 4.2). In the case of enzyme-catalysed reactions, one way by which the enzyme may bring about a lowering of the activation energy barrier is by providing an alternative reaction pathway involving the formation and breakdown of an intermediate. Depending on the depth of the energy minimum, it may be possible to isolate or trap the intermediate, e.g. by rapidly changing the pH or lowering the temperature. Characterization of the intermediate will give valuable information on the mechanism by which substrate is converted to product. The modified side chain can be identified by a mass spectrometric approach (Chapter 8, section 8.2.8.5).

It is important to demonstrate that any proposed intermediate is *kinetically competent* in the reaction, i.e. that the rate of its formation and breakdown are compatible with (that is, not greater than) the measured overall rate of the reaction. In this section, we consider some ways in which evidence for intermediate formation can be gathered.

9.9.1 Direct trapping of intermediates

A number of enzyme-catalysed reactions proceed via the reversible formation of a imine, >C=N– (also known as a Schiff base) between a carbonyl group of a substrate ($R_1R_2.C{=}O$) and the amino group on the side chain of a lysine at the active site of the enzyme:

$$R_1R_2.C{=}O + NH_2{-}(CH_2)_4{-}Enz \rightleftharpoons R_1R_2.C{=}N{-}(CH_2)_4{-}Enz + H_2O$$

The imine can be readily protonated on the N atom (>C=NH$^+$–), which markedly increases the electrophilic character of the carbon atom of the carbonyl group, facilitating subsequent reactions. Examples of enzyme-catalysed reactions which proceed via imines include aldolase, dehydroquinase (type I), decarboxylases, and transaminases.

Although the imine is unstable, it can be trapped (i.e. stabilized) by reduction with sodium borohydride to give the amine (>CH–NH–), which is stable. Radioactively labelled reagent ($NaB[^3H]_4$) can be used to determine the stoichiometry of amine formation, as demonstrated in the Worked example below. Trapping of the imine by reduction will lead to the loss of enzyme activity.

WORKED EXAMPLE 

The type I dehydroquinase from *E. coli* has a subunit molecular mass of 27.5 kDa. When 0.24 mg of the enzyme was treated with excess substrate (3-dehydroquinate), followed by addition of $NaB[^3H]_4$ (specific radioactivity 2.67×10^{10} Bq mol^{-1}), the radioactivity incorporated into the protein was 1.05×10^3 cpm. Given that the efficiency of counting is 15%, what is the stoichiometry of ^3H incorporation?

STRATEGY

The amounts (moles) of enzyme subunits and the radioactivity incorporated are calculated and then compared with each other to obtain the stoichiometry.

SOLUTION

The enzyme amounts to $0.24 \times 10^{-3}/27\,500$ mol = 8.73 nmol. The radioactivity incorporated = $(1.05 \times 10^3) \times (100/15)$ dpm = 7×10^3 dpm = 1.17×10^2 dps (Bq). This is equivalent to $(1.17 \times 10^2)/(2.67 \times 10^{10})$ mol = 4.38 nmol $NaB[^3H]_4$ = 17.5 nmol $[^3H]$ atoms (there are four H atoms per mol $NaBH_4$). Thus two H atoms are incorporated per subunit of the enzyme, as would be expected for reduction of the imine across the >C=N– double bond.

A second type of mechanism involves the formation and breakdown of a phosphoenzyme (E-P) intermediate. Examples include the ATPases involved in pumping ions such as sodium, potassium, or calcium across membranes, and phosphoglycerate mutase (PGAM), which interconverts 3- and 2-phosphoglycerates in the glycolytic pathway. The reaction of PGAM from mammalian and many fungal (e.g. *S. cerevisiae*) sources involves a second substrate, 2,3-bisphosphoglycerate

(BPG). Incubation of PGAM with [^{32}P]-BPG leads to incorporation of radioactivity into the enzyme, which corresponds to formation of the E-P intermediate (see the Self test question **ST 9.13** and Problem 9.7 at the end of this chapter).

9.9.2 Kinetic evidence for intermediate formation

In some cases it may not be possible to isolate or trap an intermediate, but its participation in the mechanism may be inferred from kinetic measurements, such as the rate of product formation. For example, if the rate of product formation is initially rapid, but then decreases to a constant (steady state) value, this would indicate the existence of an intermediate, the breakdown of which to regenerate enzyme is slow:

$$E + S_1 \rightarrow E^* + P_1 \text{ (fast)}$$

$$E^* + S_2 \rightarrow E + P_2 \text{ (slow)}$$

where E* is a modified form or distinct complex of the enzyme.

We would expect there to be a 'burst' of formation of product P_1, the amount of which would be limited by the amount of enzyme added. Subsequent production of P_1 would require the breakdown of E* to regenerate E to go through the catalytic cycle again. Clearly to observe the 'burst' we would need to add enzyme at sufficiently high concentration, so that P_1 can be detected.

Note that, as formulated in eqn. 9.8, k_1 and k_2 are pseudo-first-order rate constants (see Chapter 4, section 4.2.1). These have been derived from the true second-order rate constants by multiplying by the concentration of the substrates S_1 and S_2, respectively (these are assumed to be present in excess).

If the first and second steps are assigned rate constants k_1 and k_2, respectively, the equation for the concentration of product P_1 formed (in terms of mol/mol enzyme) at time t is given by eqn. 9.8 (Fersht, 1999a):

$$[P_1] = \left(\frac{k_1}{k_1 + k_2}\right) \times \left(\frac{k_1}{k_1 + k_2}\{1 - \exp[-(k_1 + k_2)t]\} + k_2 t\right) \qquad \textbf{9.8}$$

The form of eqn. 9.8 means that initially (at low values of t) there is a steep rise in $[P_1]$; this eventually gives way to a linear (steady state) phase of P_1 formation.

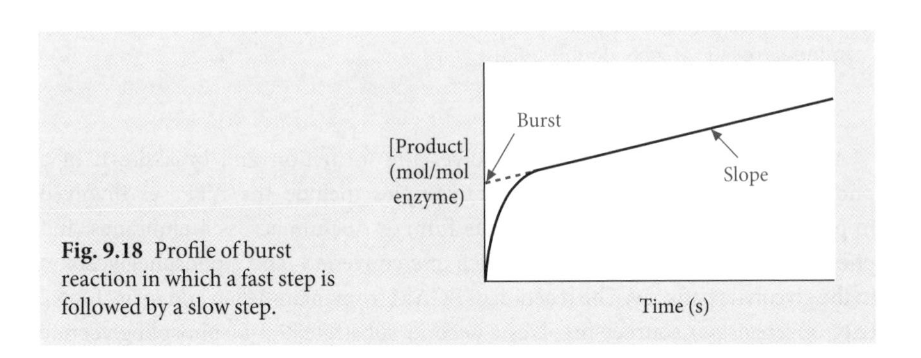

Fig. 9.18 Profile of burst reaction in which a fast step is followed by a slow step.

If the slower, steady-state part of the graph is extrapolated to the *y*-axis the intercept or burst phase (*I*) is given by eqn. 9.9:

$$I = \left(\frac{k_1}{k_1 + k_2}\right)^2 \qquad 9.9$$

and the slope (*S*) of the steady-state phase is given by eqn. 9.10:

$$S = \left(\frac{k_1 \times k_2}{k_1 + k_2}\right) \qquad 9.10$$

The size of the burst phase will therefore depend on the relative values of k_1 and k_2. If $k_1 \gg k_2$, then the burst phase will be close to 1 mol P_1 per mol enzyme; the closer the value of k_2 to k_1, the smaller the size of the burst phase.

Examples of the use of the kinetic approach are described in section 9.9.3.

WORKED EXAMPLE 

If the values of *I* (extrapolated intercept) and *S* (slope) have been measured for a particular reaction, how would you calculate the values of k_1 and k_2?

STRATEGY

This involves rearranging eqns. 9.9 and 9.10 to solve for k_1 and k_2.

SOLUTION

From eqn. 9.9 $\sqrt{I} = \left(\frac{k_1}{k_1 + k_2}\right)$

Hence, using eqn. 9.10 $$\frac{S}{\sqrt{I}} = k_2 \qquad 9.11$$

Knowing k_2, k_1 can be calculated from the equation:

$$k_1 = \frac{k_2\sqrt{I}}{(1-\sqrt{I})} = \frac{S}{(1-\sqrt{I})} \qquad 9.12$$

Practice calculations of k_1 and k_2 are described in section 9.9.3.

9.9.3 The hydrolysis of ester and amide substrates catalysed by serine proteases

The action of the serine proteases chymotrypsin and trypsin on ester and amide substrates provides good examples of reactions which proceed via a modified form of the enzyme (in this case an acyl-enzyme, which is modified on a reactive Ser (195 in the case of chymotrypsin). This Ser is part of the charge relay system Ser

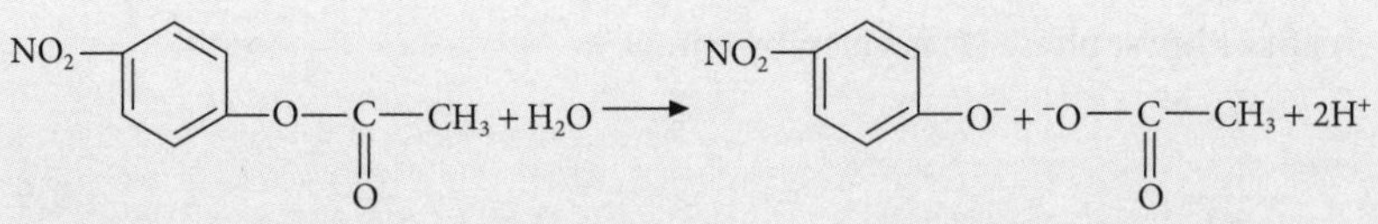

$$NO_2{-}C_6H_4{-}O{-}\overset{\overset{\large O}{\|}}{C}{-}CH_3 + H_2O \longrightarrow NO_2{-}C_6H_4{-}O^- + {}^-O{-}\overset{\overset{\large O}{\|}}{C}{-}CH_3 + 2H^+$$

Fig. 9.19 Hydrolysis of the ester substrate 4-nitrophenylacetate (4NPA).

$$R{-}C({=}O){-}NH{-}C_6H_4{-}NO_2 + H_2O \longrightarrow R{-}C({=}O){-}O^- + NH_2{-}C_6H_4{-}NO_2 + H^+$$

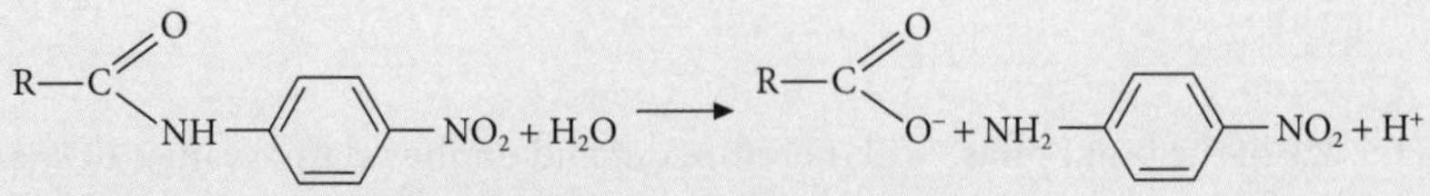

Fig. 9.20 Hydrolysis of amide substrates (BAPNA has R = N-benzoyl–Arg–; SAAPFPNA has R = N-succinyl–Ala–Ala–Pro–Phe–).

. . . His . . . Asp. The second substrate is H_2O, which hydrolyses the acyl-enzyme intermediate. In order to monitor the reactions conveniently, the substrates are designed so that the first product (P_1), an alcohol in the case of ester substrates and an amine in the case of amides, has a strong absorbance compared with that of the substrate.

9.9.3.1 Ester substrates

Experimental details are given in Practical activity 9.4, which is available for lecturers in the Online Resource Centre for this text to support teaching of this topic.

The reactions can be conveniently studied in the laboratory without needing specialized apparatus.

Typical data obtained for the reaction of 4-nitrophenyl acetate (4NPA) (0.5 mM) with chymotrypsin are shown in Fig. 9.21.

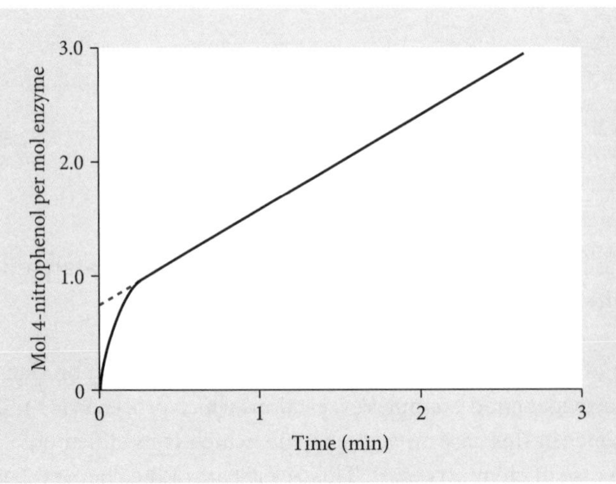

Fig. 9.21 Hydrolysis of 4NPA catalysed by chymotrypsin.

The rate of breakdown of the acyl-enzyme is markedly reduced at low pH. This makes it possible to isolate (and indeed crystallize) this intermediate by mixing chymotrypsin with an ester substrate and quenching the mixture by rapid addition of acid.

9.9.3.2 *Active site titration*

If the intermediate (acyl-enzyme) is stable, $k_2 = 0$ and the concentration of product P_1 released will reflect the concentration of active sites present. The substrate is termed an 'active site titrant'. Such a situation is observed in the reaction between trypsin and 4-nitrophenyl-4′-guanidinobenzoate (NPGB) (Fig. 9.22); by contrast the intermediate acyl-enzyme formed by reaction between chymotrypsin and NPGB is not stable.

Determination of the amount of active enzyme present is particularly important in the case of proteases, since solutions of these enzymes can self-digest in a process termed autolysis.

Active site titrants are also termed 'suicide substrates' since the enzyme is inactivated when it acts on such a substrate. There is considerable pharmaceutical interest in the design of such substrates for enzymes such as β-lactamase and monoamine oxidase (Walsh, 1983; Silverman, 1995).

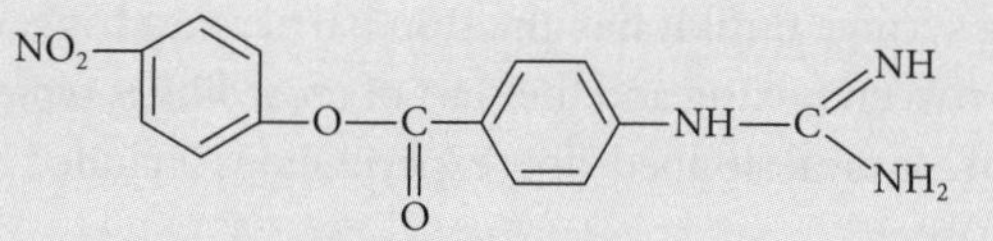

Fig. 9.22 Structure of NPGB, an active site titrant for trypsin.

Experimental details are given in Practical activity 9.4, which is available for lecturers in the Online Resource Centre for this text to support teaching of this topic.

WORKED EXAMPLE 

In the reaction between trypsin and NPGB, the value of the burst phase (I) was 0.87 mol mol^{-1} and the steady-state rate (S) was 0 mol mol^{-1} s^{-1}. In the reaction between chymotrypsin and NPGB the values of I and S were 0.74 mol mol^{-1} and 0.007 mol mol^{-1} s^{-1}, respectively. What are the values of k_1 and k_2 for the two enzymes?

STRATEGY

Eqns. 9.11 and 9.12 are used to calculate k_1 and k_2 from the values of I and S.

SOLUTION

For trypsin, $k_2 = 0$ (the acyl-enzyme is stable, because the guanidino group can lock into the active site pocket of the enzyme, which recognizes positively charged groupings). The value of k_1 cannot be determined by the approach used above (it would require rapid reaction techniques to examine the release of nitrophenol in the burst phase). The result would indicate that the preparation of trypsin contains 74% active enzyme.

In the case of chymotrypsin, the values of k_1 and k_2 are 0.049 and 0.0082 s^{-1}, respectively. The ratio of $k_1/k_2 = 6.0$. The values of the rate constants are similar to those for the reaction of chymotrypsin with 4NPA, indicating that this enzyme shows relatively little selectivity towards different ester substrates.

Active site titrants for chymotrypsin have been designed, for example 2-hydroxy-5-nitro-α-toluenesulphonic acid sultone or *trans*-cinnamoylimidazole (Kézdy and Kaiser, 1970).

9.9.3.3 Amide substrates

In the case of amide substrates there is no evidence for a burst phase, indicating that the breakdown of the acyl-enzyme intermediate is faster than its formation. We can use data on the rate of formation of the product P_1 (in this case 4-nitroaniline) to assess the relative specificities of trypsin and chymotrypsin enzymes towards different substrates (see Problem 9.13 at the end of the chapter).

9.9.4 Rapid reaction studies using the stopped flow technique

The early steps of an enzyme-catalysed reaction can often be studied in detail using the stopped-flow technique, which allows rapid mixing of solutions (usually complete within 0.1 ms) and rapid observation. The apparatus is shown schematically in Fig. 9.23.

The contents of the two syringes are mixed when the drive barrier is pushed forward, usually by gas at high pressure. After mixing, the resulting solution pushes out the stopping syringe until it hits the stop barrier and triggers data recording. The dead time between mixing and the start of recording is typically 1 ms.

The analysis of a typical stopped-flow experiment is included as Problem 9.14 at the end of the chapter.

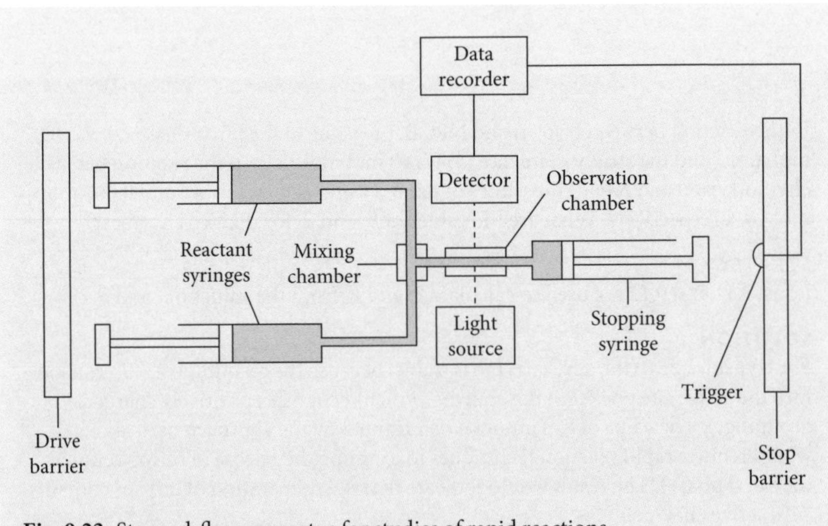

Fig. 9.23 Stopped-flow apparatus for studies of rapid reactions.

SELF TEST ?

Check that you have mastered the key concepts at the start of the section by attempting the following questions.

ST 9.13 Yeast PGAM is a tetramer of total molecular mass 110 kDa. A sample containing 0.014 mg enzyme was reacted with a large molar excess of [2,3-^{32}P] BPG of specific radioactivity 1.1 TBq mol^{-1}. The protein was then separated from excess BPG and it was found that 495 Bq of radioactivity had been incorporated. What is the stoichiometry of the incorporation?

ST 9.14 In an experiment with chymotrypsin- and trypsin-catalysed hydrolysis of 4-NPA, the values of the burst phase (I) were 0.70 and 0.12 mol mol^{-1}, respectively, and the values of the steady-state rate (S) were 0.0114 and 0.0035 mol mol^{-1} s^{-1}, respectively. What are the values of k_1 and k_2 for the two enzymes?

Answers

ST 9.13 Moles of PGAM tetramer used = $(0.014 \times 10^{-3})/110\,000 = 0.127$ nmol. Moles of ^{32}P incorporated = $495/(1.1 \times 10^{12})$ mol = 0.45 nmol. The stoichiometry is thus 0.45/0.127, i.e. 3.54 mol ^{32}P per tetramer, or 0.89 per subunit. This is in accordance with the proposed mechanism of the enzyme (see sections 9.5.3 and 9.10.1), which indicates that a phosphoenzyme intermediate is involved.

ST 9.14 Using eqns. 9.11 and 9.12, the values of k_1 and k_2 are 0.070 and 0.014 s^{-1} (chymotrypsin) and 0.004 and 0.010 s^{-1} (trypsin). The ratio of k_1/k_2 is 5.2 for chymotrypsin and 0.4 for trypsin, explaining why there is a much more marked burst phase in the former. While there is a pronounced difference in the rates of formation of the intermediate acyl-enzyme, there is little difference in the rates of its subsequent breakdown by water. This gives an interesting insight into the specificity differences shown by the two enzymes towards this substrate.

Attempt Problems 9.13 and 9.14 at the end of the chapter.

9.10 Identification of amino acid residues involved in a mechanism

KEY CONCEPTS

- Knowing the various approaches used to identify key amino acid residues
- Understanding the strengths and weaknesses of these approaches

There are three main approaches (involving bioinformatics, chemical modification, and site-directed mutagenesis) used to identify those amino acids in an enzyme which are likely to be important in the mechanism, whether in terms of binding substrate(s), playing a key covalent role in catalysis, or maintaining the precise architecture of the active site. We now consider each of these approaches in turn.

```
      138                                  175
Ec    KVVMSNHDFHKTPEAEEI..IARLRKMQSFDADTPKIALM
St    YVVMSNHDFHQTPSAEEM..VSRLRKMQALGADIPKIAVM
An    KIIASHHDPKGELSWANMSWIKFYNKALEYG.DIIKLVGV
Sc    KIIGSHHDFQGLYSWDDAEWENRFNQALTLDVDVVKFVGT
Ps    KVIVSSHNYQYTPSVEDLG..DLVARIQATGADIVKIATT
```

Ec is *Escherichia coli*, St is *Salmonella typhi*, An is *Aspergillus nidulans*, Sc is *Saccharomyces cerevisiae*, Ps is *Pisum savitum*

Fig. 9.24 Alignment of part of the sequence of type I dehydroquinases from a number of organisms. Ec, *Escherichia coli*; St, *Salmonella typhi*; An, *Aspergillus nidulans*; Sc, *Saccharomyces cerevisiae*; Ps, *Pisum savitum*.

9.10.1 Bioinformatics approach

Comparison of sequences of the enzyme from number of organisms (see Chapter 5, section 5.8.3) will highlight conserved amino acids; those that are strictly conserved are likely to play key roles, whereas those which are freely variable are unlikely to be crucial.

An example of this type of approach is provided by a comparison of the sequences of the type I dehydroquinase enzymes from a number of organisms (Fig. 9.24). For the present purpose, particular attention has been focussed on the portion from residues 138 to 175 (*E. coli* numbering), which is thought to contain important active site side chains.

In the alignment, sequences are written in the one letter code (see Chapter 1, section 1.2.1): A = Ala, C = Cys, D = Asp, E = Glu, F = Phe, G = Gly, H = His, I = Ile, K = Lys, L = Leu, M = Met, N = Asn, P = Pro, Q = Gln, R = Arg, S = Ser, T = Thr, V = Val, W = Trp, Y = Tyr. A dot indicates where a gap has been introduced in the sequence to maximize the identity of the aligned sequences.

Across the organisms, four amino acids are conserved, namely Ser 142, His 144, Asp 168, and Lys 171. Other experiments show that Lys 171 is involved in forming a Schiff base (imine) with the carbonyl group at C3 of the substrate (see section 9.9.1) and His 144 plays a role as general base abstracting a proton from C2 of the substrate to give a carbanion intermediate. The precise roles played by Ser 142 and Asp 168 in the mechanism of the enzyme have not yet been clarified.

Another example of this type of approach is the phosphoglycerate mutases from yeast and various mammalian sources where two His residues (His 8 and His 180) and several Arg residues (including Arg 7, Arg 59, Arg 113, and Arg 114) (yeast enzyme numbering) are strictly conserved. From an analysis of the structure of the enzyme and its complexes with substrate analogues, it is clear that one His (His 8) is involved in formation of a phosphoenzyme (E-P) intermediate and His 180 plays a role in acid/base catalysis. The Arg residues are involved in binding the highly negatively charged substrates of the enzyme.

The mammalian serine proteases such as trypsin, chymotrypsin, elastase, and thrombin, have very similar overall 3-dimensional structures. There is also a very high degree of conservation of sequence around the three amino acids (Ser, His, and Asp) involved in the charge relay system (see section 9.9.3):

. . . Gly–Asp–SER–Gly–Gly . . . . . . Ala–Ala–HIS–Cys . . . . . . ASP–Ile . . .

By contrast, in the serine protease subtilisin (from the Gram-positive bacterium *Bacillus amyloliquefaciens*, see section 9.10.3.1), the sequences adjacent to the residues of the catalytic triad are very different from those in the mammalian enzymes. In addition, there is essentially no discernible similarity between the overall three-dimensional structure of subtilisin and the mammalian serine proteases:

. . . Gly–Thr–SER–Met–Ala . . . . . . Asn–Ser–HIS–Gly . . . . . . ASP–Ser . . .

More direct evidence for the involvement of particular amino acids in a mechanism can be obtained at either the protein (section 9.10.2) or gene (section 9.10.3) level. The principle underlying these approaches is that if modification or substitution of a particular amino acid results in loss of enzyme activity, then that amino acid must be important for the activity.

9.10.2 Chemical modification of amino acid side chains

In this approach, the aim is to react the enzyme with a reagent so that only one type of amino acid side chain (and ideally that only one or a very small number of that type of side chain) is modified. This poses considerable problems of experimental design since it can be difficult to distinguish between, for example nucleophilic side chains such as Lys, Cys, His, and Tyr. In addition, a number of side chains are essentially unreactive, e.g. the aliphatic side chains of Ala, Val, Leu, Ile, and Pro. Nevertheless, the technique has proved of value in identifying important amino acid side chains of several enzymes.

The first approach to developing specificity in chemical modification is to exploit chemical principles. For example, the thiol group in Cys is well known as undergoing thiol/disulphide interchange reactions or reacting with mercury derivatives. The activated aromatic ring in the Tyr side chain will readily undergo electrophilic substitution, for example nitration using tetranitromethane or iodination using I_2. There are also differences in selectivity between the different nucleophilic side chains which can be exploited. Thus towards acylating agents such as iodacetamide the order of reactivity is Cys > Tyr > His > Lys, whereas towards arylating agents such as 2,4,6-trinitrobenzene sulphonate (TNBS) the order is Cys > Lys > Tyr > His. Thus, TNBS would be expected to be a better reagent for modification of Lys side chains. Changing the pH can also be used to affect selectivity since the deprotonated forms of the side chains are much more effective nucleophiles.

Using these types of ideas, a number of reagents which show reasonable degrees of specificity for particular amino acid side chains have been developed. Some of these are listed in Table 9.14. Details of these reagents can be found in the book by Lundblad, 1995.

A second approach to developing specificity is to exploit the fact that many enzymes have super-reactive amino acid side chains, so that specificity can be achieved

DIFP is an extremely toxic volatile compound, which is a nerve gas because of its action in inhibiting acetylcholinesterase, an enzyme which catalyses the hydrolysis of the neurotransmitter acetylcholine.

Table 9.14 Typical reagents used for modification of amino acid side chains

Amino acid side chain	Reagent(s) used
Cysteine	Mercurials, e.g. 4-chloromercuribenzoate Disulphides, e.g. 5,5′-dithiobis-(2-nitrobenzoic acid) Iodoacetamide Maleimide derivatives, e.g. *N*-ethylmaleimide
Lysine	2,4,6-Trinitrobenzenesulphonate Pyridoxal phosphate (± reducing agent such as $NaBH_4$)
Histidine	Diethylpyrocarbonate Photo-oxidation
Arginine	Phenylglyoxal 2,3-Butanedione
Tyrosine	Tetranitromethane *N*-Acetylimidazole Iodine
Tryptophan	*N*-Bromosuccinimide
Aspartic acid or glutamic acid	Water-soluble carbodiimide plus nucleophile, e.g. glycine methylester

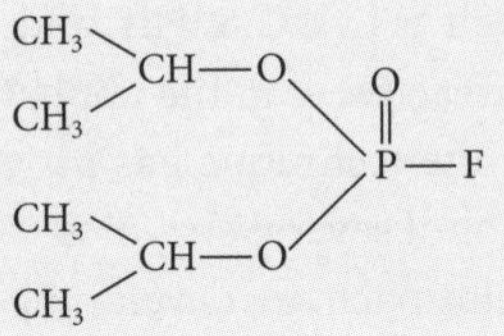

$$(CH_3)_2CH{-}O{-}P(=O)(F){-}O{-}CH(CH_3)_2$$

Fig. 9.25 Structure of diisopropyl fluorophosphate (DIFP).

serendipitously. For example, there are 28 Ser side chains in bovine chymotrypsin but only one of these (Ser 195) is reactive towards electrophilic reagents such as diisopropyl fluorophosphate (DIFP) (Fig. 9.25) or phenylmethylsulphonyl fluoride. In addition, the side chain of the free amino acid Ser does not react with DIFP.

The reason for the greatly enhanced reactivity of Ser 195 is the 'charge relay system' (or 'catalytic triad') of Ser 195, His 57, and Asp 102 in which negative charge from Asp 102 is transferred to Ser 195, thereby greatly enhancing its nucleophilic character (see section 9.9.3).

A second example is provided by the enhanced reactivity of Lys 126 in bovine liver glutamate dehydrogenase (GDH) towards TNBS, compared with the other 32 Lys in the polypeptide chain. This enhanced reactivity can be ascribed to the fact that the side chain of Lys 126 has a unusually low pK_a (estimated as 7.7–8.0, compared with 10.5 for free Lys), thereby greatly increasing the proportion of the reactive unprotonated form at neutral or mildly alkaline pH values. It is thought that Lys 126 is involved in the catalytic mechanism of the enzyme by forming a Schiff base (imine) with the carbonyl group of the substrate 2-oxoglutarate.

Experimental details of the modification of GDH by pyridoxal-5′-phosphate in the absence and presence of $NaBH_4$ are given in Practical activity 9.5, which is available for lecturers in the Online Resource Centre for this text to support teaching of this topic.

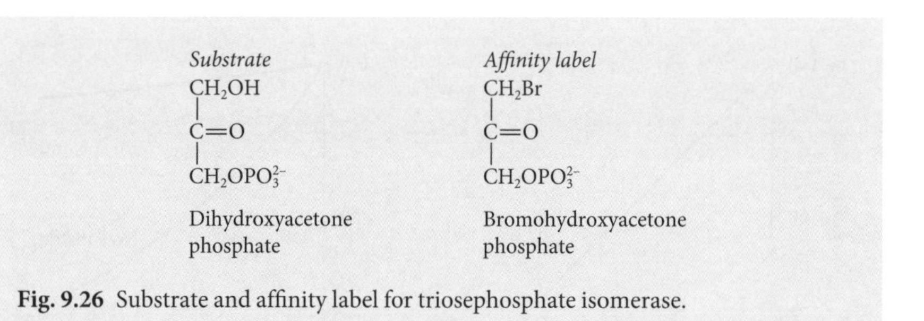

Fig. 9.26 Substrate and affinity label for triosephosphate isomerase.

Although in many cases (including the two examples described above) the super-reactive side chains are involved in the catalytic mechanism, this is not always the case. For example, the very reactive Cys side chain (Cys 282) in muscle creatine kinase is not required for catalytic activity. By contrast, Glu 35 of lysozyme, which acts as a general acid catalyst by donating a proton to the O atom of the glycosidic bond to be cleaved, is unreactive without prior denaturation of the enzyme.

The third approach to modifying specific amino acids is to build locational specificity into the structure of the modifying species reagent, creating 'active-site directed reagents' or 'affinity labels'. The example of triosephosphate isomerase illustrates the principle; the substrate and the affinity label are shown in Fig. 9.26.

The close structural relationship between the substrate and the affinity label will ensure that the latter will bind at the active site and any suitably positioned nucleophilic side chain will attack the C atom carrying the Br atom, releasing the Br^- leaving group. The site of modification is Glu 165, and it is proposed that this side chain plays a key role in catalysis by acting as a general base in proton transfer steps in the mechanism. (It should be noted that after the initial modification of Glu 165 in triosephosphate isomerase, the label can migrate to the neighbouring Tyr 164; this transfer can be prevented by reduction of the initial adduct with $NaBH_4$.)

Experimental details for studying the affinity labelling of chymotrypsin and trypsin with *N*-tosyl-L-phenylalanine chloromethyl ketone (TPCK) and Nα-tosyl-L-lysine chloromethyl ketone (TLCK) are given in Practical activity 9.4, which is available for lecturers in the Online Resource Centre for this text to support teaching of this topic.

Although chemical modification can provide some valuable insights into the nature of amino acid side chains involved in the mechanisms of enzymes, there are certain precautions which should be observed in interpreting the data. It is important to establish that the modification has not resulted in significant structural perturbation of the enzyme which might lead to loss of activity in an accidental manner. This can be conveniently demonstrated by use of circular dichroism (Chapter 5, section 5.2.2, and Chapter 8, section 8.4.2.2). In addition, two basic criteria to be satisfied are those of (a) stoichiometry of modification (i.e. modification of one side chain per active site leads to total inactivation) and (b) protection (i.e. the inclusion of substrate or competitive inhibitor protects against inactivation). These criteria are illustrated in Fig. 9.27.

However, these should only be regarded as minimal criteria and the interpretation needs to supported by other data such as site-directed mutagenesis (section 9.10.3) and detailed structural information.

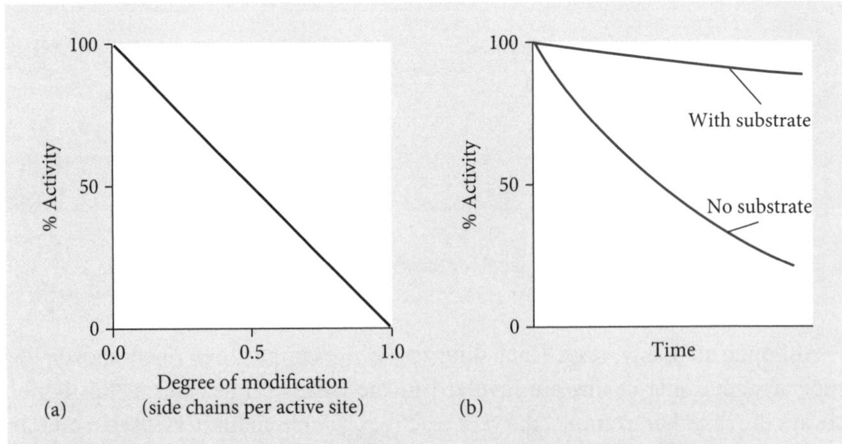

Fig. 9.27 Minimal criteria for modification at the active site: (a) stoichiometric inactivation; (b) protection by substrate or competitive inhibitor.

9.10.3 Site-directed mutagenesis of amino acid side chains

It is also possible to use a type of site-directed mutagenesis to introduce non-natural amino acids such as Se–Cys and Se–Met into proteins. Replacement of the normal S atom by the heavier Se can assist structure determination by X-ray crystallography.

Site-directed mutagenesis is a means of modifying an amino acid side chain at the gene level. In essence, the gene encoding a particular protein is modified so that at a chosen position the amino acid found in the native protein is replaced by a selected amino acid. This mutated gene is then expressed (see Chapter 5, section 5.4.2) in a suitable host system and the mutant protein purified. Clearly it is essential to have the appropriate gene and expression system available.

In principle, astronomical numbers of potential mutant proteins could be produced by this means. The number of mutants can be calculated by remembering that each of the amino acids could be replaced by any of the other 19 amino acids, thus in the case of a 140 amino acid protein, this would give $140 \times 19 = 2660$ mutants. If a second amino acid is to be mutated, there are $139 \times 19 = 2641$ mutants with this additional mutation. Thus there are $2660 \times 2641 = 6\,972\,240$ (or about 7 million) double mutants. If three amino acids are mutated there are over 1.8×10^{10} (18 billion) mutants.

The technique of directed evolution involues generating all possible mutants. Identification of mutants of interest requires a rapid screening method (Kuchner and Arnold, 1997).

Given the potential of the technique, it is important to consider which are the most useful mutations to make. It is most useful to have three-dimensional structural information about the enzyme and its complexes with substrates or inhibitors available; this can identify particular amino acids which might be playing key catalytic or structural roles in the enzyme. Information from sequence comparisons can also be useful in identifying such amino acids (see section 9.10.1 and Chapter 5, section 5.8.3).

Having identified which amino acids are worthwhile mutating, the next stage of the process is to consider which replacements should be made. Ideally, if the role of an amino acid in catalysis is to be explored the overall size and polarity of the

Table 9.15 Common amino acid replacements in site-directed mutagenesis

Amino acid	Replacement
Ala	Ser, Thr, Gly
Arg	Lys, His, Gln
Asn	Asp, Gln, Glu, Ser, His, Lys
Asp	Asn, Gln, Glu, His
Cys	Ser
His	Asn, Asp, Gln, Glu, Arg
Leu	Met, Ile, Val, Phe
Lys	Arg, Gln, Asn
Ser	Ala, Thr, Asn, Gly
Tyr	Phe, His, Trp
Trp	Phe, Tyr

amino acid side chain should be maintained but one key chemical feature should be altered. For example, replacement of Glu by Gln removes the negative charge. In general, it is better to replace a bulky side chain by a smaller one, since this causes less disturbance to the overall structure than the reverse type of substitution. It is always important to check that the overall structure and folding of the enzyme has not been perturbed by the mutation introduced; again circular dichroism (see Chapter 5, section 5.5.2 and Chapter 8, section 8.4.2.2) is an ideal technique to check this. A list of common amino acid replacements to investigate the function of side chains is shown in Table 9.15.

Site-directed mutagenesis has become a very widely used technique to explore the roles of amino acid side chains in proteins, primarily because it allows the role of all amino acids to be explored in a systematic fashion. However, it should be remembered that chemical modification can introduce a wider range of functionalities into a protein. For example, many proteins can be tagged by chemical derivatives of dyes such as fluorescein or rhodamine to provide convenient means of tracking them in cells using specialized optical techniques such as confocal microscopy.

Many examples of the use of site-directed mutagenesis to explore the contributions of particular amino acid side chains to the mechanism of enzymes have been described (see, Fersht, 1999c). These include the following enzymes.

9.10.3.1 Subtilisin, a serine protease

Subtilisin is a protease isolated from *B. amyloliquefaciens* which possesses a catalytic triad of Ser . . . His . . . Asp residues but has little similarity to the mammalian serine proteases such as trypsin and chymotrypsin in terms of its sequence or three-dimensional structure. The contribution of the three components of the triad (Ser 221, His 64, and Asp 32 in the case of subtilisin) to the catalytic efficiency was assessed by mutating each in turn to Ala; in addition, the triple mutant was made. Although the catalytic efficiency of each mutant was reduced by a large factor

Substitution of any amino acid by Ala is considered 'safe' in that it is unlikely to cause any structural distortion of the protein.

(between 10^4- and 10^6-fold) the mutants could still bring about a modest (10^3-fold) degree of rate enhancement. This was presumably because a water molecule could be activated by other side chains in the enzyme to hydrolyse the appropriately bound substrate molecule. Note that although the roles of Ser 195 and His 57 in chymotrypsin had been established by chemical modification (DIFP and TPCK, respectively), no modification experiments had been carried out on Asp 102.

9.10.3.2 Carboxypeptidase, a metalloprotease

There are in fact two forms of carboxypeptidase, A and B. A will remove all amino acids except Lys and Arg, B will remove only Lys and Arg. In the B enzyme there is an appropriately positioned side chain (Asp 255) to bind to the positive charge of the side chains of Lys and Arg. This side chain is not present in the A form of the enzyme.

Carboxypeptidase catalyses the hydrolysis of the C-terminal amide bond in a protein to release the C-terminal amino acids in turn.

Reaction of carboxypeptidase A with tetranitromethane leads to a loss of over 90% of the activity; this inactivation can be protected by inclusion of the competitive inhibitor β-phenylpropionate in the reaction mixture. The site of modification was shown to be Tyr 248, and in the X-ray structure this side chain was located close to the active site so that it appeared that the Tyr could donate a proton to make the –NH– of the peptide bond a better leaving group (Riordan *et al.*, 1967).

The involvement of Tyr 248 in the mechanism was examined by site-directed mutagenesis by making the Tyr 248 → Phe mutant; in this mutant residue 248 would be unable to act as a proton donor. Somewhat surprisingly, the mutant was active with k_{cat} unchanged but K_m raised. This suggested that Tyr 248 does not play a crucial role as a proton donor, but is instead involved in the binding of substrates (Gardell *et al.*, 1985).

9.10.3.3 Lysozyme, a glycosidase

Lysozyme catalyses the hydrolysis of the polysaccharide component of the bacterial cell wall of Gram-positive bacteria. From the X-ray structure of the enzyme, two amino acid side chains were identified as essential for activity, namely Glu 35 and Asp 52. The variation of activity with pH indicates that Glu 35 has an unusually high pK_a (6.0) because it is in a non-polar environment; the protonated form of the side chain donates a proton to the glycosidic O atom. The negative charge on the side chain of Asp 52 can stabilize the developing positive charge on the carbocation intermediate. The location of Asp 52 at the active site was confirmed by using an affinity label (the 2′,3′-epoxypropyl-β-glycoside derivative of $(NAG)_2$) which led to complete inactivation by reaction with Asp 52 (Eshdat *et al.*, 1973). However, it was not possible to modify Glu 35 without denaturing the enzyme. Site-directed mutagenesis was used to confirm the importance of these two side chains for catalysis. Whereas, the Glu 35 → Gln mutant was essentially completely inactive, the Asp 52 → Asn mutant retained a small amount (about 5%) of the activity of wild-type enzyme. However, both mutants were able to bind the competitive inhibitor chitotriose with affinities similar to that of the wild type, showing that the overall architecture of the active site had not been perturbed (Malcolm *et al.*, 1989; Matsumura and Kirsch, 1996).

SELF TEST ?

Check that you have mastered the key concepts at the start of the section by attempting the following question.

ST 9.15 One form of LDH from human muscle contains 332 amino acids in the polypeptide chain. How many single and double mutants of this enzyme could be made?

Answer

ST 9.15 The number of single mutants of the lactate dehydrogenase is $332 \times 19 = 6308$. The number of double mutants is $6308 \times 6289 = 39\,671\,012$ (i.e. nearly 40 million).

Attempt Problem 9.15 at the end of the chapter

9.11 Concluding comments about mechanisms

This chapter has described some of the main approaches used to gain an understanding of the mechanism of enzyme action. As indicated at the start of this chapter a complete description of a mechanism involves both kinetic and structural aspects and hence it can be a long process to achieve this goal. High-resolution structural studies (of both the free enzyme and appropriate complexes with substrate or inhibitors) can provide information on the amino acids that make up the active site and play roles in binding the substrate(s) and facilitating the bond breaking and making of covalent bonds during conversion to product(s). The structural studies can also indicate those parts of the enzyme that undergo conformational changes during the catalytic cycle. Proposals about the roles of particular amino acids can be tested by methods such as site-directed mutagenesis or specific chemical modification. Because high-resolution structural information cannot generally be obtained on the timescale at which catalytic events occur (ms or less), a certain amount of 'inspired guesswork' can be involved in linking the kinetic and structural aspects of mechanisms. Nevertheless, enormous progress has been made in understanding how enzymes work and, in a semi-quantitative manner at least, in being able to account for the degree of rate enhancement they can bring about.

This understanding of mechanisms has led to the prospect that new enzyme activities might be generated, which will be able to catalyse reactions of great commercial and environmental importance, for example in bioremediation, digestion of cellulose-based waste, functionalization of otherwise unreactive hydrocarbons, and stereospecific synthesis of pharmaceutical compounds. An early successful attempt to change the specificity of an enzyme was the 'conversion' of LDH to malate dehydrogenase (see the worked example below).

✓ WORKED EXAMPLE

The enzymes LDH and malate dehydrogenase (MDH) catalyse-related reactions (i.e. reduction of α-ketoacids by NADH), and have very similar secondary and tertiary structures (although it should be noted that LDH is a tetramer and MDH a dimer).

The two reactions catalysed are:

LDH lactate + $NAD^+ \rightleftharpoons$ pyruvate + NADH + H^+

MDH malate + $NAD^+ \rightleftharpoons$ oxaloacetate + NADH + H^+

From a detailed consideration of the architecture of the active sites of the two enzymes, Holbrook and colleagues (Clarke *et al.*, 1989a,b) decided that selected site-directed mutations could alter the specificity of LDH to that of MDH. Three amino acids were mutated (Asp197 → Asn, Thr246 → Gly and Gln102 → Arg) so as to increase the volume and change the charge characteristics of the active site.

The kinetic properties of the wild-type and mutant enzymes are shown in Table 9.16. Note that mutant 1 is the single mutant Asp197 → Asn, mutant 2 is the double mutant Asp197 → Asn, Thr246 → Gly, and mutant 3 is the triple mutant Asp197 → Asn, Thr246 → Gly, Gln102 → Arg.

Table 9.16 Kinetic data for wild-type lactate dehydrogenase and mutants

Enzyme	Pyruvate		Oxaloacetate	
	k_{cat} (s^{-1})	K_m (mM)	k_{cat} (s^{-1})	K_m (mM)
Wild type	250	0.06	6.0	1.5
Mutant 1	90	0.66	0.50	0.15
Mutant 2	16	13.0	0.94	0.20
Mutant 3	0.9	1.8	250	0.06

By considering the reactions catalysed by these two enzymes, explain why each of the mutations listed might help to achieve the required change in specificity. For each enzyme (wild type and mutants), evaluate the specificity constant (k_{cat}/K_m, in units of M^{-1} s^{-1}) for each substrate and hence the ratio $\{(k_{cat}/K_m)_{oxal}/(k_{cat}/K_m)_{pyr}\}$. Would you consider that the authors achieved their stated goal of changing LDH into MDH?

STRATEGY

The k_{cat}/K_m ratios should be calculated for each form of the enzyme. It is important to make sure that these ratios are expressed in the correct units.

SOLUTION

Compared with pyruvate, oxaloacetate has one of the methyl hydrogen atoms replaced by $-CH_2-CO_2^-$, so the substrate for the MDH reaction is larger and carries an extra negative charge. The three mutations are designed to allow the enzyme to accommodate the new substrate. Replacing Asp 197 by Asn removes negative charge from the active site and replacing Thr 246 by Gly increases the volume of the active site. Finally, the replacement of Gln 102 by Arg adds positive charge to the active site.

From the kinetic data, the values in Table 9.17 can be calculated.

Table 9.17 Specificity constants for wild-type lactate dehydrogenase and mutants

Enzyme	k_{cat}/K_m (M^{-1} s^{-1}) pyruvate	k_{cat}/K_m (M^{-1} s^{-1}) oxaloacetate	Ratio (Oxal/Pyr)
Wild type	4.17×10^6	4.0×10^3	9.6×10^{-4}
Mutant 1	1.36×10^5	3.33×10^3	2.4×10^{-2}
Mutant 2	1.23×10^3	4.7×10^3	3.8
Mutant 3	5×10^2	4.17×10^6	8.3×10^3

Each successive mutation helps to drive the enzyme from LDH activity towards MDH activity. Thus, whereas wild-type LDH favours pyruvate over oxaloacetate by a factor of 1040-fold, the triple mutant favours oxaloacetate over pyruvate by an even larger factor (8300-fold). On this basis, the authors can be said to have achieved their stated goal.

From this work and other studies of this type, it is clear that successful redesign of enzymes can be achieved but is more likely to be successful when (a) detailed structures of the starting and target enzyme and appropriate complexes are available and (b) the reactions catalysed are of a similar chemical nature to each other. It is also worth pointing out that the effects of amino acid replacements are not always predictable, and so it is advisable to explore the properties of a number of mutations.

9.12 Problems

Full solutions to odd-numbered problems are available to all in the student section of the Online Resource Centre at www.oxfordtextbooks.co.uk/orc/price/. Full solutions to even-numbered problems are available to lecturers only in the lecturer section of the Online Resource Centre.

9.1 The enzyme histidine decarboxylase catalyses the formation of histamine from histidine. How would you assay this enzyme?

9.2 RNA polymerase catalyses the synthesis of RNA using one of the strands of double-stranded DNA as a template. The assay system includes UTP, GTP, CTP, and [^{14}C]-ATP each at 0.15 mM in a total volume of 0.25 mL. The specific radioactivity of the ATP is 37 GBq mol^{-1}. Of a partially purified preparation of RNA polymerase, 0.085 mg was assayed. After 10-min incubation the reaction was stopped by addition of trichloroacetic acid and the radioactivity incorporated into the RNA was measured as 1980 cpm; the counting efficiency was 75%. One unit of RNA polymerase activity corresponds to the incorporation of 1 nmol ATP into RNA in 10 min. Calculate how many units of activity were present in the assay and the specific activity of the RNA polymerase preparation.

9.3 Creatine kinase catalyses the reaction creatine + ATP ⇌ phosphocreatine + ADP. How might you assay this enzyme and how would you check that the assay gave reliable results?

9.4 Dehydroquinase (DHQase) catalyses a step on the shikimate pathway for the biosynthesis of aromatic amino acids. There are two distinct types of DHQases (types I and II), which operate by very different chemical mechanisms. Kinetic parameters for the enzymes from a number of sources are shown in the table below. What can you conclude about the catalytic efficiencies of these enzymes?

Enzyme	k_{cat} (s^{-1})	K_m (μM)
Type I		
E. coli	135	16
Type II		
Aspergillus nidulans	1300	150
S. coelicolor	125	99
M. tuberculosis	5.2	24

9.5 Fructose bisphosphatase catalyses the hydrolysis of fructose-1,6-bisphosphate (FBP) to fructose-6-phosphate and phosphate; the reaction is part of the gluconeogenic pathway which results in the synthesis of glucose. The enzyme is inhibited by AMP and the following data were obtained. What type of inhibition is being observed?

[FBP] (μM)	Rate (μmol min^{-1} mg^{-1})		
	No AMP	+70 μM AMP	+140 μM AMP
0.5	19.5	10.4	7.1
1.0	33.8	18.0	12.3
2.0	53.2	28.4	19.4
4.0	74.6	39.8	27.2
6.0	86.2	46.0	31.4
8.0	93.5	49.9	34.0
10.0	98.4	52.5	35.8

9.6 β-Galactosidase catalyses the hydrolysis of the disaccharide lactose to yield the two constituent monosaccharides, i.e. glucose and galactose. Lactose is a major constituent of milk (about 50 g L^{-1} (145 mM) in cow's milk) and is broken down by β-galactosidase (lactase) located in the brush border of the small intestine. The breakdown of lactose is known to be inhibited by dietary monosaccharides such as glucose, fructose, and galactose. The inhibition of β-galactosidase from *E. coli* was studied using a model substrate (2-nitrophenylgalactoside, 2NPGal) which gives a convenient absorbance change on reaction because of the formation of 2-nitrophenol. The following data were obtained:

[2NPGal] (mM)	Rate (μmol min^{-1} mg^{-1})		
	No inhibitor	+30 mM galactose	+25 mM glucose
0.03	96	50	96
0.05	140	78	101
0.10	215	132	134
0.15	260	173	151
0.20	290	205	161
0.30	329	250	172
0.40	353	281	178

Determine the kinetic parameters (K_m and V_{max}) in each case and the type of inhibition caused by galactose and glucose.

9.7 PGAM catalyses the interconversion of 3-phosphoglycerate (3PGA) and 2-phosphoglycerate (2PGA). The enzyme from yeast is a member of one class of PGAMs, which depend on 2,3BPG for activity, i.e. effectively the enzyme catalyses a two-substrate reaction:

$$3PGA + 2,3BPG \rightleftharpoons 2PGA + 2,3BPG$$

The following data were obtained for PGAM from yeast:

[3PGA] (μM)	Rate (μmol min^{-1} mg^{-1})			
	2 μM 2,3BPG	5 μM 2,3BPG	10 μM 2,3BPG	20 μM 2,3BPG
400	22.6	27.8	39.6	58.1
800	33.3	40.7	57.0	78.5
1500	50.7	60.3	80.3	108.2
2500	70.8	82.7	103.1	137.7

Analyse the data to determine the kinetic parameters. What can you conclude about the mechanism of the enzyme?

9.8 Aspartate carbamoyltransferase catalyses the reaction between L-aspartate and carbamoyl phosphate to give *N*-carbamoylaspartate. This is the first step in the synthesis of pyrimidine nucleotides. The enzyme is inhibited in the presence of CTP, the end product of the pathway. The following kinetic data were obtained when the concentration of L-aspartate was varied in the presence and absence of CTP. In both cases the V_{max} for the preparation was 50 μmol min^{-1} mg^{-1}. Analyse the data using the Hill equation and comment on the results.

[L-aspartate] (mM)	Rate (μmol min^{-1} mg^{-1})	
	No CTP	+0.4 mM CTP
5.0	9.8	4.1
7.5	17.8	9.1
10.0	26.4	14.6
12.5	32.3	20.0
15.0	36.8	25.4
20.0	41.6	34.7
25.0	44.6	40.1
30.0	46.3	44.0

9.9 The mechanism of the type II dehydroquinases is thought to involve a tyrosine side chain (Tyr 28) which acts as a base, abstracting a proton from the substrate. It is estimated that the pK_a of this tyrosine is lowered to 8.0 (compared with 10.5 for the free amino acid) by the presence of a neighbouring arginine side chain (Arg 23). Explain why the pK_a of the Tyr should be lowered by the presence of the arginine. Use the relationship between the standard free energy change and equilibrium constant (see Chapter 4, section 4.1) to evaluate the free energy change associated with this shift in pK_a at 25°C.

9.10 The hydrolysis of 4NPP can be catalysed by hydroxide ions or (much more effectively) by intestinal alkaline phosphatase. The following data were obtained for the temperature dependence of the rates of these two reactions.

Catalysis by OH^- ions	
Temperature (°C)	**Rate (ΔA_{400} min^{-1})**
39	7.15×10^{-5}
50	2.34×10^{-4}
61	7.50×10^{-4}
70	2.11×10^{-3}

Catalysis by alkaline phosphatase	
Temperature (°C)	**Rate (ΔA_{400} min^{-1})**
2	3.40×10^{-2}
12	6.00×10^{-2}
22	1.10×10^{-1}
32	1.67×10^{-1}

Evaluate the activation energy for each reaction. What would you expect to happen to the rate of the alkaline phosphatase at higher temperatures?

9.11 The activity of a preparation of the membrane-bound enzyme Na^+-, K^+-activated ATPase (which acts as an ion pump) from rabbit kidney was measured as a function of temperature. What can you conclude from the Arrhenius plot of these data?

Temperature (°C)	Rate (μmol min^{-1} mg^{-1})
10	0.082
12	0.127
14	0.203
16	0.273
18	0.407
20	0.549
25	0.872
30	1.34
37	2.28

9.12 A number of mutants of yeast ADH have been studied in detail. The kinetic properties of one of these (Trp 93 → Ala) are shown below. What do these data indicate about the active site of this mutant?

Substrate	k_{cat} (s^{-1})	K_m (mM)
Ethanol	66	1170
Propanol	9.1	150
Pentanol	20	13
2-Methyl-1-propanol	0.34	110
Benzyl alcohol	0.26	15

9.13 The values of the steady-state rates (mol mol^{-1} s^{-1}) for the action of chymotrypsin on the amide substrates BAPNA and SAAPFPNA are 0.00083 and 47.9, respectively, and for trypsin on BAPNA and SAAPFPNA 1.3 and 0.1, respectively. Comment on the specificities of the two enzymes towards these substrates.

9.14 The reaction catalysed by horse liver ADH was studied using the stopped-flow method. One syringe contained enzyme (8 mg mL^{-1}), the other a mixture of NAD^+ and ethanol in 20-fold molar excess over enzyme. Equal volumes from the two syringes were mixed and the formation of NADH monitored by absorbance measurements at 340 nm; the pathlength of the observation chamber was 2 mm. The absorption coefficient for NADH at 340 nm is 6220 M^{-1} cm^{-1}; NAD^+ does not absorb at the wavelength. The enzyme consists of two subunits each of molecular mass 40 kDa.

Time (ms)	A_{340}
0	0
10	0.0591
20	0.0933
40	0.1256
60	0.1356
80	0.1400
100	0.1449
120	0.1499
140	0.1555

What can you conclude from these data?

9.15 The possible involvement of Tyr 28 in the mechanism of the type II dehydroquinase from *S. coelicolor* was tested by mutating this residue to Phe. The kinetic parameters of the enzymes are listed below.

Enzyme	k_{cat} (s^{-1})	K_m (µM)
Wild type	125	99
Tyr 28 → Phe	0.026	100

What can you conclude from these data and how would you check your conclusions?

References for Chapter 9

Clarke, A.R., Atkinson, T., and Holbrook, J.J. (1989a) *Trends Biochem. Sci.* **14**, 101–05.

Clarke, A.R., Atkinson, T., and Holbrook, J.J. (1989b) *Trends Biochem. Sci.* **14**, 145–48.

Cornish-Bowden, A.C. (2004) *Fundamentals of Enzyme Kinetics*, 3rd edn. Portland Press, London, Chapter 7.

Dickinson, F.M. (1996) *Enzyme Kinetics*. In: *Enzymology Labfax* (Engel, P.C., ed.). Bios Scientific Publishers, Oxford, Chapter 3B.

Dixon, M. (1953) *Biochem. J.* **55**, 170–71.

Eshdat, Y., McKelvy, J.F., and Sharon, N. (1973) *J. Biol. Chem.* **248**, 5892–98.

Fernley, H.N. and Walker, P.G. (1970) *Biochem. J.* **116**, 543–44.

Fersht, A.R. (1999a) *Structure and Mechanism in Protein Science*, 3rd edn. Freeman, New York, Chapter 4.

Fersht, A.R. (1999b) *Structure and Mechanism in Protein Science*, 3rd edn. Freeman, New York, Chapter 10.

Fersht, A.R. (1999c) *Structure and Mechanism in Protein Science*, 3rd edn. Freeman, New York, Chapter 16.

Gardell, S.J., Craik, C.S., Hilvert, D., Urdea, M.S., and Rutter, W.J. (1985) *Nature* **317**, 551–55.

Ghosh, N.K. and Fishman, W.H. (1966) *J. Biol. Chem.* **241**, 2516–22.

Green, D.W., Sun, H.-W., and Plapp, B.V. (1993) *J. Biol. Chem.* **268**, 7792–98.

Kézdy, F. and Kaiser, E.T. (1970) *Methods Enzymol.* **19**, 3–20.

Kuchner, O. and Arnold, F.H. (1997) *Trends Biotech.* **15**, 523–30.

Lundblad, R.L. (1995) *Techniques in Protein Modification*. CRC Press, Boca Raton, FL, USA, 288 pp.

Malcolm, B.A., Rosenberg, S., Corey, M.J., Allen, J.S., de Baetselier, A., and Kirsch, J.F. (1989) *Proc. Natl. Acad. Sci.* **86**, 133–37.

Matsumura, I. and Kirsch, J.F. (1996) *Biochemistry* **35**, 1881–89.

Pon, N.G. and Bondar, P.J. (1967) *Analyt. Biochem.* **19**, 272–79.

Price, N.C. and Newman, B.L. (2000) *Biochem. Mol. Biol. Educ.* **28**, 207–10.

Price, N.C. and Stevens, L. (1999a) *Fundamentals of Enzymology*, 3rd edn. Oxford University Press, Oxford, Chapter 4.

Price, N.C. and Stevens, L. (1999b) *Fundamentals of Enzymology*, 3rd edn. Oxford University Press, Oxford, Chapter 6.

Riordan, J.F., Sokolovsky, M., and Vallee, B.L. (1967) *Biochemistry* **6**, 3609–17.

Sheehan, D. (2000) *Physical Biochemistry: Principles and Applications*. John Wiley and Sons, Chichester, Chapter 6.

Silverman, R.B. (1995) *Methods Enzymol.* **249**, 240–83.

Xue, Q. and Yeung, E.S. (1995) *Nature* **373**, 681–83.

Walsh, C.T. (1983) *Trends Biochem. Sci.* **8**, 254–57.

10 Binding of ligands to proteins

10.1 Introduction

Certain aspects of the binding of ligands to proteins have already been described in this book. Chapter 4, section 4.4, deals with the theoretical background in terms of the equilibria involved and how the data obtained can be analysed to yield the key parameters (strength of interaction, numbers of binding sites, and extent of any interactions between these sites). In Chapter 6, section 6.4, some of the principal experimental techniques used to study ligand binding have been outlined; these fall into two main groups: those which rely on a separation of the protein–ligand complex from the free components and those which rely on measuring a change in some property that distinguishes the complex from the free components.

In this chapter, we discuss the application of some of the experimental methods described. Worked examples are used to illustrate the results obtained and how the results are analysed. In some cases, relatively simple laboratory practical exercises can be performed to illustrate these approaches; these are described in Practical activity 10.1, which is available for lecturers in the On-line Resource Centre for this text to support teaching of this topic.

10.2 Analysis of binding data

KEY CONCEPTS

- Understanding the principal equations describing binding of ligands to proteins
- Analysing binding data using the appropriate graphical plots
- Appreciating the significance of the parameters obtained from the plots

In this section, we shall review the equations used to describe ligand binding and the methods used to analyse experimental binding data, since these are widely used across the range of techniques covered.

10.2.1 Equations describing ligand binding

The principal equations that describe the binding of a ligand (L) to a protein (P) have been described in Chapter 4, sections 4.3 and 4.4.

If the protein has a single binding site for the ligand, the expression for the fraction (Y) of the protein P which is saturated by ligand L is given by eqn. 10.1:

$$Y = \frac{[L]}{K_d + [L]} \qquad 10.1$$

where $[L]$ is the concentration of free ligand and K_d is the dissociation constant.

The value of Y can range from 0 to 1; a graph indicating how Y varies with $[L]$ in a hyperbolic fashion is shown in Fig. 10.1.

Importantly, the value of K_d corresponds to the concentration of free L at which half-saturation of the protein has occurred. That is, when $K_d = [L]$, $Y = 0.5$.

If instead of a single binding site on the protein, there are n sites for binding the ligand and these are assumed to be equivalent and independent of each other:

$$P + nL \rightleftharpoons PL_n$$

then the saturation equation is analogous to eqn. 10.1, taking into account the number of binding sites; see eqn. 10.2:

$$r = \frac{n \times [L]}{K_d + [L]} \qquad 10.2$$

where r is the average number of molecules of ligand bound per molecule of protein, n is the number of binding sites, $[L]$ is the concentration of free ligand, and K_d is the dissociation constant.

A plot of r against $[L]$ is shown in Fig. 10.2. Note that r can take values ranging from 0 to n.

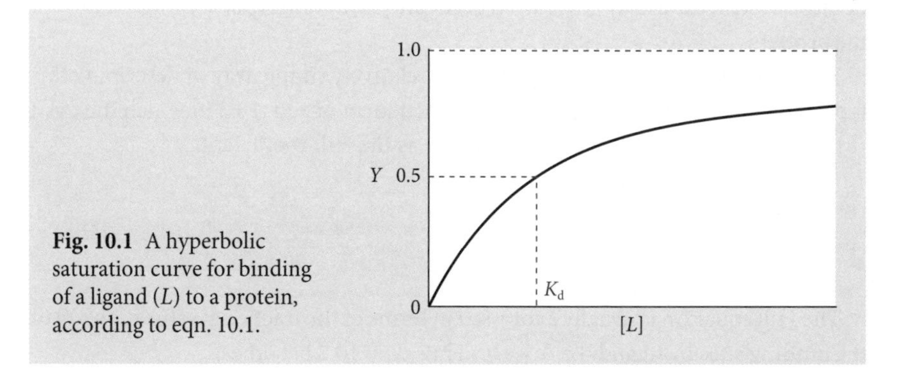

Fig. 10.1 A hyperbolic saturation curve for binding of a ligand (L) to a protein, according to eqn. 10.1.

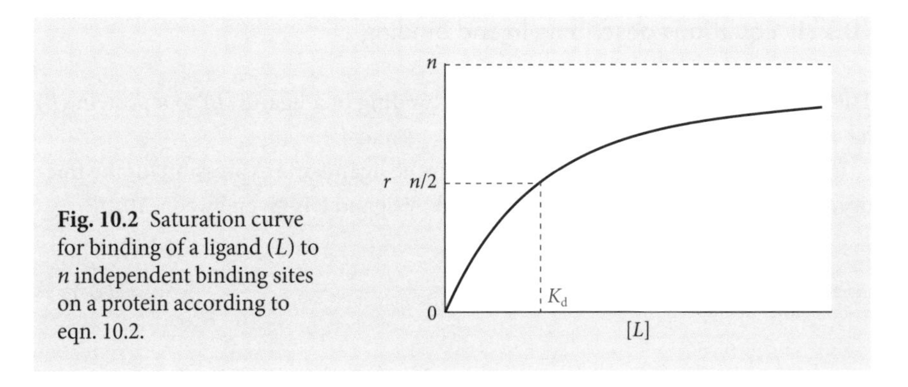

Fig. 10.2 Saturation curve for binding of a ligand (L) to n independent binding sites on a protein according to eqn. 10.2.

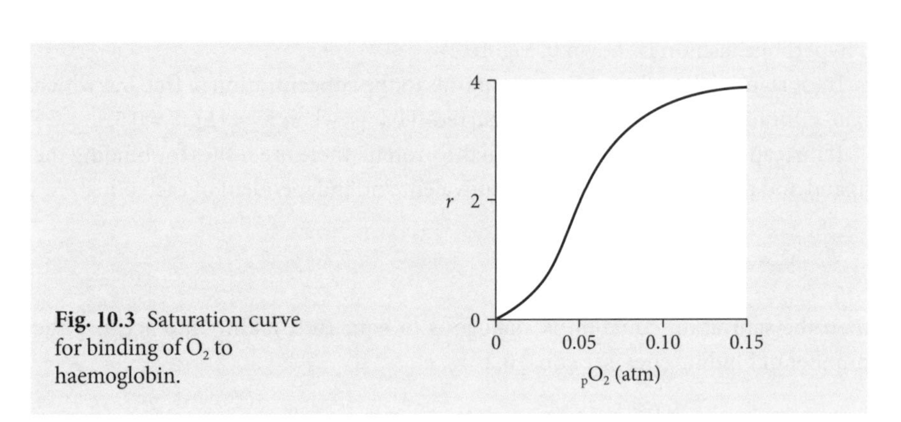

Fig. 10.3 Saturation curve for binding of O_2 to haemoglobin.

Just as in the case of a single binding site, the value of K_d is equal to the concentration of free ligand required to bring about 50% saturation of the available sites. This can be shown by substituting $r = 0.5\ n$ into eqn. 10.2, which then gives $[L] = K_d$. Thus, when there are interactions between multiple binding sites on a protein, the shape of the saturation curve will differ. For example, the binding of oxygen to haemoglobin (four binding sites) shows positive cooperativity, with the binding at the first site making it easier to bind to the subsequent sites. The saturation curve in this case is sigmoidal (see Fig. 10.3). This behaviour arises because of structural changes relayed between the subunits (polypeptide chains) of the protein.

The Hill equation (eqn. 10.3) provides a relatively simple way of describing this type of behaviour. The equation is a modified form of eqn. 10.2 in which the term $[L]$ is raised to a power h, where h is known as the Hill coefficient.

$$r = \frac{n \times [L]^h}{K + [L]^h} \qquad \textbf{10.3}$$

The Hill equation is usually expressed in terms of the fractional saturation (Y) of the binding sites by ligand, i.e. $Y = r/n$. Thus eqn. 10.3 becomes:

$$Y = \frac{[L]^h}{K + [L]^h} \qquad 10.4$$

When $h = 1$, there are no interactions between the multiple ligand binding sites on the macromolecule. A value of $h > 1$ indicates positive cooperativity. The value of h can range up to n, the number of binding sites, and its magnitude gives a measure of the extent of cooperativity between the sites. The Hill coefficient does not have to be an integer; for example the value of h for the binding of oxygen to haemoglobin is typically about 2.8. (A value of $h < 1$ is said to indicate negative cooperativity, where the binding of the first ligand molecule hinders binding of subsequent molecules.)

10.2.2 Analysis of binding data

Binding data can be analysed to determine the number of ligand binding sites on the protein and the strength of the ligand–protein interaction. We can use either direct, computer-based, fitting to the saturation curve (non-linear regression; see Chapter 2, section 2.6.5) or transformations of the saturation equations (eqns. 10.1, 10.2, and 10.4) to give straight-line equations (see Chapter 2, section 2.5.2), which can be plotted as Hughes–Klotz, Scatchard and Hanes–Woolf plots. From the slopes and intercepts of these plots (Figs. 10.4, 10.5, and 10.6) the parameters

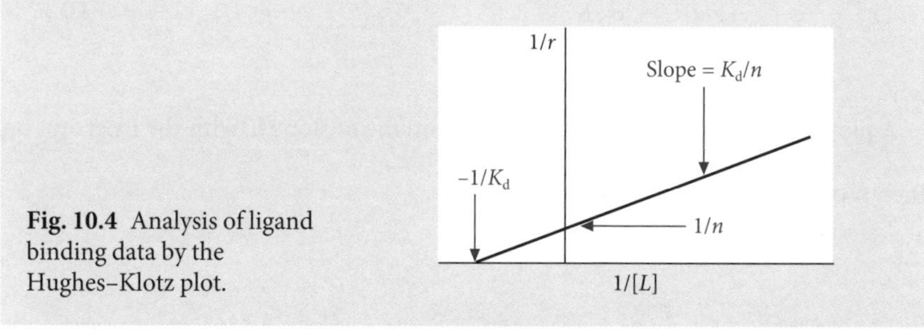

Fig. 10.4 Analysis of ligand binding data by the Hughes–Klotz plot.

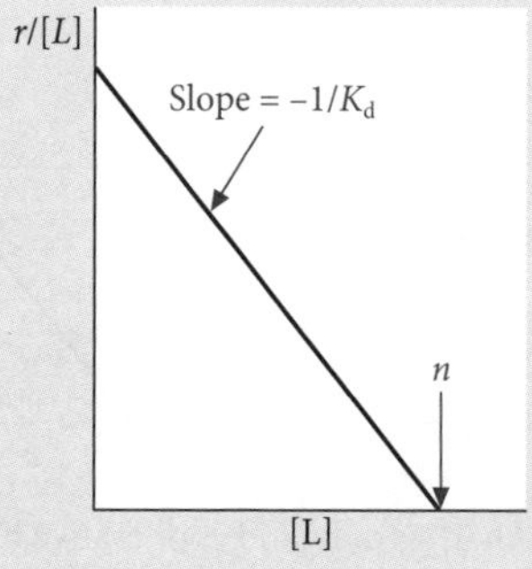

r/[L]
Slope = $-1/K_d$
n
[L]

Fig. 10.5 Analysis of ligand binding data using the Scatchard plot.

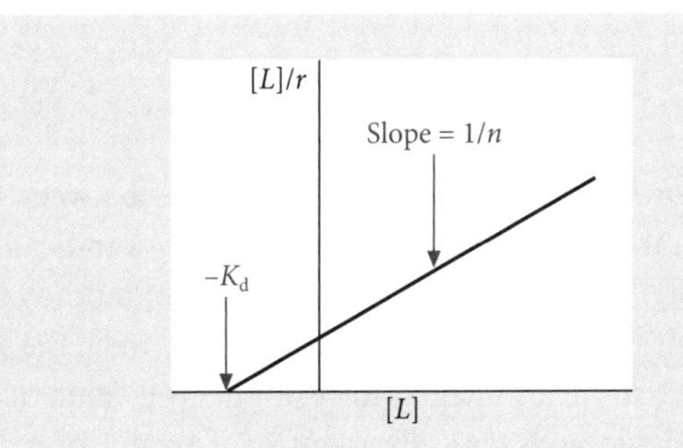

Fig. 10.6 Analysis of ligand binding data using the Hanes-Woolf plot.

(K_d and n) can be obtained. The Scatchard plot is preferred by many workers for two reasons: (i) it provides a good test of the quality of the experimental data; poor data will give a clearly unsatisfactory Scatchard plot, i.e. one in which the experimental points are widely scattered, rather than falling on a straight line, and (ii) there is a reasonably uniform error distribution across the saturation range, so the best straight line can be reliably obtained using the least-squares approach (Chapter 2, section 2.6.4).

For cooperative ligand binding, the value of the Hill coefficient can be derived by taking logarithms of both sides of eqn. 10.4 and then rearranging to give eqn. 10.5.

$$\log\left(\frac{Y}{1-Y}\right) = h\log[L] - \log K \qquad \textbf{10.5}$$

A plot of $\log\left(\frac{Y}{1-Y}\right)$ vs log $[L]$ is a straight line of slope h, with the intercept on the $y = 0$ axis equal to $(\log K)/h$ (Fig. 10.7).

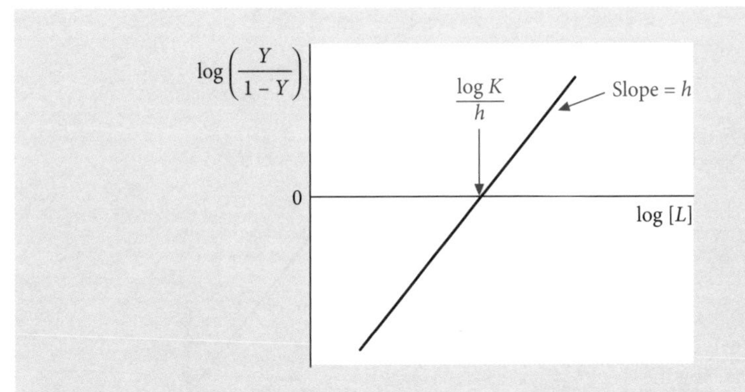

Fig. 10.7 Analysis of ligand cooperative binding data using the Hill plot.

10.2.3 Free energy changes of ligand–protein interactions

The value of K_d can be related to the free energy change involved in ligand–protein interactions.

$$-\Delta G^0 = RT\ln K_d \quad (10.6)$$

Thus, if $K_d = 1\ \mu M$ at 310K (37°C), $\Delta G^0 = -8.31 \times 310 \times \ln(1 \times 10^{-6})\ J\ mol^{-1} = 35\,600\ J\ mol^{-1} = 35.6\ kJ\ mol^{-1}$. The energy change can be compared with typical values for the weak, non-covalent forces described in Chapter 1, section 1.7. Table 1.4 in Chapter 1 (section 1.7.6) lists some typical K_d values for interactions involving proteins and the corresponding values of ΔG^0_{310}.

Note that K_d is expressed in terms of molar concentration, since this is referred to the standard state of the solution of ligand, i.e. a 1 M solution (see Chapter 4, section 4.1).

In this example, the value of ΔG^0 is large and positive, telling us that, under standard state conditions (i.e. 1 M PL) there would be little tendency for the reaction to proceed from left to right, i.e. for PL to dissociate. Dissociation of the complex would be promoted by lowering the concentration of PL (which would be the case in living systems, where concentrations of macromolecules are likely to be in the micromolar range, or even lower).

The ΔG^0 involved in an interaction can be estimated very easily if we remember that at 310K each factor of 10 in the value of K_d corresponds to an increment in ΔG^0 of approximately 6 kJ mol^{-1} (more accurately 5.93 kJ mol^{-1}; see Chapter 2, section 2.3.6). For example, a K_d of 10^{-9} M (1 nM) for a complex corresponds to a ΔG^0 of approximately 54 kJ mol^{-1} (accurately 53.4 kJ mol^{-1}) for dissociation of that complex at 310K. This value is some 18 kJ mol^{-1} larger than for the case where $K_d = 1\ \mu M$ (see above), reflecting the 1000-fold (10^3-fold) change in the value of K_d.

SELF TEST ?

Check that you have mastered the key concepts at the start of this section by attempting the following question.

ST 10.1 A thermophilic bacterium grows optimally at 65°C. At this temperature the dissociation constant for a protein–ligand interaction is 5 μM. What is the ΔG^0 for this dissociation process under these conditions?

Answer

ST 10.1 Using the equation $-\Delta G^0 = RT\ln K$, putting $K = 5 \times 10^{-6}$ (M) and $T = 338K$, the value of ΔG^0 for the dissociation process = 34.3 kJ mol^{-1}.

Attempt Problem 10.1 at the end of the chapter.

10.3 Spectroscopic methods to study binding

KEY CONCEPTS

- Understanding the basis of the various spectroscopic techniques which can be used to monitor ligand binding to proteins
- Analysing experimental data to derive the binding parameters

The binding of the dye ANS to bovine serum albumin (BSA) provides a good example of a case where some of the binding sites are spectroscopically 'silent' (see section 10.3.2).

As explained in Chapter 6, section 6.4.2, spectroscopic methods are widely used to monitor the binding of ligands to proteins; the only requirement is that there is a spectroscopic signal (of either the ligand or the protein) which changes on complex formation. In the case of multiple ligand binding sites on a protein, it is assumed that the extent of the spectroscopic signal change is proportional to the extent of complex formation; if the sites interact significantly (as in the case of cooperative binding), this assumption may not be valid.

10.3.1 Changes in absorbance

Generally, the changes in absorbance of a protein on binding a ligand are small (of the order of a few per cent), although there can be larger changes in the absorbance of the ligand on formation of the protein–ligand complex. In this section, we shall outline two examples where the changes in absorbance are sufficiently large to allow accurate experimental data to be obtained and analysed.

Binding of 2′-CMP to bovine pancreatic ribonuclease

WORKED EXAMPLE

2′-CMP acts as an inhibitor of bovine pancreatic ribonuclease (RNase). The binding of the ligand to the enzyme can be studied by the technique of difference spectroscopy, by using the fact that the absorbance in the near-UV region changes on formation of the complex. Previous work had established that the largest changes occurred at 262 nm and that a 1:1 complex between enzyme and ligand was formed. Formation of this complex gave a decrease in the absorbance coefficient ($\Delta\varepsilon$) of 2550 M^{-1} cm^{-1} at pH 7.0 compared with the absorbances of the free components (Anderson *et al.*, 1968).

In view of the relatively small changes in absorbance which occur in this case, the binding studies were performed using 'tandem' cells in difference spectroscopy. These are 'split' cells consisting of two compartments, each of which is of 0.5 cm pathlength. In the reference cell, the ligand and protein are in separate compartments; in the sample cell, the components are mixed to allow complex formation to occur (Fig. 10.8). Using a double-beam spectrophotometer the difference in absorbance (sample *minus* reference) can be measured.

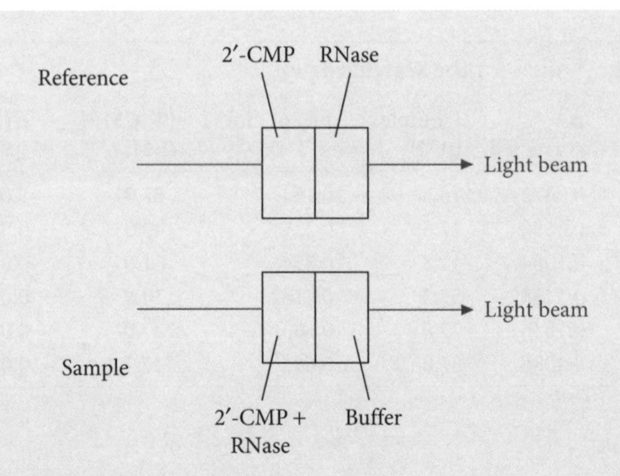

Fig. 10.8 Use of tandem cells in difference spectroscopy.

A series of measurements were made with different total concentrations of 2′-CMP added; the concentration of RNase was fixed at 120 μM. The data in Table 10.1 were obtained.

Table 10.1 Binding of 2′-CMP to RNase

[2′-CMP] added (μM)	ΔA_{262}
45.5	0.0236
70.0	0.0344
123.5	0.0542
192.0	0.0729
294.0	0.0918
399.0	0.1043

Analyse the data to obtain the K_d for the interaction between RNase and 2′-CMP.

STRATEGY

The concentration of the RNase–2′-CMP complex can be found using the observed change in absorbance and the known value of the absorption coefficient on complex formation. The concentration of free 2′-CMP can be found by difference and hence the data required for the Scatchard plot calculated.

SOLUTION

It is important to remember that the observed changes in absorbance at 262 nm have to be multiplied by 2 to correct for the pathlength of each compartment being 0.5 cm in the tandem cell. The corrected absorbance changes can then be converted to concentration of the complex formed using the value of 2550 M^{-1} cm^{-1} for the absorption coefficient on formation of the complex. The value of r is obtained by dividing the concentration of the complex by the RNase concentration (120 μM). The concentration of free 2′-CMP can be obtained by subtracting from the total concentration of 2′-CMP. This gives the values in Table 10.2.

The Scatchard plot ($r/[2'\text{-CMP}]_{free}$ vs r) (Fig. 10.9) is linear with the intercept on the x-axis = 1.0, i.e. there is one binding site on the enzyme. This confirms that the ligand forms a 1:1 complex with the enzyme. The slope of the plot = −0.00675 μM^{-1}; hence the value of K_d = 1/0.00675 μM = 148 μM.

Table 10.2 Values for the Scatchard plot

$[2'\text{-CMP}]_{total}$ (μM)	ΔA_{262} corrected	[complex] (μM)	r (mol mol^{-1})	$[2'\text{-CMP}]_{free}$ (μM)	$r/[2'\text{-CMP}]_{free}$ (μM)$^{-1}$
45.5	0.0472	18.5	0.154	27.0	0.00570
70.0	0.0688	27.0	0.225	43.0	0.00523
123.5	0.1084	42.5	0.354	81.0	0.00437
192.0	0.1458	57.2	0.476	134.8	0.00353
294.0	0.1836	72.0	0.600	222.0	0.00270
399.0	0.2086	81.8	0.682	317.2	0.00215

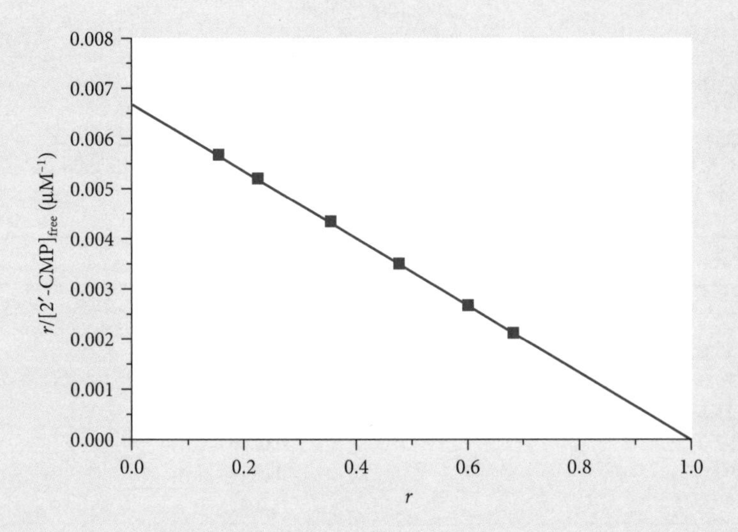

Fig. 10.9 Scatchard plot for binding of 2′-CMP to RNase.

Binding of Coomassie Brilliant Blue to proteins

✓ WORKED EXAMPLE

The spectral changes that occur on binding of Coomassie Brilliant Blue (CBB) to proteins provide the basis of a very widely used method of protein estimation (see Chapter 6, section 6.1.1). The interaction between the dye and a number of well-characterized proteins has been studied by Congdon *et al.* (1993). Under conditions where the dye was in a modest molar excess over the protein, binding only occurred at the high-affinity sites. The binding of CBB to BSA led to a marked increase in the absorption coefficient of the dye at 620 nm from 8800 M^{-1} cm^{-1} for the free dye to 48000 M^{-1} cm^{-1} for the bound dye. A titration was performed in which increasing concentrations of CBB were added to a fixed concentration of BSA (15 μM). From the observed A_{620}, the concentration of the bound dye could be evaluated, as shown in Table 10.3.

Table 10.3 Binding of Coomassie Brilliant Blue to BSA

$[CBB]_{total}$ (μM)	A_{620}	$[CBB]_{bound}$ (μM)
7.0	0.2498	4.8
15.1	0.5210	9.9
23.1	0.7678	14.4
32.1	1.0116	18.6
40.3	1.2249	22.2
52.3	1.495	26.4
69.8	1.8138	30.6

Evaluate the number of binding sites on BSA for the dye and the K_d for the interaction.

STRATEGY

The data in Table 10.3 allow values of r to be calculated; under each set of conditions, the value of free [CBB] can be found by subtracting the bound [dye] from the total [dye]. The data for the Scatchard plot are then readily calculated.

SOLUTION

From the data, the values of r (mol CBB mol BSA^{-1}) are obtained by dividing the concentration of bound dye by the concentration of BSA present (15 μM), and the concentration of free CBB by subtracting the bound dye from the total dye. This gives the values in Table 10.4.

Table 10.4 Values for the Scatchard plot

$[CBB]_{total}$ (μM)	$[CBB]_{bound}$ (μM)	$[CBB]_{free}$ (μM)	r	$r/[CBB]_{free}$ (μM^{-1})
7.0	4.8	2.2	0.32	0.145
15.1	9.9	5.2	0.66	0.127
23.1	14.4	8.7	0.96	0.110
32.1	18.6	13.5	1.24	0.092
40.3	22.2	18.1	1.48	0.082
52.3	26.4	25.9	1.76	0.068
69.8	30.6	39.2	2.04	0.052

A Scatchard plot ($r/[CBB]_{free}$ vs r) gives a straight line (Fig. 10.10); the intercept on the x-axis gives the number of binding sites as 3.0. The slope of the plot = −0.054 μM^{-1}, giving a value for K_d of 18.5 μM.

The studies by Congdon *et al.* (1993) show that although there are relatively few binding sites for CBB on BSA under these conditions, when the ratio of CBB/BSA was increased, binding occurred at a large number (approximately 100) of additional weak binding sites. The number of weak sites in different proteins seems to be correlated with their content of basic amino acids (Lys plus Arg).

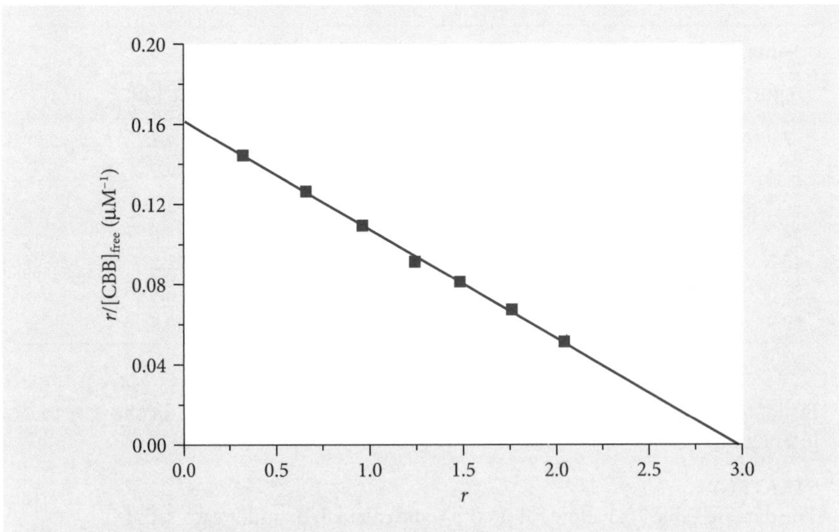

Fig. 10.10 Scatchard plot for binding of CBB to BSA.

10.3.2 Changes in fluorescence

As mentioned in Chapter 6, section 6.4.2.2, the changes in fluorescence which can occur on complex formation are usually much greater than changes in absorbance. This underlies the widespread use of fluorescence changes to monitor binding, as illustrated in the following examples.

Binding of ANS to bovine serum albumin

The dye ANS shows very weak fluorescence in aqueous solution with an emission maximum at 515 nm. In non-polar solvents the fluorescence is markedly enhanced and shifted to lower wavelengths (462 nm in the case of octanol). ANS binds to BSA with a large increase in fluorescence (some 100-fold greater than in aqueous solution) and an emission maximum at 471 nm.

The characteristics of the interaction were studied by titrating BSA (0.75 µM) with increasing concentrations of ANS and monitoring the fluorescence intensity at 471 nm. The concentration of bound ligand was calculated knowing the limiting increase corresponding to the BSA–ANS complex (Takehara *et al.*, 2006). The results in Table 10.5 were obtained.

How many binding sites for ANS are present on BSA and what is the K_d for the interaction?

STRATEGY

The values of r are obtained by dividing the $[ANS]_{bound}$ by the concentration of BSA (0.75 µM). This allows the values required for the Scatchard plot to be calculated.

Table 10.5 Binding of ANS to BSA

$[ANS]_{total}$ (μM)	$[ANS]_{bound}$ (μM)	$[ANS]_{free}$ (μM)
0.47	0.37	0.10
1.02	0.77	0.25
1.72	1.22	0.50
2.27	1.52	0.75
2.72	1.72	1.00
3.49	1.99	1.50
4.17	2.17	2.00
5.37	2.37	3.00

SOLUTION

The data points for the Scatchard plot are set out in Table 10.6.

Table 10.6 Values for the Scatchard plot

$[ANS]_{total}$ (μM)	$[ANS]_{bound}$ (μM)	$[ANS]_{free}$ (μM)	r (mol mol^{-1})	$r/[ANS]_{free}$ (μM)$^{-1}$
0.47	0.37	0.10	0.493	4.93
1.02	0.77	0.25	1.03	4.12
1.72	1.22	0.50	1.63	3.26
2.27	1.52	0.75	2.03	2.71
2.72	1.72	1.00	2.29	2.29
3.49	1.99	1.50	2.65	1.77
4.17	2.17	2.00	2.89	1.45
5.37	2.37	3.00	3.16	1.05

The Scatchard plot ($r/[ANS]_{free}$ vs r) (Fig. 10.11) is a straight line with an intercept on the x-axis of 3.9, i.e. there are four sites for binding of the dye to BSA. The slope of the line is −1.42 μM^{-1}, giving a K_d of 0.7 μM.

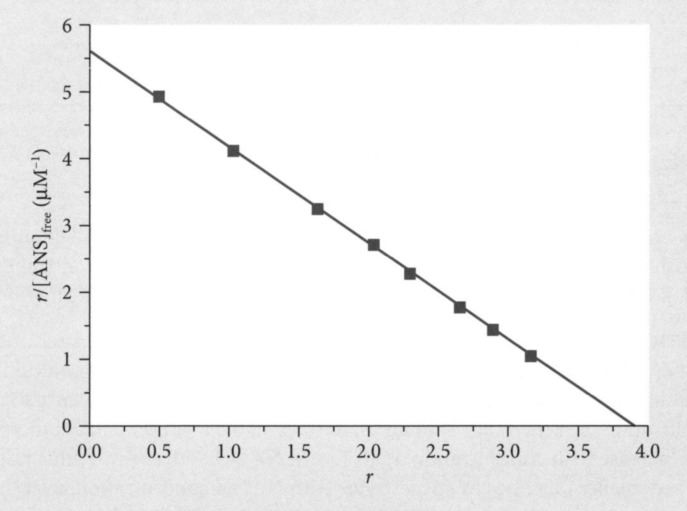

Fig. 10.11 Scatchard plot for binding of ANS to BSA.

Using equilibrium dialysis many more binding sites for ANS (at least 15 in total) were observed (Kolb and Weber, 1975). It would appear that these additional sites (which have a much weaker affinity for ANS) are sufficiently polar to give no significant fluorescence enhancement. BSA has been shown to have a large number (16) of binding sites for the non-polar dye methyl orange (see section 10.3.2).

This experiment indicates that there are four ANS binding sites on the protein detectable by fluorescence changes.

Binding of NADH to rabbit muscle lactate dehydrogenase

 WORKED EXAMPLE

The binding of NADH to rabbit muscle lactate dehydrogenase (LDH) leads to an enhancement of the fluorescence of NADH (Velick, 1958). The fluorescence was excited at 340 nm and emission was observed at 440 nm. The results in Table 10.7 were obtained when aliquots of a solution of NADH were added to a cuvette containing 1 mL of a solution of LDH (0.64 mg mL^{-1}) in buffer and to a cuvette containing a control solution consisting of 1 mL buffer. The NADH concentrations shown are the total concentrations added to the cuvettes. LDH consists of four subunits each of molecular mass 35 kDa.

Table 10.7 Binding of NADH to lactate dehydrogenase

[NADH] (μM)	Fluorescence (LDH)	Fluorescence (control)
0	0	0
2	85	30
4	169	58
8	338	112
12	508	170
16	625	223
20	720	280
24	803	332
28	862	381
32	919	440

What can you conclude about the interaction of NADH with LDH?

STRATEGY

In this example the free NADH contributes (albeit weakly) to the overall fluorescence. Hence the difference in fluorescence between the experimental solution and the control solution must be used to assess the extent of LDH–NADH complex formation.

SOLUTION

A plot of fluorescence intensity vs [NADH] (Fig. 10.12A) for the two solutions shows that in the control solution there is a linear increase in fluorescence with concentration over the range studied. In the case of the solution of LDH there is a linear increase with concentration up to 12 μM NADH added. Above this value, there are smaller increases in fluorescence with NADH concentration; over the

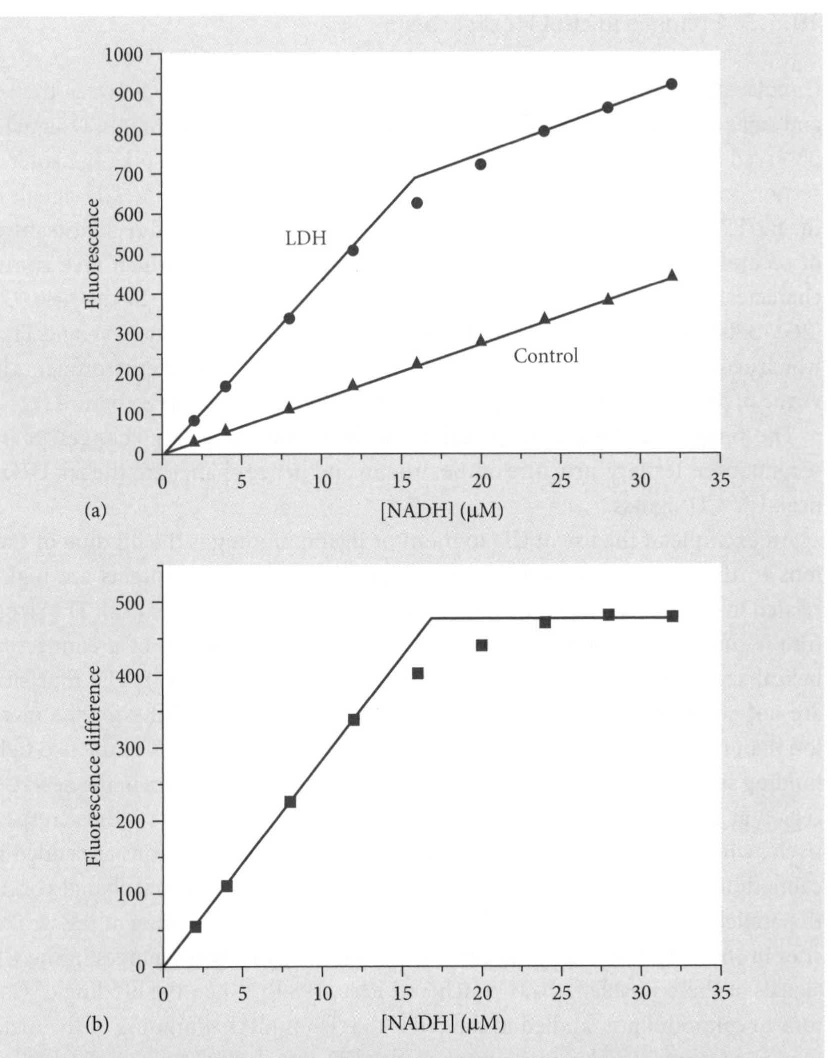

Fig. 10.12 Binding of NADH to LDH: (a) LDH and control solutions, (b) Fluorescence difference (LDH minus control).

range 24–32 μM NADH the increase in fluorescence is parallel to that of the control solution. This is seen more clearly if the difference between the fluorescence of the two solutions is plotted vs [NADH] (Fig. 10.12B).

The data can be interpreted as follows. As NADH is initially added to the solution of LDH, there is essentially complete formation of the LDH–NADH complex with enhanced fluorescence. At 24 μM NADH and above, essentially all the added ligand is free. Thus, under these conditions, binding of NADH to LDH is tight, i.e. there is little tendency to dissociate; the K_d value is much smaller than the concentration of binding sites present (18.3 μM subunits). The two linear portions of the graph intersect at 16.8 μM NADH. The concentration of LDH is 0.64 mg mL^{-1}, which corresponds to 4.57 μM tetramer (molecular mass 140 kDa). There are thus 3.7 binding sites per tetramer or 0.93 (i.e. effectively 1) sites per polypeptide chain.

10.3.3 Changes in circular dichroism

Circular dichroism (CD) measures the difference in absorbance between the left and right circularly polarized components of plane polarized light. A CD signal is observed in the spectral region where a chiral chromophore absorbs (Chapter 5, section 5.2.2, and Chapter 8, section 8.4.2.2). In the case of proteins, CD signals in the far-UV (240 nm and below) are due to the peptide bonds and regular features of secondary structure such as α-helix, β-sheet, or β-turns, which give rise to characteristic signals since these features are inherently chiral. In the near-UV (260–320 nm) CD signals arise from the aromatic amino acids (Phe, Tyr, and Trp), which acquire significant chirality when they are placed in chiral environments by virtue of the folding of the polypeptide chain in its native tertiary structure.

The binding of ligands to proteins can be accompanied by changes in the secondary or tertiary structure of the protein and hence changes in the far-UV or near-UV CD signals.

An example of the use of CD to monitor ligand binding is the binding of Ca^{2+} ions to the proteins calmodulin and troponin-C. The two proteins are highly related in structural terms with about 50% of the amino acids identical. The structure is that of a dumbbell with globular domains at each end of a connecting helical segment; each domain has two sites for Ca^{2+} (Fig. 10.13). The four sites are not equivalent, with two having a significantly higher affinity for the metal ion than the other two. In the case of calmodulin, binding of Ca^{2+} to the two tight binding sites in the C-terminal domain is accompanied by changes in the near-UV region at 279 and 268 nm arising from the Tyr 138 and Phe 65 residues, respectively, which are close to these metal binding sites. When Ca^{2+} ions are added to calmodulin, up to two metal ions per protein molecule, the spectral changes occur in parallel with the addition of the metal ions; subsequent binding at the weaker sites in the N-terminal domain cause only very small further changes in the CD signals at these wavelengths (Crouch and Klee, 1980). When the binding of Ca^{2+} ions to calmodulin is studied in the far-UV, it is found that binding of the metal ions is accompanied by an increase in the CD signal at 222 nm, indicating an increase in α-helical content. The changes in this signal are about two-fold greater for the binding to the two high-affinity sites compared with the two low-affinity sites (Klee, 1977).

The most striking difference between troponin-C and calmodulin is the presence of a 14-residue α-helix at the N-terminus of the former (Fig. 10.13). The binding of Ca^{2+} ions to troponin-C could be monitored by changes in the far-UV CD signals, indicating an increase in α-helical character. As in the case of calmodulin, binding of metal ions to the two high-affinity sites led to rather greater changes in the signal than binding to the low-affinity sites. The CD method could also be used to assess the effect of removal of the residues 1–14 from troponin-C (Smith *et al.*, 1994).

However, in many cases of protein–ligand interactions, only small changes are seen in the CD of the protein and it can be often more convenient to measure

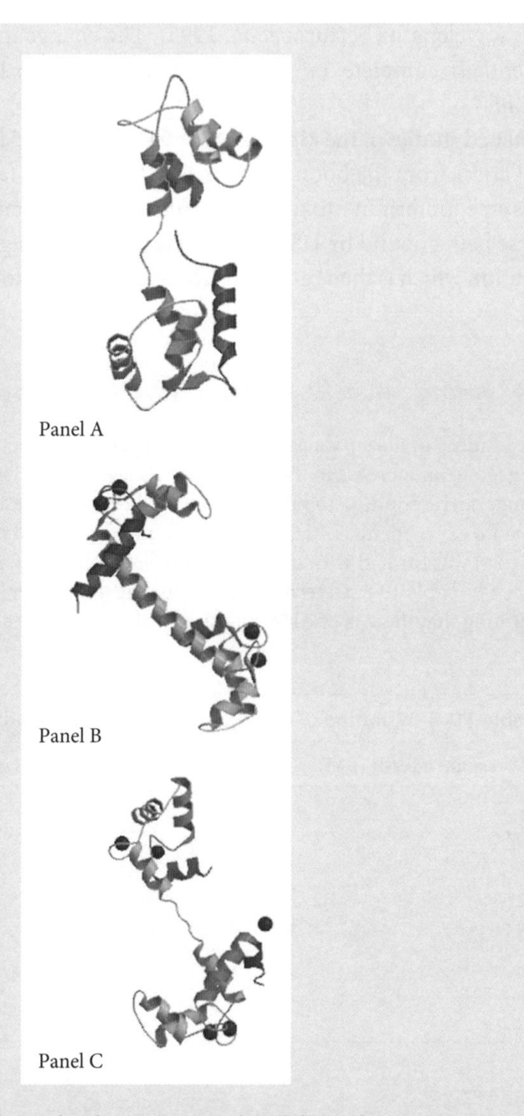

Fig. 10.13 Structures of the Ca-binding proteins calmodulin and troponin-C. Panels A, B, and C represent Ca-free calmodulin (PDB code 1cfd; Kubinowa *et al.*, 1995), Ca-bound calmodulin (PDB code 1cll; Chattopadhyaya *et al.*, 1992) and Ca-bound troponin-C (PDB code 1tcf; Soman *et al.*, 1999) respectively. Helices are shown as ribbons and Ca ions as filled circles.

changes in the spectral region where the ligand absorbs. A good example is provided by the binding of drugs to human serum albumin (HSA); this interaction represents a major means of transporting drugs to key sites of action in the tissues. The binding of diazepam (a powerful muscle relaxant, widely used in the treatment of anxiety) to HSA generates a signal at 315 nm, which can be used to assess the extent of binding of the drug to the protein; the free drug does not give a signal at this wavelength (Bertucci *et al.*, 1995). The change in the CD signal at 315 nm is essentially complete, i.e. all sites are saturated, at a 1:1 molar ratio of drug to protein.

Detailed studies of the kinetics of the CD changes at 315 nm have shown that the signal arises from the bound drug in a rather unusual fashion. Diazepam in solution is an equilibrium mixture of two mirror-image conformations, with only one of these being bound by HSA. This preferential binding perturbs the equilibrium in solution which is then re-established over a time period of a few seconds (Drake, 2001).

 WORKED EXAMPLE

The binding of diazepam to HSA was studied by CD spectroscopy. Binding of the drug led to an increase in the (negative) CD signal at 315 nm. The limiting signal change corresponding to saturation of all the protein present was determined in separate experiment as −9.9 millidegrees (mdeg). A fixed concentration of HSA (25 μM) was titrated with increasing concentrations of diazepam, with the results shown in Table 10.8. Determine the K_d for the interaction under these conditions, assuming that there is one binding site for the drug on the protein.

Table 10.8 Binding of diazepam to human serum albumin

[Diazepam added] (μM)	CD signal at 315 nm (mdeg)
7.3	−2.29
10.8	−3.30
16.4	−4.69
22.5	−5.94
29.7	−6.99
35.1	−7.54
45.8	−8.25
62.2	−8.80

STRATEGY

In order to analyse the binding using the methods outlined in section 10.2, it is necessary to evaluate the concentration of the free diazepam at each point in the titration. This can be done by calculating the fraction of HSA that is complexed with the drug.

SOLUTION

At each [diazepam], the observed signal can be used to calculate the fraction of the HSA that is complexed with the drug, since the total change is known to be -9.9 mdeg. The total [HSA] is known to be 25 μM, so that we can calculate the concentration of bound protein; this will be equal to the concentration of bound drug, since a 1:1 complex is formed. The concentration of free drug can be found by subtracting this from the total concentration of drug (Table 10.9). The Scatchard plot ($r/[\text{diazepam}]_{\text{free}}$ vs r) gives a straight line (Fig. 10.14); from the slope the value for K_d is 5.1 μM.

Table 10.9 Values for the Scatchard plot

Diazepam added (μM)	Fraction HSA bound (r)	[HSA] bound (μM)	Diazepam bound (μM)	Diazepam free (μM)	$r/[\text{Diazepam}]_{free}$ (μM^{-1})
7.3	0.23	5.8	5.8	1.5	0.153
10.8	0.33	8.3	8.3	2.5	0.132
16.4	0.47	11.8	11.8	4.6	0.102
22.5	0.60	15.0	15.0	7.5	0.080
29.7	0.71	17.7	17.7	12.0	0.059
35.1	0.76	19.0	19.0	16.1	0.047
45.8	0.83	20.8	20.8	25.0	0.033
62.2	0.89	22.2	22.2	40.0	0.022

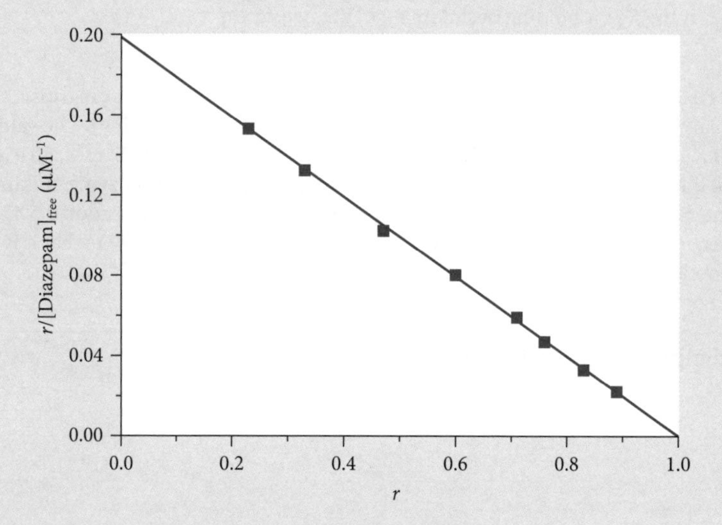

Fig. 10.14 Scatchard plot for binding of diazepam to BSA.

SELF TEST 

Check that you have mastered the key concepts at the start of this section by attempting the following question.

ST 10.2 Shikimate kinase catalyses the reaction: shikimate + ATP ⇌ shikimate-3-phosphate + ADP, which is part of the pathway of biosynthesis of the aromatic amino acids Phe, Tyr, and Trp. The enzyme from the plant pathogen *Erwinia chrysanthemi* has been studied in detail; it consists of a single polypeptide chain with one Trp. The fluorescence of this Trp is quenched by the binding of shikimate and this provides a convenient means of monitoring the interaction of the substrate with the enzyme.

The following data were obtained when the binding of shikimate was studied in the absence of ADP and in the presence of 2 mM ADP. In each case the quenching is expressed relative to that observed before shikimate is added.

[Shikimate] (mM)	% quenching (no ADP)	% quenching (+2 mM ADP)
0	0	0
0.05	7.0	12.5
0.1	12.6	21.7
0.2	21.5	34.2
0.5	37.5	52.2
1.0	49.7	63.5
1.5	55.7	68.2
2.0	59.3	71.0

What is the K_d for binding of shikimate in the two cases?

Answer

ST 10.2 We can apply one of the graphical methods or non-linear regression analysis to determine the values of K_d and Q_{lim} (limiting quenching). The values of K_d and Q_{lim} are: (no ADP) 0.48 mM, 73.6%; (+2 mM ADP) 0.27 mM, 80.6%. In the presence of ADP there is a modest strengthening of the enzyme/shikimate interaction, suggesting a synergistic interaction (see Chapter 9, section 9.5.4) between the binding sites for the two substrates of the enzyme. Note that it is not possible to study the binding of shikimate in the presence of ATP since the enzyme-catalysed reaction would proceed.

Attempt Problems 10.2–10.4 at the end of the chapter.

10.4 Equilibrium dialysis to study binding

KEY CONCEPTS

- Understanding the basis of the equilibrium dialysis technique
- Analysing experimental data to derive the binding parameters

The principles of equilibrium dialysis have been described in Chapter 6, section 6.4.1.1. In this section, we will describe some applications of the approach.

Binding of L*-Phe to pyruvate kinase*

WORKED EXAMPLE 

The binding of the amino acid L-Phe to rabbit muscle pyruvate kinase can be studied by equilibrium dialysis using radioactively labelled amino acids. The data in Table 10.10 were obtained when the binding of [^{14}C]L-Phe was studied. The binding of [^{14}C]L-Phe was studied under three different conditions: in buffer only, in buffer containing 2.5 mM Mg^{2+}, and in buffer containing 2.5 mM Mg^{2+} plus 5 mM L-Ala. The stoichiometry of bound ligand is expressed in terms of mol mol^{-1} enzyme, where the molecular mass corresponds to that of the tetramer (237 kDa).

Table 10.10 Binding of Phe to pyruvate kinase

$[Phe]_{free}$ (mM)	r (no divalent cation) (mol mol^{-1})	r (2.5 mM Mg^{2+}) (mol mol^{-1})	r (2.5 mM Mg^{2+} plus 5 mM Ala) (mol mol^{-1})
0.09	0.70	0.08	<0.02
0.2	1.27	0.30	<0.02
0.3	1.66	0.57	<0.02
0.5	2.16	1.12	<0.02
1.0	2.84	2.06	0.06
2.0	3.24	2.88	0.12
3.0	n.d.	3.23	0.18

n.d., not determined.

What can you conclude regarding the binding of L-Phe to the enzyme?

STRATEGY

The data can be conveniently analysed by evaluating the values of $r/[\text{L-Phe}]_{free}$ and constructing Scatchard plots ($r/[\text{L-Phe}]_{free}$ vs r).

SOLUTION

Inspection of the data shows that the presence of Mg^{2+} inhibits the binding of L-Phe; this effect is much greater at low L-[Phe]. The presence of L-Ala greatly decreases the binding of Phe.

The values used to construct the Scatchard plots for these data are shown in Table 10.11.

Table 10.11 Values for the Scatchard plot

$[Phe]_{free}$ (mM)	No divalent cation		+2.5 mM Mg^{2+}	
	r	$r/[Phe]_{free}$ (mM^{-1})	r	$r/[Phe]_{free}$ (mM^{-1})
0.09	0.70	7.78	0.08	0.89
0.2	1.27	6.35	0.30	1.48
0.3	1.66	5.53	0.57	1.88
0.5	2.16	4.32	1.12	2.24
1.0	2.84	2.84	2.06	2.06
2.0	3.24	1.62	2.88	1.44
3.0	n.d.	—	3.23	1.08

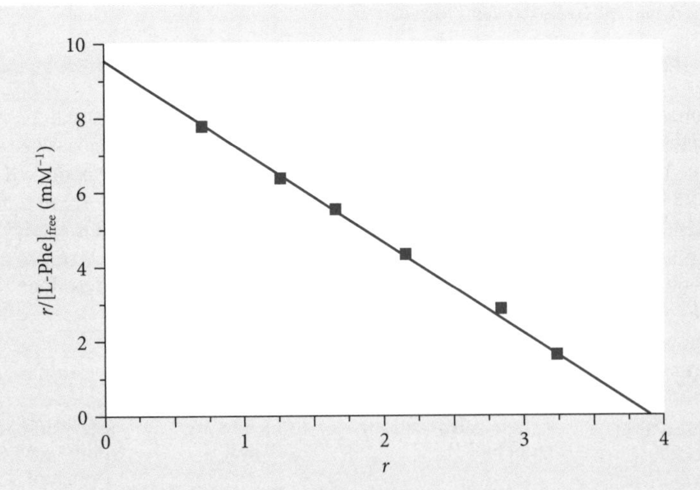

Fig. 10.15 Scatchard plot for binding of L-Phe to pyruvate kinase in the absence of divalent metal ions.

The Scatchard plot of binding of L-Phe (Fig. 10.15) in the absence of divalent cations is a straight line with an x-axis intercept of 4.0, corresponding to 4.0 binding sites per tetramer or 1 per subunit. The slope is -2.38 mM^{-1}, giving a K_d of 0.42 mM. This indicates that under these conditions the four sites on the enzyme are equivalent and independent.

In the presence of 2.5 mM Mg^{2+} the Scatchard plot (Fig. 10.16) shows pronounced curvature convex to the x-axis, indicating that the binding of the ligand displays positive cooperativity. At high values of r, the curve extrapolates to four binding sites per tetramer, i.e. 1 per subunit. The extent of cooperativity can be assessed by constructing a Hill plot of the data (see section 10.2.2 and Fig. 10.7); this gives a reasonable straight line with a slope (Hill coefficient) = 1.6. Given that there are four binding sites, this indicates a modest degree of cooperativity.

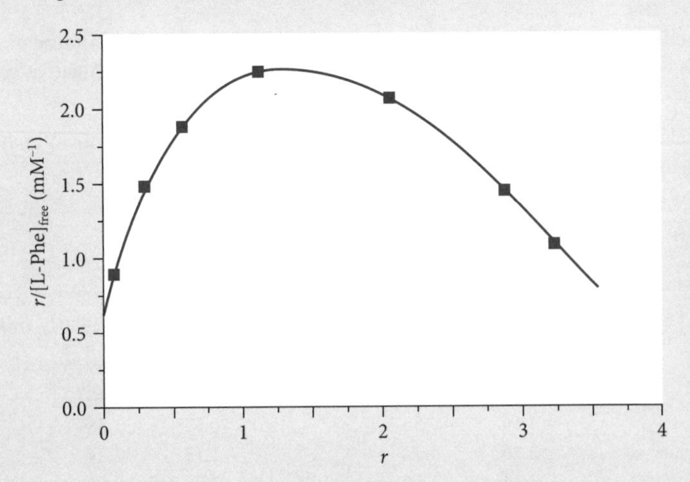

Fig. 10.16 Scatchard plot for binding of L-Phe to pyruvate kinase in buffer +2.5 mM Mg^{2+}. Note that at low values of r, the slope of the graph is positive but that at values of r above 2, the graph is a straight line of negative slope. The shape of the graph is characteristic of positive cooperativity between the binding sites on the protein.

Binding of methyl orange to bovine serum albumin

WORKED EXAMPLE 

Methyl orange (4-(*p*-[dimethylamino]phenylazo) benzene sulphonate) is a commonly used indicator, changing from pink to yellow over the pH range 3.0–4.4. The binding of this anionic dye to BSA was studied by equilibrium dialysis (Shikama, 1968). In these experiments a series of incubation mixtures were set up in which dialysis bags containing a known concentration of the protein were immersed in methyl orange solutions of different concentrations. After equilibrium had been achieved the concentration of the dye outside each bag was determined by absorbance measurements at 464 nm. By subtracting the amount of dye outside the bag from the known amount initially present in each mixture, the amount and hence the concentration inside the bag could be determined. Spectrophotometry could not be used to determine the concentration inside the bag since formation of the BSA–dye complex markedly altered the absorption spectrum of the dye. The data in Table 10.12 were obtained when the BSA concentration was 5 mg mL^{-1}; the molecular mass of BSA is 66 kDa. What can you conclude about the binding of methyl orange to BSA?

Table 10.12 Binding of methyl orange to BSA

[Methyl orange]$_{outside}$ (μM)	[Methyl orange]$_{inside}$ (μM)
53	257
97	428
174	660
254	855
368	1080
520	1330
746	1645

STRATEGY

The concentration of BSA is calculated in molar terms. The concentrations of the dye (methyl orange) outside the bag and inside the bag correspond to the free [dye] and the total (i.e.free + bound) [dye], respectively. Hence the bound [dye] can be determined by difference, and the appropriate parameters calculated to construct a Scatchard plot (r/[dye]$_{free}$ vs r).

SOLUTION

The molar concentration of BSA = 5/66 000 M = 75.8 μM. The data required for the Scatchard plot are shown in Table 10.13.

Table 10.13 Values for the Scatchard plot

[Dye]$_{outside}$ (μM)	[Dye]$_{inside}$ (μM)	[Dye]$_{bound}$ (μM)	r (mol mol^{-1})	r/[dye]$_{free}$ $(\mu M)^{-1}$
53	257	204	2.69	0.0508
97	428	331	4.37	0.0450
174	660	486	6.41	0.0368
254	855	601	7.93	0.0312
368	1080	712	9.39	0.0255
520	1330	810	10.69	0.0206
746	1645	899	11.86	0.0159

Note how in the case of this relatively weak interaction, it is necessary to use quite a high concentration of protein, in the range of the K_d, to analyse the binding accurately. In the case of the much stronger interaction between ANS and BSA (section 10.3.2) a much lower concentration of protein is used.

The Scatchard plot (Fig. 10.17) is linear with an intercept on the x-axis of 16, i.e. there are 16 binding sites for methyl orange on the protein. The slope of the plot $= -0.0605/16\ \mu M^{-1} = -0.00378\ \mu M^{-1}$; hence $K_d = 1/0.00378\ \mu M = 264\ \mu M$.

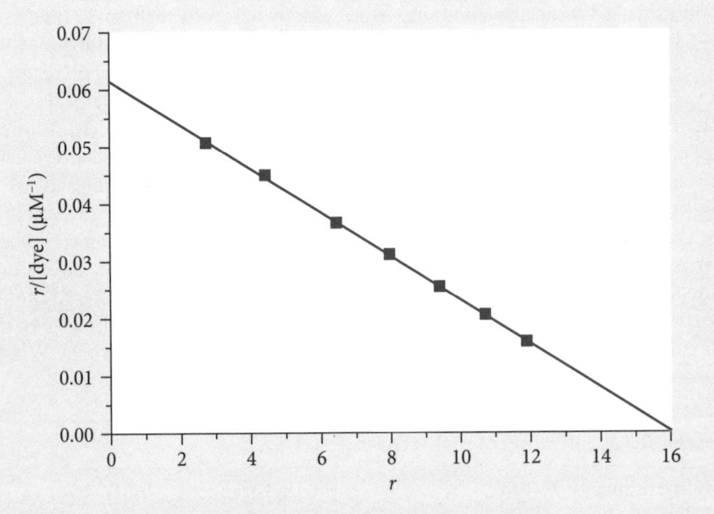

Fig. 10.17 Scatchard plot for binding of methyl orange to BSA.

? SELF TEST

Check that you have mastered the key concepts at the start of this section by attempting the following question.

ST 10.3 A study of the binding of [^{3}H] L-Ala to the tetrameric enzyme pyruvate kinase gave the following data. The values of r are given in terms of mol ligand bound per mol tetramer. How many binding sites are there for L-Ala, and what is the K_d for the interaction?

$[\text{L-Ala}]_{free}$ (mM)	r (mol mol^{-1})
0.04	0.58
0.07	0.89
0.1	1.09
0.2	1.74
0.3	2.25
0.5	2.66
0.7	2.86
1.0	3.06

Answer

ST 10.3 The binding data are analysed by constructing a Scatchard plot ($r/[\text{Ala}]_{free}$ vs r). This is a good straight line with the intercept on the x-axis (3.9) showing that there are 3.9 binding sites per tetramer or 0.975 (i.e. 1) per subunit. These sites are equivalent and independent. The slope of the line is $-4.18\ mM^{-1}$; thus the $K_d = 0.24$ mM.

Attempt Problem 10.5 at the end of the chapter.

10.5 Other methods separating free and bound ligands

KEY CONCEPTS

- Understanding the basis of the filtration and electrophoretic mobility shift assay (EMSA) techniques
- Analysing experimental data to derive the binding parameters

In this section, we shall illustrate the use of two other methods to study binding in which there is a physical separation of the free ligand and the protein–ligand complex; these methods have been already mentioned in Chapter 6, section 6.4.1.

10.5.1 Filtration to study ligand binding to adrenoreceptors

The catecholamine hormones (adrenaline and noradrenaline) bind to adrenergic receptors on cells. These receptors are divided into two main types (α and β). Binding to α-adrenergic receptors leads to contraction of smooth muscle; binding to β-receptors leads to relaxation of smooth muscle and cardiac stimulation. The types are further subdivided (α1, α2, β1, β2, etc.) depending on the action of specific agonists and antagonists which mimic or inhibit the actions of the hormones, respectively.

The binding of an antagonist to β-adrenoreceptors is described in the worked example below.

WORKED EXAMPLE 

As mentioned in section 6.4.1.1, filtration through an appropriate membrane can be used to separate free and bound ligand. A good example of the use of this approach is the binding of ligands to β-adrenoreceptors. These receptors are of particular interest as they are the sites of action of the β-blocker series of drugs, which are used to lower heart rate and blood pressure. Dihydroalprenolol (DHA) acts as an antagonist towards the natural ligand (adrenaline) and hence is a typical β-blocker.

The number of adrenergic receptors in most cells is small which would make a detailed study of the binding of DHA difficult. It is more convenient to increase the number of receptors by transfecting a cancer cell line (mouse neuroblastoma cells) to overexpress the human β2-adrenoreceptor. Radioactive ([^{3}H]-labelled) DHA is used to study the binding. The bound antagonist can be separated from free DHA by simple filtration through a glass fibre disc. The radioactivity retained on the disc after washing with buffer corresponds to the amount of DHA bound to the cells.

This experiment illustrates a number of features associated with the study of binding of ligands to systems which are more complex than purified proteins. For example, a problem with studying the binding to proteins in whole cells rather than to isolated proteins is that of non-specific weak binding to some of the many hundreds of other proteins in the cell membrane. This can be taken into account by measuring the amount of radioactive ligand bound in the presence of a large molar excess of non-radioactive ligand. This makes the calculations more involved than in the examples discussed in sections 10.3 and 10.4.

A second problem is that we cannot readily express the saturation of the binding sites in terms of the moles of ligand bound per mole of receptor protein, since the concentration (or amount) of the latter is not usually known. In such cases we modify the binding equations (section 10.2) to express the saturation in terms of a measurable parameter, such as the amount of total protein in the preparation. The saturation equation (eqn. 10.2)

$$r = \frac{n \times [L]}{K_d + [L]}$$

is modified by substituting B (mol ligand bound) for r, and B_{lim} for n (representing the limiting amount of B bound), to give eqn. 10.7:

$$B = \frac{B_{lim} \times [L]}{K_d + [L]} \quad \textbf{10.7}$$

which can be rearranged to give eqn. 10.8:

$$\frac{B}{[L]} = \frac{B_{lim}}{K_d} - \frac{B}{K_d} \quad \textbf{10.8}$$

Since B_{lim} and K_d are constants, eqn. 10.8 is of the form $y = mx + c$, where $y = B/[L]$, $x = B$, $m = -1/K_d$, and $c = B_{lim}/K_d$. Thus, a plot of $B/[L]$ vs $[B]$ (which is the appropriate Scatchard plot in this case) gives a line of slope $-1/K_d$ and an intercept on the x-axis of B_{lim}.

Incubation mixtures contained (in a total volume of 300 μL) 50 μL of membrane suspension, 30 μL of [^{3}H]DHA, 30 μL [DHA] of concentrations varying from 10 pM to 10 nM and buffer. After 30-min incubation, each mixture was filtered through a glass fibre disc which was washed with buffer, before being counted for radioactivity in a scintillation counter. The results obtained are shown in Table 10.14. Interconversion between the concentration (nM) (column 2) and

Table 10.14 Binding of dihydroalprenolol to β-adrenoreceptors

[^{3}H]DHA added (nM)	Total DHA (nM)	Total DHA (fmol)	Total DHA bound (dpm)	Specific DHA bound (dpm)
0.165	0.165	49.5	1833	1726
0.165	0.175	52.5	2006	1899
0.165	0.465	139.5	1624	1517
0.165	0.765	229.5	1412	1305
0.165	1.165	349.5	1198	1091
0.165	3.165	949.5	608	501
0.165	6.165	1 849.5	376	269
0.165	10.165	3 049.5	226	119
0.165	100.165	30 049.5	107	0

amount (fmol) of DHA (column 3) is made by noting that the volume is 300 μL. The dpm bound at a total [DHA] of 100.165 nM is used to indicate the amount of non-specific binding of ligand and hence to correct the total DHA bound (dpm in column 4) to give the specific DHA bound (dpm in column 5). In these experiments, the specific radioactivity of the stock solution of [^{3}H]DHA was found to be 196 dpm fmol^{-1}, and the amount of protein in each incubation mixture was 0.01 mg.

What can you conclude about the strength of the DHA–receptor interaction and how might you estimate the number of receptors per neuroblastoma cell?

STRATEGY

The radioactivity (dpm) of the specific DHA bound has to be converted to amounts in mol (B), using the specific radioactivity value appropriate to that incubation mixture. This will allow the data required for the Scatchard plot ($B/[L]$ vs B) to be calculated and hence the values of K_d and B_{lim} to be determined. B_{lim} can be used to calculate the number of receptors per cell, provided we know the protein content of each cell.

SOLUTION

The data are manipulated so as to be able to plot the results in the form of a Scatchard plot (bound DHA/[DHA]$_{free}$) vs bound DHA. The bound DHA is expressed in terms of fmol mg protein^{-1}, and the [DHA]$_{free}$ is expressed in nM.

The specific radioactivity of the DHA varies in the different mixtures because the concentration of non-radioactive DHA added varies.

The specific DHA bound (dpm) can be converted to fmol by using eqn. 10.9:

$$\text{specific DHA bound (fmol)} = \frac{\text{dpm specific DHA bound}}{\text{specific activity stock [}^3\text{H]DHA}} \times \frac{\text{total [DHA] concentration (column 2)}}{\text{[}^3\text{H]DHA concentration (column 1)}} \quad 10.9$$

Eqn. 10.9 expresses the fact that calculation of the amount of specific DHA bound involves division of the dpm-specific DHA bound by the specific radioactivity of the DHA actually present. The latter is derived by dividing the specific radioactivity of the stock DHA by the total DHA concentration (i.e. radioactive DHA plus non-radioactive DHA).

(Note that the specific activity of the stock [^{3}H]DHA = 196 dpm fmol^{-1}.)

Taking into account that the protein present in the 50 μL cell sample added to each mixture is 0.01 mg, we can construct Table 10.15:

Table 10.15 Values for the Scatchard plot

[Total DHA] (nM)	Specific DHA bound (fmol)	Specific DHA bound (B) (fmol mg protein^{-1})	Free DHA (fmol)	[Free DHA] (F) (nM)	B/F (fmol mg protein^{-1} nM^{-1})
0.165	8.85	885	40.7	0.136	6352
0.175	10.32	1032	42.2	0.141	7344
0.465	21.94	1268	117.6	0.392	5597
0.765	31.07	2194	198.4	0.661	4697
1.165	39.59	3959	309.9	1.033	3833
3.165	49.77	4977	899.7	2.99	1659
6.165	52.67	5267	1797	5.99	879
10.165	39.64	3964	3009	10.03	395

The Scatchard plot (B/F vs B) gives a reasonably good straight line (Fig. 10.18), provided that the point at 10.165 nM total DHA is omitted. The slope of the plot is -1.25 nM^{-1}; hence $K_d = 0.8$ nM. The intercept on the x-axis (B_{lim}) is 6400 fmol mg protein^{-1}, which corresponds to the limiting amount of DHA that can be bound.

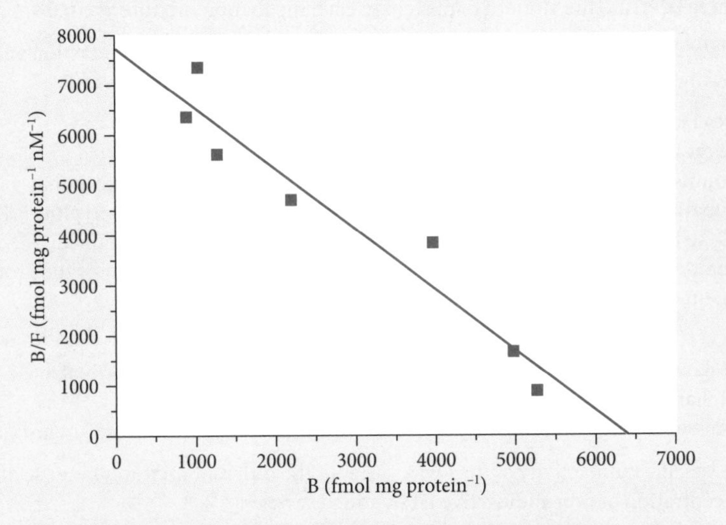

Fig. 10.18 Scatchard plot of DHA binding to β2-adrenergic receptors.

We can use the value of B_{lim} to estimate the number of binding sites (i.e. the number of receptors) on each cell. To do this, we need to know how much protein is there in each cell. In a separate experiment, it was shown that 10^6 cells contain 0.1 mg protein.

The total DHA bound = 6400 fmol mg protein^{-1}

Now 1 mg protein = 10^7 cells

Hence DHA bound = 6400 fmol 10^7 cells^{-1}

We can express the amount of DHA bound in terms of the number of molecules by noting that the Avogadro number is 6.02×10^{23} mol^{-1}.

In most cell types the number of receptors for a particular hormone is in the range 10 000–40 000 per cell, but can often be much lower. For example, human erythrocytes have 2000 receptors for insulin (Gambhir *et al.*, 1978).

Hence DHA bound = $6400 \times 10^{-15} \times 6.02 \times 10^{23}$ molecules per 10^7 cells
= 3.85×10^5 molecules cell^{-1}

Hence the number of receptors per cell is about 4×10^5. This value is rather greater than the typical number of receptors found in cells because the cancer cell line used (mouse neuroblastoma cells) has been transfected to overexpress the human β2-adrenoreceptor.

10.5.2 Electrophoretic mobility shift assay (EMSA) to study DNA–protein binding

The principles underlying the electrophoretic mobility shift assay (EMSA) technique for studying the binding of proteins to nucleic acids or fragments of nucleic acids have been outlined in Chapter 6, section 6.4.1.3. An example of the use of this approach is described in the Worked example below.

WORKED EXAMPLE 

Molybdenum is a trace element required for the activity of a number of oxidoreductase enzymes that catalyse reactions at carbon, sulphur, and nitrogen atoms and play a key role in anaerobic metabolism. Bacteria such as *Escherichia coli* contain a transport and uptake system for molybdenum (in the form of molybdate, MoO_4^{2-}) consisting of four proteins that are encoded by a single operon *modABCD*. A regulatory protein ModE is responsible for the molybdenum-dependent repression of transcription of the *modABCD* operon. The binding of ModE to the appropriate portion of the operon can be studied by mobility shift assays.

A 127-base pair fragment of the *modABCD* operon was amplified by the polymerase chain reaction. This fragment was labelled by reaction with [γ-^{32}P] ATP using T4 DNA kinase; each reaction mixture contained 5 ng in a reaction volume of 10 μL.

These mixtures were incubated with increasing concentrations of the ModE protein and then analysed by electrophoresis on polyacrylamide gels. The intensities of the bands corresponding to free DNA and the DNA–ModE complex were determined by autoradiography. Additional experiments showed that the ModE dimer formed a 1:1 complex with the DNA. It can be assumed that each nucleotide contributes 330 Da to the mass (see Chapter 2, section 2.1).

The data in Table 10.16 were obtained.

Table 10.16 Binding of DNA to ModE

$[\text{ModE}]_{added}$ (nM)	Free DNA (%)	DNA–ModE complex (%)
1	96.6	3.4
5	84	16
10	73	27
25	50	50
50	32	68
100	19	81
200	10	90

Estimate the K_d for the DNA–ModE interaction.

STRATEGY

It is necessary to express the [DNA] in molar terms which requires the mass of the fragment to be estimated from the chain length. We can then calculate the concentrations of bound DNA and hence the concentration of bound ModE. Subtraction from the total [ModE] gives the free [ModE] and hence allows the data for the Scatchard plot to be calculated.

As mentioned in Chapter 4, section 4.6, it is usual to mix a radioactive compound with an excess of the non-radioactive compound because the product of the radioactive decay can possess very different chemical properties.

SOLUTION

The concentration of the DNA fragment in the mixture is 5 ng per 10 μL; this is equivalent to $5 \times 10^{-6}/10^{-2} = 5 \times 10^{-4}$ mg mL^{-1}. The molar concentration can be obtained knowing the molecular mass of the DNA fragment (see Chapter 3, section 3.4.1). Using the rule that each nucleotide contributes 330 Da to the mass, the molecular mass of the fragment is $127 \times 2 \times 330$ Da = 83 820 Da. The molar concentration is thus $5 \times 10^{-4}/83\,820$ M = 5.97 nM.

The concentration of bound DNA at each [ModE] can be calculated by noting that 100% complex corresponds to 5.97 nM bound ModE. This concentration is subtracted from the total [ModE] to give the free [ModE] (Table 10.17).

Table 10.17 Values for the Scatchard plot

[ModE] added (nM)	$[DNA]_{bound}$ (nM)	$[ModE]_{free}$ (nM)	r (mol mol^{-1})	$r/[ModE]_{free}$ (nM^{-1})
1	0.20	0.80	0.034	0.0425
5	0.96	4.04	0.16	0.040
10	1.61	8.39	0.27	0.032
25	2.99	23.01	0.50	0.022
50	4.06	45.94	0.68	0.015
100	4.84	95.16	0.81	0.009
200	5.37	194.63	0.90	0.005

The slope of the Scatchard plot (Fig. 10.19) is −0.045 nM^{-1}, hence the K_d is 22 nM. This experiment indicates the need to use concentrations of the components of the same order as the K_d. At these low concentrations of DNA, detection requires the sensitivity provided by radioactivity. In this experiment, the specific radioactivity of the [γ-^{32}P]ATP is 5000 Ci mmol^{-1}, i.e. $5000 \times 3.7 \times 10^{10} \times 10^{3} = 1.85 \times 10^{17}$ dps mol^{-1}. 5 ng DNA corresponds to $5 \times 10^{-9}/83\,820$ mol = 5.97×10^{-14} mol. Hence, if the labelling with ATP is 100% efficient, the radioactivity in the DNA = 11 040 dps, which is easily detectable by autoradiography.

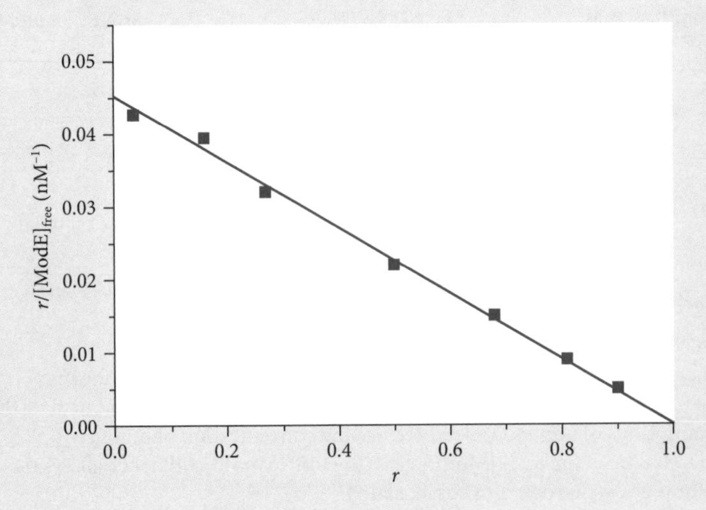

Fig. 10.19 Scatchard plot for binding of ModE to DNA.

SELF TEST ?

Check that you have mastered the calculations associated with the procedures involved in studying protein–ligand binding by attempting the following questions.

ST 10.4 A sample of [^{3}H]adrenaline was purchased with a specific activity of 65 Curies (Ci) mol^{-1}. A solution for binding studies was prepared by adding 5 µCi of this sample to 1.5 mL of 2 mM non-radioactive adrenaline. What is the specific radioactivity of this latter solution?

ST 10.5 In an EMSA experiment, it is important to check the specificity of complex formation. If the DNA being studied is radioactively (^{32}P) labelled, this can be done by adding a large excess of the same (but unlabelled) DNA. In an experiment, a 230 base pair length of DNA was used and the concentration of ^{32}P-labelled DNA was 3.2 nM. What is the concentration (in terms of mg mL^{-1}) of unlabelled DNA if it is added at 200-fold molar excess over labelled DNA? It can be assumed that each nucleotide contributes 330 Da to the molecular mass.

Answers

ST 10.4 The amount of [^{3}H]adrenaline added = $5 \times 10^{-6}/65$ mol = 0.077 µmol. The amount of non-radioactive adrenaline to which this was added = $1.5 \times 10^{-3} \times 2 \times 10^{-3}$ mol = 3 µmol. Hence the total adrenaline present = 3.077 µmol, and the specific radioactivity = 5/3.077 µC$_i$ µmol^{-1} = 1.625Ci mol^{-1}.

ST 10.5 **Answer:** The molar concentration of unlabelled DNA = 640 nM. Molecular mass of DNA = $230 \times 2 \times 330$ Da = 151.8 kDa. Hence the concentration of DNA = $640 \times 10^{-9} \times 150 \times 10^3$ mg mL^{-1} = 0.096 mg mL^{-1}.

Attempt Problems 10.6 and 10.7 at the end of the chapter.

10.6 Isothermal titration calorimetry

When a ligand binds to a protein, there will generally be a change in enthalpy; using a very sensitive microcalorimeter, it is possible to monitor these changes.

KEY CONCEPTS

- Understanding the basis of isothermal titration calorimetry (ITC)
- Using the experimental results to calculate the thermodynamic parameters for ligand binding

It should be remembered (Chapter 4, section 4.1) that both enthalpy and entropy changes contribute to the change in free energy characterizing an interaction ($\Delta G^0 = \Delta H^0 - T\Delta S^0$). In many cases of protein–ligand interactions, the contribution of the $T\Delta S^0$ term can be significant, or even dominant; it usually arises from changes in solvation which accompany complex formation.

In practice, small aliquots of ligand are added from a mixing syringe to a solution of the macromolecule in one cell in a thermally insulated chamber. The sample cell (containing the macromolecule) and a reference cell are contained within a doubly insulated chamber which is unable to exchange heat with its surroundings, i.e. it is adiabatic. As small aliquots of the ligand are added from the syringe to the sample cell, any enthalpy changes are detected by the thermopile and the heat needed to be

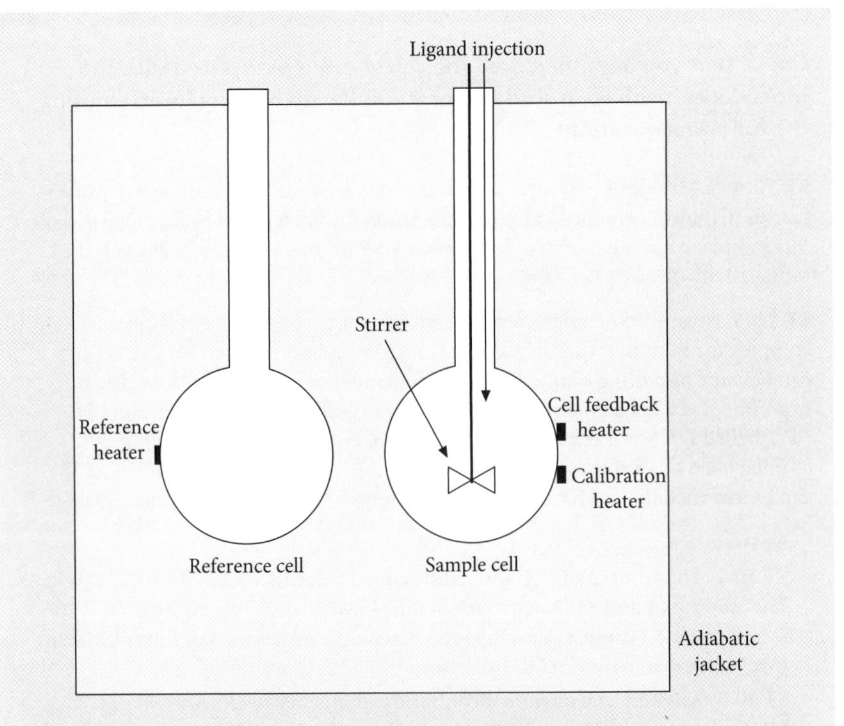

Fig. 10.20 The isothermal titration calorimetry experiment. Each cell is suspended by a narrow filling tube in an insulated adiabatic chamber, which is unable to exchange heat with its surroundings. Small aliquots of the ligand are added from the syringe to the sample cell and any enthalpy changes are detected by a sensitive thermopile. The heat needed to be supplied to the sample cell or the reference cell in order to keep the temperatures of the two cells identical is recorded.

supplied to the sample cell or the reference cell to keep the temperatures of the two cells identical is recorded (Fig. 10.20). The pattern of the heat changes during the titration can be used to generate a saturation curve for the binding of the ligand to the protein (Wiseman *et al.*, 1989).

This curve can be analysed to give the binding parameters K_d, ΔH^0 and stoichiometry of the interaction. Because the K_d can be used to derive ΔG^0, it is possible to calculate the ΔS^0 for the interaction, using the relationship $\Delta G^0 = \Delta H^0 - T\Delta S^0$ (see Chapter 4, section 4.1). From the contributions of the $T\Delta S^0$ and ΔH^0 terms to the overall ΔG^0 it is possible to reach some conclusions about the importance of various types of the weak intermolecular forces in the overall interaction. For example, hydrogen bonds and hydrophobic interactions tend to contribute to the enthalpy and entropy changes, respectively.

A typical change in enthalpy on complex formation might be 20 kJ mol^{-1}. 1 mL of a solution of a 2 mg mL^{-1} solution of protein of molecular mass 35 kDa would contain 0.057 μmol protein. Thus, the enthalpy change on complex formation would be 1.14 mJ. In order to measure the enthalpy changes during the titration, a calorimeter with a sensitivity of about 10 μJ would be required. The enthalpy changes arising from complex formation are of a similar magnitude to those due to stirring, mixing, dilution, neutralization, etc., hence careful attention to experimental conditions and use of appropriate control solutions are required. Despite these drawbacks, ITC has been extensively used to examine the interactions between proteins and small ligands, as illustrated in the Worked example below.

To appreciate how small the enthalpy changes measured in ITC are, it should be noted that the human body is estimated to radiate about 100 watts of heat (i.e. 100 J s^{-1}). A typical change in enthalpy observed in an ITC experiment is thus about 10^5-fold less than the heat you would radiate in 1 s!

WORKED EXAMPLE 

ITC was used to study the binding of ε-aminocaproic acid to the so-called kringle 4 domain of human plasminogen (the precursor of the enzyme plasmin which dissolves blood clots). At 25°C, the values obtained for the dissociation of the complex were $K_d = 26.2$ μM and $\Delta H^0 = +18.8$ kJ mol^{-1} (Sehl and Castellino, 1990). What is the value of ΔS^0 for the dissociation of the complex?

STRATEGY

The value of ΔG^0 can be calculated from the K_d value; ΔS^0 can then be calculated using the equation $\Delta G^0 = \Delta H^0 - T\Delta S^0$.

SOLUTION

For this interaction, the ΔG^0 for the dissociation can be calculated by using the equation $-\Delta G^0 = RT \ln K_d$; this gives $\Delta G^0 = 26.13$ kJ mol^{-1} for the dissociation process. From the values of ΔH^0 and ΔG^0, ΔS^0 for the dissociation is −24.5 J K^{-1} mol^{-1}. It is thus clear that both enthalpy and entropy changes contribute to the driving force for complex formation, pointing to the importance of both hydrogen bonding and hydrophobic interactions in this process.

The term 'kringle' comes from the Old Norse word 'kringla' meaning ring or circle. In the USA, it refers to a knotted-shaped pastry, derived from Scandinavia. In proteins, kringle domains are characterized by their entangled appearance.

ITC is now being increasingly applied to the study of protein–protein interactions, partly as a result of the ability of over-expression techniques (see Chapter 5, section 5.4.2) to provide the quantities of pure proteins required. A study has been made of the interactions between some of the component enzymes of the large pyruvate dehydrogenase complex from the bacterium *Bacillus stearothermophilus* (Jung *et al.*, 2002). The so-called peripheral subunit-binding domain (PSBD) of the E2 enzyme (dihydrolipoyl acetyltransferase) binds to both the E1 (pyruvate decarboxylase) and the E3 (dihydrolipoyl dehydrogenase) enzymes, but in a mutually exclusive fashion. The PSBD interactions with E1 and E3 at 25°C were found to be of similar affinity (with K_d values of 0.32 and 0.59 nM, respectively), but the ΔH^0 values for the dissociation processes were very different (35.1 kJ mol^{-1} and −9.2 kJ mol^{-1}, respectively).

Using the standard equations (see Chapter 4, section 4.1), the values of the thermodynamic parameters for the dissociation of the two complexes can be derived as shown in Table 10.18.

The very distinct ways in which combinations of enthalpy and entropy changes can give similar free energy changes for these interactions has been discussed in

Table 10.18 Thermodynamic parameters for dissociation of PSBD complexes

Interaction	ΔH^0 (kJ mol^{-1})	ΔG^0 (kJ mol^{-1})	ΔS^0 (J K^{-1} mol^{-1})
PSBD-E1	35.1	54.1	−63.8
PSBD-E3	−9.2	52.6	−207

structural terms (Frank *et al.*, 2005). Although the precise origin of the enthalpy and entropy contributions remains uncertain, it is clear that the large entropy change in the PDSB–E3 interaction reflects the release of water molecules on formation of the complex.

? SELF TEST

Check that you have mastered the key concepts at the start of this section by attempting the following question.

ST 10.6 The binding of the inhibitor 2′-CMP to bovine ribonuclease (RNase) was studied by ITC. At 28°C, the value of K_d was 12.1 μM and the ΔH^0 for the dissociation process was 51.4 kJ mol^{-1}. What is the ΔS^0 for the dissociation process?

Answer

ST 10.6 Using the equation $-\Delta G^0 = RT \ln K_d$, the ΔG^0 for the dissociation process = 28.3 kJ mol^{-1}. Using the equation $\Delta G^0 = \Delta H^0 - T\Delta S^0$, $\Delta S^0 = 76.7$ J/K/mol. The interaction of 2′-CMP with RNase is very much enthalpy dominated, suggesting that hydrophobic interactions do not play a major role in driving complex formation.

Attempt Problem 10.8 at the end of the chapter.

10.7 Surface plasmon resonance

KEY CONCEPTS

- Understanding the basis of surface plasmon resonance (SPR)
- Using the experimental results to calculate the dissocation constant for ligand binding

This relatively recently developed technique depends on the reflection of plane polarized light from the surface at a metal–liquid interface. The metal is a layer of gold (50 nm thickness), to which one component of the interacting pair (known as the ligand) is attached. The interaction of the light wave with the metal sets up electromagnetic waves, known as plasmons, which can penetrate a short distance (a few hundred nm, of the order of the wavelength of the light) into the medium containing the immobilized ligand. A solution containing the other component of the pair (known as the analyte) flows past the sensor. The polarized light is directed

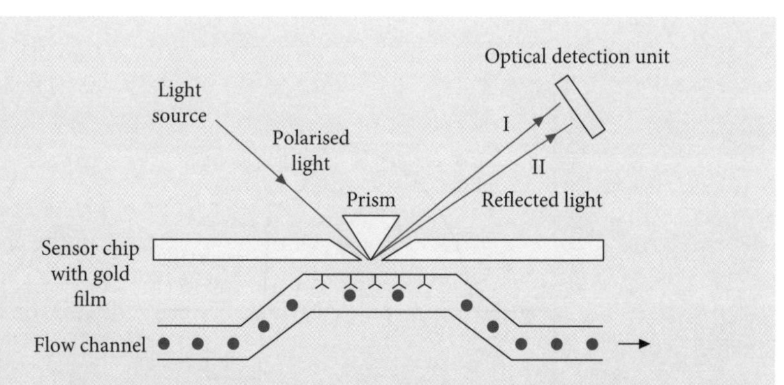

Fig. 10.21 The surface plasmon resonance experiment. One component of the complex (inverted Y) is immobilized on the sensor chip and the other component (filled circle) is passed through the flow channel. Formation of a complex is accompanied by a change in the angle (from I to II) of reflection of the polarized light. By measuring the reflected light at a chosen angle, complex formation will be detected by a change in intensity.

at the side of the sensor not in contact with the sample and the intensity of the reflected light is measured.

When binding occurs, there is a change in refractive index, leading to a change in the SPR signal and hence a change in the reflected light intensity. Changes in the SPR signal are proportional to the change in mass and hence respond directly to the extent of complex formation (Fig. 10.21).

A variety of approaches can be used to immobilize the ligand on the gold layer; these include chemical linking to specific functional groups (or amino acid side chains), or the use of strong non-covalent interactions, such as those, for example, between the protein avidin linked to the gold layer and a biotin-conjugated ligand.

The interaction between the protein avidin (from egg white) and biotin is one of the strongest known in biological systems; the K_d is estimated to be about 10^{-15} M.

In the SPR technique the approach is to flow the solution containing the analyte past the immobilized ligand and observe the increase in SPR signal with time; this is the association phase. When the signal reaches (or approaches) its limiting value, the solution of analyte is replaced by buffer, leading to a time-dependent decrease in the signal; this is the dissociation phase (Fig. 10.22). The sensor can then be regenerated by treatment with a solution which will lead to complete removal of the analyte (usually a strongly alkaline solution or a solution containing a high concentration of a denaturant such as urea).

Analysis of the association and dissociation phases in terms of an assumed model of binding will yield the rate constants (k_a and k_d) for the two phases; the equilibrium constant will then be the ratio of these two rate constants, i.e. $K_d = k_d/k_a$ (see Chapter 4, section 4.3.1) (Morton and Myszka, 1998).

The correctness of the model and the reliability of the rate constants derived can be tested by approaches such as those described in Chapter 2, section 2.6.5 to test non-linear regression analysis.

Although the SPR equipment is rather specialized, the technique has proved very valuable in analysing protein–ligand interactions, for ligands of both low and high molecular masses. For example, SPR has been used to examine (a) the interaction of the HIV protease with low molecular mass inhibitors (Gossas and

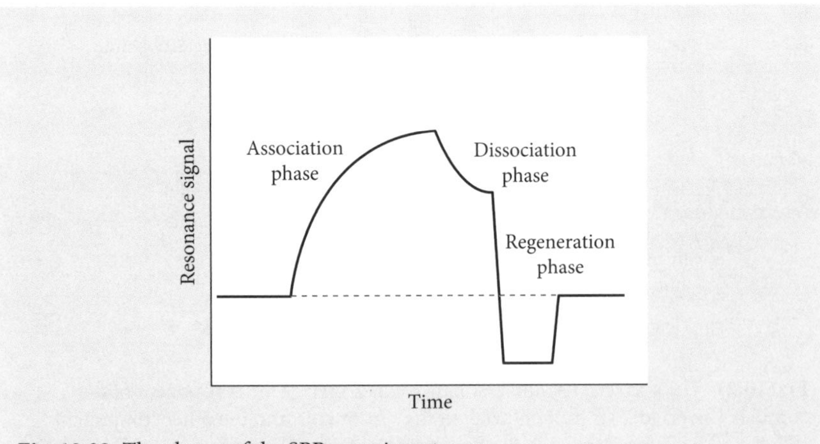

Fig. 10.22 The phases of the SPR experiment.

The SPR experiment can provide valuable information for pharmacological studies. For example, if a certain drug has a low rate of dissociation from its receptor, this may well indicate that relatively low doses of the drug are needed to maintain effective saturation of the receptor and thus bring about the desired physiological response.

Danielson, 2003) and (b) the interaction of immunoglobulin E and its fragments with its high-affinity receptor (Henry *et al.*, 1997). It is possible to programme the SPR equipment to perform a number of association/dissociation cycles under a defined range of experimental conditions allowing a large amount of information to be gathered in an efficient manner.

✓ WORKED EXAMPLE

Interleukin-2 (IL-2) is a locally acting protein hormone involved in the immune response. The IL-2 receptor (IL-2R) has three types of cell surface subunits (α, β, and γ). The binding of IL-2 to IL-2Rα and IL-2Rβ was studied by SPR and the results in Table 10.19 obtained.

Table 10.19 Rate constants in the IL-2-receptor interaction

	IL-2Rα	IL-2Rβ
k_a (M^{-1} s^{-1})	7.8×10^6	1.3×10^6
k_d (s^{-1})	0.24	0.66

What is the K_d for the interaction between IL-2 and these receptor subunits?

STRATEGY

The values of K_d are given by the ratio of the rate constants k_d/k_a in each case.

SOLUTION

The K_d for interaction with the IL-2Rα subunit = 31 nM; the K_d for the interaction with the IL-2Rβ subunit is considerably larger (510 nM). There is only a very weak interaction between IL-2 and the IL-2Rγ subunit. Interestingly, if both α and β subunits are immobilized on the surface, a high-affinity site for the ligand (K_d = 1 nM) is generated. This suggests that the two types of subunit can each contribute structural features to generate this high-affinity site (Myszka *et al.*, 1996).

SELF TEST ?

Check that you have mastered the key concepts at the start of this section by attempting the following question.

ST 10.7 The interaction of immunoglobulin E (IgE) with its high-affinity receptor is a key part of the development of the allergic response. The following data were obtained for the interaction of the receptor with wild-type IgE and an IgE mutant in which Arg 334 had been replaced by Ser.

	Wild type	Arg334→Ser
k_a (M^{-1} s^{-1})	2.5×10^5	2.9×10^5
k_d (s^{-1})	6.5×10^{-5}	1.5×10^{-3}

What is the value of K_d for each interaction, and what can you conclude regarding the importance of Arg 334?

Answer

ST 10.7 The value of K_d is given by the ratio of the rate constants k_d/k_a. For wild-type IgE, $K_d = 2.6 \times 10^{-10}$ M (0.26 nM); for the Arg334 → Ser mutant, $K_d = 5.2 \times 10^{-9}$ M (5.2 nM). The replacement of the Arg has a significant effect on the affinity, principally because of the much faster rate of dissociation of the complex. It is likely that this Arg side chain in IgE is closely involved in the interaction with the receptor.

10.8 Competition assays

KEY CONCEPTS

- Understanding the basis of competition assays
- Using experimental results to calculate the dissociation constant for ligand binding

The principles of competition assays have been described in Chapter 6, section 6.4.3. They are especially valuable in monitoring protein–ligand interactions under two sets of conditions: (i) when the binding of the ligand does not give rise to a convenient change in a measurable parameter, and (ii) when the dissociation constant for the interaction is outside a conveniently measurable range.

An example of (ii) would be where there is a very tight interaction between protein and ligand (for example, the interaction of a hormone with its receptor), so that very low concentrations of protein and ligand would be required to achieve significant dissociation of the complex and provide data for detailed analysis of the binding. If the concentration of protein were very low, it may not be possible to perform a convenient assay.

Spectroscopic methods for monitoring binding are suited to particular ranges of K_d values, reflecting the sensitivities of the techniques. Fluorescence, absorbance, and nuclear magnetic resonance approaches are best in the approximate ranges 10^{-9} to 10^{-6} M, 10^{-6} to 10^{-3} M, and 10^{-4} to 10^{-1} M, respectively.

Analysis of competition assay data depends on a good understanding of the relationship between the modes of binding of the natural ligand and the competitor ligand. The simplest situation (and the one most frequently employed) is where the two ligands bind to the same site on the protein. There will then be strict competition between the two ligands and the analysis used will be analogous to that used to describe competitive inhibition of enzymes (see Chapter 4, sections 4.3.4 and 4.4). The K_d for the natural ligand will be increased by a factor of $(1 + ([L]/K_L))$ where $[L]$ is the concentration of the competitor ligand and K_L is the dissociation constant of the protein–competitor ligand complex. Competitive binding is indicated if addition of increasing concentrations of the competitor ligand leads to complete displacement of the natural ligand, and the validity of the analysis can be confirmed by showing that the constant values of K_L are obtained when different values of $[L]$ are used in the experiments.

Strictly speaking $[L]$ is the concentration of the free competitor ligand; however, it will be assumed that this can be equated with the total concentration of ligand added.

A number of other possible modes of binding can be envisaged, for example where the competitor ligand binds to a site distinct from that where the natural ligand is bound. If it is possible to detect a change in the mode of binding of the natural ligand under these circumstances, the value of K_L can be determined using methods analogous to those used to analyse non-competitive inhibition of enzymes (see Chapter 4, sections 4.3.4 and 4.4).

Two examples of the use of competition assays are described below as worked examples.

Binding of ibuprofen to human serum albumin (HSA)

WORKED EXAMPLE

HSA is able to bind many types of drug molecules. The interaction of HSA with diazepam can be conveniently studied by CD, see section 10.5.3. However, addition of ibuprofen to HSA does not give rise to any significant CD signal. When ibuprofen is added to the HSA–diazepam complex the CD signal at 315 nm, characteristic of that complex, is abolished. Under certain conditions, the K_d for HSA–diazepam is 5 μM; in the presence of 40 μM ibuprofen, the K_d is increased to 22 μM. What is the K_d for the interaction of HSA with ibuprofen?

STRATEGY

Since we know the increase in K_d for diazepam in the presence of ibuprofen we can apply the standard formula for competitive binding.

SOLUTION

Ibuprofen is very likely to act as a competitive ligand with respect to diazepam since it abolishes the CD signal characteristic of the HSA–diazepam complex. The increase in K_d in the presence of ibuprofen is 4.4-fold; this is equal to $(1 + ([L]/K_L))$. Since $[L] = 40$ μM, the value of the K_d for interaction with ibuprofen (K_L) = 11.8 μM.

Binding of substrate to lactate dehydrogenase (LDH)

WORKED EXAMPLE 

The binding of the substrate NADH to LDH can be monitored by the increase in fluorescence of the substrate on complex formation (see section 10.5.2). When a large excess of NAD^+ is added to the LDH–NADH complex the fluorescence is reduced to that of free NADH. Under certain conditions the K_d for the LDH–NADH complex is 4 μM. In the presence of 0.6-mM NAD^+, the K_d is increased to 11 μM. What is the K_d for the interaction of LDH with NAD^+?

STRATEGY

Since we know the increase in K_d for NADH in the presence of NAD^+, we can apply the standard formula for competitive binding.

SOLUTION

NAD^+ is very likely to act as a competitive ligand with respect to NADH since it abolishes the fluorescence signal of the LDH–NADH complex. The increase in K_d in the presence of NAD^+ is 2.75-fold; this is equal to $(1 + ([L]/K_L))$. Since $[L] = 0.6$ mM, the value of the K_d for interaction with NAD^+ $(K_L) = 0.34$ mM.

SELF TEST 

Check that you have mastered the key concepts at the start of this section by attempting the following question.

ST 10.8 Alprenolol acts as an antagonist with respect to the binding of adrenaline to β-adrenergic receptors. The K_d for binding of adrenaline to the receptors was found to be 0.5 μM. In the presence of 10-nM alprenolol, the K_d for adrenaline was 2.1 μM. What is the K_d for the interaction of alprenolol to the receptors?

Answer

ST 10.8 The increase in $K_L = (1 + ([L]/K_L))$-fold, where $[L]$ is the concentration of the competing ligand. The increase in K_d in the presence of 10 nM alprenolol = (2.1/0.5)-fold, i.e. 4.2-fold. From the equation, $K_L = 3.13$ nM. The very tight binding of alprenolol to the receptor indicates that it could be a therapeutically useful β-blocker. It has been used to treat conditions such as hypertension, angina, and cardiac arrhythmia.

Attempt Problems 10.9 and 10.10 at the end of the chapter.

10.9 Problems

Full solutions to odd-numbered problems are available to all in the student section of the Online Resource Centre at www.oxfordtextbooks.co.uk/orc/price/. Full solutions to even-numbered problems are available to lecturers only in the lecturer section of the Online Resource Centre.

10.1 Haemoglobin Howick is a mutant form of the protein in which the β37 amino acid in normal protein (Trp) is replaced by Gly. Under certain conditions, the binding of O_2 to the protein gives the following results:

pO_2 (mm Hg)	Fractional saturation (Y)
0.25	0.115
0.5	0.242
1.0	0.441
2.0	0.660
3.0	0.767
4.0	0.827

Determine the values of the Hill coefficient (h) and K in the Hill equation (eqn. 10.3) for haemoglobin Howick. Compare these values with those for normal haemoglobin ($h = 2.8$, $K = 9160$).

10.2 The binding of Coomassie Brilliant Blue (CBB) to α-lactalbumin was studied using an approach analogous to that described for the binding of the dye to BSA (section 10.3.1). The following results were obtained when a solution of 83 μM α-lactalbumin was titrated with CBB. (It should be noted that, because the observed values of A_{620} are high, the absorbance values were read using short pathlength (0.2-cm cuvettes). Analyse the data to estimate the number of binding sites for the dye and the K_d for the interaction.

$[CBB]_{total}$ (μM)	$[CBB]_{bound}$ (μM)
23.4	14.9
63.3	38.2
106.5	59.8
159.7	81.3
238.6	104.6
326.8	121.2

10.3 The dimeric protein ModE is involved in the regulation of the expression of the operon encoding molybdenum uptake and transport system in the bacterium *E. coli* (see section 10.5.2). When molybdate is added to ModE, there is a marked quenching (48% reduction at saturation) of the fluorescence of the protein. A titration at high [ModE] (20 μM) confirmed that there are two sites for molybdate per dimer, but this was unable to give an accurate value for K_d since dissociation of ModE–molybdate complex was not significant under these conditions. The following data were obtained in a titration of ModE at a lower concentration (0.6 μM dimer) with molybdate. What is the value of K_d for the interaction?

$[\text{Molybdate}]_{total}$ (μM)	% Maximum change in fluorescence
0.25	12.1
0.50	22.4
0.82	34.4
1.22	45.6
1.64	55.0
2.32	65.5
3.16	72.4
3.96	77.9

10.4 The trp repressor protein is involved in controlling the expression of the genes encoding key enzymes in the synthesis of the amino acid L-Trp. In the absence of L-Trp, the repressor binds only weakly to the operator stretch of the DNA involved (the *trp* operon); however, in the presence of L-Trp, binding is much tighter. The interaction between the repressor protein from *E. coli* (a protein of molecular mass 25 kDa consisting of two identical subunits) and L-Trp was studied by CD; there is a marked change in the CD signal at 284 nm on complex formation (Lane, 1986). The following results were obtained using a solution of 0.51 mg mL^{-1} repressor protein.

$[\text{L-Trp}]_{total}$ (μM)	$[\text{L-Trp}]_{bound}$ (μM)
6.4	4.3
15.1	9.6
26.0	15.3
34.0	18.8
43.2	22.1
55.3	25.5
71.4	28.8
92.1	31.6

Analyse the data to obtain the number of binding sites for L-Trp and the K_d for the interaction.

10.5 The enzyme 5-aminolaevulinate dehydratase catalyses a key step in the synthesis of the pyrrole building blocks of the haem group found in several proteins such as haemoglobin, myoglobin, and cytochromes. The binding of the essential metal ion Zn^{2+} to the enzyme was studied using the isotope ^{65}Zn, which emits γ-radiation (Gibbs *et al.*, 1985). The enzyme has a molecular mass of 280 kDa and consists of eight identical subunits. The following results were obtained in an equilibrium dialysis experiment; the concentration of enzyme inside each dialysis bag was 2.2 mg mL^{-1}. How many binding sites for Zn^{2+} are present in the enzyme and what is the K_d for the interaction?

$[Zn^{2+}]_{outside}$ (μM)	$[Zn^{2+}]_{inside}$ (μM)
23.8	38.3
31.3	49.1
38.5	59.2
50.1	74.7
71.4	101.7
126.1	165.4
201.2	245.8
252.1	300.1

10.6 A preparation of hepatoma cells (cancer cells from liver) contain 1.5×10^6 cells mL^{-1}. The binding of ^{125}I-labelled insulin to these cells was studied. When ^{125}I-labelled insulin (100 nM final concentration) was added to 1 mL of the cell suspension, the amount of radioactive insulin bound was 1.05×10^{-13} mol (105 fmol). When the experiment was repeated in the presence of 10-μM unlabelled insulin, the amount of radioactive insulin was 22.5 fmol. How many receptors for insulin are there on each hepatoma cell? (Avogadro's number is 6.02×10^{23} mol^{-1}).

10.7 The EMSA technique was used to examine the binding of catabolite activator protein (CAP) to the *lac* promoter (the region of DNA adjacent to the start point for transcription of the *lac* genes). The concentrations of CAP and promoter DNA were 0.9 and 0.2 μM, respectively. Under these conditions, there is no evidence for a CAP–promoter complex. In the presence of 20 μM cyclic AMP (cAMP) which binds to CAP, a complex was observed. By measuring the intensities of the bands, it was estimated that 85% of the DNA was present in the complex. If the L8 mutant promoter DNA was used, there was no evidence for complex formation with CAP in the absence or presence of cAMP. Interpret these data and estimate the K_d value of any complex formed; it can be assumed that the stoichiometry of complex formation is 1:1.

10.8 The binding of a monoclonal antibody (MAb2B5) to cytochrome *c* was studied by ITC. At 25°C the value of K_d was 0.5 nM and the ΔH^0 for the dissociation was 87.8 kJ mol^{-1}. What is the ΔS^0 for the dissociation process?

10.9 L-Phe acts as an uncompetitive inhibitor of alkaline phosphatase. The V_{max} of the enzyme from human placenta was reduced to 52 and 35% of the uninhibited value in the presence of 0.5 and 1 mM L-Phe, respectively. What is the dissociation constant for the complex between L-Phe and the enzyme–substrate intermediate?

10.10 The opiates morphine and heroin (3,6-diacetylmorphine) both bind to the opioid receptors to produce a number of physiological effects, including pain relief. The K_d for interaction of each compound with the opioid receptors in a preparation from rat brain was found to be 2 nM. Naloxone acts as an antagonist to morphine and heroin (i.e. it binds to the receptors without eliciting physiological effects). In the presence of 20 pM naloxone, the K_d for morphine–receptor interaction was increased to 7.5 nM. What is the K_d for the naloxone–receptor interaction?

References for Chapter 10

Anderson, D.G., Hammes, G.G., and Walz, F.G., Jr. (1968) *Biochemistry* 7, 1637–45.

Bertucci, C., Viegi, A., Ascoli, G., and Salvadori, P. (1995) *Chirality* 7, 57–61.

Chattopadhyaya, R., Meador, W.E., Means, A.R., and Quiocho, F.A. (1992) *J. Mol. Biol.* **228**, 1177–92.

Congdon, R.W., Muth, G.W., and Splittgerber, A.G. (1993) *Analyt. Biochem.* **213**, 407–13.

Crouch, T.H. and Klee, C.B. (1980) *Biochemistry* **19**, 3692–98.

Drake, A.F. (2001) Circular Dichroism. In: *Protein–Ligand Interactions: Structure and Spectroscopy*, eds Harding, S.E. and Chowdhry, B.Z. Oxford University Press, Oxford, pp. 123–67.

Frank, R.A.W., Pratap, J.V., Pei, X.Y., Perham, R.N., and Luisi, B.F. (2005) *Structure* **13**, 1119–30.

Gambhir, K.K., Archer, J.A., and Bradley, C.J. (1978) *Diabetes* **27**, 701–08.

Gibbs, P.N., Gore, M.G., and Jordan, P.M. (1985) *Biochem. J.* **225**, 573–80.

Gossas, T. and Danielson, U.H. (2003) *J. Mol. Recog.* **16**, 203–12.

Henry, A.J., Cook, J.P., McDonnell, J.M., Mackary, G.A., Shi, J., Sutton, B.J., and Gould, H.J. (1997) *Biochemistry* **36**, 15568–78.

Jung, H-I., Bowden, S.J., Cooper, A., and Perham, R.N. (2002) *Protein Science* **11**, 1091–100.

Klee, C.B. (1977) *Biochemistry* **16**, 1017–24.

Kolb, D.A. and Weber, G. (1975) *Biochemistry* **14**, 4476–81.

Kubinowa, H., Tjandra, N., Grzesiek, S., Ren, H., Klee, C.B., and Bax, A. (1995) *Nat. Struct. Biol.* **2**, 768–76.

Lane, A.N. (1986) *Europ. J. Biochem.* **157**, 405–13.

Morton, T.A. and Myszka, D.G. (1998) *Methods Enzymol.* **295**, 268–95.

Myszka, D.G., Arulanantham, P.R., Sana, T., Wu, Z., Morton, T.A., and Ciardelli, T.L. (1996) *Protein Sci.* **5**, 2468–78.

Sehl, L.C. and Castellino, F.J. (1990) *J. Biol. Chem.* **265**, 5482–86.

Shikama, K. (1968) *J. Biochem.* **64**, 55–63.

Smith, L., Greenfield, N.J., and Hitchcock-DeGregori, S.E. (1994) *J. Biol. Chem.* **269**, 9857–63.

Soman, J., Tao, T., and Phillips, Jr., G.N. (1999) *Proteins* **37**, 510–11.

Takehara, K., Morinaga, Y., Nakashima, S., Matsuoka, S., Kamaya, H., and Ueda, I. (2006) *Analyt. Sci.* **22**, 1571–75.

Velick, S.F. (1958) *J. Biol. Chem.* **233**, 1455–67.

Wiseman, T., Williston, S., Brandts, J.F., and Lin, L.-N. (1989) *Analyt. Biochem.* **179**, 131–37.

11 How to report experimental work

Many of the points made in this chapter are based on those described in the companion volume *Research Methods for the Biosciences* (Holmes *et al.*, 2006) with some modifications to reflect the differences in emphasis for the molecular biosciences. We are grateful to the authors and publishers for permission to use this material.

11.1 Introduction

KEY CONCEPTS

- Knowing the different ways in which scientific work can be reported
- Appreciating the need for good communication skills when reporting work

Even if you are highly skilled as an experimental scientist in the laboratory (or sitting in front of the computer screen), your work will have little impact unless you can communicate it to other scientists or to the wider public. This can be done by oral presentations either in a relatively small-scale setting, such as a group meeting or a departmental seminar, or, if you really are one of the international superstars, at a large conference in some exotic location with an audience of several thousand hanging on your every word (and picture) and with television and newspaper correspondents in close attendance. However, for most scientists the primary means of communication is via the written word in the form of a report (e.g. for an undergraduate laboratory practical, an honours year research project), a thesis (for a postgraduate degree), or a scientific paper. Papers are published in journals which can be aimed at a general scientific audience (e.g. *Nature*, *Science*, or the *Proceedings of the National Academy of Sciences of the USA* (usually abbreviated to PNAS)), or show varying degrees of specialization. Thus the journals, *Biochemistry*, *Biochemical Journal*, and *European Journal of Biochemistry* all cover biochemical topics in a broad sense, but journals such as *Journal of Endocrinology* or the *American Journal of Cancer* clearly have a more defined scope and readership. The important point is that reports, theses, and papers all have the same basic structure, so that you should aim to start looking at primary journals early in your studies so that you can see how work is reported. You may well be able to access many of these journals directly from your computer either using the publishers' websites (for example http://www.jbc.org for the *Journal of Biological Chemistry*, or http://www.biochemj.org for the *Biochemical Journal*) or using database searching tools such as those in PubMed (http://www.pubmed.gov) or Web of

Another very commonly used means of communication at scientific meetings is the poster presentation, where you would have typically an area of about 1 m^2 to display a poster describing your work. Posters are discussed further in section 11.5.

Primary journals (such as those named) report new research work. Secondary journals, such as *Annual Reviews of Biochemistry*, review work published in a certain area over a period of time and assess the significance of the various findings reported.

Science (portal.isiknowledge.com (accessed in the UK via wok.mimas.ac.uk)). A number of scientific papers reporting studies on proteins are available on the website linked with this book.

There has been an explosive growth in the scale of publishing in the biosciences, both in terms of the numbers of journals and the sizes of the journals themselves. For example, the *Journal of Biological Chemistry* consisted of 905 pages in 1950; this has approximately doubled each decade, so that in the year 2000, there were over 41 000 pages of the journal published!

11.2 The structure of a scientific report

KEY CONCEPTS

- Understanding the function of each part of a scientific report
- Appreciating the importance of good writing skills when reporting work

If you look at a scientific report you will see that it consists of a number of sections:

- Title
- List of authors and affiliations
- List of keywords
- List of abbreviations
- Acknowledgements
- Abstract
- Introduction
- Experimental section (materials and methods)
- Results
- Discussion
- References.

Sometimes the results and discussion sections are combined (see section 11.2.9); in such cases there may be a short conclusions section.

The report may also include one or more appendices in which additional information or sample calculations are given. In some journals there is extreme pressure on space and additional primary data may be given in a supplementary information section, which is usually accessible online.

All of these sections are not necessarily required in every type of report; for example lists of authors and keywords, and acknowledgements are not generally required in an undergraduate laboratory report.

There may be a suggested overall word (or page) limit for the report and it is essential to bear this in mind when writing. You may well be penalized if your laboratory practical or honours project is too long (or too short!). In the case of a paper submitted for publication, your work may be rejected without detailed scrutiny if it is deemed too long for the amount of new information contained. *You should always aim to be clear and concise in your writing.*

The report is generally written in the *third person* (i.e. avoiding the use of I, we, and you) and in the *past tense*. This is certainly true for the experimental and the results sections. However, it is permissible to use the first person (I or we) at the end of the introduction section where the aims of the work are explained, and in the discussion section where the importance of the results is described and evaluated.

In undergraduate laboratory reports you are often asked not to repeat sets of instructions given in the manual; in such cases the experimental section may be very short, stating that procedures were described in the manual and only giving details of any significant modifications made to these.

The two most important parts of a report are the experimental and results sections. The experimental section describes how the work was carried out, thus allowing others to carry out the work and check that the results you obtained are correct. The results section describes the new findings which have been made, i.e. it records the advances in knowledge. It is important that within the overall word limit there is a good balance so that these two sections are not squeezed by those of lesser importance. In papers, it would be usual to find that the introduction and discussion sections constitute between 20 and 30% of the overall length; but this proportion may be higher in a postgraduate thesis, for example.

The assessment of any report will usually involve a number of stated criteria. In the case of undergraduate practical reports and honours projects, there should be clear marking schemes available. Assessment of honours project reports and postgraduate theses will usually involve an oral examination as well as a grading of the report itself. In the case of scientific papers, the editorial board of a journal will seek advice from reviewers (usually two or three) who are chosen for their expertise in the appropriate area (a process known as peer review). The reviewers will be asked to comment on the findings presented and confirm that they are properly documented and represent significant advances in the field of knowledge. There may then be a need for modification of the paper before it is accepted for publication. In some of the more prestigious journals, the rejection rate is very high (over 90% of papers submitted). In such cases the most common reason for rejection is that the work is not considered to represent a significant enough advance in the field.

The essence of scientific practice is that the work of one team of scientists can be replicated in other laboratories. Despite the careful scrutiny of papers before publication, there are occasional examples where the work cannot be replicated. In many of these cases the authors are asked to retract the paper, or choose to do so themselves.

11.2.1 Title

The title of the report should aim to capture the attention of the reader. In the case of an undergraduate practical report, the title may already be supplied in the manual for the experiment. Try to avoid the use of extra words, thus the title 'An investigation of the effect of insulin on glucose uptake into muscle tissue in mice' could be shortened to 'The effect of insulin on glucose uptake into muscle tissue in mice'. On the other hand, the title has to be informative, so that the reader knows it is worth exploring the paper in some detail. In the case of journal papers, authors are asked to supply in addition a short title (known as the running title) which appears at the top of all pages apart from the first one.

11.2.2 List of authors and affiliations

Most undergraduate laboratory reports and honours project reports (and all postgraduate theses) are assessed on an individual basis, unless, of course, they are specifically set as group exercises to foster such skills as team working. However, it is now very unusual to find single author research reports and papers. In part this reflects the fact that most research involves the application of a number of experimental approaches to tackle a specific problem or question, and therefore large teams or collections of teams are needed to make rapid progress. It is important that all those who have contributed to the work are duly recognized, usually by being included among the list of authors, or at least by appropriate acknowledgements. Deciding on the order in which authors are listed can be a delicate matter; a convention often followed is that the person who has done the bulk of the experimental work is first author, and that the senior person in the laboratory who may have formulated the ideas and obtained funding for the research is the last author and is designated as 'corresponding' author. The order of listing any other authors will usually reflect their relative contributions to the obtaining, interpreting, or reporting of data, but this is by no means always the case. Some journals now insist that the contribution made by each author is specified. The address of each author should be given.

Some of the recent papers reporting large-scale genome sequencing projects have involved several hundred authors. This can create its own problems in terms of citing references correctly!

In some cases it may be difficult to say which author has made the largest contribution, and a number of authors are stated as having made equal contributions to the work.

11.2.3 List of keywords

When submitting papers to journals, authors are usually required to submit a number (usually in the range 6–10) of keywords or phrases. These are then used for indexing purposes in databases so that other scientists can be readily directed to the paper. The keywords will generally include the names of the specific systems (e.g. organisms, cell types, or proteins) studied, the techniques employed, and the precise field of study (e.g. catalytic mechanism, protein folding, ligand recognition, etc.).

11.2.4 List of abbreviations

With the increasing complexity of biological systems under experimental investigation, it has become almost routine to use abbreviations especially for terms which are referred to frequently. Any substantial report, paper or postgraduate thesis should explain the abbreviation the first time it is used and also provide a complete list of them, usually near the beginning of the report. Examples of abbreviations would include terms such as: ANS, 8-anilino-1-naphthalene-sulphonic acid; NBD, nucleotide binding domain; SK, shikimate kinase.

Many abbreviations have become accepted as standard biochemical nomenclature. These include, for example ATP (adenosine 5′-triphosphate), EDTA (ethylenediaminetetra-acetate), NADH (reduced form of nicotinamide-adenine dinucleotide), and the abbreviations for amino acids and nucleosides (e.g. Ala, alanine; Trp, Tryptophan; A, adenosine; dA, 2′-deoxyribosyladenine). A full list of accepted abbreviations is included in the Instructions to authors given by the *Biochemical Journal* (http://www.biochemj.org).

11.2.5 Acknowledgements

Acknowledgements are important in a postgraduate thesis or a scientific paper. They are used to indicate the source(s) of funding for the work and to thank individuals who may have provided reagents or expertise to assist the work. In a paper the style of the acknowledgements is formal but in a postgraduate thesis you are allowed to be more personal and you may wish to thank friends and family for their support both in- and outside the laboratory.

11.2.6 Abstract

Many scientists who encounter your paper during a database search will read only the abstract (or summary). The job of the abstract is a difficult one; in the space of 200 words, or whatever limit is suggested or required, it has to convey the important messages of the paper and thereby encourage the reader to study the whole paper. You may find the following guidelines useful in writing an abstract.

1) Within the word limit you must try to give a balanced summary covering all the aspects required. You do not want to produce something that covers some points in detail but then runs out of words.
2) Remember that the summary is meant to be a brief, but self-contained, account of the work or the topic. Someone reading it should be able to work out what has gone on without needing to look at the main paper.
3) There are four main elements which the summary should contain:
   - A statement of the purpose of the work
   - A brief mention of the principal techniques/approaches employed in the study
   - An account of the principal results obtained; this would generally mention some specific results (e.g. actual values of kinetic parameters or numbers of binding sites)
   - A brief indication of the significance of these results in the context of what is known about the system and/or related systems.

4) From this you can see that it is not an easy thing to produce a good summary and you should look at actual scientific journals to see how it is done. Whatever you write, ask yourself whether it meets the points above or not. If necessary, revise it until it does.

11.2.7 Introduction

The aim of the introduction is to set the scene for the work by describing the system under investigation, reviewing earlier work on it and indicating the areas of uncertainty which your studies are intended to address. It must include a clear statement of the aims of the work. In some cases, the introduction may include a reference to the general nature of the results obtained, for example that a particular hypothesis has been confirmed or rejected.

The introduction should establish the correct nomenclature for the protein, and, in the case of enzymes, give the Enzyme Commission (EC) number. The online version of the EC recommendations is available at: http://www.chem.qmul.ac.uk/iubmb/enzyme/.

In writing the introduction, especially of a scientific paper, it is important to follow two general principles: (i) be concise and describe only the key points of the system and approaches employed and (ii) give proper recognition of the work of other scientists by giving appropriate references (section 11.2.11). It is not acceptable practice to copy sections of text and claim them as your own; this is plagiarism (section 11.3). *You should always aim to express the key findings of other scientists in your own words and give the appropriate reference(s).*

In a postgraduate thesis there is considerably more scope for providing greater detail in the introduction.

11.2.8 Experimental section (materials and methods)

The aim of this section is to provide the reader with the information to be able to replicate the experiments (and hopefully confirm the findings!). It is therefore essential that adequate details about the materials and procedures employed are given. Because of pressures of space, journals usually insist that authors of papers are as concise as possible and, wherever possible, give reference to published work rather than details of the procedures. However, in postgraduate theses it would be more usual to provide these details either in the main text or in an appendix. In all cases, it is important to describe any significant modifications made to published procedures.

The descriptions of the materials used should include the names of chemicals and reagents, and a note of the suppliers and the quality or purity of the materials. Most chemicals used are of analytical grade; for biological materials such as proteins, it would be important to specify the purity and specific activity of the preparation, if the data are available.

In the case of biological materials which can be propagated, including animals, cell lines, DNA clones, and microorganisms, it is important to give full information about the material with strain names or numbers as appropriate. The material may have been obtained commercially in which case details of the supplier should be given, or obtained from a national collection centre, whose details should be supplied. Many journals now insist that newly created strains, etc. are deposited at such centres so that they are available to other scientists.

As mentioned in section 11.2, in undergraduate laboratory practicals the methods are often set out in detail for you. In your report, you are not usually expected to repeat this information, but state that you followed the procedures outlined in the manual unless, of course, significant modifications were made during the course of the experiments.

The methods part of this section should mention any standard (i.e. well-accepted) methods used with appropriate references to published work. In the case of specialized techniques, for example spectroscopic procedures, you should provide details of the instruments used (manufacturer, model number, details of calibration procedures, etc.). You should also describe the methods of data analysis and a note of any statistical procedures employed to assess the significance of the results (see Chapter 2, section 2.6).

11.2.9 Results

In this section, you are essentially providing a narrative about the findings you have made. This means that the order of the various parts of this section does not necessarily reflect the precise order in which you carried out the experiments. For instance, you might only have realized the importance of performing a particular control experiment towards the end of your work, but it should probably be described early on in this section.

The results section will involve text and (usually) figures and tables to help the reader understand the findings you have made. The text is crucial in linking the various parts of this section and in highlighting key features of the figures and tables. Do not make the results section long, unless, of course, you have undertaken substantial amounts of experimental work. In general, it is better to assess the significance of particular results in a separate discussion section (see section 11.2.10). However, in some cases it may be necessary to discuss the results as you go along, for example to explain why the following sets of experiments were undertaken; in such cases there would be a combined results and discussion section.

If you have very large amounts of primary data, it is best to provide them in the form of an appendix to the report. In the case of a published paper, they can be deposited as supplementary information, which is made available online via the publisher's website.

Many scientific observations are more clearly understood by the use of figures and tables. The figures could be actual illustrations, for example micrograph images of cells obtained by either light or electron microscopy. These should carry a scale bar and the main features should be appropriately labelled. In most of the studies you will undertake on proteins, numerical data will be represented in a graphical format. Guidelines for drawing graphs are given in Box 11.1, and it is recommended that you study them carefully before plotting your data. *Although the guidelines have been drawn up with manual plotting of graphs in mind, the same considerations (choice of axes and scale, labelling of axes, title and legend, symbols for plotting, trend lines, or joining experimental points) apply equally well to computer-generated graphs. In the proof reading stages of report preparation (section 11.4) you should check that the graph is actually conveying to the reader what it should do!*

Colour illustrations are becoming more common in reports and papers. They are especially useful, for example when depicting the structures of large molecules such as proteins or cellular structures in which particular proteins have been tagged with fluorescent groups. In some cases, such as showing how muscle fibres contract, it is possible to depict real-time changes by depositing suitable video files to be accessed online.

Tables are a convenient way of listing a large amount of (often complex) numerical data. They should be carefully and clearly set out with a title, explanation (or legend), and careful attention to the units of quantities listed in the table.

It is important to realise that each figure and table should be reasonably self-contained, i.e. the title and legend should contain sufficient information so that the message conveyed can be understood without extensive reference to the text of the report.

When submitting papers, it is usual for the figure legends to be collected together in a separate section, although in the published version each figure will appear with its appropriate legend.

The title of a figure or table should not start with phrases such as 'Figure showing . . .' or 'Table listing values of . . .'. Rather it should state the actual subject matter, for example 'The variation of the rate of the chymotrypsin-catalysed hydrolysis of casein with temperature'. Have a look at the actual titles and legends used in a number of journal articles to see what is appropriate.

BOX 11.1

PRESENTATION OF GRAPHS

The purpose of plotting a graph is to make the relationship between two sets of variables clearer.

Does your graph achieve this objective? If it is badly drawn or untidy it will not.

You may wish to plot a rough draft of your graph in pencil. However, unless you are told otherwise, the final version of any work which is to be submitted for assessment should be drawn in pen (a medium thickness black ball point pen is particularly suitable).

It is not possible to give a set of hard and fast rules for presentation of graphs which apply under all circumstances. However, the following guidelines apply in the vast majority of cases. In the appendix to this chapter some examples of well-drawn and badly drawn graphs are illustrated and discussed.

1) Choice of axes
   - The horizontal axis (x-axis or abscissa) is used to plot the quantity which you vary in the experiment, e.g. time, temperature, pH, concentration, etc.
   - The vertical axis (y-axis or ordinate) is used to plot the quantity which you measure as a result of that change, e.g. absorbance, rate, velocity, etc.

   Thus:
   - The x-axis is used for the experimental (or independent) variable
   - The y-axis is used for the observed (or dependent) variable.
2) Choice of scale
   - Choose a scale for each axis which allows you to *represent the data over the range you have collected.* In many cases the *origin* (0, 0) should be included, but in others this is not appropriate (e.g. there is no such thing as zero wavelength)
   - *Make sure that the scale is uniform,* i.e. that each cm along the axis represents the same change in the particular variable (0–10, 10–20, 20–30 units, etc.)
   - The scale should be *sensible*—do not make 2 squares equal 3 units or 7 units, for example since it is difficult to read values off such a graph

- The choice of scale should, in general, allow the graph to occupy half to two-thirds of a page of A4
- *Do not put the axes at the edge of the sheet of graph paper*; leave a margin of about 2 cm to allow labelling of the *y*-axis, and a gap of several cm below the *x*-axis and a suitable legend (see point 3 below).

3) Labelling of the graph
   - The axes should be labelled clearly showing the variable plotted (e.g. concentration) and the units or dimensions of that variable (e.g. molar, which can also be denoted as mol/litre, mol L^{-1}, mol dm^{-3} or M)
   - Add a *few* values along each axis to indicate that scale, e.g. 0, 5, 10, 15 . . . or 0, 4, 8, 12 . . . , etc. *Do not label every division on the paper* (0, 1, 2, 3, 4, 5, 6, etc.); this makes the graph look cluttered
   - All graphs must include a *title*, which should be placed below the graph. The title should be *informative* and *specific* without being verbose and should *allow the reader to appreciate what is being plotted in the graph*. The title should not start with the words 'A graph showing . . .', etc.
   - It may be necessary to add additional information (legend) perhaps giving details of the important experimental conditions involved or a key identifying different sets of data shown
   - In a set of figures, number them consecutively (Fig. 1, Fig. 2, etc.).

4) Plotting experimental data
   - Make sure that the data points are clearly designated. *Use clear symbols* (•, X, ♦, etc.) *not small dots* (.) to represent the points
   - If you plot different sets of data on the same graph, use different symbols and different lines, for instance broken (- - -), continuous (——), or dotted (····) to make sure that they are clearly differentiated
   - Remember, if you use different colours for the different sets, you will not be able to distinguish them in a photocopy!

5) Should the points be joined or should a smooth curve or straight line be drawn? This question is a difficult one to answer unequivocally
   - As a general rule, if you expect there to be a straight line or smooth curve relationship between the variables then it is best to draw the best line or curve to go through (or near) the points. Deviations from the line or curve are likely to be due to random errors in making the experimental measurements
   - Do not extend the line or curve beyond the limits of your data without good reason. (One reason might be to determine the value of an intercept)
   - If you do not expect any simple relationship between the variables, then it is appropriate to join the points
   - Do not obliterate the symbols with the line.

   The properties of straight line graphs have been described in Chapter 2, section 2.5.1.

11.2.10 Discussion

The discussion section should aim to evaluate the significance of the results obtained and set them in the context of the current state of knowledge of the field. It is entirely appropriate to indicate any potential limitations or difficulties with your experiments, provided these are not of your own making (e.g. you were careless in carrying out the procedures or analysing the data). It is best to aim to be concise; while it is important to provide a good balance in the coverage of the results, you do not need to discuss every experiment you performed in great detail. In setting the results in context you will need to refer to other published work, from either yourself or other scientists. Many discussion sections tend to 'ramble' and leave the reader confused as to what the real message is. It is therefore a good idea to have a closing paragraph which reiterates the main conclusions and points the way to possible future studies.

If you have combined the results and discussion sections, it is useful to have a separate conclusions section, to emphasize the main points of the work.

11.2.11 References

Scientific work is seen as a community activity in which the results from one group or individual are used to inform and guide the work in other laboratories. It is therefore essential that due credit is given to the work of others when writing a report. This is done by indicating the citation at the appropriate place in the text and then giving full details of the citations in a special section at the end of the paper or report. Rather annoyingly, there is no universally agreed format for citations and references; different journals have their own 'house styles'. There are two main systems in use:

1) The numeric system in which references are cited in the text by sequential number, in some journals as superscripts (^{1}, $^{2–5}$, etc.) in others in square brackets ([1], [2–5], etc.). In the references section, the full details are given for each cited reference in the numerical order.
2) The author–date system (often called the 'Harvard system'). In the text, the citation will list the author or authors (if there are two) by surname, followed by the date, e.g. (Smith, 2002) or (Smith and Jones, 2003). If there are three or more authors the citation indicates the first author name followed by *et al.* (note the italics to show that this is an abbreviation of the Latin phrase *et alia*, meaning 'and others'). In the references section, the full details of each cited reference are given in alphabetical order.

The precise way in which the details of the references are given varies between journals and you should check the format required for the journal in question. The format for the *Biochemical Journal* (http://www.biochemj.org) is as follows:

For a journal article:

Number of reference, names and initials of all authors, date, title of paper, name of journal, volume number, page numbers. (Note that the names of journals are abbreviated in a style agreed by most major journals.) For example:

Price, N.C. and Radda, G.K. (1969) Desensitization of glutamate dehydrogenase by reaction of tyrosine residues. Biochem. J. **114**, 419–427.

In the case of papers with very large numbers of authors, it is standard practice to list only the first few (perhaps 6) and then add '*et al.*'.

For a book or chapter from a book:

Number of reference, names and initials of all authors, date, title of chapter and title of book, names of editors, page numbers, publishers, and place of publication. For example:

Price, N.C. (1995) Circular Dichroism in Protein Analysis in Molecular Biology and Biotechnology: a Comprehensive Desk Reference (Meyers, R.A., ed.), pp. 179–185, VCH Press, New York.

For a URL (web address):

Many of the major journals actively discourage citation of URLs for two main reasons: (i) the reliability of much of the information on websites is variable (there is no systematic use of peer review, for example) and (ii) many websites are not maintained and cannot always be accessed. In the case of the *Biochemical Journal*, for instance, URLs may only be cited in the text (and then only when a literature reference is not sufficient) and may not be listed in the references. If a particular website referred to is not one maintained by an internationally recognized organization, such as a university or a government agency or department, authors are required to provide written assurance from the webmaster that the site and the relevant information will be maintained.

Several reference manager programs (such as Endnote) are available which can be used to help you keep track of citations as you write a report or paper. However, during the proof reading stage (section 11.4), you should check that the reference information is both accurate in content and in the correct format. Failure to give accurate reference information is a major cause of delay in the publishing of papers.

11.3 Plagiarism

KEY CONCEPTS

- Understanding what is meant by plagiarism and why it incurs penalties
- Knowing how to give due credit to the work of other investigators and authors

In section 11.2.7 we mentioned the importance of giving due credit to the work of other scientists and making sure that you did not try to pass off the work of others as your own (a practice known as plagiarism, i.e. cheating). The availability of so much information via web-based resources and the ability to copy and paste makes this a very tempting prospect, especially when you are under time pressure to finish an assignment. However, you should be aware that those who are assessing your work know the tell-tale signs to look for, such as inconsistencies in style between sections of a piece of work or grammatical constructions which seem out of character. In addition, search engines are now well developed so that it is often relatively easy to trace the source of any such copied material.

In academic institutions there is a range of penalties for plagiarism, such as loss of marks or, in extreme cases, suspension or even expulsion. In other environments, there will be different penalties, but disciplinary proceedings are very likely to result.

The instructions to authors provided by the *Biochemical Journal* make it clear that the onus is on the authors of a paper to try to cite all previous publications which are particularly relevant to the work described. Failure to acknowledge this 'prior art' (a term used in the process of patenting) may result in the publication of an erratum (i.e. a note about errors in a paper). Publication of such a note is likely to damage the reputation of the authors involved.

11.4 How to set about writing a report

KEY CONCEPTS

- Understanding the order in which the sections of a report should be compiled
- Appreciating that conveying the results in a meaningful way is the key function of a report
- Appreciating the importance of careful checking of a report for accuracy and consistency

In this chapter, we have explained the structure of a scientific report and the functions of the various sections. However, what about the actual task of writing a report—how do you go about it? Unlike the advice in the song from the *Sound of Music*, when it comes to writing reports 'the very beginning is *not* a very good place to start'!

Ultimately, the order in which the various sections of a report is written is a matter of personal choice, but most scientists agree that a process something like that set out below works best in practice. The use of word-processing programs means that you can modify sections at will during the process.

If you are producing an individual report (as would be the case for most undergraduate project reports, for example) you will have to write the report on your own. However, if you are writing as part of a group it is a good idea to consult with your colleagues along the way; you do not want to give a 'finished' report to another member of the group only to have it returned with suggestions for extensive revision!

- Define the key results
  It is best to start by sorting out in your own mind (or with colleagues if you are part of a group) the important data which you wish to communicate in the report. This will help you to focus on those parts of other sections which are

relevant to those results, and prevent wandering off into other areas. It may be a good idea to think about the key figures or tables of data you will use in the results section

- Write the experimental section
  Now that you have sorted out the focus of the report, you should aim to start writing. The experimental section is the easiest to tackle first, as it tends to be fairly 'mechanical' in style and content and does not require too much synthesis on your part. It is surprising how much better you feel when you have actually written something (even if you have to revise it later—it is 'in the bank'!)

- Write the results section, including figures and tables
  The next section to tackle is the results. If you have already sorted out the figures and tables that you will use in this section, then the main task is to write the text which links them together and directs the reader to the key features of the data. If you are in a group it is a good idea to get a colleague to read through your attempts so far and take his/her comments on board. Make sure that the titles and legends for the figures and tables are appropriate

- Write the discussion and/or conclusions sections
  By now you will begin to see the shape of the overall report emerging and you should be able to attempt to evaluate the results and set them in context. You will need to make sure that you have copies of the relevant literature to hand when you write this section

- Write the introduction
  The last major section to be written is the introduction and you need to describe the background to the system and the approaches employed in a concise fashion. Again, it is essential to have the relevant literature available as you are writing this section

- Write the abstract
  The guidelines for this have been already been given (section 11.2.6), and your improved writing skills will now make this task much easier

- Compile the references and other material
  Now it is a case of compiling and sorting all the other information required to put the finishing touches to the report. Among the final tasks is the choice of a suitable title for the report

- Proofreading and checking
  You must check the entire document over for correct style, format, and consistency. Make sure that what you say is shown in the figures is actually there. Do not underestimate how long careful checking can take; you should build this time into your schedule. However, the satisfaction of producing your own impressive-looking report makes it all worthwhile!

When using a computer to produce your report it is very good practice to make sure that your work is saved and backed up at regular intervals, for example you might save the file on the hard drive of the computer, on a portable storage device (floppy disk, compact disk, or a memory stick), and possibly on a server as well. This should prevent last minute emergencies, such as a disk failure just before a piece of work is due to be submitted.

The mechanism(s) used for file storage may need to take security issues into account, for example in a company it is not a good idea to have confidential information stored in an insecure manner.

11.5 Poster presentations

KEY CONCEPT

- Understanding the factors which help to maximize the impact of a poster presentation

As mentioned in section 11.1, poster presentations are commonly used at scientific conferences; at large meetings there may well be several hundred posters presented. These will usually be displayed in a separate location near lecture theatres, often with refreshments available for conference delegates. Authors are usually requested to be in attendance near their posters between stated times during the conference. Poster sessions therefore have both a scientific and a social function. You will want your poster to attract attention and some careful planning is usually necessary to achieve this goal.

Poster presentations will contain most of the features of a scientific paper or report, i.e. title, list of authors, summary, introduction, experimental, results, discussion (or conclusion), references, and acknowledgements. Given the limited space available for the poster (typically about 1 m^2) and the need to make as strong an impact as possible, it is important to minimize the amount of text and to make more use of figures and tables to convey the important points. It is important not to try to cram too much into the space available and to avoid using too small a font size. Delegates should be able to 'navigate' their way around the poster and be able to understand what you are trying to say from a distance of about 1 m. You may well need to call on the services of a specialized graphics department who will be able to print and laminate large sheets of card, but the finished results can look very impressive indeed, and this will help to get your poster noticed!

Reference for Chapter 11

Holmes, D., Moody, P. and Dine, D. (2006) *Research Methods for the Biosciences.* Oxford University Press.

Appendix

Appendix 11.1 Examples of badly drawn and well-drawn graphs

Many of the points made in the guidelines given in Box 11.1 can be illustrated by the following examples. Suppose that we have obtained data on the velocity of the reaction catalysed by lactate dehydrogenase assayed in the direction of pyruvate reduction (i.e. pyruvate + NADH + H^+ → lactate + NAD^+). The reaction can be monitored by the change in absorbance at 340 nm, since NADH absorbs at this wavelength but NAD^+ does not. The data obtained when the concentration of pyruvate was varied over the range 0–200 μM, with the concentration of NADH being kept constant, are shown in Table A11.1. In the absence of pyruvate, the velocity is zero.

In plotting the data, we should note that the experimental variable (i.e. the concentration of substrate) is plotted on the *x*-axis, and the measured variable (i.e. the velocity of the reaction) is plotted on the *y*-axis.

Graphs A11.1A and A11.1B illustrate plots of velocity against [pyruvate].

Graph A11.1A contains several errors. The most important of these is that the scale on the *x*-axis is non-uniform; the 1-cm divisions on the graph paper correspond to successive increments of concentration of 10, 10, 20, 40, 40, and 80 μM. The values of concentrations used have been plotted with equal spacing (note, this can be the default setting of some computer graphics programs). As a consequence, the shape of the graph is distorted so that it is difficult to discern the appropriate trend in the data points. In the graph the points have merely been joined together by a series of straight lines. The second major error is that the units of the variables have not been shown on the axis labels.

Table A11.1 Variation of the velocity of the lactate dehydrogenase-catalysed reaction with [pyruvate]

[Pyruvate] (μM)	Velocity (ΔA_{340} min^{-1})
10	0.033
20	0.045
40	0.071
80	0.090
120	0.110
200	0.118

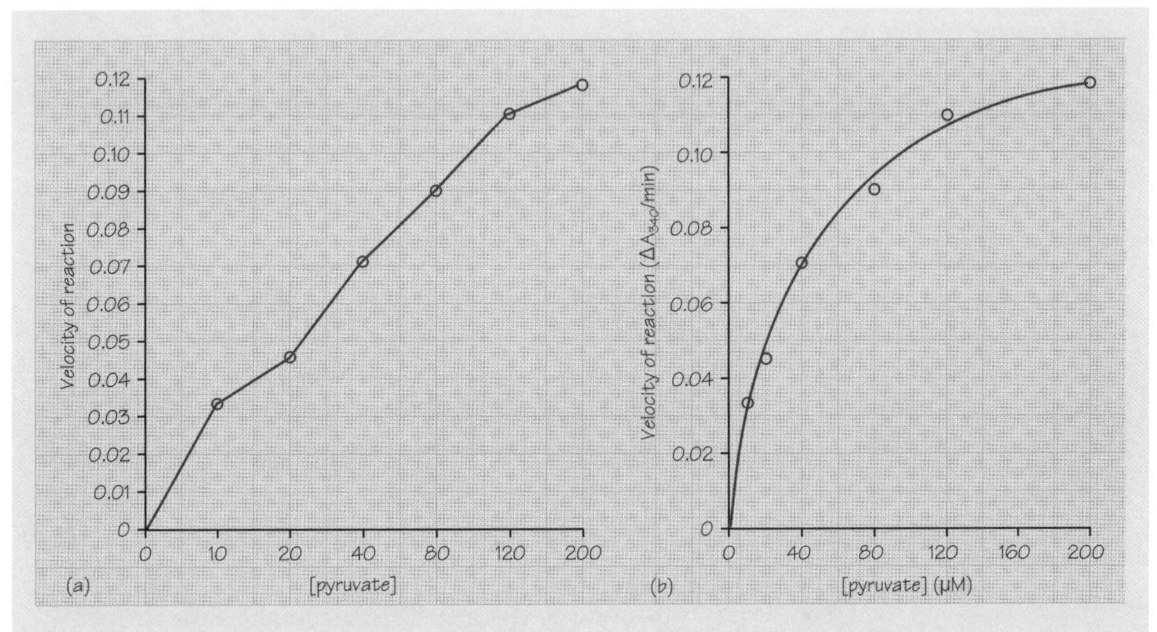

Figs. A11.1A and B The variation of the velocity of the lactate dehydrogenase-catalysed reaction with [pyruvate].

In Graph A11.1B these errors have been corrected. The x-axis scale is uniform so that the trend of the data points is shown correctly; the curve drawn represents the rectangular hyperbola characteristic of most enzyme-catalysed reactions (Chapter 4, section 4.3.3). The labels on the axes show the units of the quantities plotted.

In this example, Graph A11.1 is given a short but informative title. There should also be a more detailed legend provided, giving important experimental details such as the pH, buffer system and temperature used, the volume of the assay mixture, the concentration of NADH, and the quantity of enzyme added.

Graph A11.2 is similar to Graph A11.1B, except that the choice of scales on the axes is inconvenient (3 cm correspond to increments of 40 μM on the x-axis and 0.02 ΔA_{340} min^{-1} on the y-axis. This makes it more difficult both to plot the experimental values and to read off any derived values accurately. It is better to use more convenient scales shown in Graph A11.1B.

Graphs A11.3A and A11.3B depict the effect of including 50 μM oxamate in each assay. The experimental data points obtained in the presence of oxamate are listed in Table A11.2.

In Graph A11.3A, the experimental points are depicted by small dots. This can make them difficult to discern, so that it is sometimes hard to tell what is an experimental point and what is merely an aid to drawing the line shown. In addition, the two sets of data are not distinguished from each other in terms of either the symbols for the experimental points or the nature of the curve drawn (although they do each carry a label). By contrast in Graph A11.3B, the data sets are distinguished by the use of empty circles with a solid line and crossed circles with

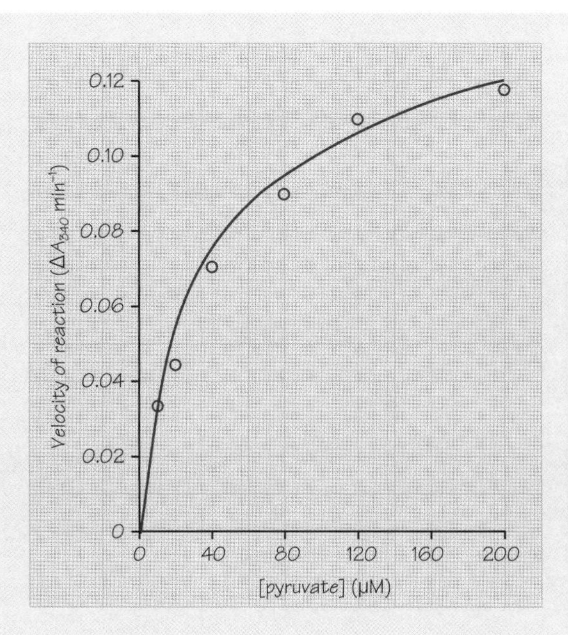

Fig. A11.2 The variation of the velocity of the lactate dehydrogenase-catalysed reaction with [pyruvate].

Table A11.2 Variation of the velocity of the lactate dehydrogenase-catalysed reaction with [pyruvate] in the presence of 50 μM oxamate

[Pyruvate] (μM)	Velocity (ΔA_{340} min^{-1})
10	0.014
20	0.030
40	0.050
80	0.075
120	0.080
200	0.100

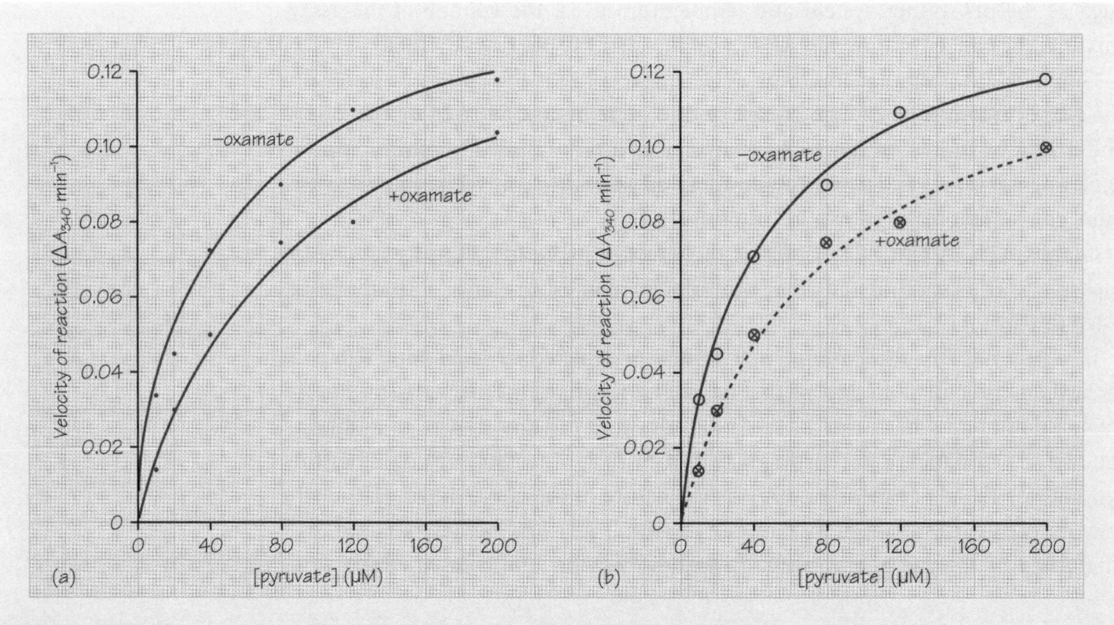

Figs. A11.3A and B The effect of oxamate on the lactate dehydrogenase-catalysed reaction.

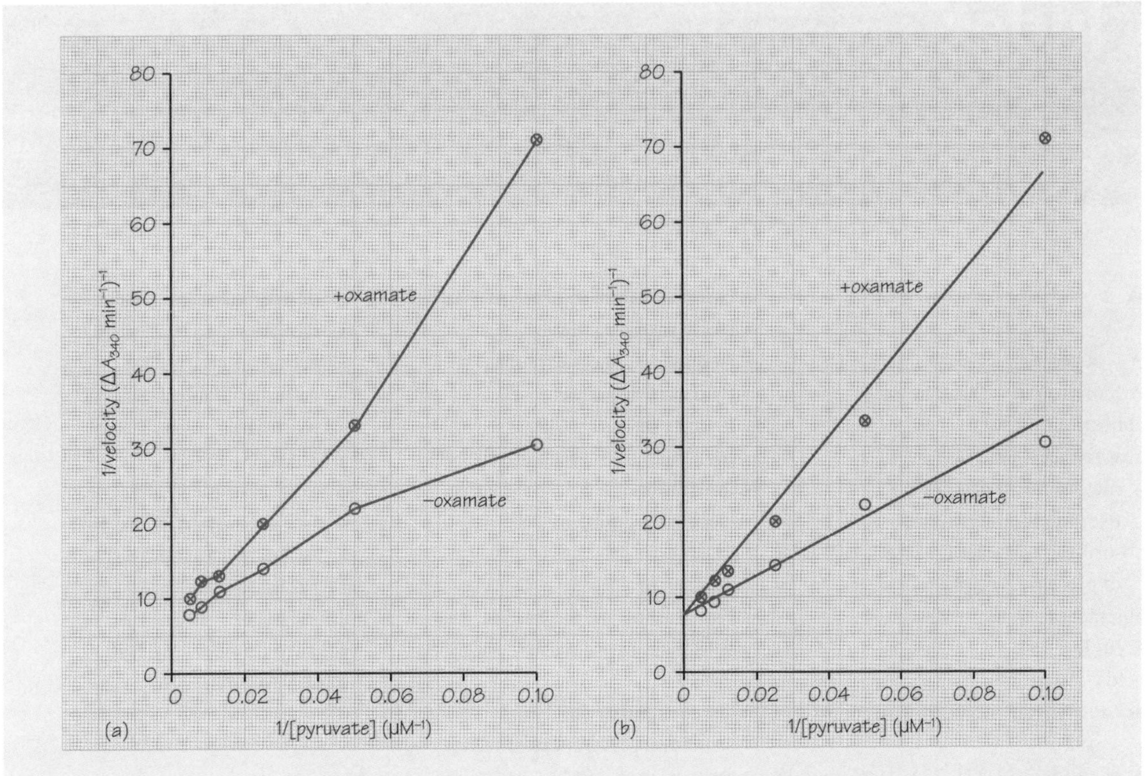

Figs. A11.4A and B Lineweaver–Burk plots showing the effect of oxamate on the lactate dehydrogenase-catalysed reaction.

a dashed line. The title chosen for Graph A11.3A is rather long and contains information which should more properly be given in the accompanying legend; Graph A11.3B carries a more suitable concise title.

In Graphs A11.4A and A11.4B, the data plotted in Graphs A11.3A and A11.3B have been transformed using the method of Lineweaver and Burk (Chapter 4, section 4.4), so that 1/velocity is plotted against 1/[substrate]. In Graph A11.4A, the points have been joined together by straight lines; this would make it difficult to extrapolate the data to infinite [substrate], i.e. when 1/[substrate] = 0. In Graph A11.4B, the best-fit straight lines (Chapter 2, section 2.6.4) have been drawn through each set of data points. From the slope and *y*-axis intercepts of these lines (Chapter 4, section 4.4) the values of V_{max} in the absence or presence of oxamate are 0.130 ΔA_{340} min^{-1}; the K_m value in the absence of oxamate is 32 μM and in the presence of oxamate 70 μM. We can thus conclude that oxamate behaves as a competitive inhibitor towards pyruvate, and from the change in K_m, the inhibitor constant K_{EI} can be calculated to be 42 μM (Chapter 4, section 4.4).

Index

A

B

I

J

K

L

M